$$\begin{bmatrix} a_1 & b_1 & c_1 \\ a_2 & b_2 & c_2 \end{bmatrix}$$ Two-by-three matrix

$$\begin{vmatrix} a_1 & b_1 \\ a_2 & b_2 \end{vmatrix}$$ Determinant

a_n nth term of a sequence

S_n Sum of n terms of a sequence

$\sum\limits_{i=1}^{n}$ Summation from $i=1$ to $i=n$

S_∞ Infinite sum

$n!$ n factorial

$P(n,n)$ Permutations of n things taken n at a time

$P(n,r)$ Permutations of n things taken r at a time

$C(n,r)$ Combinations of n things taken r at a time or r-element subsets taken from a set of n elements

$P(E)$ Probability of an event E

$n(E)$ Number of elements in the event space E

$n(S)$ Number of elements in the sample space S

E' The complement of set E

E_v Expected value

$P(E|F)$ Conditional probability of E given F

Set notation	Interval notation
$\{x|x>a\}$	(a,∞)
$\{x|x<b\}$	$(-\infty,b)$
$\{x|a<x<b\}$	(a,b)
$\{x|x\geq a\}$	$[a,\infty)$
$\{x|x\leq b\}$	$(-\infty,b]$
$\{x|a<x\leq b\}$	$(a,b]$
$\{x|a\leq x<b\}$	$[a,b)$
$\{x|a\leq x\leq b\}$	$[a,b]$

COLLEGE ALGEBRA

COLLEGE ALGEBRA

Jerome E. Kaufmann
Western Illinois University

PWS-KENT Publishing Company
Boston

PWS-KENT
Publishing Company

20 Park Plaza
Boston, Massachusetts 02116

PWS-KENT Publishing Company is a division of Wadsworth, Inc.

Portions of this book also appear in *College Algebra and Trigonometry* by Jerome E. Kaufmann, copyright © 1987 by PWS Publishers.

Library of Congress Cataloging-in-Publication Data

Kaufmann, Jerome E.
    College algebra.

    "Portions of this book also appear in College algebra and trigonometry"--T.p. verso.
    Includes index.
    1. Algebra.   I. Kaufmann, Jerome E.   College algebra and trigonometry.   II. Title.
QA154.2.K365   1987   512.9   86-25477
ISBN 0-87150-109-0

Printed in the United States of America

89 90 91 — 10 9 8 7 6 5 4

Sponsoring Editor: David Pallai
Production Coordinator: Susan Graham
Production: Deborah Schneider/Schneider & Company
Manuscript Editor: Deborah Schneider/Schneider & Company
Interior Design: Susan Graham
Cover Design: Julia Gecha
Cover Photo: © Greg Mancuso/Stock Boston, Inc.
Interior Illustration: J & R Art Services
Typesetting: Syntax International
Cover Printing: New England Book Components
Printing and Binding: Halliday Lithograph

To
William C. Lowry

With great respect and admiration I
want to say thank you for all that you
have done for me and many others as an
adviser and true friend.

Preface

This text was written for those college students who need a college algebra course to (a) serve as a prerequisite for the standard calculus sequence, (b) serve as a prerequisite for a finite mathematics or business calculus course, or (c) satisfy a liberal arts requirement. An example of a course outline for each of these classifications of students is included at the end of this preface.

The basic concepts of college algebra are presented in a simple and straightforward manner. The concepts are often motivated by examples, and they are continuously reinforced by additional examples. Algebraic ideas are developed in a logical sequence and in an easy-to-read manner without excessive technical vocabulary and formality.

Four major ideas serve as unifying themes, namely, (1) solving equations and inequalities, (2) solving problems, (3) developing graphing techniques, and (4) the concept of a function. These are four vital areas for precalculus students.

Specific Comments about Some of the Chapters

1. Chapter 1, with the possible exception of binomial expansions in Section 1.3 and complex numbers in Section 1.8, is a review of some basic concepts of intermediate algebra. This material has been written so that it can be reviewed with a minimum of instructor help.

2. Chapter 2 pulls together and expands upon a variety of equality and inequality solving processes. The development of problem-solving techniques is highlighted throughout the chapter.

3. Chapter 3 was written on the premise that students at this level need more work with coordinate geometry concepts—specifically graphing techniques—*before* being introduced to the idea of a function. In Chapter 3 various ellipses and hyperbolas (centers at the origin) are graphed that are a result of varying the coefficients of the equation $Ax^2 + By^2 = F$. Then in Chapter 9 the standard approach of developing basic forms from the definitions is used.

4. Chapters 4, 5, and 6 are tied together by the function concept. A straightforward approach to the function concept is presented in Chapter 4. The entire chapter is devoted to functions and the issue is not clouded by jumping back and forth between functions and relations that are not functions. Chapter 5 presents a modern-day version

of exponents and logarithms. The emphasis is on making the concepts and their applications understood, with the calculator used as a computational tool.

5. Chapters 7 and 8 have been written to again provide the instructor with some flexibility as to choice of topics. Chapter 7 is devoted entirely to solving systems of linear equations, including the use of matrices and determinants. Then in Chapter 8 the algebra of matrices is the focal point.

6. Problem solving is the unifying theme of Chapters 10 and 11. In contrast to most college algebra books, Chapter 11 contains a significant amount of probability.

Other Special Features of the Book

1. The problem sets have been very carefully constructed on an even/odd basis; that is, all variations of skill-development exercises are contained in both the even- and the odd-numbered problems. Thus, a meaningful assignment can be given using either the "evens" or the "odds" and a double dosage is available by assigning all of them. I also considered the fact that some skills need more drill than others.

2. Many of the problem sets contain a special section called *Miscellaneous Problems*. These problems encompass a variety of ideas. Some of them are proofs, some exhibit different approaches to topics covered in the text, some bring in supplementary topics and relationships, and some are just more difficult problems. All of them could be omitted without breaking the continuity pattern of the text; however, I feel that they do add another flexibility feature. Chapter 1, being basically a review chapter, does not contain any miscellaneous problems.

3. There is a *Review Problem Set* at the end of each chapter. These sets are designed to help students pull together the ideas presented in the chapter. For example, in Chapter 2 each section presents a different type of equation or inequality to solve. Then the review problem set contains a mixture of the various types of equations and inequalities. There is a *Cumulative Review Problem Set* at the end of Chapter 3. *All* of the answers to review problem sets are given in the back of the book.

4. I have tried to make *Chapter Summaries* truly useful from a student's viewpoint. No standard format for each chapter has been used. At the end of each chapter I asked myself the question "What is the most effective way of summarizing the big ideas of this chapter?" In Chapter 3, for example, I felt that organizing the summary around the two basic types of coordinate geometry problems was effective.

5. I tried to assign the calculator its rightful place in the study of mathematics—a tool, useful at times, unnecessary at other times. No special problems have been created just so that we can use the calculator. Instead it is used as needed. For example, Chapter 5, on exponents and logarithms, lends itself to the use of a calculator.

The following are sample outlines for a college algebra course that (a) serves as a prerequisite for the standard calculus sequence, (b) serves as a prerequisite for a finite mathematics or business calculus course, or (c) satisfies a liberal arts requirement.

Precalculus	*Business*	*Liberal Arts*
1.1–1.8 Some Basic Concepts of Algebra	*1.1–1.7* Some Basic Concepts of Algebra	*1.1–1.6* Some Basic Concepts of Algebra
2.1–2.7 Equations, Inequalities, and Problem Solving	*2.1–2.7* Equations, Inequalities, and Problem Solving	*2.1–2.4 and 2.6* Equations, Inequalities, and Problem Solving
3.1–3.5 Coordinate Geometry and Graphing Techniques	*3.1–3.4* Coordinate Geometry and Graphing Techniques	*3.1–3.4* Coordinate Geometry and Graphing Techniques
4.1–4.6 Functions	*4.1–4.6* Functions	*4.1–4.5* Functions
5.1–5.5 Exponential and Logarithmic Functions	*5.1–5.5* Exponential and Logarithmic Functions	*5.1–5.5* Exponential and Logarithmic Functions
6.1–6.7 Polynomial and Rational Functions	*6.1–6.5* Polynomial and Rational Functions	*7.1–7.3* Systems of Equations
9.1–9.3 Conic Sections	*7.1–7.5* Systems of Equations	*8.1 and 8.2* Matrices
10.1–10.3 Sequences	*8.4* Linear Programming	*10.1–10.3* Sequences
11.6 The Binomial Theorem	*10.1–10.3* Sequences	*11.1–11.4* Counting Techniques and Probability
	11.1–11.5 Counting Techniques and Probability	

For the precalculus course any remaining days could be spent on topics from Chapters 7 and 8 that would supplement any future linear algebra course.

I would like to take this opportunity to thank all of the people who served as reviewers for this manuscript and to extend my special thanks to

James E. Arnold, Jr.
University of Wisconsin

Jerome Bloomberg
Essex Community College

Patrick Costello
Eastern Kentucky University

Hugh B. Easler
College of William & Mary

Christopher Ennis
University of Minnesota

Naomi Halpern
Brooklyn College of City University of New York

Ruth Hunt
Brooklyn College of City University of New York

Toni Kasper
Borough of Manhattan Community College

Jimmie G. Lakin
University of Wisconsin

Robert Lohman
Kent State University

Samuel Lynch
Southwest Missouri State University

Lois Miller
Golden West College

Stephen P. Peterson
University of Notre Dame

William D. Popejoy
University of Northern Colorado

Susan Prazak
College of Charleston

Howard Sherwood
University of Central Florida

Dorothy Sulock
University of North Carolina

William F. Ward
Indian River Community College

Again I am very grateful to Dave Pallai, my editor, and to the staff of Prindle, Weber & Schmidt for their continuous cooperation and assistance throughout this project. In particular, I would like to thank Susan Graham for her complete cooperation in her role as production editor. A special word of thanks is due to Debbie Schneider.

Special thanks again are due to my wife, Arlene, who continues to spend numerous hours proofreading and typing the manuscript along with the answer book, solutions manual, and accompanying sets of tests.

Jerome E. Kaufmann

Contents

CHAPTER 1

Some Basic Concepts of Algebra   **1**

1.1	Some Basic Ideas	**2**
1.2	Exponents	**13**
1.3	Polynomials	**20**
1.4	Factoring Polynomials	**27**
1.5	Rational Expressions	**35**
1.6	Radicals	**44**
1.7	The Relationship between Exponents and Roots	**52**
1.8	Complex Numbers	**57**
	Chapter Summary	**64**
	Review Problem Set	**66**

CHAPTER 2

Equations, Inequalities, and Problem Solving   **68**

2.1	Linear Equations and Problem Solving	**69**
2.2	More Equations and Applications	**76**
2.3	Quadratic Equations	**88**
2.4	Applications of Linear and Quadratic Equations	**98**
2.5	Miscellaneous Equations	**107**
2.6	Inequalities	**112**
2.7	Inequalities Involving Quotients and Absolute Value	**119**
	Chapter Summary	**125**
	Review Problem Set	**126**

CHAPTER 3

Coordinate Geometry and Graphing Techniques   **129**

3.1	Coordinate Geometry	**130**
3.2	Graphing Techniques: Linear Equations and Inequalities	**138**
3.3	Determining the Equation of a Line	**142**
3.4	More on Graphing	**152**
3.5	Circles, Ellipses, and Hyperbolas	**157**
	Chapter Summary	**164**
	Review Problem Set	**165**
	Cumulative Review Problem Set	**166**

CHAPTER 4

Functions   **168**

4.1	The Concept of a Function	**169**
4.2	Linear and Quadratic Functions	**177**
4.3	More about Functions and Problem Solving	**183**
4.4	Combining Functions	**191**
4.5	Inverse Functions	**195**
4.6	Direct and Inverse Variations	**202**
	Chapter Summary	**209**
	Review Problem Set	**211**

CHAPTER 5

Exponential and Logarithmic Functions   **213**

5.1	Exponents and Exponential Functions	**214**
5.2	Applications of Exponential Functions	**218**
5.3	Logarithms	**226**
5.4	Logarithmic Functions	**233**
5.5	Exponential and Logarithmic Equations; Problem Solving	**241**
5.6	Computation with Common Logarithms (Optional)	**250**
	Chapter Summary	**255**
	Review Problem Set	**257**

CHAPTER 6

Polynomial and Rational Functions

259

6.1	Dividing Polynomials	**260**
6.2	The Remainder and Factor Theorems	**265**
6.3	Polynomial Equations	**270**
6.4	Graphing Polynomial Functions	**279**
6.5	Graphing Rational Functions	**287**
6.6	More on Graphing Rational Functions	**293**
6.7	Partial Fractions	**297**
	Chapter Summary	**302**
	Review Problem Set	**304**

CHAPTER 7

Systems of Equations

305

7.1	Systems of Two Linear Equations in Two Variables	**306**
7.2	Systems of Three Linear Equations in Three Variables	**315**
7.3	A Matrix Approach to Solving Systems	**322**
7.4	Determinants	**332**
7.5	Cramer's Rule	**342**
	Chapter Summary	**348**
	Review Problem Set	**350**

CHAPTER 8

The Algebra of Matrices

353

8.1	The Algebra of 2×2 Matrices	**354**
8.2	Multiplicative Inverses	**360**
8.3	$m \times n$ Matrices	**365**
8.4	Systems Involving Linear Inequalities: Linear Programming	**374**
	Chapter Summary	**383**
	Review Problem Set	**385**

CHAPTER 9

Conic Sections   387

9.1	Parabolas	388
9.2	Ellipses	394
9.3	Hyperbolas	400
9.4	Systems Involving Nonlinear Equations	407
	Chapter Summary	412
	Review Problem Set	413

CHAPTER 10

Sequences and Mathematical Induction   414

10.1	Arithmetic Sequences	415
10.2	Geometric Sequences	423
10.3	Another Look at Problem Solving	430
10.4	Mathematical Induction	436
	Chapter Summary	440
	Review Problem Set	442

CHAPTER 9

CHAPTER 11

Counting Techniques, Probability, and the Binomial Theorem   444

11.1	The Fundamental Principle of Counting	445
11.2	Permutations and Combinations	450
11.3	Probability	458
11.4	Some Properties of Probability; Expected Values	465
11.5	Conditional Probability; Dependent and Independent Events	476
11.6	The Binomial Theorem	484
	Chapter Summary	488
	Review Problem Set	490

Answers to Odd-numbered Exercises   A1

Index   A29

Some Basic Concepts of Algebra

1

1.1 Some Basic Ideas

1.2 Exponents

1.3 Polynomials

1.4 Factoring Polynomials

1.5 Rational Expressions

1.6 Radicals

1.7 The Relationship between Exponents and Roots

1.8 Complex Numbers

Algebra is often described as generalized arithmetic. That description may not tell the whole story, but it does indicate an important idea, namely, that a good understanding of arithmetic provides a sound basis for the study of algebra. Furthermore, a good understanding of some basic algebraic concepts provides an even better basis for the study of more advanced algebraic ideas. *Be sure* that you can work effectively with the algebraic concepts reviewed in this first chapter.

1.1

Some Basic Ideas

Let's begin by pulling together some basic ideas that are needed in the study of algebra. In arithmetic, symbols such as 6, $\frac{2}{3}$, 0.27, and π are used to represent numbers. The operations of addition, subtraction, multiplication, and division are commonly indicated by the symbols $+$, $-$, $\times$, and $\div$, respectively. Using these symbols, specific **numerical expressions** can be formed. For example, the indicated sum of six and eight can be written $6 + 8$.

In algebra, the concept of a variable provides the basis for generalizing arithmetic ideas. For example, by using x and y to represent *any* two numbers, the expression $x + y$ can be used to represent the indicated sum of *any* two numbers. The x and y in such an expression are called **variables** and the phrase $x + y$ is called an **algebraic expression**.

Many of the notational agreements made in arithmetic are extended to algebra, with a few modifications. The following chart summarizes these notational agreements pertaining to the four basic operations.

OPERATION	ARITHMETIC	ALGEBRA	VOCABULARY
addition	$4 + 6$	$x + y$	The sum of x and y
subtraction	$14 - 10$	$a - b$	The difference of a and b
multiplication	7×5 or $7 \cdot 5$	$a \cdot b$ or $a(b)$ or $(a)b$ or $(a)(b)$ or ab	The product of a and b
division	$8 \div 4$ or $\frac{8}{4}$ or $8/4$ or $4\,\overline{)8}$	$x \div y$ or $\dfrac{x}{y}$ or x/y or $y\,\overline{)x}$ $(y \neq 0)$	The quotient of x divided by y

Note the different ways of indicating a product, including the use of parentheses. The *ab* form is the simplest and probably most widely used form. Expressions such as *abc*, $6xy$, and $14xyz$ all indicate multiplication. We also call your attention to the various forms used to indicate division. In algebra, the fractional forms $\dfrac{x}{y}$ or x/y are usually used (x/y is frequently seen in textbooks), although the other forms do serve a purpose at times.

The Use of Sets

Some of the vocabulary and symbolism associated with the concept of sets can be effectively used in the study of algebra. A **set** is a collection of objects and the objects are called **elements** or **members** of the set. The use of capital letters to name sets and set braces, { }, to enclose the elements or a description of the elements provides a convenient way to communicate about sets. For example, a set A consisting of the vowels of the alphabet can be represented as follows.

$$A = \{\text{vowels of the alphabet}\} \qquad \text{Word description}$$

or $\quad A = \{a, e, i, o, u\} \qquad\qquad\qquad$ List or roster description

or $\quad A = \{x \,|\, x \text{ is a vowel}\} \qquad\qquad$ Set builder notation

A set consisting of no elements is called the **null** or **empty set** and is written $\varnothing$.

The **set-builder notation** combines the use of braces and the concept of a variable. For example, $\{x \,|\, x \text{ is a vowel}\}$ is read, "The set of all x such that x is a vowel." Note that the vertical line is read "such that."

Two sets are said to be **equal** if they contain exactly the same elements. For example, $\{1, 2, 3\} = \{2, 1, 3\}$ because both sets contain exactly the same elements; the order in which the elements are listed doesn't matter. A slash mark through an equality symbol denotes "not equal to." Thus if $A = \{1, 2, 3\}$ and $B = \{3, 6\}$, we can write $A \neq B$, which is read, "Set A is not equal to set B."

Real Numbers

The following terminology is commonly used to classify different types of numbers.

$\{1, 2, 3, 4, \ldots\}$	Natural numbers, counting numbers, positive integers
$\{0, 1, 2, 3, \ldots\}$	Whole numbers, nonnegative integers
$\{\ldots, -3, -2, -1\}$	Negative integers
$\{\ldots, -3, -2, -1, 0\}$	Nonpositive integers
$\{\ldots, -2, -1, 0, 1, 2, \ldots\}$	Integers

A **rational number** is defined as any number that can be expressed in the form a/b, where a and b are integers and b is not zero. The following are examples of rational numbers:

$\frac{2}{3}$; $\quad -\frac{3}{4}$; $\quad 6$ because $6 = \frac{6}{1}$; $\quad -4$ because $-4 = \frac{-4}{1} = \frac{4}{-1}$;

0 because $0 = \frac{0}{1} = \frac{0}{2} = \frac{0}{3}$, etc.; $\quad 0.3$ because $0.3 = \frac{3}{10}$;

$6\frac{1}{2}$ because $6\frac{1}{2} = \frac{13}{2}$.

A rational number can also be defined in terms of a decimal representation. Before doing so, let's briefly review the different possibilities for decimal representations. Decimals can be classified as *terminating*, *repeating*, or *nonrepeating*. Here are some examples of each.

$$\left.\begin{array}{l} 0.3 \\ 0.46 \\ 0.789 \\ 0.2143 \end{array}\right\} \text{Terminating decimals}$$

$$\left.\begin{array}{l} 0.3333\ldots \\ 0.141414\ldots \\ 0.712712712\ldots \\ 0.24171717\ldots \\ 0.9675283283283\ldots \end{array}\right\} \text{Repeating decimals}$$

$$\left.\begin{array}{l} 0.472195631\ldots \\ 0.21411711191111\ldots \\ 0.752389433215333\ldots \end{array}\right\} \text{Nonrepeating decimals}$$

A **repeating decimal** has a block of digits that repeats indefinitely. This repeating block of digits may be of any size and may or may not begin immediately after the decimal point. A small horizontal bar is commonly used to indicate the repeating block. Thus, 0.3333 ... can be expressed as $0.\overline{3}$ and 0.24171717 ... as $0.24\overline{17}$.

In terms of decimals, a rational number is defined to be a number that has either a terminating or a repeating decimal representation. The following examples illustrate some rational numbers written in $\frac{a}{b}$ form and in the equivalent decimal form.

$$\tfrac{3}{4} = 0.75; \quad \tfrac{3}{11} = 0.\overline{27}; \quad \tfrac{1}{8} = 0.125; \quad \tfrac{1}{7} = 0.\overline{142857}; \quad \tfrac{1}{3} = 0.\overline{3}$$

An **irrational number** is defined to be a number that cannot be expressed in $\frac{a}{b}$ form, where a and b are integers and b is not zero. Furthermore, an irrational number has a nonrepeating decimal representation. Following are some examples of irrational numbers and a partial decimal representation for each.

$$\sqrt{2} = 1.414213562373095\ldots$$
$$\sqrt{3} = 1.73205080756887\ldots$$
$$\pi = 3.14159265358979\ldots$$

The entire set of **real numbers** is composed of the rational numbers along with the irrationals. The following tree diagram can be used to summarize the various classifications of the real number system.

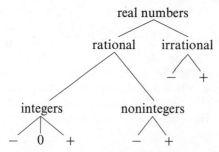

Any real number can be traced down through the tree. For example,

7 is real, rational, an integer, and positive;

$-\frac{2}{3}$ is real, rational, a noninteger, and negative;

$\sqrt{7}$ is real, irrational, and positive;

0.59 is real, rational, a noninteger, and positive.

The concept of subset is convenient to use at this time. A set A is a **subset** of another set B if and only if every element of A is also an element of B. For example, if $A = \{1, 2\}$ and $B = \{1, 2, 3\}$, then A is a subset of B. This is written $A \subseteq B$ and is read, "A is a subset of B." The slash mark can also be used here to denote negation. If $A = \{1, 2, 4, 6\}$ and $B = \{2, 3, 7\}$, we can say, "A is not a subset of B" by writing $A \nsubseteq B$. The following kinds of statements can be made using the subset vocabulary and symbolism.

1. The set of whole numbers is a subset of the set of integers.

$$\{0, 1, 2, 3, \ldots\} \subseteq \{\ldots, -2, -1, 0, 1, 2, \ldots\}$$

2. The set of integers is a subset of the set of rational numbers.

$$\{\ldots, -2, -1, 0, 1, 2, \ldots\} \subseteq \{x \mid x \text{ is a rational number}\}$$

3. The set of rational numbers is a subset of the set of real numbers.

$$\{x \mid x \text{ is a rational number}\} \subseteq \{y \mid y \text{ is a real number}\}$$

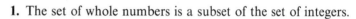

The Real Number Line and Absolute Value

It is often convenient to have a geometric representation of the set of real numbers in front of us, as indicated in Figure 1.1. Such a representation, called the **real number line**, indicates a "one-to-one correspondence" between the set of real numbers and the points on a line. That is to say, to each real number there corresponds one and only one point on the line, and to each point on the line there corresponds one and only one real number. The number that corresponds to a particular point on the line is called the **coordinate** of that point.

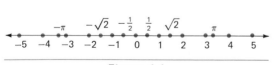

Figure 1.1

Figure 1.2

Many operations, relations, properties, and concepts pertaining to real numbers can be given a geometric interpretation on the number line. For example, the addition problem $(-1) + (-2)$ can be interpreted on the number line as in Figure 1.2.

The inequality relations also have a geometric interpretation. The statement $a > b$ (read "a is greater than b") means that a is to the right of b, and the statement $c < d$ (read "c is less than d") means that c is to the left of d (see Figure 1.3).

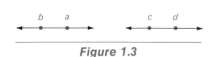

Figure 1.3

The property $-(-x) = x$ can be "pictured" on the number line in a sequence of steps:

1. Choose a point having a coordinate of x.

2. Locate its opposite (written as $-x$) on the other side of zero.

3. Locate the opposite of $-x$ (written as $-(-x)$) on the other side of zero.

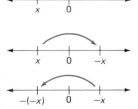

Therefore, we conclude that **the opposite of the opposite of any real number is the number itself**, and we symbolically express this by $-(-x) = x$.

> **Remark:** The symbol -1 can be read **negative one**, the **negative of one**, the **opposite of one**, or the **additive inverse of one**. The "opposite of" and "additive inverse of" terminology is especially meaningful when working with variables. For example, the symbol $-x$, read "the opposite of x" or "the additive inverse of x," emphasizes an important issue. Since x can be any real number, $-x$ (opposite of x) can be zero, positive, or negative. If x is positive, then $-x$ is negative. If x is negative, then $-x$ is positive. If x is zero, then $-x$ is zero.

The concept of absolute value can be interpreted on the number line. Geometrically, the **absolute value** of any real number is the distance between that number and zero on the number line. For example, the absolute value of 2 is 2, the absolute value of -3 is 3, and the absolute value of zero is zero (see Figure 1.4). Symbolically, absolute value is denoted with vertical bars. Thus, we write $|2| = 2$, $|-3| = 3$, and $|0| = 0$.
More formally, the concept of absolute value is defined as follows.

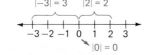

Figure 1.4

DEFINITION 1.1

> For all real numbers a,
>
> 1. If $a \geq 0$, then $|a| = a$;
> 2. If $a < 0$, then $|a| = -a$.

According to Definition 1.1, we obtain

$$|6| = 6 \qquad \text{by applying part 1;}$$
$$|0| = 0 \qquad \text{by applying part 1;}$$
$$|-7| = -(-7) = 7 \quad \text{by applying part 2.}$$

Notice that the absolute value of a positive number is the number itself, but the absolute value of a negative number is its opposite. Thus, the absolute value of any number except zero is positive, and the absolute value of zero is zero. Together, these facts indicate that the absolute value of any real number is equal to the absolute value of its opposite. All of these ideas are summarized in the following properties.

Properties of
Absolute Value

> a and b represent any real number.
>
> **1.** $|a| \geq 0$
> **2.** $|a| = |-a|$
> **3.** $|a - b| = |b - a|$ $a - b$ and $b - a$ are opposites of each other.

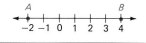

Figure 1.5

In Figure 1.5 we have indicated points A and B at -2 and 4, respectively. The distance between A and B is six units and can be calculated by using either $|-2 - 4|$ or $|4 - (-2)|$. In general, if two points have coordinates x_1 and x_2, the distance between the two points is determined by using either $|x_2 - x_1|$ or $|x_1 - x_2|$, since they are the same quantity by property 3.

Properties of the Real Numbers

As you work with the set of real numbers, the basic operations, and the relations of equality and inequality, the following properties will guide your study. Be sure that you understand these properties, for not only do they facilitate manipulations with real numbers, but they also serve as a basis for many algebraic computations. The variables a, b, and c represent real numbers.

Properties of
the Real Numbers

Closure properties	$a + b$ is a real number. ab is a real number.
Commutative properties	$a + b = b + a$ $ab = ba$
Associative properties	$(a + b) + c = a + (b + c)$ $(ab)c = a(bc)$
Identity properties	There exists a real number 0 such that $a + 0 = 0 + a = a$. There exists a real number 1 such that $a(1) = 1(a) = a$.
Inverse properties	For every real number a, there exists a real number $-a$ such that $a + (-a) = (-a) + a = 0$. For every nonzero real number a, there exists a real number $\frac{1}{a}$ such that $$a\left(\frac{1}{a}\right) = \frac{1}{a}(a) = 1.$$
Multiplication property of zero	$a(0) = 0(a) = 0$
Multiplication property of negative one	$a(-1) = -1(a) = -a$
Distributive property	$a(b + c) = ab + ac$

Let's make a few comments about the previous list of properties. The set of real numbers is said to be **closed** with respect to addition and multiplication. That is to say, the sum of two real numbers is a real number and the product of two real numbers is a real number. **Closure** plays an important role when proving additional properties pertaining to real numbers.

Addition and multiplication are said to be **commutative** operations. This means that the order in which you add or multiply two real numbers does not affect the result. For example, $6 + (-8) = -8 + 6$ and $(-4)(-3) = (-3)(-4)$. It is also important to realize that *subtraction and division are not commutative operations*; order does make a difference. For example, $3 - 4 = -1$, but $4 - 3 = 1$. Likewise, $2 \div 1 = 2$, but $1 \div 2 = \frac{1}{2}$.

Addition and multiplication are **associative** operations. The associative properties are grouping properties. For example, $(-8 + 9) + 6 = -8 + (9 + 6)$; changing the grouping of the numbers does not affect the final sum. Likewise, for multiplication, $[(-4)(-3)](2) = (-4)[(-3)(2)]$. *Subtraction and division are not associative operations*. For example, $(8 - 6) - 10 = -8$, but $8 - (6 - 10) = 12$. An example showing that division is not associative is $(8 \div 4) \div 2 = 1$, but $8 \div (4 \div 2) = 4$.

Zero is the **identity element for addition**. This means that the sum of any real number and zero is identically the same real number. For example, $-87 + 0 = 0 + (-87) = -87$.

One is the **identity element for multiplication**: the product of any real number and 1 is identically the same real number. For example, $(-119)(1) = (1)(-119) = -119$.

The real number $-a$ is called the **additive inverse** of a or the **opposite** of a. The sum of a number and its additive inverse is the identity element for addition. For example, 16 and -16 are additive inverses and their sum is zero. The additive inverse of zero is zero.

The real number $1/a$ is called the **multiplicative inverse** or **reciprocal** of a. The product of a number and its multiplicative inverse is the identity element for multiplication. For example, the reciprocal of 2 is $\frac{1}{2}$ and $2(\frac{1}{2}) = \frac{1}{2}(2) = 1$.

The product of any real number and zero is zero. For example, $(-17)(0) = (0)(-17) = 0$. The product of any real number and -1 is the opposite of the real number. For example, $(-1)(52) = (52)(-1) = -52$.

The distributive property ties together the operations of addition and multiplication. We say that *multiplication distributes over addition*. For example, $7(3 + 8) = 7(3) + 7(8)$. Furthermore, because $b - c = b + (-c)$, it follows that *multiplication also distributes over subtraction*. This can be symbolically expressed $a(b - c) = ab - ac$. For example, $6(8 - 10) = 6(8) - 6(10)$.

Algebraic Expressions

Algebraic expressions such as

$$2x, \quad 8xy, \quad -3xy, \quad -4abc, \quad \text{and} \quad z$$

are called **terms**. A term is an indicated product and may have any number of factors. The variables involved in a term are called **literal factors** and the numerical factor is called the **numerical coefficient**. Thus in $8xy$, the x and y

are literal factors and 8 is the numerical coefficient. Since $1(z) = z$, the numerical coefficient of the term z is understood to be 1. Terms having the same literal factors are called **similar** or **like** terms. The distributive property in the form $ba + ca = (b + c)a$ provides the basis for simplifying algebraic expressions by *combining similar terms*, as illustrated in the following examples.

$$3x + 5x = (3 + 5)x \qquad -6xy + 4xy = (-6 + 4)xy \qquad 4x - x = 4x - 1x$$
$$= 8x \qquad\qquad\qquad = -2xy \qquad\qquad\qquad = (4 - 1)x$$
$$= 3x$$

Sometimes an algebraic expression can be simplified by applying the distributive property to remove parentheses and then by combining similar terms, as the next examples illustrate.

$$4(x + 2) + 3(x + 6) = 4(x) + 4(2) + 3(x) + 3(6)$$
$$= 4x + 8 + 3x + 18$$
$$= 7x + 26$$

$$-5(y + 3) - 2(y - 8) = -5(y) - 5(3) - 2(y) - 2(-8)$$
$$= -5y - 15 - 2y + 16$$
$$= -7y + 1$$

An algebraic expression takes on a numerical value whenever each variable in the expression is replaced by a real number. For example, if x is replaced by 5 and y by 9, the algebraic expression $x + y$ becomes the numerical expression $5 + 9$, which is equal to 14. We say that $x + y$ has a value of 14 when x equals 5 and y equals 9.

Consider the following examples, which illustrate the process of finding a value of an algebraic expression. The process is commonly referred to as **evaluating an algebraic expression**.

EXAMPLE 1

Find the value of $3xy - 4z$ when $x = 2$, $y = -4$, and $z = -5$.

Solution

$$3xy - 4z = 3(2)(-4) - 4(-5) \qquad \text{when } x = 2, y = -4, \text{ and } z = -5$$
$$= -24 + 20$$
$$= -4$$

EXAMPLE 2

Find the value of $a - [4b - (2c + 1)]$ when $a = -8$, $b = -7$, and $c = 14$.

Solution

$$a - [4b - (2c + 1)] = -8 - [4(-7) - (2(14) + 1)]$$
$$= -8 - [-28 - 29]$$
$$= -8 - [-57]$$
$$= 49$$

EXAMPLE 3

Evaluate $\dfrac{a - 2b}{3c + 5d}$ when $a = 14$, $b = -12$, $c = -3$, and $d = -2$.

Solution

$$\frac{a - 2b}{3c + 5d} = \frac{14 - 2(-12)}{3(-3) + 5(-2)}$$

$$= \frac{14 + 24}{-9 - 10}$$

$$= \frac{38}{-19} = -2$$

Look back at the previous examples and notice that the following *order of operations* was followed when simplifying numerical expressions.

1. Perform the operations inside the symbols of inclusion (parentheses, brackets, and braces) and above and below each fraction bar. Start with the innermost inclusion symbol.

2. Perform all multiplications and divisions in the order in which they appear, from left to right.

3. Perform all additions and subtractions in the order in which they appear, from left to right.

We should also realize that first simplifying by combining similar terms can sometimes aid in the process of evaluating algebraic expressions. The last example of this section illustrates this idea.

EXAMPLE 4

Evaluate $2(3x + 1) - 3(4x - 3)$ when $x = -5$.

Solution

$$2(3x + 1) - 3(4x - 3) = 2(3x) + 2(1) - 3(4x) - 3(-3)$$

$$= 6x + 2 - 12x + 9$$

$$= -6x + 11$$

Now substituting -5 for x, we obtain

$$-6x + 11 = -6(-5) + 11$$

$$= 30 + 11$$

$$= 41.$$

Problem Set 1.1

Identify each of the following as *true* or *false*.

1. Every rational number is a real number.

2. Every irrational number is a real number.

3. Every real number is a rational number.

4. If a number is real, then it is irrational.

5. Some irrational numbers are also rational numbers.

6. All integers are rational numbers.

7. The number zero is a rational number.

8. Zero is a positive integer.

9. Zero is a negative number.

10. All whole numbers are integers.

In the set of numbers $\{0, \sqrt{5}, -\sqrt{2}, \frac{7}{8}, -\frac{10}{13}, 7\frac{1}{8}, 0.279, 0.4\overline{67}, -\pi, -14, 46, 6.75\}$, list those elements that belong to each of the following sets. (Problems 11–18)

11. The natural numbers   **12.** The whole numbers

13. The integers   **14.** The rational numbers

15. The irrational numbers   **16.** The nonnegative integers

17. The nonpositive integers   **18.** The real numbers

For Problems 19–32, use the following set designations.

$$N = \{x \mid x \text{ is a natural number}\}$$
$$W = \{x \mid x \text{ is a whole number}\}$$
$$I = \{x \mid x \text{ is an integer}\}$$
$$Q = \{x \mid x \text{ is a rational number}\}$$
$$H = \{x \mid x \text{ is an irrational number}\}$$
$$R = \{x \mid x \text{ is a real number}\}$$

Place $\subseteq$ or $\nsubseteq$ in each blank to make a true statement.

19. N _____ R **20.** R _____ N

21. N _____ I **22.** I _____ Q

23. H _____ Q **24.** Q _____ H

25. W _____ I **26.** N _____ W

27. I _____ W **28.** I _____ N

29. $\{0, 2, 4, \ldots\}$ _____ W **30.** $\{1, 3, 5, 7, \ldots\}$ _____ I

31. $\{-2, -1, 0, 1, 2\}$ _____ W **32.** $\{0, 3, 6, 9, \ldots\}$ _____ N

For Problems 33–42, list the elements of each set. For example, the elements of $\{x \mid x \text{ is a natural number less than 4}\}$ can be listed $\{1, 2, 3\}$.

33. $\{x \mid x \text{ is a natural number less than 2}\}$

34. $\{x \mid x \text{ is a natural number greater than 5}\}$

35. $\{n \mid n \text{ is a whole number less than 4}\}$

36. $\{y \mid y \text{ is an integer greater than } -3\}$

37. $\{y \mid y \text{ is an integer less than 2}\}$

38. $\{n \mid n \text{ is a positive integer greater than } -4\}$

39. $\{x \mid x \text{ is a whole number less than 0}\}$

40. $\{x \mid x \text{ is a negative integer greater than } -5\}$

41. $\{n \mid n \text{ is a nonnegative integer less than 3}\}$

42. $\{n \mid n$ is a nonpositive integer greater than $1\}$

43. Find the distance on the real number line between two points whose coordinates are as follows.

(a) 17 and 35   (b) -14 and 12

(c) 18 and -21 (d) -17 and -42

(e) -56 and -21 (f) 0 and -37

44. Evaluate each of the following if x is a nonzero real number.

(a) $\dfrac{|x|}{x}$ (b) $\dfrac{x}{|x|}$ (c) $\dfrac{|-x|}{-x}$ (d) $|x| - |-x|$

In Problems 45–58, state the property that justifies each of the statements. For example, $3 + (-4) = (-4) + 3$ because of the commutative property of addition.

45. $x(2) = 2(x)$ **46.** $(7 + 4) + 6 = 7 + (4 + 6)$

47. $1(x) = x$ **48.** $43 + (-18) = (-18) + 43$

49. $(-1)(93) = -93$ **50.** $109 + (-109) = 0$

51. $5(4 + 7) = 5(4) + 5(7)$ **52.** $-1(x + y) = -(x + y)$

53. $7yx = 7xy$ **54.** $(x + 2) + (-2) = x + [2 + (-2)]$

55. $6(4) + 7(4) = (6 + 7)(4)$ **56.** $(\frac{2}{3})(\frac{3}{2}) = 1$

57. $4(5x) = (4 \cdot 5)x$ **58.** $[(17)(8)](25) = (17)[(8)(25)]$

For Problems 59–79, evaluate each of the algebraic expressions for the given values of the variables.

59. $5x + 3y$; $x = -2$ and $y = -4$

60. $7x - 4y$; $x = -1$ and $y = 6$

61. $-3ab - 2c$; $a = -4$, $b = 7$, and $c = -8$

62. $x - (2y + 3z)$; $x = -3$, $y = -4$, and $z = 9$

63. $(a - 2b) + (3c - 4)$; $a = 6$, $b = -5$, and $c = -11$

64. $3a - [2b - (4c + 1)]$; $a = 4$, $b = 6$, and $c = -8$

65. $\dfrac{-2x + 7y}{x - y}$; $x = -3$ and $y = -2$

66. $\dfrac{x - 3y + 2z}{2x - y}$; $x = 4$, $y = 9$, and $z = -12$

67. $(5x - 2y)(-3x + 4y)$; $x = -3$ and $y = -7$

68. $(2a - 7b)(4a + 3b)$; $a = 6$ and $b = -3$

69. $5x + 4y - 9y - 2y$; $x = 2$ and $y = -8$

70. $5a + 7b - 9a - 6b$; $a = -7$ and $b = 8$

71. $-5x + 8y + 7y + 8x$; $x = 5$ and $y = -6$

72. $|x - y| - |x + y|$; $x = -4$ and $y = -7$

73. $|3x + y| + |2x - 4y|$; $x = 5$ and $y = -3$

74. $\left|\dfrac{x - y}{y - x}\right|$; $x = -6$ and $y = 13$

75. $\left|\dfrac{2a - 3b}{3b - 2a}\right|$; $a = -4$ and $b = -8$

76. $5(x - 1) + 7(x + 4);$ $x = 3$

77. $2(3x + 4) - 3(2x - 1);$ $x = -2$

78. $-4(2x - 1) - 5(3x + 7);$ $x = -1$

79. $5(a - 3) - 4(2a + 1) - 2(a - 4);$ $a = -3$

 80. You should be able to do calculations like those in Problems 59–79 *with* and *without* a calculator. Different types of calculators handle the "priority of operations" issue in different ways. Be sure you can do Problems 59–79 with *your* calculator.

1.2

Exponents

Positive integers are used as **exponents** to indicate repeated multiplication. For example, $4 \cdot 4 \cdot 4$ can be written 4^3, where the "raised 3" indicates that 4 is to be used as a factor three times. The following general definition is helpful.

DEFINITION 1.2

> If n is a positive integer and b is any real number, then
>
> $$b^n = \underbrace{bbb \cdots b}_{n \text{ factors of } b}$$

The number b is referred to as the **base** and n is called the **exponent**. The expression b^n can be read "b to the nth power." The terms **squared** and **cubed** are commonly associated with exponents of 2 and 3, respectively. For example, b^2 is read "b squared" and b^3 as "b cubed." An exponent of 1 is usually not written, so b^1 is simply written b.

The following examples illustrate Definition 1.2.

$$2^3 = 2 \cdot 2 \cdot 2 = 8; \qquad (\tfrac{1}{2})^5 = \tfrac{1}{2} \cdot \tfrac{1}{2} \cdot \tfrac{1}{2} \cdot \tfrac{1}{2} \cdot \tfrac{1}{2} = \tfrac{1}{32}$$

$$3^4 = 3 \cdot 3 \cdot 3 \cdot 3 = 81; \qquad (0.7)^2 = (0.7)(0.7) = 0.49$$

$$(-5)^2 = (-5)(-5) = 25; \qquad -5^2 = -(5 \cdot 5) = -25$$

We especially want to call your attention to the last two examples. Note that $(-5)^2$ means -5 is the base and is to be used as a factor twice. However, -5^2 means that 5 is the base and that after it is squared, we take the opposite of that result.

Properties of Exponents

In a previous algebra course, you have seen some properties pertaining to the use of positive integers as exponents. Those properties can be summarized as follows.

PROPERTY 1.1
Properties of
Exponents

If a and b are real numbers and m and n are positive integers, then

1. $b^n \cdot b^m = b^{n+m}$;

2. $(b^n)^m = b^{mn}$;

3. $(ab)^n = a^n b^n$;

4. $\left(\dfrac{a}{b}\right)^n = \dfrac{a^n}{b^n}$, $b \neq 0$;

5. $\dfrac{b^n}{b^m} = b^{n-m}$ when $n > m$, $b \neq 0$;

$\dfrac{b^n}{b^m} = 1$ when $n = m$, $b \neq 0$;

$\dfrac{b^n}{b^m} = \dfrac{1}{b^{m-n}}$ when $n < m$, $b \neq 0$.

Each part of Property 1.1 can be justified by using Definition 1.2. For example, to justify part 1 we can reason as follows:

$$b^n \cdot b^m = \underbrace{(bbb \cdots b)}_{\substack{n \text{ factors} \\ \text{of } b}} \cdot \underbrace{(bbb \cdots b)}_{\substack{m \text{ factors} \\ \text{of } b}}$$

$$= \underbrace{bbb \cdots b}_{\substack{n + m \text{ factors} \\ \text{of } b}}$$

$$= b^{n+m}.$$

Similar reasoning can be used to verify the other parts of Property 1.1.

The following examples illustrate the use of Property of 1.1 along with the commutative and associative properties of the real numbers. The steps enclosed in the dashed boxes might be performed mentally.

EXAMPLE 1

$$(3x^2 y)(4x^3 y^2) = \boxed{3 \cdot 4 \cdot x^2 \cdot x^3 \cdot y \cdot y^2}$$
$$= \boxed{12x^{2+3} y^{1+2}} \qquad b^n \cdot b^m = b^{n+m}$$
$$= 12x^5 y^3$$

EXAMPLE 2

$$(-2y^3)^5 = \boxed{(-2)^5 (y^3)^5} \qquad (ab)^n = a^n b^n$$
$$= -32y^{15} \qquad\qquad (b^n)^m = b^{mn}$$

EXAMPLE 3

$$\left(\frac{a^2}{b^4}\right)^7 = \frac{[(a^2)^7]}{[(b^4)^7]} \qquad \left(\frac{a}{b}\right)^n = \frac{a^n}{b^n}$$

$$= \frac{a^{14}}{b^{28}} \qquad (b^n)^m = b^{mn}$$

EXAMPLE 4

$$\frac{-56x^9}{7x^4} = \boxed{-8x^{9-4}} \qquad \frac{b^n}{b^m} = b^{n-m} \quad \text{when } n > m$$

$$= -8x^5$$

Zero and Negative Integers as Exponents

Now we can extend the concept of an exponent to include the use of zero and negative integers. First let's consider the use of zero as an exponent. We want to use zero in a way that Property 1.1 will continue to hold. For example, if $b^n \cdot b^m = b^{n+m}$ is to hold, then $x^4 \cdot x^0$ should equal x^{4+0}, which equals x^4. In other words, x^0 *acts like* 1 because $x^4 \cdot x^0 = x^4$. This line of reasoning suggests the following definition.

DEFINITION 1.3

If b is a nonzero real number, then

$$b^0 = 1.$$

Therefore, according to Definition 1.3, the following statements are all true:

$$5^0 = 1; \qquad (-413)^0 = 1; \qquad (\tfrac{3}{11})^0 = 1;$$

$$(x^3y^4)^0 = 1, \qquad \text{if } x \neq 0 \text{ and } y \neq 0.$$

A similar line of reasoning can be used to motivate a definition for the use of negative integers as exponents. Consider the example $x^4 \cdot x^{-4}$. If $b^n \cdot b^m = b^{n+m}$ is to hold, then $x^4 \cdot x^{-4}$ should equal $x^{4+(-4)}$, which equals $x^0 = 1$. Therefore x^{-4} must be the reciprocal of x^4, since their product is 1. That is to say, $x^{-4} = 1/x^4$. This suggests the following definition.

DEFINITION 1.4

If n is a positive integer and b is a nonzero real number, then

$$b^{-n} = \frac{1}{b^n}.$$

According to Definition 1.4, the following statements are true:

$$x^{-5} = \frac{1}{x^5}; \qquad 2^{-4} = \frac{1}{2^4} = \frac{1}{16};$$

$$\left(\frac{3}{4}\right)^{-2} = \frac{1}{(\frac{3}{4})^2} = \frac{1}{\frac{9}{16}} = \frac{16}{9}; \qquad \frac{2}{x^{-3}} = \frac{2}{\frac{1}{x^3}} = 2x^3.$$

The first four parts of Property 1.1 hold true *for all integers*. Furthermore, we do not need all three equations in part 5 of Property 1.1 The first equation,

$$\frac{b^n}{b^m} = b^{n-m},$$

can be used for *all integral exponents*. Let's restate Property 1.1 as it holds for all integers. We will include "name tags" for easy reference.

PROPERTY 1.2

If m and n are integers and a and b are real numbers, with $b \neq 0$ whenever it appears in a denominator, then

1. $b^n \cdot b^m = b^{n+m}$ Product of two powers

2. $(b^n)^m = b^{mn}$ Power of a power

3. $(ab)^n = a^n b^n$ Power of a product

4. $\left(\dfrac{a}{b}\right)^n = \dfrac{a^n}{b^n}$ Power of a quotient

5. $\dfrac{b^n}{b^m} = b^{n-m}$ Quotient of two powers

Having the use of all integers as exponents allows us to work with a large variety of numerical and algebraic expressions. Let's consider some examples illustrating the various parts of Property 1.2.

EXAMPLE 5

Evaluate each of the following numerical expressions.

(a) $(2^{-1} \cdot 3^2)^{-1}$ **(b)** $\left(\dfrac{2^{-3}}{3^{-2}}\right)^{-2}$

Solutions

(a) $(2^{-1} \cdot 3^2)^{-1} = (2^{-1})^{-1}(3^2)^{-1}$ Power of a product

$\qquad\qquad\qquad\;\; = (2^1)(3^{-2})$ Power of a power

$\qquad\qquad\qquad\;\; = (2)\left(\dfrac{1}{3^2}\right)$

$\qquad\qquad\qquad\;\; = 2(\tfrac{1}{9}) = \tfrac{2}{9}$

(b) $\left(\dfrac{2^{-3}}{3^{-2}}\right)^{-2} = \dfrac{(2^{-3})^{-2}}{(3^{-2})^{-2}}$ Power of a quotient

$\qquad\qquad = \dfrac{2^6}{3^4}$ Power of a power

$\qquad\qquad = \dfrac{64}{81}$

EXAMPLE 6

Find the indicated products and quotients, expressing final results with positive integral exponents only.

(a) $(3x^2y^{-4})(4x^{-3}y)$ **(b)** $\dfrac{12a^3b^2}{-3a^{-1}b^5}$ **(c)** $\left(\dfrac{15x^{-1}y^2}{5xy^{-4}}\right)^{-1}$

Solutions

(a) $(3x^2y^{-4})(4x^{-3}y) = 12x^{2+(-3)}y^{-4+1}$ Product of powers

$\qquad\qquad\qquad\quad = 12x^{-1}y^{-3}$

$\qquad\qquad\qquad\quad = \dfrac{12}{xy^3}$

(b) $\dfrac{12a^3b^2}{-3a^{-1}b^5} = -4a^{3-(-1)}b^{2-5}$ Quotient of powers

$\qquad\qquad\quad = -4a^4b^{-3}$

$\qquad\qquad\quad = -\dfrac{4a^4}{b^3}$

(c) $\left(\dfrac{15x^{-1}y^2}{5xy^{-4}}\right)^{-1} = (3x^{-1-1}y^{2-(-4)})^{-1}$ First simplify inside parentheses

$\qquad\qquad\qquad\quad = (3x^{-2}y^6)^{-1}$ Power of a product

$\qquad\qquad\qquad\quad = 3^{-1}x^2y^{-6}$

$\qquad\qquad\qquad\quad = \dfrac{x^2}{3y^6}$

The final examples of this section illustrate the simplification of numerical and algebraic expressions involving sums and differences. In such cases, Definition 1.4 is used to change from negative to positive exponents so that we can proceed in the usual ways.

EXAMPLE 7

Simplify $2^{-3} + 3^{-1}$.

Solution

$$2^{-3} + 3^{-1} = \frac{1}{2^3} + \frac{1}{3^1}$$

$$= \frac{1}{8} + \frac{1}{3}$$

$$= \frac{3}{24} + \frac{8}{24} = \frac{11}{24}$$

EXAMPLE 8

Simplify $(4^{-1} - 3^{-2})^{-1}$.

Solution

$$(4^{-1} - 3^{-2})^{-1} = \left(\frac{1}{4^1} - \frac{1}{3^2}\right)^{-1}$$

$$= \left(\frac{1}{4} - \frac{1}{9}\right)^{-1}$$

$$= \left(\frac{9}{36} - \frac{4}{36}\right)^{-1}$$

$$= \left(\frac{5}{36}\right)^{-1}$$

$$= \frac{1}{\left(\frac{5}{36}\right)^1} = \frac{36}{5}$$

EXAMPLE 9

Express $a^{-1} + b^{-2}$ as a single fraction involving positive exponents only.

Solution

$$a^{-1} + b^{-2} = \frac{1}{a^1} + \frac{1}{b^2}$$

$$= \left(\frac{1}{a}\right)\left(\frac{b^2}{b^2}\right) + \left(\frac{1}{b^2}\right)\left(\frac{a}{a}\right)$$

$$= \frac{b^2}{ab^2} + \frac{a}{ab^2}$$

$$= \frac{b^2 + a}{ab^2}$$

Remark: The expression $(n)(10)^k$, where n is a number between 1 and 10, written in decimal form, and k is an integer, is commonly called **scientific**

notation or the **scientific form** of a number. Very large and very small numbers can be conveniently expressed in scientific notation. For example, 950,000,000,000 can be written $(9.5)(10)^{11}$ and 0.00000049 is $(4.9)(10)^{-7}$.

Problem Set 1.2

Evaluate each of the following numerical expressions.

1. 2^{-3}
2. 3^{-2}
3. -10^{-3}
4. 10^{-4}

5. $\dfrac{1}{3^{-3}}$
6. $\dfrac{1}{2^{-5}}$
7. $(\frac{1}{2})^{-2}$
8. $-(\frac{1}{3})^{-2}$

9. $(-\frac{2}{3})^{-3}$
10. $(\frac{5}{6})^{-2}$
11. $(-\frac{1}{5})^{0}$
12. $\dfrac{1}{(\frac{3}{5})^{-2}}$

13. $\dfrac{1}{(\frac{4}{5})^{-2}}$
14. $(\frac{4}{5})^{0}$
15. $2^{5} \cdot 2^{-3}$
16. $3^{-2} \cdot 3^{5}$

17. $10^{-6} \cdot 10^{4}$
18. $10^{6} \cdot 10^{-9}$
19. $10^{-2} \cdot 10^{-3}$
20. $10^{-1} \cdot 10^{-5}$

21. $(3^{-2})^{-2}$
22. $((-2)^{-1})^{-3}$
23. $(4^{2})^{-1}$
24. $(3^{-1})^{3}$

25. $(3^{-1} \cdot 2^{2})^{-1}$
26. $(2^{3} \cdot 3^{-2})^{-2}$
27. $(4^{2} \cdot 5^{-1})^{2}$
28. $(2^{-2} \cdot 4^{-1})^{3}$

29. $\left(\dfrac{2^{-2}}{5^{-1}}\right)^{-2}$
30. $\left(\dfrac{3^{-1}}{2^{-3}}\right)^{-2}$
31. $\left(\dfrac{3^{-2}}{8^{-1}}\right)^{2}$
32. $\left(\dfrac{4^{2}}{5^{-1}}\right)^{-1}$

33. $\dfrac{2^{3}}{2^{-3}}$
34. $\dfrac{2^{-3}}{2^{3}}$
35. $\dfrac{10^{-1}}{10^{4}}$
36. $\dfrac{10^{-3}}{10^{-7}}$

37. $3^{-2} + 2^{-3}$
38. $2^{-3} + 5^{-1}$
39. $(\frac{2}{3})^{-1} - (\frac{3}{4})^{-1}$
40. $3^{-2} - 2^{3}$

41. $(2^{-4} + 3^{-1})^{-1}$
42. $(3^{-2} - 5^{-1})^{-1}$

Simplify each of the following; express final results without using zero or negative integers as exponents.

43. $x^{3} \cdot x^{-7}$
44. $x^{-2} \cdot x^{-3}$
45. $a^{2} \cdot a^{-3} \cdot a^{-1}$

46. $b^{-3} \cdot b^{5} \cdot b^{-4}$
47. $(a^{-3})^{2}$
48. $(b^{5})^{-2}$

49. $(x^{3}y^{-4})^{-1}$
50. $(x^{4}y^{-2})^{-2}$
51. $(ab^{2}c^{-1})^{-3}$

52. $(a^{2}b^{-1}c^{-2})^{-4}$
53. $(2x^{2}y^{-1})^{-2}$
54. $(3x^{4}y^{-2})^{-1}$

55. $\left(\dfrac{x^{-2}}{y^{-3}}\right)^{-2}$
56. $\left(\dfrac{y^{4}}{x^{-1}}\right)^{-3}$
57. $\left(\dfrac{2a^{-1}}{3b^{-2}}\right)^{-2}$

58. $\left(\dfrac{3x^{2}y}{4a^{-1}b^{-3}}\right)^{-1}$
59. $\dfrac{x^{-5}}{x^{-2}}$
60. $\dfrac{a^{-3}}{a^{5}}$

61. $\dfrac{a^{2}b^{-3}}{a^{-1}b^{-2}}$
62. $\dfrac{x^{-1}y^{-2}}{x^{3}y^{-1}}$

Find the indicated products and quotients; express results using positive integral exponents only.

63. $(2x^{-1}y^{2})(3x^{-2}y^{-3})$
64. $(4x^{-2}y^{3})(-5x^{3}y^{-4})$

65. $(-6a^{5}y^{-4})(-a^{-7}y)$
66. $(-8a^{-4}b^{-5})(-6a^{-1}b^{8})$

67. $\dfrac{24x^{-1}y^{-2}}{6x^{-4}y^{3}}$
68. $\dfrac{56xy^{-3}}{8x^{2}y^{2}}$
69. $\dfrac{-35a^{3}b^{-2}}{7a^{5}b^{-1}}$

70. $\dfrac{27a^{-4}b^{-5}}{-3a^{-2}b^{-4}}$
71. $\left(\dfrac{14x^{-2}y^{-4}}{7x^{-3}y^{-6}}\right)^{-2}$
72. $\left(\dfrac{24x^{5}y^{-3}}{-8x^{6}y^{-1}}\right)^{-3}$

Express each of the following as a single fraction involving positive exponents only.

73. $x^{-1} + x^{-2}$

74. $x^{-2} + x^{-4}$

75. $x^{-2} - y^{-1}$

76. $2x^{-1} - 3y^{-3}$

77. $3a^{-2} + 2b^{-3}$

78. $a^{-2} + a^{-1}b^{-2}$

79. $x^{-1}y - xy^{-1}$

80. $x^2y^{-1} - x^{-3}y^2$

Find each of the following products and quotients. Assume that all variables appearing as exponents represent integers. For example,

$$(x^{2b})(x^{-b+1}) = x^{2b+(-b+1)} = x^{b+1}.$$

81. $(3x^a)(4x^{2a+1})$

82. $(5x^{-a})(-6x^{3a-1})$

83. $(x^a)(x^{-a})$

84. $(-2y^{3b})(-4y^{b+1})$

85. $\dfrac{x^{3a}}{x^a}$

86. $\dfrac{4x^{2a+1}}{2x^{a-2}}$

87. $\dfrac{-24y^{5b+1}}{6y^{-b-1}}$

88. $(x^a)^{2b}(x^b)^a$

89. $\dfrac{(xy)^b}{y^b}$

90. $\dfrac{(2x^{2b})(-4x^{b+1})}{8x^{-b+2}}$

1.3

Polynomials

Recall that algebraic expressions such as $5x$, $-6y^2$, $2x^{-1}y^{-2}$, $14a^2b$, $5x^{-4}$, and $-17ab^2c^3$ are called **terms**. Terms containing variables with only non-negative integers as exponents are called **monomials**. Of the previously listed terms, $5x$, $-6y^2$, $14a^2b$, and $-17ab^2c^3$ are monomials. The **degree** of a monomial is the sum of the exponents of the literal factors. For example, $7xy$ is of degree 2, while $14a^2b$ is of degree 3, and $-17ab^2c^3$ is of degree 6. If the monomial contains only one variable, then the exponent of that variable is the degree of the monomial. For example, $5x^3$ is of degree 3 and $-8y^4$ is of degree 4. Any nonzero constant term, such as 8, is of degree zero.

A **polynomial** is a monomial or a finite sum of monomials. Thus

$$4x^2, \quad 3x^2 - 2x - 4, \quad 7x^4 - 6x^3 + 5x^2 - 2x - 1,$$
$$3x^2y + 2y, \quad \tfrac{1}{5}a^2 - \tfrac{2}{3}b^2, \quad \text{and} \quad 14$$

are examples of polynomials. In addition to calling a polynomial with one term a monomial, we also classify polynomials with two terms as **binomials** and those with three terms, **trinomials**.

The **degree of a polynomial** is the degree of the term with the highest degree in the polynomial. The following examples illustrate some of this terminology.

The polynomial $4x^3y^4$ is a monomial in two variables of degree 7.

The polynomial $4x^2y - 2xy$ is a binomial in two variables of degree 3.

The polynomial $9x^2 - 7x - 1$ is a trinomial in one variable of degree 2.

Addition and Subtraction of Polynomials

Both adding and subtracting polynomials rely on basically the same ideas. The commutative, associative, and distributive properties provide the basis for rearranging, regrouping, and combining similar terms. Consider the following addition problems.

$$(4x^2 + 5x + 1) + (7x^2 - 9x + 4) = (4x^2 + 7x^2) + (5x - 9x) + (1 + 4)$$
$$= 11x^2 - 4x + 5$$

$$(5x - 3) + (3x + 2) + (8x + 6) = (5x + 3x + 8x) + (-3 + 2 + 6)$$
$$= 16x + 5$$

The definition of subtraction as "adding the opposite" $(a - b = a + (-b))$ extends to polynomials in general. The opposite of a polynomial can be formed by taking the opposite of each term. For example, the opposite of $3x^2 - 7x + 1$ is $-3x^2 + 7x - 1$. Symbolically this is expressed

$$-(3x^2 - 7x + 1) = -3x^2 + 7x - 1.$$

You can also think in terms of the property $-x = -1(x)$ and the distributive property. Therefore,

$$-(3x^2 - 7x + 1) = -1(3x^2 - 7x + 1) = -3x^2 + 7x - 1.$$

Now consider the following subtraction problems.

$$(7x^2 - 2x - 4) - (3x^2 + 7x - 1) = (7x^2 - 2x - 4) + (-3x^2 - 7x + 1)$$
$$= (7x^2 - 3x^2) + (-2x - 7x) + (-4 + 1)$$
$$= 4x^2 - 9x - 3$$

$$(4y^2 + 7) - (-3y^2 + y - 2) = (4y^2 + 7) + (3y^2 - y + 2)$$
$$= (4y^2 + 3y^2) + (-y) + (7 + 2)$$
$$= 7y^2 - y + 9$$

Multiplying Polynomials

The distributive property is usually stated as $a(b + c) = ab + ac$, but it can be extended as follows:

$$a(b + c + d) = ab + ac + ad;$$
$$a(b + c + d + e) = ab + ac + ad + ae;$$

$$\text{etc.}$$

The commutative and associative properties, the properties of exponents, and the distributive property work together to form a basis for finding the product of a monomial and a polynomial. The following example illustrates this idea.

$$3x^2(2x^2 + 5x + 3) = 3x^2(2x^2) + 3x^2(5x) + 3x^2(3)$$
$$= 6x^4 + 15x^3 + 9x^2$$

Extending the method of finding the product of a monomial and a polynomial to finding the product of two polynomials is again based on the distributive property.

$$(x + 2)(y + 5) = x(y + 5) + 2(y + 5)$$
$$= x(y) + x(5) + 2(y) + 2(5)$$
$$= xy + 5x + 2y + 10$$

Notice that each term of the first polynomial multiplies each term of the second polynomial.

$$(x - 3)(y + z + 3) = x(y + z + 3) - 3(y + z + 3)$$
$$= xy + xz + 3x - 3y - 3z - 9$$

Frequently, multiplying polynomials will produce similar terms that can be combined, simplifying the resulting polynomial.

$$(x + 5)(x + 7) = x(x + 7) + 5(x + 7)$$
$$= x^2 + 7x + 5x + 35$$
$$= x^2 + 12x + 35$$

$$(x - 2)(x^2 - 3x + 4) = x(x^2 - 3x + 4) - 2(x^2 - 3x + 4)$$
$$= x^3 - 3x^2 + 4x - 2x^2 + 6x - 8$$
$$= x^3 - 5x^2 + 10x - 8$$

In a previous algebra course, you may have developed a shortcut for multiplying binomials, as illustrated by the following example.

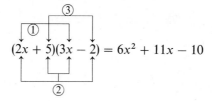

$$(2x + 5)(3x - 2) = 6x^2 + 11x - 10$$

Step ①: Multiply $(2x)(3x)$.

Step ②: Multiply $(5)(3x)$ and $(2x)(-2)$ and combine.

Step ③: Multiply $(5)(-2)$.

Remark: Shortcuts can be very helpful for certain manipulations in mathematics. But a word of caution: do not lose the understanding of what you are doing. Always be able to do the manipulation without the shortcut.

Exponents can also be used to indicate repeated multiplication of polynomials. For example, $(3x - 4y)^2$ means $(3x - 4y)(3x - 4y)$, and $(x + 4)^3$ means $(x + 4)(x + 4)(x + 4)$. Therefore, raising a polynomial to a power is merely another multiplication problem.

$$(3x - 4y)^2 = (3x - 4y)(3x - 4y) = 9x^2 - 24xy + 16y^2$$

(When squaring a binomial, be careful not to forget the middle term. That is to say, $(x + 5)^2 \neq x^2 + 25$; instead, $(x + 5)^2 = x^2 + 10x + 25$.)

$$(x + 4)^3 = (x + 4)(x + 4)(x + 4)$$
$$= (x + 4)(x^2 + 8x + 16)$$
$$= x(x^2 + 8x + 16) + 4(x^2 + 8x + 16)$$
$$= x^3 + 8x^2 + 16x + 4x^2 + 32x + 64$$
$$= x^3 + 12x^2 + 48x + 64$$

Special Patterns

When multiplying binomials, some special patterns occur that you should learn to recognize. These patterns can be used to find products, and some of them are very helpful later when factoring polynomials.

$$(a + b)^2 = a^2 + 2ab + b^2$$
$$(a - b)^2 = a^2 - 2ab + b^2$$
$$(a + b)(a - b) = a^2 - b^2$$
$$(a + b)^3 = a^3 + 3a^2b + 3ab^2 + b^3$$
$$(a - b)^3 = a^3 - 3a^2b + 3ab^2 - b^3$$

The following three examples illustrate the first three patterns, respectively.

$$(2x + 3)^2 = (2x)^2 + 2(2x)(3) + (3)^2$$
$$= 4x^2 + 12x + 9$$

$$(5x - 2)^2 = (5x)^2 - 2(5x)(2) + (2)^2$$
$$= 25x^2 - 20x + 4$$

$$(3x + 2y)(3x - 2y) = (3x)^2 - (2y)^2 = 9x^2 - 4y^2$$

In the first two examples, the resulting trinomial is called a **perfect-square trinomial**; it is the result of squaring a binomial. In the third example, the resulting binomial is called the **difference of two squares**. We will use both of these patterns extensively later when factoring polynomials.

The "cubing of a binomial" patterns are helpful primarily when multiplying. These patterns can shorten your work when cubing a binomial, as the next two examples illustrate.

$$(3x + 2)^3 = (3x)^3 + 3(3x)^2(2) + 3(3x)(2)^2 + (2)^3$$
$$= 27x^3 + 54x^2 + 36x + 8$$

$$(5x - 2y)^3 = (5x)^3 - 3(5x)^2(2y) + 3(5x)(2y)^2 - (2y)^3$$
$$= 125x^3 - 150x^2y + 60xy^2 - 8y^3$$

Keep in mind that these multiplying patterns are useful shortcuts, but if you forget them, simply revert to applying the distributive property.

The Binomial Expansion Pattern

It is possible to write the expansion of $(a + b)^n$, where n is *any* positive integer, without showing all of the intermediate steps of multiplying and combining similar terms. To do this, let's observe some patterns in the following examples, each of which can be verified by direct multiplication.

$$(a + b)^1 = a + b$$
$$(a + b)^2 = a^2 + 2ab + b^2$$
$$(a + b)^3 = a^3 + 3a^2b + 3ab^2 + b^3$$
$$(a + b)^4 = a^4 + 4a^3b + 6a^2b^2 + 4ab^3 + b^4$$
$$(a + b)^5 = a^5 + 5a^4b + 10a^3b^2 + 10a^2b^3 + 5ab^4 + b^5$$

First, note the patterns of the exponents for a and b on a term-by-term basis. The exponents of a begin with the exponent of the binomial and decrease by 1, term-by-term, until the last term, which has $a^0 = 1$. The exponents of b begin with zero ($b^0 = 1$) and increase by 1, term-by-term, until the last term, which contains b to the power of the original binomial. In other words, the variables in the expansion of $(a + b)^n$ have the pattern

$$a^n, \quad a^{n-1}b, \quad a^{n-2}b^2, \quad \ldots, \quad ab^{n-1}, \quad b^n,$$

where for each term, the *sum* of the exponents of a and b is n.

Next let's arrange the *coefficients* in a triangular formation: this yields an easy-to-remember pattern.

$$
\begin{array}{ccccccccc}
 & & & & 1 & & 1 & & \\
 & & & 1 & & 2 & & 1 & \\
 & & 1 & & 3 & & 3 & & 1 \\
 & 1 & & 4 & & 6 & & 4 & & 1 \\
1 & & 5 & & 10 & & 10 & & 5 & & 1
\end{array}
$$

Row number n in the formation contains the coefficients of the expansion of $(a + b)^n$. For example, the fifth row contains 1 5 10 10 5 1, and these numbers are the coefficients of the terms in the expansion of $(a + b)^5$. Furthermore, each row can be formed from the previous row as follows:

1. Start and end each row with 1.

2. All other entries result from adding the two numbers in the row immediately above, one number to the left and one number to the right.

Thus, from row 5 we can form row 6.

row 5: 1   5   10   10   5   1

add add add add add

row 6: 1   6   15   20   15   6   1

Now we can use these 7 coefficients and our discussion about the exponents to write out the expansion for $(a + b)^6$:

$$(a + b)^6 = a^6 + 6a^5b + 15a^4b^2 + 20a^3b^3 + 15a^2b^4 + 6ab^5 + b^6.$$

Remark: The triangular formation of numbers that we have been discussing is often referred to as *Pascal's triangle*. This is in honor of Blaise Pascal, a seventeenth century mathematician, to whom the discovery of this pattern is attributed.

Let's consider two more examples using Pascal's triangle and the exponent relationships.

EXAMPLE 1

Expand $(a - b)^4$.

Solution We can treat $a - b$ as $a + (-b)$ and use the fourth row of Pascal's triangle to obtain the coefficients.

$$[a + (-b)]^4 = a^4 + 4a^3(-b) + 6a^2(-b)^2 + 4a(-b)^3 + (-b)^4$$
$$= a^4 - 4a^3b + 6a^2b^2 - 4ab^3 + b^4$$

EXAMPLE 2

Expand $(2x + 3y)^5$.

Solution Let $2x = a$ and $3y = b$. The coefficients come from the fifth row of Pascal's triangle.

$$(2x + 3y)^5 = (2x)^5 + 5(2x)^4(3y) + 10(2x)^3(3y)^2 + 10(2x)^2(3y)^3$$
$$+ 5(2x)(3y)^4 + (3y)^5$$
$$= 32x^5 + 240x^4y + 720x^3y^2 + 1080x^2y^3 + 810xy^4 + 243y^5$$

Dividing Polynomials by Monomials

In Section 1.5 we will review the addition and subtraction of rational expressions using the properties

$$\frac{a}{b} + \frac{c}{b} = \frac{a + c}{b} \qquad \text{and} \qquad \frac{a}{b} - \frac{c}{b} = \frac{a - c}{b}.$$

These properties can also be viewed as

$$\frac{a + c}{b} = \frac{a}{b} + \frac{c}{b} \qquad \text{and} \qquad \frac{a - c}{b} = \frac{a}{b} - \frac{c}{b}.$$

Together with our knowledge of dividing monomials, they provide the basis for dividing polynomials by monomials. Consider the following examples.

$$\frac{18x^3 + 24x^2}{6x} = \frac{18x^3}{6x} + \frac{24x^2}{6x} = 3x^2 + 4x$$

$$\frac{35x^2y^3 - 55x^3y^4}{5xy^2} = \frac{35x^2y^3}{5xy^2} - \frac{55x^3y^4}{5xy^2} = 7xy - 11x^2y^2$$

Therefore, to divide a polynomial by a monomial, we divide each term of the polynomial by the monomial. As with many skills, once you feel comfortable with the process you may then choose to perform some of the steps mentally. Your work could take the following format.

$$\frac{40x^4y^5 + 72x^5y^7}{8x^2y} = 5x^2y^4 + 9x^3y^6$$

$$\frac{36a^3b^4 - 48a^3b^3 + 64a^2b^5}{-4a^2b^2} = -9ab^2 + 12ab - 16b^3$$

Problem Set 1.3

In Problems 1–10, perform the indicated operations.

1. $(5x^2 - 7x - 2) + (9x^2 + 8x - 4)$ **2.** $(-9x^2 + 8x + 4) + (7x^2 - 5x - 3)$

3. $(14x^2 - x - 1) - (15x^2 + 3x + 8)$ **4.** $(-3x^2 + 2x + 4) - (4x^2 + 6x - 5)$

5. $(3x - 4) - (6x + 3) + (9x - 4)$ **6.** $(7a - 2) - (8a - 1) - (10a - 2)$

7. $(8x^2 - 6x - 2) + (x^2 - x - 1) - (3x^2 - 2x + 4)$

8. $(12x^2 + 7x - 2) - (3x^2 + 4x + 5) + (-4x^2 - 7x - 2)$

9. $5(x - 2) - 4(x + 3) - 2(x + 6)$ **10.** $3(2x - 1) - 2(3x + 4) - 4(5x - 1)$

In Problems 11–54, find the indicated products. Remember the special patterns that were discussed in this section.

11. $3xy(4x^2y + 5xy^2)$ **12.** $-2ab^2(3a^2b - 4ab^3)$

13. $6a^3b^2(5ab - 4a^2b + 3ab^2)$ **14.** $-xy^4(5x^2y - 4xy^2 + 3x^2y^2)$

15. $(x + 8)(x + 12)$ **16.** $(x - 9)(x + 6)$ **17.** $(n - 4)(n - 12)$

18. $(n + 6)(n - 10)$ **19.** $(s - t)(x + y)$ **20.** $(a + b)(c + d)$

21. $(3x - 1)(2x + 3)$ **22.** $(5x + 2)(3x + 4)$ **23.** $(4x - 3)(3x - 7)$

24. $(4n + 3)(6n - 1)$ **25.** $(x + 4)^2$ **26.** $(x - 6)^2$

27. $(2n + 3)^2$ **28.** $(3n - 5)^2$

29. $(x + 2)(x - 4)(x + 3)$ **30.** $(x - 1)(x + 6)(x - 5)$

31. $(x - 1)(2x + 3)(3x - 2)$ **32.** $(2x + 5)(x - 4)(3x + 1)$

33. $(x - 1)(x^2 + 3x - 4)$ **34.** $(t + 1)(t^2 - 2t - 4)$

35. $(t - 1)(t^2 + t + 1)$ **36.** $(2x - 1)(x^2 + 4x + 3)$

37. $(3x + 2)(2x^2 - x - 1)$ **38.** $(3x - 2)(2x^2 + 3x + 4)$

39. $(x^2 + 2x - 1)(x^2 + 6x + 4)$ **40.** $(x^2 - x + 4)(2x^2 - 3x - 1)$

41. $(5x - 2)(5x + 2)$ **42.** $(3x - 4)(3x + 4)$

43. $(x^2 - 5x - 2)^2$ **44.** $(-x^2 + x - 1)^2$ **45.** $(2x + 3y)(2x - 3y)$

46. $(9x + y)(9x - y)$ **47.** $(x + 5)^3$ **48.** $(x - 6)^3$

49. $(2x + 1)^3$ **50.** $(3x + 4)^3$ **51.** $(4x - 3)^3$

52. $(2x - 5)^3$ **53.** $(5x - 2y)^3$ **54.** $(x + 3y)^3$

For Problems 55–66, use Pascal's triangle to help expand each of the following.

55. $(a + b)^7$ **56.** $(a + b)^8$ **57.** $(x - y)^5$ **58.** $(x - y)^6$

59. $(x + 2y)^4$ **60.** $(2x + y)^5$ **61.** $(2a - b)^6$ **62.** $(3a - b)^4$

63. $(x^2 + y)^7$ **64.** $(x + 2y^2)^7$ **65.** $(2a - 3b)^5$ **66.** $(4a - 3b)^3$

In Problems 67–72, perform the indicated divisions.

67. $\dfrac{15x^4 - 25x^3}{5x^2}$ **68.** $\dfrac{-48x^8 - 72x^6}{-8x^4}$

69. $\dfrac{30a^5 - 24a^3 + 54a^2}{-6a}$ **70.** $\dfrac{18x^3y^2 + 27x^2y^3}{3xy}$

71. $\dfrac{-20a^3b^2 - 44a^4b^5}{-4a^2b}$ **72.** $\dfrac{21x^5y^6 + 28x^4y^3 - 35x^5y^4}{7x^2y^3}$

In Problems 73–82, find the indicated products. Assume all variables appearing as exponents represent integers.

73. $(x^a + y^b)(x^a - y^b)$ **74.** $(x^{2a} + 1)(x^{2a} - 3)$ **75.** $(x^b + 4)(x^b - 7)$

76. $(3x^a - 2)(x^a + 5)$ **77.** $(2x^b - 1)(3x^b + 2)$ **78.** $(2x^a - 3)(2x^a + 3)$
79. $(x^{2a} - 1)^2$ **80.** $(x^{3b} + 2)^2$ **81.** $(x^a - 2)^3$ **82.** $(x^b + 3)^3$

1.4

Factoring Polynomials

If a polynomial is equal to the product of other polynomials, then each polynomial in the product is called a **factor** of the original polynomial. For example, since $x^2 - 4$ can be expressed as $(x + 2)(x - 2)$, we say that $x + 2$ and $x - 2$ are factors of $x^2 - 4$. The process of expressing a polynomial as a product of polynomials is called **factoring**. In this section we will consider methods of factoring polynomials with integer coefficients.

In general, factoring is the reverse of multiplication, so we can use our knowledge of multiplication to help develop factoring techniques. For example, we previously used the distributive property to find the product of a monomial and a polynomial, as the next examples illustrate.

$$3(x + 2) = 3(x) + 3(2) = 3x + 6$$

$$3x(x + 4) = 3x(x) + 3x(4) = 3x^2 + 12x$$

For factoring purposes, the distributive property (now in the form $ab + ac = a(b + c)$) can be used to reverse the process. (The steps indicated in the dashed boxes can be done mentally.)

$$3x + 6 = \boxed{3(x) + 3(2)} = 3(x + 2)$$

$$3x^2 + 12x = \boxed{3x(x) + 3x(4)} = 3x(x + 4)$$

Polynomials can be factored in a variety of ways. Consider some factorizations of $3x^2 + 12x$.

$$3x^2 + 12x = 3x(x + 4) \qquad \text{or} \qquad 3x^2 + 12x = 3(x^2 + 4x) \qquad \text{or}$$
$$3x^2 + 12x = x(3x + 12) \qquad \text{or} \qquad 3x^2 + 12x = \tfrac{1}{2}(6x^2 + 24x)$$

We are, however, primarily interested in the first of these factorization forms; we shall refer to it as the **completely factored form**. A polynomial with integral coefficients is in completely factored form if:

1. it is expressed as a product of polynomials with *integral coefficients*, and

2. no polynomial, other than a monomial, within the factored form can be further factored into polynomials with integral coefficients.

Do you see why only the first of the factored forms of $3x^2 + 12x$ is said to be in completely factored form? In each of the other three forms, the polynomial inside the parentheses can be further factored. Furthermore, in the last form, $\tfrac{1}{2}(6x^2 + 24x)$, the condition of using only integers is violated.

This application of the distributive property is often referred to as **factoring out the highest common monomial factor**. The following examples further illustrate the process.

$$12x^3 + 16x^2 = 4x^2(3x + 4)$$

$$8ab - 18b = 2b(4a - 9)$$

$$6x^2y^3 + 27xy^4 = 3xy^3(2x + 9y)$$

$$30x^3 + 42x^4 - 24x^5 = 6x^3(5 + 7x - 4x^2)$$

Sometimes there may be a common *binomial* factor rather than a common monomial factor. For example, each of the two terms in the expression $x(y + 2) + z(y + 2)$ has a binomial factor of $y + 2$. Thus we can factor $y + 2$ from each term and obtain the following result:

$$x(y + 2) + z(y + 2) = (y + 2)(x + z).$$

Consider a few more examples involving a common binomial factor:

$$a^2(b + 1) + 2(b + 1) = (b + 1)(a^2 + 2);$$

$$x(2y - 1) - y(2y - 1) = (2y - 1)(x - y);$$

$$x(x + 2) + 3(x + 2) = (x + 2)(x + 3).$$

It sometimes seems that a given polynomial exhibits no apparent common monomial or binomial factor. Such is the case with $ab + 3a + bc + 3c$. However, by factoring a from the first two terms and c from the last two terms, it can be expressed

$$ab + 3a + bc + 3c = a(b + 3) + c(b + 3).$$

Now a common binomial factor of $b + 3$ is obvious and we can proceed as before:

$$a(b + 3) + c(b + 3) = (b + 3)(a + c).$$

This factoring process is referred to as **factoring by grouping**. Let's consider another example of this type.

$$ab^2 - 4b^2 + 3a - 12 = b^2(a - 4) + 3(a - 4)$$
Factor b^2 from first two terms, 3 from last two.

$$= (a - 4)(b^2 + 3)$$
Factor common binomial from both terms.

The Difference of Two Squares

In Section 1.3 we called your attention to some special multiplication patterns. One of these patterns was the following:

$$(a + b)(a - b) = a^2 - b^2.$$

This same pattern, viewed as a factoring pattern,

$$a^2 - b^2 = (a + b)(a - b)$$

is referred to as the **difference of two squares**. Applying the pattern is a fairly simple process, as these next examples illustrate. Again the steps we have included in dashed boxes are usually performed mentally.

$$x^2 - 16 = \boxed{(x)^2 - (4)^2} = (x + 4)(x - 4)$$
$$4x^2 - 25 = \boxed{(2x)^2 - (5)^2} = (2x + 5)(2x - 5)$$

Since multiplication is commutative, the order of writing the factors is not important. For example, $(x + 4)(x - 4)$ can also be written $(x - 4)(x + 4)$.

You must be careful not to assume an analogous factoring pattern for the *sum* of two squares; *it does not exist*. For example, $x^2 + 4 \neq (x + 2)(x + 2)$ because $(x + 2)(x + 2) = x^2 + 4x + 4$. We say that a polynomial such as $x^2 + 4$ is **not factorable using integers**.

Sometimes the difference-of-two-squares pattern can be applied more than once, as the next example illustrates:

$$16x^4 - 81y^4 = (4x^2 + 9y^2)(4x^2 - 9y^2) = (4x^2 + 9y^2)(2x + 3y)(2x - 3y).$$

It may also happen that the squares are not just simple monomial squares. These next three examples illustrate such polynomials.

$$(x + 3)^2 - y^2 = [(x + 3) + y][(x + 3) - y] = (x + 3 + y)(x + 3 - y)$$
$$4x^2 - (2y + 1)^2 = [2x + (2y + 1)][2x - (2y + 1)]$$
$$= (2x + 2y + 1)(2x - 2y - 1)$$
$$(x - 1)^2 - (x + 4)^2 = [(x - 1) + (x + 4)][(x - 1) - (x + 4)]$$
$$= (x - 1 + x + 4)(x - 1 - x - 4)$$
$$= (2x + 3)(-5)$$

It is possible that both the technique of factoring out a common monomial factor and the pattern of the difference of two squares can be applied to the same problem. *In general, it is best to look first for a common monomial factor.* Consider the following examples.

$$2x^2 - 50 = 2(x^2 - 25)$$
$$= 2(x + 5)(x - 5)$$
$$48y^3 - 27y = 3y(16y^2 - 9)$$
$$= 3y(4y + 3)(4y - 3)$$
$$9x^2 - 36 = 9(x^2 - 4)$$
$$= 9(x + 2)(x - 2)$$

Factoring Trinomials

Expressing a trinomial as the product of two binomials is one of the most common factoring techniques used in algebra. Like before, to develop a factoring technique we first look at some multiplication ideas. Let's consider the product $(x + a)(x + b)$, using the distributive property to show how each term of the resulting trinomial is formed.

$$(x + a)(x + b) = x(x + b) + a(x + b)$$
$$= x(x) + x(b) + a(x) + a(b)$$
$$= x^2 + (a + b)x + ab$$

Notice that the coefficient of the middle term is the *sum* of a and b and the last term is the *product* of a and b. These two relationships can be used to factor trinomials. Let's consider some examples.

EXAMPLE 1

Factor $x^2 + 12x + 20$.

Solution We need two integers whose sum is 12 and whose product is 20. The numbers are 2 and 10, and we can complete the factoring as follows.

$$x^2 + 12x + 20 = (x + 2)(x + 10)$$

EXAMPLE 2

Factor $x^2 - 3x - 54$.

Solution We need two integers whose sum is -3 and whose product is -54. The integers are -9 and 6, and we can factor as follows.

$$x^2 - 3x - 54 = (x - 9)(x + 6)$$

EXAMPLE 3

Factor $x^2 + 7x + 16$.

Solution We need two integers whose sum is 7 and whose product is 16. The only possible pairs of factors of 16 are $1 \cdot 16$, $2 \cdot 8$, and $4 \cdot 4$. Since a sum of 7 is not produced by any of these pairs, the polynomial $x^2 + 7x + 16$ is *not factorable using integers*.

Trinomials of the Form $ax^2 + bx + c$

Now let's consider factoring trinomials for which the coefficient of the squared term is not 1. One approach to factoring such trinomials is based on the ideas used in the previous examples. To see the basis of this technique, consider the following product.

$$(px + r)(qx + s) = px(qx) + px(s) + r(qx) + r(s)$$
$$= (pq)x^2 + (ps + rq)x + rs$$

Notice that the product of the coefficient of x^2 and the constant term is *pqrs*. Likewise, the product of the two coefficients of x (*ps* and *rq*) is also *pqrs*. Therefore, the coefficient of x must be a sum of the form *ps* + *rq*, such that the product of the coefficient of x^2 and the constant term equals *pqrs*. Now let's see how this works in some specific examples.

EXAMPLE 4

Factor $6x^2 + 17x + 5$.

Solution

$$6x^2 + 17x + 5 \qquad \text{sum of 17}$$

product of $6 \cdot 5 = 30$

We need two integers whose sum is 17 and whose product is 30. The integers 2 and 15 satisfy these conditions. Therefore the middle term, $17x$, of the given trinomial can be expressed as $2x + 15x$ and we can proceed as follows.

$$\begin{aligned}
6x^2 + 17x + 5 &= 6x^2 + 2x + 15x + 5 \\
&= 2x(3x + 1) + 5(3x + 1) \\
&= (3x + 1)(2x + 5)
\end{aligned}$$

EXAMPLE 5

Factor $5x^2 - 18x - 8$.

Solution

$$5x^2 - 18x - 8 \qquad \text{sum of } -18$$

product of $5(-8) = -40$

We need two integers whose sum is -18 and whose product is -40. The integers -20 and 2 satisfy these conditions. Therefore the middle term, $-18x$, of the trinomial can be written $-20x + 2x$ and we can factor as follows.

$$\begin{aligned}
5x^2 - 18x - 8 &= 5x^2 - 20x + 2x - 8 \\
&= 5x(x - 4) + 2(x - 4) \\
&= (x - 4)(5x + 2)
\end{aligned}$$

EXAMPLE 6

Factor $4x^2 + 6x + 9$.

Solution

$$4x^2 + 6x + 9 \qquad \text{sum of 6}$$

product of $4 \cdot 9 = 36$

We need two integers whose sum is 6 and whose product is 36. The only possible pairs of factors of 36 are $1 \cdot 36$, $2 \cdot 18$, $3 \cdot 12$, $4 \cdot 9$, and $6 \cdot 6$. A sum of 6 is not produced by any of these pairs and therefore the original trinomial, $4x^2 + 6x + 9$, is *not factorable using integers*.

EXAMPLE 7

Factor $24x^2 + 2x - 15$.

Solution

$$24x^2 + 2x - 15 \quad \rightarrow \text{sum of 2}$$

product of $24(-15) = -360$

We need two integers whose sum is 2 and whose product is -360. To help find these integers, let's factor 360 into primes.

$$360 = 2 \cdot 2 \cdot 2 \cdot 3 \cdot 3 \cdot 5$$

Now by grouping these factors in various ways, we find that $2 \cdot 2 \cdot 5 = 20$ and $2 \cdot 3 \cdot 3 = 18$, so we can use the integers 20 and -18 to produce a sum of 2 and a product of -360. Therefore the middle term, $2x$, of the trinomial can be expressed as $20x - 18x$ and we can proceed as follows.

$$24x^2 + 2x - 15 = 24x^2 + 20x - 18x - 15$$
$$= 4x(6x + 5) - 3(6x + 5)$$
$$= (6x + 5)(4x - 3)$$

One final point should be made about factoring trinomials. We have been using a rather systematic approach; however, you may find that as you become more proficient at factoring, a less systematic, trial-and-error process may be convenient to use at times. For example, suppose that we need to factor $2n^2 + 15n + 7$. By looking at the first term, $2n^2$, and the two positive signs, we know that the binomials are of the form

$$(n + \underline{\quad})(2n + \underline{\quad}).$$

Furthermore, since the constant term of 7 can be factored only as $7 \cdot 1$, we only have two possibilities to try:

$$(n + 1)(2n + 7) \quad \text{or} \quad (n + 7)(2n + 1).$$

By checking the middle term of each product, we can quickly determine that the second product yields the correct middle term. Therefore we obtain

$$2n^2 + 15n + 7 = (n + 7)(2n + 1).$$

The Sum and Difference of Two Cubes

Earlier in this section we discussed the difference-of-squares factoring pattern. It was also pointed out that no analogous sum-of-squares pattern exists; that is to say, a polynomial such as $x^2 + 9$ is not factorable using integers. However, there do exist patterns for both the *sum* and the *difference of two cubes*. These patterns are as follows.

$$x^3 + y^3 = (x + y)(x^2 - xy + y^2)$$
$$x^3 - y^3 = (x - y)(x^2 + xy + y^2)$$

Note how these patterns are used in the next three examples.

$$x^3 + 8 = x^3 + 2^3 = (x + 2)(x^2 - 2x + 4)$$
$$8x^3 - 27y^3 = (2x)^3 - (3y)^3 = (2x - 3y)(4x^2 + 6xy + 9y^2)$$
$$8a^6 + 125b^3 = (2a^2)^3 + (5b)^3 = (2a^2 + 5b)(4a^4 - 10a^2b + 25b^2)$$

We do want to leave you with one final word of caution: **be sure to factor completely**. Sometimes more than one technique needs to be applied, or perhaps the same technique can be applied more than once. Study the following examples very carefully.

$$2x^2 - 8 = 2(x^2 - 4) = 2(x + 2)(x - 2)$$
$$3x^2 + 18x + 24 = 3(x^2 + 6x + 8) = 3(x + 4)(x + 2)$$
$$3x^3 - 3y^3 = 3(x^3 - y^3) = 3(x - y)(x^2 + xy + y^2)$$
$$a^4 - b^4 = (a^2 + b^2)(a^2 - b^2) = (a^2 + b^2)(a + b)(a - b)$$
$$x^4 - 6x^2 - 27 = (x^2 - 9)(x^2 + 3) = (x + 3)(x - 3)(x^2 + 3)$$
$$3x^4y + 9x^2y - 84y = 3y(x^4 + 3x^2 - 28)$$
$$= 3y(x^2 + 7)(x^2 - 4)$$
$$= 3y(x^2 + 7)(x + 2)(x - 2)$$

Problem Set 1.4

Factor completely each of the following. Indicate any that are not factorable using integers.

1. $6xy - 8xy^2$ **2.** $4a^2b^2 + 12ab^3$ **3.** $x(z + 3) + y(z + 3)$

4. $5(x + y) + a(x + y)$ **5.** $3x + 3y + ax + ay$ **6.** $ac + bc + a + b$

7. $ax - ay - bx + by$ **8.** $2a^2 - 3bc - 2ab + 3ac$ **9.** $9x^2 - 25$

10. $4x^2 + 9$ **11.** $1 - 81n^2$ **12.** $9x^2y^2 - 64$

13. $(x + 4)^2 - y^2$ **14.** $x^2 - (y - 1)^2$ **15.** $9s^2 - (2t - 1)^2$

16. $4a^2 - (3b + 1)^2$ **17.** $x^2 - 5x - 14$ **18.** $a^2 + 5a - 24$

19. $15 - 2x - x^2$ **20.** $40 - 6x - x^2$ **21.** $x^2 + 7x - 36$

22. $x^2 - 4xy - 5y^2$ **23.** $3x^2 - 11x + 10$ **24.** $2x^2 - 7x - 30$

25. $10x^2 - 33x - 7$

26. $8y^2 + 22y - 21$

27. $x^3 - 8$

28. $x^3 + 64$

29. $64x^3 + 27y^3$

30. $27x^3 - 8y^3$

31. $4x^2 + 16$

32. $n^3 - 49n$

33. $x^3 - 9x$

34. $12n^2 + 59n + 72$

35. $9a^2 - 42a + 49$

36. $1 - 16x^4$

37. $2n^3 + 6n^2 + 10n$

38. $x^2 - (y - 7)^2$

39. $10x^2 + 39x - 27$

40. $3x^2 + x - 5$

41. $36a^2 - 12a + 1$

42. $18n^3 + 39n^2 - 15n$

43. $8x^2 + 2xy - y^2$

44. $12x^2 + 7xy - 10y^2$

45. $2n^2 - n - 5$

46. $25t^2 - 100$

47. $2n^3 + 14n^2 - 20n$

48. $25n^2 + 64$

49. $4x^3 + 32$

50. $2x^3 - 54$

51. $x^4 - 4x^2 - 45$

52. $x^4 - x^2 - 12$

53. $2x^4y - 26x^2y - 96y$

54. $3x^4y - 15x^2y - 108y$

55. $(a + b)^2 - (c + d)^2$

56. $(a - b)^2 - (c - d)^2$

57. $x^2 + 8x + 16 - y^2$

58. $4x^2 + 12x + 9 - y^2$

59. $x^2 - y^2 - 10y - 25$

60. $y^2 - x^2 + 16x - 64$

Factor each of the following, assuming that all variables appearing as exponents represent integers.

61. $x^{2a} - 16$

62. $x^{4n} - 9$

63. $x^{3n} - y^{3n}$

64. $x^{3a} + y^{6a}$

65. $x^{2a} - 3x^a - 28$

66. $x^{2a} + 10x^a + 21$

67. $2x^{2n} + 7x^n - 30$

68. $3x^{2n} - 16x^n - 12$

69. $x^{4n} - y^{4n}$

70. $16x^{2a} + 24x^a + 9$

71. Suppose that we want to factor $x^2 + 34x + 288$. We need to complete the following with two numbers whose sum is 34 and whose product is 288:

$$x^2 + 34x + 288 = (x + \underline{\quad})(x + \underline{\quad}).$$

These numbers can be found as follows: Since we need a product of 288, let's consider the prime factorization of 288.

$$288 = 2^5 \cdot 3^2$$

Now we need to use five 2's and two 3's in the statement

$$(\quad) + (\quad) = 34.$$

Since 34 is divisible by 2 but not by 4, four factors of 2 must be in one number and one factor of 2 in the other number. Also, since 34 is not divisible by 3, both factors of 3 must be in the same number. These facts aid us in determining that

$$(2 \cdot 2 \cdot 2 \cdot 2) + (2 \cdot 3 \cdot 3) = 34$$

or

$$16 + 18 = 34.$$

Thus, we can complete the original factoring problem.

$$x^2 + 34x + 288 = (x + 16)(x + 18)$$

Use this approach to factor each of the following expressions.

(a) $x^2 + 35x + 96$

(b) $x^2 + 27x + 176$

(c) $x^2 - 45x + 504$

(d) $x^2 - 26x + 168$

(e) $x^2 + 60x + 896$

(f) $x^2 - 84x + 1728$

1.5

Rational Expressions

Indicated quotients of algebraic expressions are called **algebraic fractions**, or **fractional expressions**. The indicated quotient of two polynomials is called a **rational expression**. (This is analogous to defining a rational number as the indicated quotient of two integers.) The following are examples of rational expressions:

$$\frac{3x^2}{5}, \quad \frac{x-2}{x+3}, \quad \frac{x^2+5x-1}{x^2-9}, \quad \frac{xy^2+x^2y}{xy}, \quad \frac{a^3-3a^2-5a-1}{a^4+a^3+6}.$$

Because division by zero must be avoided, no values can be assigned to variables that will create a denominator of zero. Thus, the rational expression $\frac{x-2}{x+3}$ is meaningful for all real-number values of x except $x = -3$. Rather than making restrictions for each individual expression, we will merely assume that **all denominators represent nonzero real numbers**.

The basic properties of the real numbers can be used for working with rational expressions. For example, the property

$$\frac{a \cdot k}{b \cdot k} = \frac{a}{b},$$

which is used to reduce rational numbers, is also used to *simplify* rational expressions. Consider the following examples.

$$\frac{15xy}{25y} = \frac{3 \cdot \cancel{5} \cdot x \cdot \cancel{y}}{\cancel{5} \cdot 5 \cdot \cancel{y}} = \frac{3x}{5}$$

$$\frac{-9}{18x^2y} = -\frac{\overset{1}{\cancel{9}}}{\underset{2}{\cancel{18}}x^2y} = -\frac{1}{2x^2y}$$

Notice that slightly different formats were used in these two examples. In the first one we factored the coefficients into primes and then proceeded to simplify; however, in the second problem we simply divided a common factor of 9 out of both the numerator and the denominator. This is basically a format issue and depends upon your personal preference. Also notice that in the second example, we applied the property $\frac{-a}{b} = -\frac{a}{b}$. This is part of the general property that states

$$\frac{-a}{b} = \frac{a}{-b} = -\frac{a}{b}.$$

The factoring techniques discussed in the previous section can be used to factor numerators and denominators so that the property $(a \cdot k)/(b \cdot k) = a/b$ can be applied. Consider the following examples.

$$\frac{x^2 + 4x}{x^2 - 16} = \frac{x(x + 4)}{(x - 4)(x + 4)} = \frac{x}{x - 4}$$

$$\frac{5n^2 + 6n - 8}{10n^2 - 3n - 4} = \frac{(5n - 4)(n + 2)}{(5n - 4)(2n + 1)} = \frac{n + 2}{2n + 1}$$

$$\frac{x^3 + y^3}{x^2 + xy + 2x + 2y} = \frac{(x + y)(x^2 - xy + y^2)}{x(x + y) + 2(x + y)}$$

$$= \frac{(x + y)(x^2 - xy + y^2)}{(x + y)(x + 2)} = \frac{x^2 - xy + y^2}{x + 2}$$

$$\frac{6x^3y - 6xy}{x^3 + 5x^2 + 4x} = \frac{6xy(x^2 - 1)}{x(x^2 + 5x + 4)} = \frac{6xy(x + 1)(x - 1)}{x(x + 1)(x + 4)} = \frac{6y(x - 1)}{x + 4}$$

Note that in the last example, we left the numerator of the final fraction in factored form. This is often done if expressions other than monomials are involved. Either

$$\frac{6y(x - 1)}{x + 4} \qquad \text{or} \qquad \frac{6xy - 6y}{x + 4}$$

is an acceptable answer.

Remember that the quotient of any nonzero real number and its opposite is -1. For example, $6/-6 = -1$ and $-8/8 = -1$. Likewise, the indicated quotient of any polynomial and its opposite is equal to -1. For example,

$$\frac{a}{-a} = -1 \qquad \text{because } a \text{ and } -a \text{ are opposites,}$$

$$\frac{a - b}{b - a} = -1 \qquad \text{because } a - b \text{ and } b - a \text{ are opposites,}$$

$$\frac{x^2 - 4}{4 - x^2} = -1 \qquad \text{because } x^2 - 4 \text{ and } 4 - x^2 \text{ are opposites.}$$

The next example illustrates the use of this idea when simplifying rational expressions.

$$\frac{4 - x^2}{x^2 + x - 6} = \frac{(2 + x)}{(x + 3)} \boxed{\frac{(2 - x)}{(x - 2)}}$$

$$= (-1)\left(\frac{x + 2}{x + 3}\right) \qquad \frac{2 - x}{x - 2} = -1$$

$$= -\frac{x + 2}{x + 3} \qquad \text{or} \qquad \frac{-x - 2}{x + 3}$$

Multiplying and Dividing Rational Expressions

Multiplication of rational expressions is based on the following property:

$$\frac{a}{b} \cdot \frac{c}{d} = \frac{ac}{bd}.$$

In other words, we multiply numerators and multiply denominators and express the final product in simplified form. Study the following examples carefully and pay special attention to the formats used to organize the computational work.

$$\frac{3x}{4y} \cdot \frac{8y^2}{9x} = \frac{3 \cdot \overset{2}{\cancel{8}} \cdot \cancel{x} \cdot \overset{y}{\cancel{y^2}}}{\cancel{4} \cdot \underset{3}{\cancel{9}} \cdot \cancel{x} \cdot \cancel{y}} = \frac{2y}{3}$$

$$\frac{12x^2y}{-18xy} \cdot \frac{-24xy^2}{56y^3} = \frac{\overset{2}{\cancel{12}} \cdot \overset{\overset{8}{\cancel{24}}}{} \cdot \overset{x^2}{\cancel{x^3}} \cdot \cancel{y^3}}{\underset{\cancel{3}}{\cancel{18}} \cdot \underset{7}{\cancel{56}} \cdot \cancel{x} \cdot \underset{y}{\cancel{y^4}}} = \frac{2x^2}{7y}$$

$$\frac{12x^2y}{-18xy} = -\frac{12x^2y}{18xy} \quad \text{and} \quad \frac{-24xy^2}{56y^3} = -\frac{24xy^2}{56y^3}$$

so the product is positive.

$$\frac{y}{x^2 - 4} \cdot \frac{x + 2}{y^2} = \frac{\cancel{y}\cancel{(x+2)}}{\underset{y}{\cancel{y^2}}\cancel{(x+2)}(x-2)} = \frac{1}{y(x-2)}$$

$$\frac{x^2 - x}{x + 5} \cdot \frac{x^2 + 5x + 4}{x^4 - x^2} = \frac{x\cancel{(x-1)}\cancel{(x+1)}(x+4)}{(x+5)\underset{x}{\cancel{(x^2)}}\cancel{(x+1)}\cancel{(x-1)}} = \frac{x+4}{x(x+5)}$$

To divide rational expressions we merely apply the following property:

$$\frac{a}{b} \div \frac{c}{d} = \frac{a}{b} \cdot \frac{d}{c} = \frac{ad}{bc}.$$

That is to say, the quotient of two rational expressions is the product of the first expression times the reciprocal of the second. Consider the following examples.

$$\frac{16x^2y}{24xy^3} \div \frac{9xy}{8x^2y^2} = \frac{16x^2y}{24xy^3} \cdot \frac{8x^2y^2}{9xy} = \frac{16 \cdot 8 \cdot \overset{x^2}{\cancel{x^4}} \cdot y^3}{\underset{3}{\cancel{24}} \cdot 9 \cdot x^2 \cdot \underset{y}{\cancel{y^4}}} = \frac{16x^2}{27y}$$

$$\frac{3a^2 + 12}{3a^2 - 15a} \div \frac{a^4 - 16}{a^2 - 3a - 10} = \frac{3a^2 + 12}{3a^2 - 15a} \cdot \frac{a^2 - 3a - 10}{a^4 - 16}$$

$$= \frac{\cancel{3}(a^2 + 4)\cancel{(a-5)}\cancel{(a+2)}}{\cancel{3}a\cancel{(a-5)}\cancel{(a^2+4)}\cancel{(a+2)}(a-2)}$$

$$= \frac{1}{a(a-2)}$$

Adding and Subtracting Rational Expressions

The following two properties provide the basis for adding and subtracting rational expressions:

$$\frac{a}{b} + \frac{c}{b} = \frac{a+c}{b};$$

$$\frac{a}{b} - \frac{c}{b} = \frac{a-c}{b}.$$

These properties state that rational expressions with a common denominator can be added (or subtracted) by adding (or subtracting) the numerators and placing the result over the common denominator. Let's illustrate this idea.

$$\frac{8}{x-2} + \frac{3}{x-2} = \frac{8+3}{x-2} = \frac{11}{x-2}$$

$$\frac{9}{4y} - \frac{7}{4y} = \frac{9-7}{4y} = \frac{2}{4y} = \frac{1}{2y}$$

(Don't forget to simplify the final result.)

$$\frac{n^2}{n-1} - \frac{1}{n-1} = \frac{n^2-1}{n-1} = \frac{(n+1)(n\!\!-\!\!1)}{n\!\!-\!\!1} = n+1$$

If we need to add or subtract rational expressions that do not have a common denominator, then we apply the property $a/b = (a \cdot k)/(b \cdot k)$ to obtain equivalent fractions with a common denominator. Study the next examples and again pay special attention to the format we use to organize our work.

Remark: Remember that the **least common multiple** of a set of whole numbers is the smallest nonzero whole number divisible by each of the numbers in the set. When we add or subtract rational numbers, the least common multiple of the denominators of those numbers is the **least common denominator (LCD)**. This concept of a least common denominator can be extended to include polynomials.

EXAMPLE 1

Add $\dfrac{x+2}{4} + \dfrac{3x+1}{3}$.

Solution By inspection we see that the LCD is 12.

$$\frac{x+2}{4} + \frac{3x+1}{3} = \left(\frac{x+2}{4}\right)\left(\frac{3}{3}\right) + \left(\frac{3x+1}{3}\right)\left(\frac{4}{4}\right)$$

$$= \frac{3(x+2)}{12} + \frac{4(3x+1)}{12}$$

$$= \frac{3x+6+12x+4}{12} = \frac{15x+10}{12}$$

EXAMPLE 2

Perform the indicated operations:

$$\frac{x+3}{10} + \frac{2x+1}{15} - \frac{x-2}{18}.$$

Solution If you cannot determine the LCD by inspection, then use the prime-factored forms of the denominators:

$$10 = 2 \cdot 5, \qquad 15 = 3 \cdot 5, \qquad 18 = 2 \cdot 3 \cdot 3.$$

The LCD must contain one factor of 2, two factors of 3, and one factor of 5. Thus the LCD is $2 \cdot 3 \cdot 3 \cdot 5 = 90$.

$$\frac{x + 3}{10} + \frac{2x + 1}{15} - \frac{x - 2}{18} = \left(\frac{x + 3}{10}\right)\left(\frac{9}{9}\right) + \left(\frac{2x + 1}{15}\right)\left(\frac{6}{6}\right) - \left(\frac{x - 2}{18}\right)\left(\frac{5}{5}\right)$$

$$= \frac{9(x + 3)}{90} + \frac{6(2x + 1)}{90} - \frac{5(x - 2)}{90}$$

$$= \frac{9x + 27 + 12x + 6 - 5x + 10}{90}$$

$$= \frac{16x + 43}{90}$$

Having variables in the denominators does not create any serious difficulty; our approach remains the same. Study the following examples very carefully. For each problem notice the same basic procedure: (1) find the LCD; (2) change each fraction to an equivalent fraction having the LCD as its denominator; (3) add or subtract numerators and place this result over the LCD; and (4) look for possibilities to simplify the resulting fraction.

EXAMPLE 3

Add $\dfrac{3}{2x} + \dfrac{5}{3y}$.

Solution Using an LCD of $6xy$, we can proceed as follows:

$$\frac{3}{2x} + \frac{5}{3y} = \left(\frac{3}{2x}\right)\left(\frac{3y}{3y}\right) + \left(\frac{5}{3y}\right)\left(\frac{2x}{2x}\right)$$

$$= \frac{9y}{6xy} + \frac{10x}{6xy} = \frac{9y + 10x}{6xy}$$

EXAMPLE 4

Subtract $\dfrac{7}{12ab} - \dfrac{11}{15a^2}$.

Solution We can factor the numerical coefficients of the denominators into primes to help find the LCD.

$$\left.\begin{array}{l} 12ab = 2 \cdot 2 \cdot 3 \cdot a \cdot b \\ 15a^2 = 3 \cdot 5 \cdot a^2 \end{array}\right\} \quad \text{LCD} = 2 \cdot 2 \cdot 3 \cdot 5 \cdot a^2 \cdot b = 60a^2b$$

$$\frac{7}{12ab} - \frac{11}{15a^2} = \left(\frac{7}{12ab}\right)\left(\frac{5a}{5a}\right) - \left(\frac{11}{15a^2}\right)\left(\frac{4b}{4b}\right)$$

$$= \frac{35a}{60a^2b} - \frac{44b}{60a^2b} = \frac{35a - 44b}{60a^2b}$$

EXAMPLE 5

Add $\dfrac{8}{x^2 - 4x} + \dfrac{2}{x}$.

Solution

$$\left.\begin{array}{l} x^2 - 4x = x(x - 4) \\ x = x \end{array}\right\} \qquad \text{LCD} = x(x - 4)$$

$$\frac{8}{x(x - 4)} + \frac{2}{x} = \frac{8}{x(x - 4)} + \left(\frac{2}{x}\right)\left(\frac{x - 4}{x - 4}\right)$$

$$= \frac{8}{x(x - 4)} + \frac{2(x - 4)}{x(x - 4)}$$

$$= \frac{8 + 2x - 8}{x(x - 4)}$$

$$= \frac{2x}{x(x - 4)}$$

$$= \frac{2}{x - 4}$$

EXAMPLE 6

Add $\dfrac{3n}{n^2 + 6n + 5} + \dfrac{4}{n^2 - 7n - 8}$.

Solution

$$\left.\begin{array}{l} n^2 + 6n + 5 = (n + 5)(n + 1) \\ n^2 - 7n - 8 = (n - 8)(n + 1) \end{array}\right\} \qquad \text{LCD} = (n + 1)(n + 5)(n - 8)$$

$$\frac{3n}{n^2 + 6n + 5} + \frac{4}{n^2 - 7n - 8} = \left[\frac{3n}{(n + 5)(n + 1)}\right]\left(\frac{n - 8}{n - 8}\right)$$

$$+ \left[\frac{4}{(n - 8)(n + 1)}\right]\left(\frac{n + 5}{n + 5}\right)$$

$$= \frac{3n(n - 8)}{(n + 5)(n + 1)(n - 8)}$$

$$+ \frac{4(n + 5)}{(n + 5)(n + 1)(n - 8)}$$

$$= \frac{3n^2 - 24n + 4n + 20}{(n + 5)(n + 1)(n - 8)}$$

$$= \frac{3n^2 - 20n + 20}{(n + 5)(n + 1)(n - 8)}$$

Simplifying Complex Fractions

Fractional forms that contain rational expressions in the numerator and/or denominator are called **complex fractions**. The following examples illustrate some approaches to simplifying complex fractions.

EXAMPLE 7

Simplify $\dfrac{\dfrac{3}{x} + \dfrac{2}{y}}{\dfrac{5}{x} - \dfrac{6}{y^2}}$.

Solution A Treating the numerator as the sum of two rational expressions and the denominator as the difference of two rational expressions, we can proceed as follows.

$$\frac{\dfrac{3}{x} + \dfrac{2}{y}}{\dfrac{5}{x} - \dfrac{6}{y^2}} = \frac{\left(\dfrac{3}{x}\right)\left(\dfrac{y}{y}\right) + \left(\dfrac{2}{y}\right)\left(\dfrac{x}{x}\right)}{\left(\dfrac{5}{x}\right)\left(\dfrac{y^2}{y^2}\right) - \left(\dfrac{6}{y^2}\right)\left(\dfrac{x}{x}\right)}$$

$$= \frac{\dfrac{3y}{xy} + \dfrac{2x}{xy}}{\dfrac{5y^2}{xy^2} - \dfrac{6x}{xy^2}} = \frac{\dfrac{3y + 2x}{xy}}{\dfrac{5y^2 - 6x}{xy^2}}$$

$$= \frac{3y + 2x}{xy} \cdot \frac{\overset{y}{\cancel{xy^2}}}{5y^2 - 6x}$$

$$= \frac{y(3y + 2x)}{5y^2 - 6x}$$

Solution B The LCD of all four denominators $(x, y, x,$ and $y^2)$ is xy^2. Let's multiply the entire complex fraction by a form of 1, namely, $(xy^2)/(xy^2)$.

$$\frac{\dfrac{3}{x} + \dfrac{2}{y}}{\dfrac{5}{x} - \dfrac{6}{y^2}} = \left(\frac{\dfrac{3}{x} + \dfrac{2}{y}}{\dfrac{5}{x} - \dfrac{6}{y^2}}\right)\left(\frac{xy^2}{xy^2}\right)$$

$$= \frac{(xy^2)\left(\dfrac{3}{x}\right) + (xy^2)\left(\dfrac{2}{y}\right)}{(xy^2)\left(\dfrac{5}{x}\right) - (xy^2)\left(\dfrac{6}{y^2}\right)}$$

$$= \frac{3y^2 + 2xy}{5y^2 - 6x} \quad \text{or} \quad \frac{y(3y + 2x)}{5y^2 - 6x}$$

Certainly either approach (Solution A or Solution B) will work with a problem such as Example 7. We suggest that you study Solution B very carefully. This approach works effectively with complex fractions when the LCD of all the denominators is easy to find. Let's consider one final example to illustrate another idea.

EXAMPLE 8

Simplify $1 - \dfrac{n}{1 - \dfrac{1}{n}}$.

Solution We first simplify the complex fraction by multiplying by n/n.

$$\left(\frac{n}{1 - \frac{1}{n}}\right)\left(\frac{n}{n}\right) = \frac{n^2}{n - 1}$$

Now we can perform the subtraction.

$$1 - \frac{n^2}{n - 1} = \left(\frac{n - 1}{n - 1}\right)\left(\frac{1}{1}\right) - \frac{n^2}{n - 1}$$

$$= \frac{n - 1}{n - 1} - \frac{n^2}{n - 1}$$

$$= \frac{n - 1 - n^2}{n - 1} \quad \text{or} \quad \frac{-n^2 + n - 1}{n - 1}$$

Problem Set 1.5

Simplify each of the following rational expressions.

1. $\dfrac{14x^2y}{21xy}$

2. $\dfrac{-26xy^2}{65y}$

3. $\dfrac{-63xy^4}{-81x^2y}$

4. $\dfrac{x^2 - y^2}{x^2 + xy}$

5. $\dfrac{a^2 + 7a + 12}{a^2 - 6a - 27}$

6. $\dfrac{6x^2 + x - 15}{8x^2 - 10x - 3}$

7. $\dfrac{2x^3 + 3x^2 - 14x}{x^2y + 7xy - 18y}$

8. $\dfrac{3x - x^2}{x^2 - 9}$

9. $\dfrac{x^3 - y^3}{x^2 + xy - 2y^2}$

10. $\dfrac{ax - 3x + 2ay - 6y}{2ax - 6x + ay - 3y}$

11. $\dfrac{2y - 2xy}{x^2y - y}$

12. $\dfrac{16x^3y + 24x^2y^2 - 16xy^3}{24x^2y + 12xy^2 - 12y^3}$

Perform the following indicated operations involving rational expressions. Express final answers in simplest form.

13. $\dfrac{4x^2}{5y^2} \cdot \dfrac{15xy}{24x^2y^2}$

14. $\dfrac{5xy}{8y^2} \cdot \dfrac{18x^2y}{15}$

15. $\dfrac{-14xy^4}{18y^2} \cdot \dfrac{24x^2y^3}{35y^2}$

16. $\dfrac{6xy}{9y^4} \cdot \dfrac{30x^3y}{-48x}$

17. $\dfrac{7a^2b}{9ab^3} \div \dfrac{3a^4}{2a^2b^2}$

18. $\dfrac{9a^2c}{12bc^2} \div \dfrac{21ab}{14c^3}$

19. $\dfrac{5xy}{x+6} \cdot \dfrac{x^2-36}{x^2-6x}$

20. $\dfrac{2a^2+6}{a^2-a} \cdot \dfrac{a^3-a^2}{8a-4}$

21. $\dfrac{5a^2+20a}{a^3-2a^2} \cdot \dfrac{a^2-a-12}{a^2-16}$

22. $\dfrac{t^4-81}{t^2-6t+9} \cdot \dfrac{6t^2-11t-21}{5t^2+8t-21}$

23. $\dfrac{x^2+5xy-6y^2}{xy^2-y^3} \cdot \dfrac{2x^2+15xy+18y^2}{xy+4y^2}$

24. $\dfrac{10n^2+21n-10}{5n^2+33n-14} \cdot \dfrac{2n^2+6n-56}{2n^2-3n-20}$

25. $\dfrac{9y^2}{x^2+12x+36} \div \dfrac{12y}{x^2+6x}$

26. $\dfrac{x^2-4xy+4y^2}{7xy^2} \div \dfrac{4x^2-3xy-10y^2}{20x^2y+25xy^2}$

27. $\dfrac{2x^2+3x}{2x^3-10x^2} \cdot \dfrac{x^2-8x+15}{3x^3-27x} \div \dfrac{14x+21}{x^2-6x-27}$

28. $\dfrac{a^2-4ab+4b^2}{6a^2-4ab} \cdot \dfrac{3a^2+5ab-2b^2}{6a^2+ab-b^2} \div \dfrac{a^2-4b^2}{8a+4b}$

29. $\dfrac{x+4}{6} + \dfrac{2x-1}{4}$

30. $\dfrac{3n-1}{9} - \dfrac{n+2}{12}$

31. $\dfrac{x+1}{4} + \dfrac{x-3}{6} - \dfrac{x-2}{8}$

32. $\dfrac{x-2}{5} - \dfrac{x+3}{6} + \dfrac{x+1}{15}$

33. $\dfrac{7}{16a^2b} + \dfrac{3a}{20b^2}$

34. $\dfrac{5b}{24a^2} - \dfrac{11a}{32b}$

35. $\dfrac{1}{n^2} + \dfrac{3}{4n} - \dfrac{5}{6}$

36. $\dfrac{3}{n^2} - \dfrac{2}{5n} + \dfrac{4}{3}$

37. $\dfrac{3}{4x} + \dfrac{2}{3y} - 1$

38. $\dfrac{5}{6x} - \dfrac{3}{4y} + 2$

39. $\dfrac{3}{2x+1} + \dfrac{2}{3x+4}$

40. $\dfrac{5}{x-1} - \dfrac{3}{2x-3}$

41. $\dfrac{4x}{x^2+7x} + \dfrac{3}{x}$

42. $\dfrac{6}{x^2+8x} - \dfrac{3}{x}$

43. $\dfrac{4a-4}{a^2-4} - \dfrac{3}{a+2}$

44. $\dfrac{6a+4}{a^2-1} - \dfrac{5}{a-1}$

45. $\dfrac{3}{x+1} + \dfrac{x+5}{x^2-1} - \dfrac{3}{x-1}$

46. $\dfrac{5}{x} - \dfrac{5x-30}{x^2+6x} + \dfrac{x}{x+6}$

47. $\dfrac{5}{x^2+10x+21} + \dfrac{4}{x^2+12x+27}$

48. $\dfrac{8}{a^2-3a-18} - \dfrac{10}{a^2-7a-30}$

49. $\dfrac{5}{x^2-1} - \dfrac{2}{x^2+6x-16}$

50. $\dfrac{4}{x^2+2} - \dfrac{7}{x^2+x-12}$

51. $x - \dfrac{x^2}{x-1} + \dfrac{1}{x^2-1}$

52. $x - \dfrac{x^2}{x+7} - \dfrac{x}{x^2-49}$

53. $\dfrac{2n^2}{n^4-16} - \dfrac{n}{n^2-4} + \dfrac{1}{n+2}$

54. $\dfrac{n}{n^2+1} + \dfrac{n^2+3n}{n^4-1} - \dfrac{1}{n-1}$

55. $\dfrac{2x+1}{x^2-3x-4} + \dfrac{3x-2}{x^2+3x-28}$

56. $\dfrac{3x-4}{2x^2-9x-5} - \dfrac{2x-1}{3x^2-11x-20}$

57. Consider the addition problem $\dfrac{8}{x-2} + \dfrac{5}{2-x}$. Note that the denominators are opposites of each other. If the property $\dfrac{a}{-b} = -\dfrac{a}{b}$ is applied to the second fraction, we obtain $\dfrac{5}{2-x} = -\dfrac{5}{x-2}$. Thus we can proceed as follows:

$$\frac{8}{x-2} + \frac{5}{2-x} = \frac{8}{x-2} - \frac{5}{x-2} = \frac{8-5}{x-2} = \frac{3}{x-2}.$$

Use this approach to do the following problems.

(a) $\dfrac{7}{x-1} + \dfrac{2}{1-x}$

(b) $\dfrac{5}{2x-1} + \dfrac{8}{1-2x}$

(c) $\dfrac{4}{a-3} - \dfrac{1}{3-a}$

(d) $\dfrac{10}{a-9} - \dfrac{5}{9-a}$

(e) $\dfrac{x^2}{x-1} - \dfrac{2x-3}{1-x}$

(f) $\dfrac{x^2}{x-4} - \dfrac{3x-28}{4-x}$

Simplify each of the following complex fractions.

58. $\dfrac{\dfrac{2}{x} + \dfrac{7}{y}}{\dfrac{3}{x} - \dfrac{10}{y}}$

59. $\dfrac{\dfrac{5}{x^2} - \dfrac{3}{x}}{\dfrac{1}{y} + \dfrac{2}{y^2}}$

60. $\dfrac{\dfrac{1}{x} + 3}{\dfrac{2}{y} + 4}$

61. $\dfrac{1 + \dfrac{1}{x}}{1 - \dfrac{1}{x}}$

62. $\dfrac{3 - \dfrac{2}{n-4}}{5 + \dfrac{4}{n-4}}$

63. $\dfrac{1 - \dfrac{1}{n+1}}{1 + \dfrac{1}{n-1}}$

64. $\dfrac{\dfrac{2}{x-3} - \dfrac{3}{x+3}}{\dfrac{5}{x^2-9} - \dfrac{2}{x-3}}$

65. $\dfrac{\dfrac{-2}{x} - \dfrac{4}{x+2}}{\dfrac{3}{x^2+2x} + \dfrac{3}{x}}$

66. $\dfrac{\dfrac{-1}{y-2} + \dfrac{5}{x}}{\dfrac{3}{x} - \dfrac{4}{xy-2x}}$

67. $1 + \dfrac{x}{1 + \dfrac{1}{x}}$

68. $2 - \dfrac{x}{3 - \dfrac{2}{x}}$

69. $\dfrac{\dfrac{a}{1} + 1}{\dfrac{1}{a} + 4}$

70. $\dfrac{3a}{2 - \dfrac{1}{a}} - 1$

1.6

Radicals

Recall from our work with exponents that to **square a number** means to raise it to the second power, that is, to use the number as a factor twice. For example, $4^2 = 4 \cdot 4 = 16$ and $(-4)^2 = (-4)(-4) = 16$. A **square root of a number** is one of its two equal factors. Thus, 4 and -4 are both square roots of 16. In general, a is a square root of b if $a^2 = b$. The following statements are generalizations of these ideas.

1. Every positive real number has two square roots; one is positive and the other is negative. They are opposites of each other.

2. Negative real numbers have no real-number square roots because the square of any nonzero real number is positive.

3. The square root of zero is zero.

The symbol $\sqrt{\ }$, called a **radical sign**, is used to designate the *nonnegative* square root, which is called the **principal square root**. The number under the radical sign is called the **radicand** and the entire expression, such as $\sqrt{16}$, is referred to as a **radical**.

The following examples demonstrate the use of the square root notation.

$\sqrt{16} = 4$ $\sqrt{16}$ indicates the *nonnegative* or *principal* square root of 16.

$-\sqrt{16} = -4$ $-\sqrt{16}$ indicates the negative square root of 16.

$\sqrt{0} = 0$ Zero has only one square root. Technically, we could also write $-\sqrt{0} = -0 = 0$.

$\sqrt{-4}$ Not a real number

$-\sqrt{-4}$ Not a real number

To **cube a number** means to raise it to the third power, that is, to use the number as a factor three times. For example, $2^3 = 2 \cdot 2 \cdot 2 = 8$ and $(-2)^3 = (-2)(-2)(-2) = -8$. A **cube root of a number** is one of its three equal factors. Thus, 2 is a cube root of 8 and, as we will discuss later, it is the only real number that is a cube root of 8. Furthermore, -2 is the only real number that is a cube root of -8. In general, a is a cube root of b if $a^3 = b$. The following statements are generalizations of these ideas.

1. Every positive real number has one positive real-number cube root.

2. Every negative real number has one negative real-number cube root.

3. The cube root of zero is zero.

Remark: Every nonzero real number has three cube roots, but only one of them is a real number. The other two roots are complex numbers and will be discussed in Section 1.8.

The symbol $\sqrt[3]{\ }$ is used to designate the cube root of a number. Thus we can write

$$\sqrt[3]{8} = 2, \quad \sqrt[3]{-8} = -2, \quad \sqrt[3]{\tfrac{1}{27}} = \tfrac{1}{3}, \quad \text{and} \quad \sqrt[3]{-\tfrac{1}{27}} = -\tfrac{1}{3}.$$

The concept of root can be extended to fourth roots, fifth roots, sixth roots, and, in general, nth roots. Some generalizations can be made.

If n is an *even positive integer*, then the following statements are true:

1. Every positive real number has exactly two real nth roots, one positive and one negative. For example, the real fourth roots of 16 are 2 and -2.

2. Negative real numbers do not have real nth roots. For example, there are no real fourth roots of -16.

If n is an *odd positive integer* greater than 1, then the following statements are true:

1. Every real number has exactly one real nth root.

2. The real nth root of a positive number is positive. For example, the fifth root of 32 is 2.

3. The real nth root of a negative number is negative. For example, the fifth root of -32 is -2.

In general, the following definition is useful.

DEFINITION 1.5

$$\sqrt[n]{b} = a \quad \text{if and only if} \quad a^n = b.$$

In Definition 1.5, if n is an even positive integer, then a and b are both nonnegative. If n is an odd positive integer greater than 1, then a and b are both nonnegative or both negative. The symbol $\sqrt[n]{}$ designates the principal root.
The following examples are applications of Definition 1.5.

$$\sqrt[4]{81} = 3 \quad \text{because} \quad 3^4 = 81.$$
$$\sqrt[5]{32} = 2 \quad \text{because} \quad 2^5 = 32.$$
$$\sqrt[5]{-32} = -2 \quad \text{because} \quad (-2)^5 = -32.$$

To complete our terminology, the n in the radical $\sqrt[n]{b}$ is called the **index** of the radical. If $n = 2$, we commonly write $\sqrt{b}$ instead of $\sqrt[2]{b}$. In this text, when we use symbols such as $\sqrt[n]{b}$, $\sqrt[m]{y}$, and $\sqrt[k]{x}$, we will assume the previous agreements relative to the existence of real roots, without listing the various restrictions, unless a special restriction is needed.
From Definition 1.5 we see that if n is any positive integer greater than 1 and $\sqrt[n]{b}$ exists, then

$$(\sqrt[n]{b})^n = b.$$

For example, $(\sqrt{4})^2 = 4$, $(\sqrt[3]{-8})^3 = -8$, and $(\sqrt[4]{81})^4 = 81$. Furthermore, if $b \geq 0$ and n is any positive integer greater than 1 or if $b < 0$ and n is an odd positive integer greater than 1, then

$$\sqrt[n]{b^n} = b.$$

For example, $\sqrt{4^2} = 4$, $\sqrt[3]{(-2)^3} = -2$, and $\sqrt[5]{6^5} = 6$. *But we must be careful*, because $\sqrt{(-2)^2} \neq -2$ and $\sqrt[4]{(-2)^4} \neq -2$.

Simplest Radical Form

Let's use some examples to motivate another very useful property of radicals.

$$\sqrt{16 \cdot 25} = \sqrt{400} = 20 \qquad \text{and} \quad \sqrt{16} \cdot \sqrt{25} = 4 \cdot 5 = 20$$
$$\sqrt[3]{8 \cdot 27} = \sqrt[3]{216} = 6 \qquad \text{and} \quad \sqrt[3]{8} \cdot \sqrt[3]{27} = 2 \cdot 3 = 6$$
$$\sqrt[3]{-8 \cdot 64} = \sqrt[3]{-512} = -8 \quad \text{and} \quad \sqrt[3]{-8} \cdot \sqrt[3]{64} = -2 \cdot 4 = -8$$

In general, the following property can be stated.

PROPERTY 1.3

$$\sqrt[n]{bc} = \sqrt[n]{b}\,\sqrt[n]{c} \quad \text{if } \sqrt[n]{b} \text{ and } \sqrt[n]{c} \text{ are real numbers.}$$

Property 1.3 states that **the nth root of a product is equal to the product of the nth roots.**

The definition of *n*th root, along with Property 1.3, provides the basis for changing radicals to simplest radical form. The concept of **simplest radical form** takes on additional meaning as we encounter more complicated expressions, but for now it simply means that the radicand does not contain any perfect powers of the index. Consider the following examples of reductions to simplest radical form.

$$\sqrt{45} = \sqrt{9 \cdot 5} = \sqrt{9}\sqrt{5} = 3\sqrt{5}$$
$$\sqrt{52} = \sqrt{4 \cdot 13} = \sqrt{4}\sqrt{13} = 2\sqrt{13}$$
$$\sqrt[3]{24} = \sqrt[3]{8 \cdot 3} = \sqrt[3]{8}\sqrt[3]{3} = 2\sqrt[3]{3}$$

A variation of the technique for changing radicals with index *n* to simplest form is to factor the radicand into primes and then to look for perfect *n*th powers in exponential form. The following examples illustrate this technique.

$$\sqrt{80} = \sqrt{2^4 \cdot 5} = \sqrt{2^4}\sqrt{5} = 2^2\sqrt{5} = 4\sqrt{5}$$
$$\sqrt[3]{108} = \sqrt[3]{2^2 \cdot 3^3} = \sqrt[3]{3^3}\sqrt[3]{2^2} = 3\sqrt[3]{4}$$

The distributive property can be used to combine radicals that have the same index and the same radicand. Consider the following examples.

$$3\sqrt{2} + 5\sqrt{2} = (3 + 5)\sqrt{2} = 8\sqrt{2}$$
$$7\sqrt[3]{5} - 3\sqrt[3]{5} = (7 - 3)\sqrt[3]{5} = 4\sqrt[3]{5}$$

Sometimes it is necessary to simplify the radicals first and then to combine them by applying the distributive property.

$$3\sqrt{8} + 2\sqrt{18} - 4\sqrt{2} = 3\sqrt{4}\sqrt{2} + 2\sqrt{9}\sqrt{2} - 4\sqrt{2}$$
$$= 6\sqrt{2} + 6\sqrt{2} - 4\sqrt{2}$$
$$= (6 + 6 - 4)\sqrt{2}$$
$$= 8\sqrt{2}$$

Property 1.3 can also be viewed as $\sqrt[n]{b}\sqrt[n]{c} = \sqrt[n]{bc}$. Then, along with the commutative and associative properties of the real numbers, it provides the basis for multiplying radicals having the same index. Consider the following examples.

$$(7\sqrt{6})(3\sqrt{8}) = 7 \cdot 3 \cdot \sqrt{6} \cdot \sqrt{8}$$
$$= 21\sqrt{48}$$
$$= 21\sqrt{16}\sqrt{3}$$
$$= 21 \cdot 4 \cdot \sqrt{3}$$
$$= 84\sqrt{3}$$
$$(2\sqrt[3]{6})(5\sqrt[3]{4}) = 2 \cdot 5 \cdot \sqrt[3]{6} \cdot \sqrt[3]{4}$$
$$= 10\sqrt[3]{24}$$
$$= 10\sqrt[3]{8}\sqrt[3]{3}$$
$$= 10 \cdot 2 \cdot \sqrt[3]{3}$$
$$= 20\sqrt[3]{3}$$

The distributive property, along with Property 1.3, provides a way of handling special products involving radicals, as the next examples illustrate.

$$2\sqrt{2}(4\sqrt{3} - 5\sqrt{6}) = (2\sqrt{2})(4\sqrt{3}) - (2\sqrt{2})(5\sqrt{6})$$
$$= 8\sqrt{6} - 10\sqrt{12}$$
$$= 8\sqrt{6} - 10\sqrt{4}\sqrt{3}$$
$$= 8\sqrt{6} - 20\sqrt{3}$$

$$(2\sqrt{2} - \sqrt{7})(3\sqrt{2} + 5\sqrt{7}) = 2\sqrt{2}(3\sqrt{2} + 5\sqrt{7}) - \sqrt{7}(3\sqrt{2} + 5\sqrt{7})$$
$$= (2\sqrt{2})(3\sqrt{2}) + (2\sqrt{2})(5\sqrt{7}) - (\sqrt{7})(3\sqrt{2})$$
$$- (\sqrt{7})(5\sqrt{7})$$
$$= 6 \cdot 2 + 10\sqrt{14} - 3\sqrt{14} - 5 \cdot 7$$
$$= -23 + 7\sqrt{14}$$

$$(\sqrt{5} + \sqrt{2})(\sqrt{5} - \sqrt{2}) = \sqrt{5}(\sqrt{5} - \sqrt{2}) + \sqrt{2}(\sqrt{5} - \sqrt{2})$$
$$= (\sqrt{5})(\sqrt{5}) - (\sqrt{5})(\sqrt{2}) + (\sqrt{2})(\sqrt{5}) - (\sqrt{2})(\sqrt{2})$$
$$= 5 - \sqrt{10} + \sqrt{10} - 2$$
$$= 3$$

Pay special attention to the last example. It fits the special product pattern $(a + b)(a - b) = a^2 - b^2$. We will use that idea in a moment.

More about Simplest Radical Form

Another property of nth roots is motivated by the following examples.

$$\sqrt{\frac{36}{9}} = \sqrt{4} = 2 \qquad \text{and} \qquad \frac{\sqrt{36}}{\sqrt{9}} = \frac{6}{3} = 2$$

$$\sqrt[3]{\frac{64}{8}} = \sqrt[3]{8} = 2 \qquad \text{and} \qquad \frac{\sqrt[3]{64}}{\sqrt[3]{8}} = \frac{4}{2} = 2$$

In general, the following property can be stated.

PROPERTY 1.4

$$\sqrt[n]{\frac{b}{c}} = \frac{\sqrt[n]{b}}{\sqrt[n]{c}} \qquad \text{if } \sqrt[n]{b} \text{ and } \sqrt[n]{c} \text{ are real numbers and } c \neq 0.$$

Property 1.4 states that **the nth root of a quotient is equal to the quotient of the nth roots**.

To evaluate radicals such as $\sqrt{\frac{4}{25}}$ and $\sqrt[3]{\frac{27}{8}}$, where the numerator and denominator of the fractional radicands are perfect nth powers, we can either use Property 1.4 or rely on the definition of nth root:

$$\sqrt{\frac{4}{25}} = \frac{\sqrt{4}}{\sqrt{25}} = \frac{2}{5} \quad \text{or} \quad \sqrt{\frac{4}{25}} = \frac{2}{5} \qquad \text{because } \frac{2}{5} \cdot \frac{2}{5} = \frac{4}{25};$$

$$\sqrt[3]{\frac{27}{8}} = \frac{\sqrt[3]{27}}{\sqrt[3]{8}} = \frac{3}{2} \quad \text{or} \quad \sqrt[3]{\frac{27}{8}} = \frac{3}{2} \qquad \text{because } \frac{3}{2} \cdot \frac{3}{2} \cdot \frac{3}{2} = \frac{27}{8}.$$

Radicals such as $\sqrt{\frac{28}{9}}$ and $\sqrt[3]{\frac{24}{27}}$, where only the denominators of the radicand are perfect nth powers, can be simplified as follows:

$$\sqrt{\frac{28}{9}} = \frac{\sqrt{28}}{\sqrt{9}} = \frac{\sqrt{4}\sqrt{7}}{3} = \frac{2\sqrt{7}}{3};$$

$$\sqrt[3]{\frac{24}{27}} = \frac{\sqrt[3]{24}}{\sqrt[3]{27}} = \frac{\sqrt[3]{8}\sqrt[3]{3}}{3} = \frac{2\sqrt[3]{3}}{3}.$$

Before considering more examples, let's summarize some ideas about simplifying radicals. A radical is said to be in **simplest radical form** if the following conditions are satisfied.

1. No fraction appears within a radical sign. (Thus $\sqrt{\frac{3}{4}}$ violates this condition.)
2. No radical appears in the denominator. (So $\sqrt{2}/\sqrt{3}$ violates this condition.)
3. No radicand contains a perfect power of the index. (Therefore $\sqrt{7^2 \cdot 5}$ violates this condition.)

Now let's consider an example in which neither the numerator nor the denominator of the radicand is a perfect nth power.

$$\sqrt{\frac{2}{3}} = \frac{\sqrt{2}}{\sqrt{3}} = \frac{\sqrt{2}}{\sqrt{3}} \cdot \frac{\sqrt{3}}{\sqrt{3}} = \frac{\sqrt{6}}{3}$$

form of 1

The process used to simplify the radical in the previous example is referred to as **rationalizing the denominator**. The process of rationalizing the denominator often can be accomplished in more than one way, as illustrated by the next example.

EXAMPLE 1

Simplify $\dfrac{\sqrt{5}}{\sqrt{8}}$.

Solution A $\dfrac{\sqrt{5}}{\sqrt{8}} = \dfrac{\sqrt{5}}{\sqrt{8}} \cdot \dfrac{\sqrt{8}}{\sqrt{8}} = \dfrac{\sqrt{40}}{8} = \dfrac{\sqrt{4}\sqrt{10}}{8} = \dfrac{2\sqrt{10}}{8} = \dfrac{\sqrt{10}}{4}$

Solution B $\dfrac{\sqrt{5}}{\sqrt{8}} = \dfrac{\sqrt{5}}{\sqrt{8}} \cdot \dfrac{\sqrt{2}}{\sqrt{2}} = \dfrac{\sqrt{10}}{\sqrt{16}} = \dfrac{\sqrt{10}}{4}$

Solution C $\dfrac{\sqrt{5}}{\sqrt{8}} = \dfrac{\sqrt{5}}{\sqrt{4}\sqrt{2}} = \dfrac{\sqrt{5}}{2\sqrt{2}} = \dfrac{\sqrt{5}}{2\sqrt{2}} \cdot \dfrac{\sqrt{2}}{\sqrt{2}} = \dfrac{\sqrt{10}}{4}$

The three approaches to Example 1 again illustrate the need to think first and then to push the pencil. You may find one approach easier than another.

EXAMPLE 2

Simplify $\dfrac{4}{\sqrt{5} + \sqrt{2}}$ by rationalizing the denominator.

Solution   Remember that a moment ago we found that $(\sqrt{5} + \sqrt{2})(\sqrt{5} - \sqrt{2}) = 3$. Let's use that idea here.

$$\frac{4}{\sqrt{5} + \sqrt{2}} = \left(\frac{4}{\sqrt{5} + \sqrt{2}}\right)\left(\frac{\sqrt{5} - \sqrt{2}}{\sqrt{5} - \sqrt{2}}\right)$$

$$= \frac{4(\sqrt{5} - \sqrt{2})}{(\sqrt{5} + \sqrt{2})(\sqrt{5} - \sqrt{2})} = \frac{4(\sqrt{5} - \sqrt{2})}{3}$$

Radicals Containing Variables

Before illustrating simplification of radicals containing variables, there is one important point that should be called to your attention. Let's look at some examples to illustrate the idea.

Consider the radical $\sqrt{x^2}$ for different values of x:

Let $x = 3$;   then $\sqrt{x^2} = \sqrt{3^2} = \sqrt{9} = 3$.
Let $x = -3$;   then $\sqrt{x^2} = \sqrt{(-3)^2} = \sqrt{9} = 3$.

Thus if $x \geq 0$ then $\sqrt{x^2} = x$, but if $x < 0$ then $\sqrt{x^2} = -x$. Using the concept of absolute value, we can state that **for all real numbers, $\sqrt{x^2} = |x|$.**

Now consider the radical $\sqrt{x^3}$. Since x^3 is negative when x is negative, we need to restrict x to the nonnegative real numbers when working with $\sqrt{x^3}$. Thus we can write,

If $x \geq 0$ then $\sqrt{x^3} = \sqrt{x^2}\sqrt{x} = x\sqrt{x}$,

and no absolute value sign is needed.

Finally, let's consider the radical $\sqrt[3]{x^3}$.

Let $x = 2$;   then $\sqrt[3]{x^3} = \sqrt[3]{2^3} = \sqrt[3]{8} = 2$.
Let $x = -2$;   then $\sqrt[3]{x^3} = \sqrt[3]{(-2)^3} = \sqrt[3]{-8} = -2$.

Thus it is correct to write,

$\sqrt[3]{x^3} = x$ for all real numbers,

and again no absolute value sign is needed.

The previous discussion indicates that, technically, every radical expression with variables in the radicands needs to be analyzed individually to determine the necessary restrictions on the variables. However, to avoid having to do this on a problem-by-problem basis, we shall merely **assume that all variables repre-sent positive real numbers**.

Let's conclude this section by simplifying some radical expressions contain-ing variables.

$$\sqrt{72x^3y^7} = \sqrt{36x^2y^6}\sqrt{2xy} = 6xy^3\sqrt{2xy}$$

$$\sqrt[3]{40x^4y^8} = \sqrt[3]{8x^3y^6}\sqrt[3]{5xy^2} = 2xy^2\sqrt[3]{5xy^2}$$

$$\frac{\sqrt{5}}{\sqrt{12a^3}} = \frac{\sqrt{5}}{\sqrt{12a^3}}\cdot\frac{\sqrt{3a}}{\sqrt{3a}} = \frac{\sqrt{15a}}{\sqrt{36a^4}} = \frac{\sqrt{15a}}{6a^2}$$

$$\frac{3}{\sqrt[3]{4x}} = \frac{3}{\sqrt[3]{4x}}\cdot\frac{\sqrt[3]{2x^2}}{\sqrt[3]{2x^2}} = \frac{3\sqrt[3]{2x^2}}{\sqrt[3]{8x^3}} = \frac{3\sqrt[3]{2x^2}}{2x}$$

Problem Set 1.6

Evaluate each of the following.

1. $\sqrt{81}$ **2.** $-\sqrt{49}$ **3.** $\sqrt[3]{125}$ **4.** $\sqrt[4]{81}$

5. $\sqrt{\frac{36}{49}}$ **6.** $\sqrt{\frac{256}{64}}$ **7.** $\sqrt[3]{-\frac{27}{8}}$ **8.** $\sqrt[3]{\frac{64}{27}}$

Express each of the following in simplest radical form. All variables represent positive real numbers.

9. $\sqrt{24}$ **10.** $\sqrt{54}$ **11.** $\sqrt{112}$ **12.** $6\sqrt{28}$

13. $-3\sqrt{44}$ **14.** $-5\sqrt{68}$ **15.** $\frac{3}{4}\sqrt{20}$ **16.** $\frac{3}{8}\sqrt{72}$

17. $\sqrt{12x^2}$ **18.** $\sqrt{45xy^2}$ **19.** $\sqrt{64x^4y^7}$ **20.** $3\sqrt{32a^3}$

21. $\frac{3}{7}\sqrt{45xy^6}$ **22.** $\sqrt[3]{32}$ **23.** $\sqrt[3]{128}$ **24.** $\sqrt[3]{54x^3}$

25. $\sqrt[3]{16x^4}$ **26.** $\sqrt[3]{81x^5y^6}$ **27.** $\sqrt[4]{48x^5}$ **28.** $\sqrt[4]{162x^6y^7}$

29. $\sqrt{\frac{12}{25}}$ **30.** $\sqrt{\frac{75}{81}}$ **31.** $\sqrt{\frac{7}{8}}$ **32.** $\frac{\sqrt{35}}{\sqrt{7}}$

33. $\frac{4\sqrt{3}}{\sqrt{5}}$ **34.** $\frac{\sqrt{27}}{\sqrt{18}}$ **35.** $\frac{6\sqrt{3}}{7\sqrt{6}}$ **36.** $\sqrt{\frac{3x}{2y}}$

37. $\frac{\sqrt{5}}{\sqrt{12x^4}}$ **38.** $\frac{\sqrt{5y}}{\sqrt{18x^3}}$ **39.** $\frac{\sqrt{12a^2b}}{\sqrt{5a^3b^3}}$ **40.** $\frac{5}{\sqrt[3]{3}}$

41. $\frac{\sqrt[3]{27}}{\sqrt[3]{4}}$ **42.** $\sqrt[3]{\frac{5}{2x}}$ **43.** $\frac{\sqrt[3]{2y}}{\sqrt[3]{3x}}$ **44.** $\frac{\sqrt[3]{12xy}}{\sqrt[3]{3x^2y^5}}$

Use the distributive property to help simplify each of the following. For example,

$$3\sqrt{8} + 5\sqrt{2} = 3\sqrt{4}\sqrt{2} + 5\sqrt{2}$$
$$= 6\sqrt{2} + 5\sqrt{2}$$
$$= (6+5)\sqrt{2}$$
$$= 11\sqrt{2}$$

45. $5\sqrt{12} + 2\sqrt{3}$ **46.** $4\sqrt{50} - 9\sqrt{32}$ **47.** $2\sqrt{28} - 3\sqrt{63} + 8\sqrt{7}$

48. $4\sqrt[3]{2} + 2\sqrt[3]{16} - \sqrt[3]{54}$ **49.** $\frac{5}{6}\sqrt{48} - \frac{3}{4}\sqrt{12}$ **50.** $\frac{2}{5}\sqrt{40} + \frac{1}{6}\sqrt{90}$

51. $\dfrac{2\sqrt{8}}{3} - \dfrac{3\sqrt{18}}{5} - \dfrac{\sqrt{50}}{2}$ **52.** $\dfrac{3\sqrt[3]{54}}{2} + \dfrac{5\sqrt[3]{16}}{3}$

Multiply and express the results in simplest radical form. All variables represent non-negative real numbers.

53. $(4\sqrt{3})(6\sqrt{8})$ **54.** $(5\sqrt{8})(3\sqrt{7})$

55. $2\sqrt{3}(5\sqrt{2} + 4\sqrt{10})$ **56.** $3\sqrt{6}(2\sqrt{8} - 3\sqrt{12})$

57. $3\sqrt{x}(\sqrt{6xy} - \sqrt{8y})$ **58.** $\sqrt{6y}(\sqrt{8x} + \sqrt{10y^2})$

59. $(\sqrt{3} + 2)(\sqrt{3} + 5)$ **60.** $(\sqrt{2} - 3)(\sqrt{2} + 4)$

61. $(4\sqrt{2} + \sqrt{3})(3\sqrt{2} + 2\sqrt{3})$ **62.** $(2\sqrt{6} + 3\sqrt{5})(3\sqrt{6} + 4\sqrt{5})$

63. $(6 + 2\sqrt{5})(6 - 2\sqrt{5})$ **64.** $(7 - 3\sqrt{2})(7 + 3\sqrt{2})$

65. $(\sqrt{x} + \sqrt{y})^2$ **66.** $(2\sqrt{x} - 3\sqrt{y})^2$

67. $(\sqrt{a} + \sqrt{b})(\sqrt{a} - \sqrt{b})$ **68.** $(3\sqrt{x} + 5\sqrt{y})(3\sqrt{x} - 5\sqrt{y})$

For each of the following, rationalize the denominator and simplify. All variables represent positive real numbers.

69. $\dfrac{3}{\sqrt{5} + 2}$ **70.** $\dfrac{7}{\sqrt{10} - 3}$ **71.** $\dfrac{4}{\sqrt{7} - \sqrt{3}}$ **72.** $\dfrac{2}{\sqrt{5} + \sqrt{3}}$

73. $\dfrac{\sqrt{2}}{2\sqrt{5} + 3\sqrt{7}}$ **74.** $\dfrac{5}{5\sqrt{2} - 3\sqrt{5}}$ **75.** $\dfrac{\sqrt{x}}{\sqrt{x} - 1}$ **76.** $\dfrac{\sqrt{x}}{\sqrt{x} + 2}$

77. $\dfrac{\sqrt{x}}{\sqrt{x} + \sqrt{y}}$ **78.** $\dfrac{2\sqrt{x}}{\sqrt{x} - \sqrt{y}}$ **79.** $\dfrac{2\sqrt{x} + \sqrt{y}}{3\sqrt{x} - 2\sqrt{y}}$ **80.** $\dfrac{3\sqrt{x} - 2\sqrt{y}}{2\sqrt{x} + 5\sqrt{y}}$

1.7

The Relationship between Exponents and Roots

Recall that we used the basic properties of positive integral exponents to motivate a definition of negative integers as exponents. In this section, we shall use the properties of integral exponents to motivate definitions for rational numbers as exponents. These definitions will tie together the concepts of *exponent* and *root*. Let's consider the following comparisons.

From our study of radicals we know:

If $(b^n)^m = b^{mn}$ is to hold when n is a rational number of the form $1/p$, where p is a positive integer greater than 1, then:

$$(\sqrt{5})^2 = 5$$
$$(\sqrt[3]{8})^3 = 8$$
$$(\sqrt[4]{21})^4 = 21$$

$$(5^{1/2})^2 = 5^{2(1/2)} = 5^1 = 5$$
$$(8^{1/3})^3 = 8^{3(1/3)} = 8^1 = 8$$
$$(21^{1/4})^4 = 21^{4(1/4)} = 21^1 = 21$$

Such examples motivate the following definition.

DEFINITION 1.6

> If b is a real number, n is a positive integer greater than 1, and $\sqrt[n]{b}$ exists, then
>
> $$b^{1/n} = \sqrt[n]{b}.$$

Definition 1.6 states that $b^{1/n}$ means the nth root of b. We shall assume that b and n are chosen so that $\sqrt[n]{b}$ exists in the real number system. For example, $(-25)^{1/2}$ is not meaningful at this time because $\sqrt{-25}$ is not a real number. The following examples illustrate the use of Definition 1.6.

$$25^{1/2} = \sqrt{25} = 5; \qquad 16^{1/4} = \sqrt[4]{16} = 2$$
$$8^{1/3} = \sqrt[3]{8} = 2; \qquad (-27)^{1/3} = \sqrt[3]{-27} = -3$$

Now the following definition provides the basis for the use of *all* rational numbers as exponents.

DEFINITION 1.7

> If m/n is a rational number, where n is a positive integer greater than one and m is any integer, and if b is a real number such that $\sqrt[n]{b}$ exists, then
>
> $$b^{m/n} = \sqrt[n]{b^m} = (\sqrt[n]{b})^m.$$

In Definition 1.7, whether we use the form $\sqrt[n]{b^m}$ or $(\sqrt[n]{b})^m$ for computational purposes depends somewhat on the magnitude of the problem. Let's use both forms on two problems to illustrate this point.

$$8^{2/3} = \sqrt[3]{8^2} = \sqrt[3]{64} = 4 \qquad \text{or} \qquad 8^{2/3} = (\sqrt[3]{8})^2 = (2)^2 = 4$$
$$27^{2/3} = \sqrt[3]{27^2} = \sqrt[3]{729} = 9 \qquad \text{or} \qquad 27^{2/3} = (\sqrt[3]{27})^2 = (3)^2 = 9$$

To compute $8^{2/3}$, either form seems to work equally well. However, to compute $27^{2/3}$, the form $(\sqrt[3]{27})^2$ is much easier to handle. The following examples further illustrate Definition 1.7.

$$25^{3/2} = (\sqrt{25})^3 = 5^3 = 125$$
$$(32)^{-2/5} = \frac{1}{(32)^{2/5}} = \frac{1}{(\sqrt[5]{32})^2} = \frac{1}{2^2} = \frac{1}{4}$$
$$(-64)^{2/3} = (\sqrt[3]{-64})^2 = (-4)^2 = 16$$
$$-8^{4/3} = -(\sqrt[3]{8})^4 = -(2)^4 = -16$$

It can be shown that all of the results pertaining to integral exponents listed in Property 1.2 (on page 16) also hold for all rational exponents. Let's consider some examples to illustrate each of those results.

$$x^{1/2} \cdot x^{2/3} = x^{1/2 + 2/3} \qquad\qquad b^n \cdot b^m = b^{n+m}$$
$$= x^{3/6 + 4/6}$$
$$= x^{7/6}$$

$$(a^{2/3})^{3/2} = a^{(3/2)(2/3)} \qquad\qquad (b^n)^m = b^{nm}$$
$$= a^1 = a$$

$$(16y^{2/3})^{1/2} = (16)^{1/2}(y^{2/3})^{1/2} \qquad (ab)^n = a^n b^n$$
$$= 4y^{1/3}$$

$$\frac{y^{3/4}}{y^{1/2}} = y^{3/4 - 1/2} \qquad\qquad \frac{b^n}{b^m} = b^{n-m}$$
$$= y^{3/4 - 2/4}$$
$$= y^{1/4}$$

$$\left(\frac{x^{1/2}}{y^{1/3}}\right)^6 = \frac{(x^{1/2})^6}{(y^{1/3})^6} \qquad\qquad \left(\frac{a}{b}\right)^n = \frac{a^n}{b^n}$$
$$= \frac{x^3}{y^2}$$

The link between exponents and roots provides a basis for multiplying and dividing some radicals even if they have a different index. The general procedure is one of changing from radical to exponential form, applying the properties of exponents, and then changing back to radical form. Let's consider three examples to illustrate this process.

$$\sqrt{2}\sqrt[3]{2} = 2^{1/2} \cdot 2^{1/3} = 2^{1/2 + 1/3} = 2^{5/6} = \sqrt[6]{2^5} = \sqrt[6]{32}$$

$$\sqrt{xy}\sqrt[5]{x^2 y} = (xy)^{1/2}(x^2 y)^{1/5}$$
$$= x^{1/2}y^{1/2}x^{2/5}y^{1/5}$$
$$= x^{1/2 + 2/5}y^{1/2 + 1/5}$$
$$= x^{9/10}y^{7/10}$$
$$= (x^9 y^7)^{1/10}$$
$$= \sqrt[10]{x^9 y^7}$$

$$\frac{\sqrt{5}}{\sqrt[3]{5}} = \frac{5^{1/2}}{5^{1/3}} = 5^{1/2 - 1/3} = 5^{1/6} = \sqrt[6]{5}$$

Earlier we agreed that a radical such as $\sqrt[3]{x^4}$ is not in simplest form because the radicand contains a perfect power of the index. Thus we simplified $\sqrt[3]{x^4}$ by expressing it $\sqrt[3]{x^3}\sqrt[3]{x}$, which in turn can be written $x\sqrt[3]{x}$. Such simplification can also be done in exponential form, as follows:

$$\sqrt[3]{x^4} = x^{4/3} = x^{3/3} \cdot x^{1/3} = x \cdot x^{1/3} = x\sqrt[3]{x}.$$

Note the use of this type of simplification in the following problems.

EXAMPLE 1

Perform the indicated operations and express the answers in simplest radical form.

(a) $\sqrt[3]{x^2}\,\sqrt[4]{x^3}$ **(b)** $\sqrt{2}\sqrt[3]{4}$ **(c)** $\dfrac{\sqrt{27}}{\sqrt[3]{3}}$

Solutions

(a) $\sqrt[3]{x^2}\,\sqrt[4]{x^3} = x^{2/3} \cdot x^{3/4} = x^{2/3+3/4} = x^{17/12} = x^{12/12} \cdot x^{5/12} = x\,\sqrt[12]{x^5}$

(b) $\sqrt{2}\sqrt[3]{4} = 2^{1/2} \cdot 4^{1/3} = 2^{1/2}(2^2)^{1/3} = 2^{1/2} \cdot 2^{2/3}$

$\qquad\qquad = 2^{1/2+2/3} = 2^{7/6} = 2^{6/6} \cdot 2^{1/6} = 2\,\sqrt[6]{2}$

(c) $\dfrac{\sqrt{27}}{\sqrt[3]{3}} = \dfrac{27^{1/2}}{3^{1/3}} = \dfrac{(3^3)^{1/2}}{3^{1/3}} = \dfrac{3^{3/2}}{3^{1/3}} = 3^{3/2-1/3} = 3^{7/6}$

$\qquad\qquad = 3^{6/6} \cdot 3^{1/6} = 3\,\sqrt[6]{3}$

It should also be recognized that the process of rationalizing the denominator can sometimes be more easily handled in exponential form. Consider the following examples, which illustrate this procedure.

EXAMPLE 2

Rationalize the denominator and express the answer in simplest radical form.

(a) $\dfrac{2}{\sqrt[3]{x}}$ **(b)** $\dfrac{\sqrt[3]{x}}{\sqrt{y}}$

Solutions

(a) $\dfrac{2}{\sqrt[3]{x}} = \dfrac{2}{x^{1/3}} = \dfrac{2}{x^{1/3}} \cdot \dfrac{x^{2/3}}{x^{2/3}} = \dfrac{2x^{2/3}}{x} = \dfrac{2\,\sqrt[3]{x^2}}{x}$

(b) $\dfrac{\sqrt[3]{x}}{\sqrt{y}} = \dfrac{x^{1/3}}{y^{1/2}} = \dfrac{x^{1/3}}{y^{1/2}} \cdot \dfrac{y^{1/2}}{y^{1/2}} = \dfrac{x^{1/3} \cdot y^{1/2}}{y} = \dfrac{x^{2/6} \cdot y^{3/6}}{y} = \dfrac{\sqrt[6]{x^2 y^3}}{y}$

Note in part (b) that if we had changed back to radical form at the step $\dfrac{x^{1/3}y^{1/2}}{y}$, we would have obtained the product of two radicals, $\sqrt[3]{x}\sqrt{y}$, in the numerator. Instead we used the exponential form to find this product and express the final result with a single radical in the numerator. Finally, let's consider an example involving "the root of a root" situation.

EXAMPLE 3

Simplify $\sqrt[3]{\sqrt{2}}$.

Solution

$$\sqrt[3]{\sqrt{2}} = (2^{1/2})^{1/3} = 2^{1/6} = \sqrt[6]{2}$$

Problem Set 1.7

Evaluate each of the following.

1. $49^{1/2}$

2. $64^{1/3}$

3. $32^{3/5}$

4. $(-8)^{1/3}$

5. $-8^{2/3}$

6. $64^{-1/2}$

7. $(\frac{1}{4})^{-1/2}$

8. $(-\frac{27}{8})^{-1/3}$

9. $16^{3/2}$

10. $(0.008)^{1/3}$

11. $(0.01)^{3/2}$

12. $(\frac{1}{27})^{-2/3}$

13. $64^{-5/6}$

14. $-16^{5/4}$

15. $(\frac{1}{8})^{-1/3}$

16. $(-\frac{1}{8})^{-2/3}$

Perform the indicated operations and simplify. Express final answers using positive exponents only.

17. $(3x^{1/4})(5x^{1/3})$

18. $(2x^{2/5})(6x^{1/4})$

19. $(y^{2/3})(y^{-1/4})$

20. $(2x^{1/3})(x^{-1/2})$

21. $(4x^{1/4}y^{1/2})^3$

22. $(5x^{1/2}y)^2$

23. $\dfrac{24x^{3/5}}{6x^{1/3}}$

24. $\dfrac{18x^{1/2}}{9x^{1/3}}$

25. $\dfrac{56a^{1/6}}{8a^{1/4}}$

26. $\dfrac{48b^{1/3}}{12b^{3/4}}$

27. $\left(\dfrac{2x^{1/3}}{3y^{1/4}}\right)^4$

28. $\left(\dfrac{6x^{2/5}}{7y^{2/3}}\right)^2$

29. $\left(\dfrac{x^2}{y^3}\right)^{-1/2}$

30. $\left(\dfrac{a^3}{b^{-2}}\right)^{-1/3}$

31. $\left(\dfrac{4a^2x}{2a^{1/2}x^{1/3}}\right)^3$

32. $\left(\dfrac{3ax^{-1}}{a^{1/2}x^{-2}}\right)^2$

Perform the indicated operations and express the answers in simplest radical form.

33. $\sqrt{2}\sqrt[4]{2}$

34. $\sqrt[3]{3}\sqrt{3}$

35. $\sqrt[3]{x}\sqrt[4]{x}$

36. $\sqrt[3]{x^2}\sqrt[5]{x^3}$

37. $\sqrt{xy}\sqrt[4]{x^3y^5}$

38. $\sqrt[3]{x^2y^4}\sqrt[4]{x^3y}$

39. $\sqrt[3]{a^2b^2}\sqrt[4]{a^3b}$

40. $\sqrt{ab}\sqrt[3]{a^4b^5}$

41. $\sqrt[3]{4}\sqrt{8}$

42. $\sqrt[3]{9}\sqrt{27}$

43. $\dfrac{\sqrt{2}}{\sqrt[3]{2}}$

44. $\dfrac{\sqrt{9}}{\sqrt[3]{3}}$

45. $\dfrac{\sqrt[3]{8}}{\sqrt[4]{4}}$

46. $\dfrac{\sqrt[3]{16}}{\sqrt[6]{4}}$

47. $\dfrac{\sqrt[4]{x^9}}{\sqrt[3]{x^2}}$

48. $\dfrac{\sqrt[5]{x^7}}{\sqrt[3]{x}}$

Rationalize the denominators and express the final answers in simplest radical form.

49. $\dfrac{5}{\sqrt[3]{x}}$

50. $\dfrac{3}{\sqrt[3]{x^2}}$

51. $\dfrac{\sqrt{x}}{\sqrt[3]{y}}$

52. $\dfrac{\sqrt[4]{x}}{\sqrt{y}}$

53. $\dfrac{\sqrt[4]{x^3}}{\sqrt[5]{y^3}}$

54. $\dfrac{2\sqrt{x}}{3\sqrt[3]{y}}$

55. $\dfrac{5\sqrt[3]{y^2}}{4\sqrt[4]{x}}$

56. $\dfrac{\sqrt{xy}}{\sqrt[3]{a^2b}}$

57. Simplify each of the following, expressing the final result as one radical. For example,

$$\sqrt{\sqrt{3}} = (3^{1/2})^{1/2} = 3^{1/4} = \sqrt[4]{3}.$$

(a) $\sqrt{\sqrt[3]{2}}$

(b) $\sqrt[3]{\sqrt[4]{3}}$

(c) $\sqrt[3]{\sqrt{x^3}}$

(d) $\sqrt{\sqrt[3]{x^4}}$

 58. If your calculator has $\boxed{y^x}$ and $\boxed{1/x}$ keys, then they can be used to evaluate cube roots, fourth roots, fifth roots, and so on. For example, $\sqrt[3]{4913}$ can be evaluated as follows.

$$4913 \,\boxed{y^x}\, 3 \,\boxed{1/x}\, = 17$$

Use your calculator to evaluate each of the following.

(a) $\sqrt[3]{1728}$

(b) $\sqrt[3]{5832}$

(c) $\sqrt[4]{2401}$

(d) $\sqrt[4]{65536}$

(e) $\sqrt[5]{161051}$

(f) $\sqrt[5]{6436343}$

59. In Definition 1.7 we stated that $b^{m/n} = \sqrt[n]{b^m} = (\sqrt[n]{b})^m$. Use your calculator to verify each of the following.

(a) $\sqrt[3]{27^2} = (\sqrt[3]{27})^2$ (b) $\sqrt[3]{8^5} = (\sqrt[3]{8})^5$ (c) $\sqrt[4]{16^3} = (\sqrt[4]{16})^3$

(d) $\sqrt[3]{16^2} = (\sqrt[3]{16})^2$ (e) $\sqrt[5]{9^4} = (\sqrt[5]{9})^4$ (f) $\sqrt[3]{12^4} = (\sqrt[3]{12})^4$

60. Use your calculator to evaluate each of the following.

(a) $16^{5/2}$ (b) $25^{7/2}$ (c) $16^{9/4}$

(d) $27^{5/3}$ (e) $343^{2/3}$ (f) $512^{4/3}$

61. Use your calculator to estimate each of the following to the nearest thousandth.

(a) $7^{4/3}$ (b) $10^{4/5}$ (c) $12^{2/5}$

(d) $19^{2/5}$ (e) $7^{3/4}$ (f) $10^{5/4}$

1.8

Complex Numbers

So far we have dealt only with real numbers. However, as we get ready to solve equations in the next chapter, there is a need for "more numbers." There are some very simple equations that do not have solutions within the set of real numbers. For example, the equation $x^2 + 1 = 0$ has no solutions among the real numbers. Therefore, to be able to solve such equations, we need to extend the real number system. In this section we will introduce a set of numbers that contains some numbers whose squares are negative real numbers. Then, in the next chapter and in Chapter 6, we will see that this set of numbers, called the **complex numbers**, provides solutions not only for equations such as $x^2 + 1 = 0$, but also for *any* polynomial equation in general.

Let's begin by defining a number i such that

$$i^2 = -1.$$

The number i is not a real number and is often called the **imaginary unit**, but the number i^2 is the real number -1. The imaginary unit i is used to define a complex number as follows.

DEFINITION 1.8

> A **complex number** is any number that can be expressed in the form
>
> $a + bi,$
>
> where a and b are real numbers.

The form $a + bi$ is called the **standard form** of a complex number. The real number a is called the **real part** of the complex number and b is called the **imaginary part**. (Note that b is a real number even though it is called the imaginary part.) Each of the following represents a complex number.

$6 + 2i$ It is already expressed in the form $a + bi$. Traditionally, complex numbers for which $a \neq 0$ and $b \neq 0$ have been called imaginary numbers.

$5 - 3i$ It can be written $5 + (-3i)$ even though the form $5 - 3i$ is often used.

$-8 + i\sqrt{2}$ It can be written $-8 + \sqrt{2}i$. It is easy to mistake $\sqrt{2}i$ for $\sqrt{2i}$. Thus, we commonly write $i\sqrt{2}$ instead of $\sqrt{2}i$ to avoid any difficulties with the radical sign.

$-9i$ It can be written $0 + (-9i)$. Complex numbers such as $-9i$, for which $a = 0$ and $b \neq 0$, traditionally have been called **pure imaginary numbers**.

5 It can be written $5 + 0i$.

The set of real numbers is a subset of the set of complex numbers. The following diagram indicates the organizational format of the complex number system.

Complex numbers

$a + bi$, where a and b
are real numbers

Real numbers   **Imaginary numbers**

$a + bi$, where $b = 0$ $a + bi$, where $b \neq 0$

Pure imaginary numbers

$a + bi$, where $a = 0$ and $b \neq 0$

Two complex numbers $a + bi$ and $c + di$ are said to be *equal* if and only if $a = c$ and $b = d$. In other words, two complex numbers are equal if and only if their real parts are equal and their imaginary parts are equal.

Adding and Subtracting Complex Numbers

The following definition provides the basis for adding complex numbers.

$$(a + bi) + (c + di) = (a + c) + (b + d)i$$

We can use this definition to find the sum of two complex numbers.

$$(4 + 3i) + (5 + 9i) = (4 + 5) + (3 + 9)i = 9 + 12i$$

$$(-6 + 4i) + (8 - 7i) = (-6 + 8) + (4 - 7)i = 2 - 3i$$

$$(\tfrac{1}{2} + \tfrac{3}{4}i) + (\tfrac{2}{3} + \tfrac{1}{5}i) = (\tfrac{1}{2} + \tfrac{2}{3}) + (\tfrac{3}{4} + \tfrac{1}{5})i$$
$$= (\tfrac{3}{6} + \tfrac{4}{6}) + (\tfrac{15}{20} + \tfrac{4}{20})i = \tfrac{7}{6} + \tfrac{19}{20}i$$

$$(3 + i\sqrt{2}) + (-4 + i\sqrt{2}) = [3 + (-4)] + (\sqrt{2} + \sqrt{2})i = -1 + 2i\sqrt{2}$$

Note the form for writing $2\sqrt{2}i$.

The set of complex numbers is **closed with respect to addition**; that is, the sum of two complex numbers is a complex number. Furthermore, the commutative and associative properties of addition hold for all complex numbers. The additive identity element is $0 + 0i$, or simply the real number 0. The additive inverse of $a + bi$ is $-a - bi$ because

$$(a + bi) + (-a - bi) = [a + (-a)] + [b + (-b)]i = 0.$$

Therefore, to *subtract* $c + di$ from $a + bi$, we add the additive inverse of $c + di$.

$$(a + bi) - (c + di) = (a + bi) + (-c - di)$$
$$= (a - c) + (b - d)i$$

The following examples illustrate the subtraction of complex numbers.

$$(9 + 8i) - (5 + 3i) = (9 - 5) + (8 - 3)i = 4 + 5i$$
$$(3 - 2i) - (4 - 10i) = (3 - 4) + [-2 - (-10)]i = -1 + 8i$$
$$\left(-\tfrac{1}{2} + \tfrac{1}{3}i\right) - \left(\tfrac{3}{4} + \tfrac{1}{2}i\right) = \left(-\tfrac{1}{2} - \tfrac{3}{4}\right) + \left(\tfrac{1}{3} - \tfrac{1}{2}\right)i = -\tfrac{5}{4} - \tfrac{1}{6}i$$

Multiplying and Dividing Complex Numbers

Since $i^2 = -1$, the number i is a square root of -1; so we write $i = \sqrt{-1}$. It should also be evident that $-i$ is a square root of -1 because

$$(-i)^2 = (-i)(-i) = i^2 = -1.$$

Therefore, in the set of complex numbers, -1 has two square roots, namely, i and $-i$. This is symbolically expressed

$$i = \sqrt{-1} \qquad \text{and} \qquad -i = -\sqrt{-1}.$$

Let's extend the definition so that in the set of complex numbers, every negative real number has two square roots. For any positive real number b,

$$(i\sqrt{b})^2 = i^2(b) = -1(b) = -b.$$

Therefore, let's denote the **principal square root of $-b$** by $\sqrt{-b}$ and define it to be

$$\sqrt{-b} = i\sqrt{b},$$

where b is any positive real number. In other words, the principal square root of any negative real number can be represented as the product of a real number and the imaginary unit i. Consider the following examples.

$$\sqrt{-4} = i\sqrt{4} = 2i$$
$$\sqrt{-17} = i\sqrt{17}$$
$$\sqrt{-24} = i\sqrt{24} = i\sqrt{4}\sqrt{6} = 2i\sqrt{6}$$

Note that we simplified the radical $\sqrt{24}$ to $2\sqrt{6}$.

We should also observe that $-\sqrt{-b}$, where $b > 0$, is a square root of $-b$ because

$$(-\sqrt{-b})^2 = (-i\sqrt{b})^2 = i^2(b) = (-1)b = -b.$$

Thus, in the set of complex numbers, $-b$ (where $b > 0$) has two square roots, $i\sqrt{b}$ and $-i\sqrt{b}$. These are expressed

$$\sqrt{-b} = i\sqrt{b} \qquad \text{and} \qquad -\sqrt{-b} = -i\sqrt{b}.$$

We must be careful with the use of the symbol $\sqrt{-b}$, where $b > 0$. Some properties that are true in the set of real numbers involving the square root symbol do not hold if the square root symbol does not represent a real number. For example, $\sqrt{a}\sqrt{b} = \sqrt{ab}$ *does not hold* if a and b are both negative numbers.

Correct: $\sqrt{-4}\sqrt{-9} = (2i)(3i) = 6i^2 = 6(-1) = -6$

Incorrect: $\sqrt{-4}\sqrt{-9} = \sqrt{(-4)(-9)} = \sqrt{36} = 6$

To avoid difficulty with this idea, you should rewrite all expressions of the form $\sqrt{-b}$, where $b > 0$ in the form $i\sqrt{b}$ *before* doing any computations. The following examples further illustrate this point.

$$\sqrt{-5}\sqrt{-7} = (i\sqrt{5})(i\sqrt{7}) = i^2\sqrt{35} = (-1)\sqrt{35} = -\sqrt{35}$$

$$\sqrt{-2}\sqrt{-8} = (i\sqrt{2})(i\sqrt{8}) = i^2\sqrt{16} = (-1)(4) = -4$$

$$\sqrt{-2}\sqrt{8} = (i\sqrt{2})(\sqrt{8}) = i\sqrt{16} = 4i$$

$$\sqrt{-6}\sqrt{-8} = (i\sqrt{6})(i\sqrt{8}) = i^2\sqrt{48} = i^2\sqrt{16}\sqrt{3} = 4i^2\sqrt{3} = -4\sqrt{3}$$

$$\frac{\sqrt{-2}}{\sqrt{3}} = \frac{i\sqrt{2}}{\sqrt{3}} = \frac{i\sqrt{2}}{\sqrt{3}} \cdot \frac{\sqrt{3}}{\sqrt{3}} = \frac{i\sqrt{6}}{3}$$

$$\frac{\sqrt{-48}}{\sqrt{12}} = \frac{i\sqrt{48}}{\sqrt{12}} = i\sqrt{\frac{48}{12}} = i\sqrt{4} = 2i$$

Since complex numbers have a *binomial form*, we can find the product of two complex numbers in the same way that we find the product of two binomials. Then, by replacing i^2 with -1, we can simplify and express the final product in the standard form of a complex number. Consider the following examples.

$$\begin{aligned}
(2 + 3i)(4 + 5i) &= 2(4 + 5i) + 3i(4 + 5i) \\
&= 8 + 10i + 12i + 15i^2 \\
&= 8 + 22i + 15(-1) \\
&= 8 + 22i - 15 \\
&= -7 + 22i
\end{aligned}$$

$$\begin{aligned}
(1 - 7i)^2 &= (1 - 7i)(1 - 7i) \\
&= 1(1 - 7i) - 7i(1 - 7i) \\
&= 1 - 7i - 7i + 49i^2 \\
&= 1 - 14i + 49(-1) \\
&= 1 - 14i - 49 \\
&= -48 - 14i
\end{aligned}$$

$$(2 + 3i)(2 - 3i) = 2(2 - 3i) + 3i(2 - 3i)$$
$$= 4 - 6i + 6i - 9i^2$$
$$= 4 - 9(-1)$$
$$= 4 + 9$$
$$= 13$$

Remark: Don't forget that the multiplication patterns

$$(a + b)^2 = a^2 + 2ab + b^2$$
$$(a - b)^2 = a^2 - 2ab + b^2$$
$$(a + b)(a - b) = a^2 - b^2$$

can also be used when multiplying complex numbers.

The last example illustrates an important idea. The complex numbers $2 + 3i$ and $2 - 3i$ are called conjugates of each other. In general, the two complex numbers $a + bi$ and $a - bi$ are called **conjugates** of each other and **the product of a complex number and its conjugate is a real number**. This can be shown as follows.

$$(a + bi)(a - bi) = a(a - bi) + bi(a - bi)$$
$$= a^2 - abi + abi - b^2i^2$$
$$= a^2 - b^2(-1)$$
$$= a^2 + b^2$$

Conjugates are used to simplify an expression such as $3i/(5 + 2i)$, which *indicates the quotient of two complex numbers.* To eliminate i in the denominator and to change the indicated quotient to the standard form of a complex number, we can multiply both the numerator and denominator by the conjugate of the denominator.

$$\frac{3i}{5 + 2i} = \frac{3i}{5 + 2i} \cdot \frac{5 - 2i}{5 - 2i}$$

$$= \frac{3i(5 - 2i)}{(5 + 2i)(5 - 2i)}$$

$$= \frac{15i - 6i^2}{25 - 4i^2}$$

$$= \frac{15i - 6(-1)}{25 - 4(-1)}$$

$$= \frac{6 + 15i}{29}$$

$$= \frac{6}{29} + \frac{15}{29}i$$

The following examples further illustrate the process of dividing complex numbers.

$$\frac{2-3i}{4-7i} = \frac{2-3i}{4-7i} \cdot \frac{4+7i}{4+7i}$$

$$= \frac{(2-3i)(4+7i)}{(4-7i)(4+7i)}$$

$$= \frac{8 + 14i - 12i - 21i^2}{16 - 49i^2}$$

$$= \frac{8 + 2i - 21(-1)}{16 - 49(-1)}$$

$$= \frac{29 + 2i}{65} = \frac{29}{65} + \frac{2}{65}i$$

$$\frac{4-5i}{2i} = \frac{4-5i}{2i} \cdot \frac{-2i}{-2i}$$

$$= \frac{(4-5i)(-2i)}{(2i)(-2i)}$$

$$= \frac{-8i + 10i^2}{-4i^2}$$

$$= \frac{-8i + 10(-1)}{-4(-1)}$$

$$= \frac{-10 - 8i}{4} = -\frac{5}{2} - 2i$$

For a problem such as the last one, in which the denominator is a pure imaginary number, we can change to standard form by choosing a multiplier other than the conjugate of the denominator. Consider the following alternate approach.

$$\frac{4-5i}{2i} = \frac{4-5i}{2i} \cdot \frac{i}{i}$$

$$= \frac{(4-5i)(i)}{(2i)(i)}$$

$$= \frac{4i - 5i^2}{2i^2}$$

$$= \frac{4i - 5(-1)}{2(-1)}$$

$$= \frac{5 + 4i}{-2}$$

$$= -\frac{5}{2} - 2i$$

Problem Set 1.8

Add or subtract as indicated.

1. $(5 + 2i) + (8 + 6i)$ 2. $(-9 + 3i) + (4 + 5i)$
3. $(8 + 6i) - (5 + 2i)$ 4. $(-6 + 4i) - (4 + 6i)$
5. $(-7 - 3i) + (-4 + 4i)$ 6. $(6 - 7i) - (7 - 6i)$
7. $(-2 - 3i) - (-1 - i)$ 8. $(\frac{1}{3} + \frac{2}{5}i) + (\frac{1}{2} + \frac{1}{4}i)$
9. $(-\frac{3}{4} - \frac{1}{4}i) + (\frac{3}{5} + \frac{2}{3}i)$ 10. $(\frac{5}{8} + \frac{1}{2}i) - (\frac{7}{8} + \frac{1}{5}i)$
11. $(\frac{3}{10} - \frac{3}{4}i) - (-\frac{2}{5} + \frac{1}{6}i)$ 12. $(4 + i\sqrt{3}) + (-6 - 2i\sqrt{3})$
13. $(5 + 3i) + (7 - 2i) + (-8 - i)$ 14. $(5 - 7i) - (6 - 2i) - (-1 - 2i)$

Write each of the following in terms of i and simplify. For example, $\sqrt{-20} = i\sqrt{20} = i\sqrt{4}\sqrt{5} = 2i\sqrt{5}$.

15. $\sqrt{-9}$ 16. $\sqrt{-49}$ 17. $\sqrt{-19}$ 18. $\sqrt{-31}$
19. $\sqrt{-\frac{4}{9}}$ 20. $\sqrt{-\frac{25}{36}}$ 21. $\sqrt{-8}$ 22. $\sqrt{-18}$
23. $\sqrt{-27}$ 24. $\sqrt{-32}$ 25. $\sqrt{-54}$ 26. $\sqrt{-40}$
27. $3\sqrt{-36}$ 28. $5\sqrt{-64}$ 29. $4\sqrt{-18}$ 30. $6\sqrt{-8}$

Write each of the following in terms of i, perform the indicated operations, and simplify. For example,

$$\sqrt{-9}\sqrt{-16} = (i\sqrt{9})(i\sqrt{16}) = (3i)(4i) = 12i^2 = 12(-1) = -12$$

31. $\sqrt{-4}\sqrt{-16}$ 32. $\sqrt{-25}\sqrt{-9}$ 33. $\sqrt{-2}\sqrt{-3}$ 34. $\sqrt{-3}\sqrt{-7}$
35. $\sqrt{-5}\sqrt{-4}$ 36. $\sqrt{-7}\sqrt{-9}$ 37. $\sqrt{-6}\sqrt{-10}$ 38. $\sqrt{-2}\sqrt{-12}$

39. $\sqrt{-8}\sqrt{-7}$ 40. $\sqrt{-12}\sqrt{-5}$ 41. $\dfrac{\sqrt{-36}}{\sqrt{-4}}$ 42. $\dfrac{\sqrt{-64}}{\sqrt{-16}}$

43. $\dfrac{\sqrt{-54}}{\sqrt{-9}}$ 44. $\dfrac{\sqrt{-18}}{\sqrt{-3}}$

Find each of the following products and express the answers in standard form.

45. $(3i)(7i)$ 46. $(-5i)(8i)$ 47. $(4i)(3 - 2i)$
48. $(5i)(2 + 6i)$ 49. $(3 + 2i)(4 + 6i)$ 50. $(7 + 3i)(8 + 4i)$
51. $(4 + 5i)(2 - 9i)$ 52. $(1 + i)(2 - i)$ 53. $(-2 - 3i)(4 + 6i)$
54. $(-3 - 7i)(2 + 10i)$ 55. $(6 - 4i)(-1 - 2i)$ 56. $(7 - 3i)(-2 - 8i)$
57. $(3 + 4i)^2$ 58. $(4 - 2i)^2$ 59. $(-1 - 2i)^2$
60. $(-2 + 5i)^2$ 61. $(8 - 7i)(8 + 7i)$ 62. $(5 + 3i)(5 - 3i)$
63. $(-2 + 3i)(-2 - 3i)$ 64. $(-6 - 7i)(-6 + 7i)$

Find each of the following quotients expressing answers in standard form.

65. $\dfrac{4i}{3 - 2i}$ 66. $\dfrac{3i}{6 + 2i}$ 67. $\dfrac{2 + 3i}{3i}$ 68. $\dfrac{3 - 5i}{4i}$

69. $\dfrac{3}{2i}$ 70. $\dfrac{7}{4i}$ 71. $\dfrac{3 + 2i}{4 + 5i}$ 72. $\dfrac{2 + 5i}{3 + 7i}$

73. $\dfrac{4 + 7i}{2 - 3i}$ 74. $\dfrac{3 + 9i}{4 - i}$ 75. $\dfrac{3 - 7i}{-2 + 4i}$ 76. $\dfrac{4 - 10i}{-3 + 7i}$

77. $\dfrac{-1 - i}{-2 - 3i}$ 78. $\dfrac{-4 + 9i}{-3 - 6i}$

79. Using $a + bi$ and $c + di$ to represent two complex numbers, verify the following properties.

(a) The conjugate of the sum of two complex numbers is equal to the sum of the conjugates of the two numbers.

(b) The conjugate of the product of two complex numbers is equal to the product of the conjugates of the numbers.

Chapter 1 Summary

Be sure of the following key concepts from this chapter: set, null set, equal sets, subset, natural numbers, whole numbers, integers, rational numbers, irrational numbers, real numbers, complex numbers, absolute value, similar terms, exponent, monomial, binomial, polynomial, degree of a polynomial, perfect-square trinomial, factoring polynomials, rational expression, least common denominator, radical, simplest radical form, root, and conjugate of a complex number.

The following properties of the real numbers provide a basis for arithmetic and algebraic computation: closure for addition and multiplication, commutativity for addition and multiplication, associativity for addition and multiplication, identity properties for addition and multiplication, inverse properties for addition and multiplication, multiplication property of zero, multiplication property of -1, and distributive property.

The following properties of absolute value are useful.

1. $|a| \geq 0$

2. $|a| = |-a|$ a and b are real numbers

3. $|a - b| = |b - a|$

The following properties of exponents provide the basis for much of our computational work with polynomials.

1. $b^n \cdot b^m = b^{n+m}$

2. $(b^n)^m = b^{mn}$

3. $(ab)^n = a^n b^n$ m and n are rational numbers and a and b are real numbers,

4. $\left(\dfrac{a}{b}\right)^n = \dfrac{a^n}{b^n}$ except $b \neq 0$ whenever it appears in a denominator

5. $\dfrac{b^n}{b^m} = b^{n-m}$

The following product patterns are helpful to recognize when multiplying polynomials.

1. $(a + b)^2 = a^2 + 2ab + b^2$

2. $(a - b)^2 = a^2 - 2ab + b^2$

3. $(a + b)(a - b) = a^2 - b^2$

4. $(a + b)^3 = a^3 + 3a^2b + 3ab^2 + b^3$

5. $(a - b)^3 = a^3 - 3a^2b + 3ab^2 - b^3$

Be sure of the following factoring techniques:

1. Factoring out the highest common monomial factor;
2. Factoring by grouping;
3. Factoring a trinomial into the product of two binomials;
4. Recognizing some basic factoring patterns, namely,

$$a^2 + 2ab + b^2 = (a + b)^2,$$
$$a^2 - 2ab + b^2 = (a - b)^2,$$
$$a^2 - b^2 = (a + b)(a - b),$$
$$a^3 + b^3 = (a + b)(a^2 - ab + b^2),$$
$$a^3 - b^3 = (a - b)(a^2 + ab + b^2).$$

Be sure that you can simplify, add, subtract, multiply, and divide rational expressions using the following properties and definitions.

1. $\dfrac{a \cdot k}{b \cdot k} = \dfrac{a}{b}$

2. $\dfrac{-a}{b} = \dfrac{a}{-b} = -\dfrac{a}{b}$

3. $\dfrac{a}{b} \cdot \dfrac{c}{d} = \dfrac{ac}{bd}$

4. $\dfrac{a}{b} \div \dfrac{c}{d} = \dfrac{a}{b} \cdot \dfrac{d}{c} = \dfrac{ad}{bc}$

5. $\dfrac{a}{c} + \dfrac{b}{c} = \dfrac{a + b}{c}$

6. $\dfrac{a}{c} - \dfrac{b}{c} = \dfrac{a - b}{c}$

Be sure that you can simplify, add, subtract, multiply, and divide radicals using the following definitions and properties:

1. $\sqrt[n]{b} = a$ if and only if $a^n = b$;
2. $\sqrt[n]{bc} = \sqrt[n]{b}\,\sqrt[n]{c}$;
3. $\sqrt[n]{\dfrac{b}{c}} = \dfrac{\sqrt[n]{b}}{\sqrt[n]{c}}$.

The following definition provides the link between exponents and roots:

$$b^{m/n} = \sqrt[n]{b^m} = (\sqrt[n]{b})^m.$$

This link, along with the properties of exponents, allows us (1) to multiply and divide some radicals with different indices, (2) to change to simplest radical form while in exponential form, and (3) to simplify expressions that are roots of roots.

Be sure that you can add, subtract, multiply, and divide complex numbers.

Chapter 1 Review Problem Set

Evaluate each of the following.

1. 5^{-3}

2. -3^{-4}

3. $(\frac{3}{4})^{-2}$

4. $\dfrac{1}{(\frac{1}{3})^{-2}}$

5. $-\sqrt{64}$

6. $\sqrt[3]{\frac{27}{8}}$

7. $\sqrt[5]{-\frac{1}{32}}$

8. $36^{-1/2}$

9. $(\frac{1}{8})^{-2/3}$

10. $-32^{3/5}$

Perform the indicated operations and simplify. Express the final answers using positive exponents only.

11. $(3x^{-2}y^{-1})(4x^4y^2)$

12. $(5x^{2/3})(-6x^{1/2})$

13. $(-8a^{-1/2})(-6a^{1/3})$

14. $(3x^{-2/3}y^{1/5})^3$

15. $\dfrac{64x^{-2}y^3}{16x^3y^{-2}}$

16. $\dfrac{56x^{-1/3}y^{2/5}}{7x^{1/4}y^{-3/5}}$

17. $\left(\dfrac{-8x^2y^{-1}}{2x^{-1}y^2}\right)^2$

18. $\left(\dfrac{36a^{-1}b^4}{-12a^2b^5}\right)^{-1}$

Perform the indicated operations.

19. $(-7x - 3) + (5x - 2) + (6x + 4)$

20. $(12x + 5) - (7x - 4) - (8x + 1)$

21. $3(a - 2) - 2(3a + 5) + 3(5a - 1)$

22. $(4x - 7)(5x + 6)$

23. $(-3x + 2)(4x - 3)$

24. $(7x - 3)(-5x + 1)$

25. $(x + 4)(x^2 - 3x - 7)$

26. $(2x + 1)(3x^2 - 2x + 6)$

27. $(5x - 3)^2$

28. $(3x + 7)^2$

29. $(2x - 1)^3$

30. $(3x + 5)^3$

31. $(x^2 - 2x - 3)(x^2 + 4x + 5)$

32. $(2x^2 - x - 2)(x^2 + 6x - 4)$

33. $\dfrac{24x^3y^4 - 48x^2y^3}{-6xy}$

34. $\dfrac{-56x^2y + 72x^3y^2}{8x^2}$

Factor each of the following polynomials completely. Indicate any that are not factorable using integers.

35. $9x^2 - 4y^2$

36. $3x^3 - 9x^2 - 120x$

37. $4x^2 + 20x + 25$

38. $(x - y)^2 - 9$

39. $x^2 - 2x - xy + 2y$

40. $64x^3 - 27y^3$

41. $15x^2 - 14x - 8$

42. $3x^3 + 36$

43. $2x^2 - x - 8$

44. $3x^3 + 24$

45. $x^4 - 13x^2 + 36$

46. $4x^2 - 4x + 1 - y^2$

Perform the following indicated operations involving rational expressions. Express final answers in simplest form.

47. $\dfrac{8xy}{18x^2y} \cdot \dfrac{24xy^2}{16y^3}$

48. $\dfrac{-14a^2b^2}{6b^3} \div \dfrac{21a}{15ab}$

49. $\dfrac{x^2 + 3x - 4}{x^2 - 1} \cdot \dfrac{3x^2 + 8x + 5}{x^2 + 4x}$

50. $\dfrac{9x^2 - 6x + 1}{2x^2 + 8} \cdot \dfrac{8x + 20}{6x^2 + 13x - 5}$

51. $\dfrac{3x - 2}{4} + \dfrac{5x - 1}{3}$

52. $\dfrac{2x - 6}{5} - \dfrac{x + 4}{3}$

53. $\dfrac{3}{n^2} + \dfrac{4}{5n} - \dfrac{2}{n}$

54. $\dfrac{5}{x^2 + 7x} - \dfrac{3}{x}$

55. $\dfrac{3x}{x^2 - 6x - 40} + \dfrac{4}{x^2 - 16}$

56. $\dfrac{2}{x - 2} - \dfrac{2}{x + 2} - \dfrac{4}{x^3 - 4x}$

Simplify each of the following complex fractions.

57. $\dfrac{\dfrac{3}{x} - \dfrac{2}{y}}{\dfrac{5}{x^2} + \dfrac{7}{y}}$

58. $\dfrac{3 - \dfrac{2}{x}}{4 + \dfrac{3}{x}}$

Express each of the following in simplest radical form. All variables represent positive real numbers.

59. $5\sqrt{48}$

60. $3\sqrt{24x^3}$

61. $\sqrt[3]{32x^4 y^5}$

62. $\dfrac{3\sqrt{8}}{2\sqrt{6}}$

63. $\sqrt{\dfrac{5x}{2y^2}}$

64. $\dfrac{3}{\sqrt{2} + 5}$

65. $\dfrac{4\sqrt{2}}{3\sqrt{2} + \sqrt{3}}$

66. $\dfrac{3\sqrt{x}}{\sqrt{x} - 2\sqrt{y}}$

Perform the indicated operations and express the answers in simplest radical form.

67. $\sqrt{5}\sqrt[3]{5}$

68. $\sqrt[3]{x^2}\sqrt[4]{x}$

69. $\sqrt{x^3}\sqrt[3]{x^4}$

70. $\sqrt{xy}\sqrt[3]{x^3 y^2}$

71. $\dfrac{\sqrt{5}}{\sqrt[3]{5}}$

72. $\dfrac{\sqrt[3]{x^2}}{\sqrt[4]{x^3}}$

Perform the indicated operations and express the resulting complex number in standard form.

73. $(-7 + 3i) + (-4 - 9i)$

74. $(2 - 10i) - (3 - 8i)$

75. $(-1 + 4i) - (-2 + 6i)$

76. $(3i)(-7i)$

77. $(2 - 5i)(3 + 4i)$

78. $(-3 - i)(6 - 7i)$

79. $(4 + 2i)(-4 - i)$

80. $(5 - 2i)(5 + 2i)$

81. $\dfrac{5}{3i}$

82. $\dfrac{2 + 3i}{3 - 4i}$

83. $\dfrac{-1 - 2i}{-2 + i}$

84. $\dfrac{-6i}{5 + 2i}$

Write each of the following in terms of i and simplify.

85. $\sqrt{-100}$

86. $\sqrt{-40}$

87. $4\sqrt{-80}$

88. $(\sqrt{-9})(\sqrt{-16})$

89. $(\sqrt{-6})(\sqrt{-8})$

90. $\dfrac{\sqrt{-24}}{\sqrt{-3}}$

2

Equations, Inequalities, and Problem Solving

2.1 Linear Equations and Problem Solving

2.2 More Equations and Applications

2.3 Quadratic Equations

2.4 Applications of Linear and Quadratic Equations

2.5 Miscellaneous Equations

2.6 Inequalities

2.7 Inequalities Involving Quotients and Absolute Value

A common thread throughout precalculus algebra courses is one of developing algebraic skills, then using the skills to solve equations and inequalities, and then using equations and inequalities to solve applied problems. In this chapter we shall review and extend a variety of concepts pertaining to that thread.

2.1

Linear Equations and Problem Solving

An algebraic equation such as $5x + 2 = 12$ is neither true nor false as it stands; it is sometimes referred to as an **open sentence**. Each time that a number is substituted for x, the algebraic equation $5x + 2 = 12$ becomes a **numerical statement** that is either true or false. For example, if $x = 5$ then $5x + 2 = 12$ becomes $5(5) + 2 = 12$, which is a false statement. If $x = 2$ then $5x + 2 = 12$ becomes $5(2) + 2 = 12$, which is a true statement. **Solving an equation** refers to the process of finding the number (or numbers) that makes an algebraic equation a true numerical statement. Such numbers are called the **solutions** or **roots** of the equation and are said to **satisfy the equation**. The set of all solutions of an equation is called its **solution set**. Thus, $\{2\}$ is the solution set of $5x + 2 = 12$.

An equation that is satisfied by all numbers that can meaningfully replace the variable is called an **identity**. For example,

$$3(x + 2) = 3x + 6, \quad x^2 - 4 = (x + 2)(x - 2), \quad \text{and} \quad \frac{1}{x} + \frac{1}{2} = \frac{2 + x}{2x}$$

are all identities. In the last identity x cannot equal zero; thus the statement

$$\frac{1}{x} + \frac{1}{2} = \frac{2 + x}{2x}$$

is true for all real numbers except zero. An equation that is true for some but not all permissible values of the variable is called a **conditional equation**. Thus the equation $5x + 2 = 12$ is a conditional equation.

Equivalent equations are equations that have the same solution set. For example,

$$7x - 1 = 20, \quad 7x = 21, \quad \text{and} \quad x = 3$$

are all equivalent equations because $\{3\}$ is the solution set of each. The general procedure for solving an equation is to continue replacing the given equation with equivalent but simpler equations until an equation of the form "variable = constant" or "constant = variable" is obtained. Thus, in the example above, $7x - 1 = 20$ has been simplified to $7x = 21$, which has been further simplified to $x = 3$, which gives us the solution set, $\{3\}$.

Techniques for solving equations are centered around properties of equality. The following list summarizes some basic properties of equality.

PROPERTY 2.1
(Properties
of Equality)

For all real numbers a, b, and c,

1. $a = a$;   Reflexive property
2. If $a = b$ then $b = a$;   Symmetric property
3. If $a = b$ and $b = c$, then $a = c$;   Transitive property

> **4.** If $a = b$, then a may be replaced by b, or b may be replaced by a, in any statement without changing the meaning of the statement; Substitution property
>
> **5.** $a = b$ if and only if $a + c = b + c$;   Addition property
>
> **6.** $a = b$ if and only if $ac = bc$, where $c \neq 0$.   Multiplication property

The addition property of equality states that any number can be added to both sides of an equation and an equivalent equation is produced. The multiplication property of equality states that an equivalent equation is produced whenever both sides of an equation are multiplied by the same nonzero real number.

Now let's consider how these properties of equality can be used to solve a variety of linear equations. A **linear equation** in the variable x is one that can be written in the form

$$ax + b = 0,$$

where a and b are real numbers and $a \neq 0$.

EXAMPLE 1

Solve the equation $-4x - 3 = 2x + 9$.

Solution

$$-4x - 3 = 2x + 9$$
$$-4x - 3 + (-2x) = 2x + 9 + (-2x) \qquad \text{Add } -2x \text{ to both sides.}$$
$$-6x - 3 = 9$$
$$-6x - 3 + 3 = 9 + 3 \qquad \text{Add 3 to both sides.}$$
$$-6x = 12$$
$$-\tfrac{1}{6}(-6x) = -\tfrac{1}{6}(12) \qquad \text{Multiply both sides by } -\tfrac{1}{6}.$$
$$x = -2$$

Check: To check an apparent solution we can substitute it into the original equation and see if a true numerical statement is obtained.

$$-4x - 3 = 2x + 9$$
$$-4(-2) - 3 \stackrel{?}{=} 2(-2) + 9$$
$$8 - 3 \stackrel{?}{=} -4 + 9$$
$$5 = 5$$

Now we know that the solution set is $\{-2\}$.

EXAMPLE 2

Solve $4(n - 2) - 3(n - 1) = 2(n + 6)$.

Solution First let's use the distributive property to remove parentheses and combine similar terms.

$$4(n - 2) - 3(n - 1) = 2(n + 6)$$
$$4n - 8 - 3n + 3 = 2n + 12$$
$$n - 5 = 2n + 12$$

Now we can apply the addition property of equality.

$$n - 5 + (-n) = 2n + 12 + (-n)$$
$$-5 = n + 12$$
$$-5 + (-12) = n + 12 + (-12)$$
$$-17 = n$$

Check:

$$4(n - 2) - 3(n - 1) = 2(n + 6)$$
$$4(-17 - 2) - 3(-17 - 1) \overset{?}{=} 2(-17 + 6)$$
$$4(-19) - 3(-18) \overset{?}{=} 2(-11)$$
$$-76 + 54 \overset{?}{=} -22$$
$$-22 = -22$$

The solution set is $\{-17\}$.

As you study our examples of solving equations, pay special attention to the steps being shown in the solutions. Certainly, there are no rules as to which steps should be performed mentally; this is an individual decision. We would suggest that you show enough steps so that the "flow" of the process is understood and so that the chances of making careless computational errors are minimized. Furthermore, we shall discontinue showing the check for each problem, but remember that *checking an answer* is the *only way* of being sure of your result.

EXAMPLE 3

Solve $\dfrac{2y - 3}{3} + \dfrac{y + 1}{2} = 3$.

Solution

$$\frac{2y - 3}{3} + \frac{y + 1}{2} = 3$$

$$6\left(\frac{2y - 3}{3} + \frac{y + 1}{2}\right) = 6(3) \qquad \text{Multiply both sides by the LCD.}$$

$$6\left(\frac{2y - 3}{3}\right) + 6\left(\frac{y + 1}{2}\right) = 6(3) \qquad \begin{array}{l}\text{Apply the distributive property} \\ \text{on the left side.}\end{array}$$

$$2(2y - 3) + 3(y + 1) = 18$$
$$4y - 6 + 3y + 3 = 18$$
$$7y - 3 = 18$$
$$7y = 21$$
$$y = 3$$

The solution set is $\{3\}$. (Check it!)

EXAMPLE 4

Solve $\dfrac{4x - 1}{10} - \dfrac{5x + 2}{4} = -3$.

Solution

$$\frac{4x - 1}{10} - \frac{5x + 2}{4} = -3$$

$$20\left(\frac{4x - 1}{10} - \frac{5x + 2}{4}\right) = 20(-3)$$

$$20\left(\frac{4x - 1}{10}\right) - 20\left(\frac{5x + 2}{4}\right) = -60$$

$$2(4x - 1) - 5(5x + 2) = -60$$

$$8x - 2 - 25x - 10 = -60$$

$$-17x - 12 = -60$$

$$-17x = -48$$

$$x = \frac{48}{17}$$

The solution set is $\{\frac{48}{17}\}$.

Problem Solving

The ability to use the tools of algebra to solve problems requires that we be able to translate the English language into the language of algebra. More specifically, at this time we need to translate *English sentences* into *algebraic equations* so that we can use our equation-solving skills. Let's work through an example and then comment on some of the problem-solving aspects of it.

PROBLEM 1

If 2 is subtracted from five times a certain number, the result is 28. Find the number.

Solution Let n represent the number to be found. The sentence, "If 2 is subtracted from five times a certain number, the result is 28" translates into the equation $5n - 2 = 28$. Solving this equation, we obtain

$$5n - 2 = 28$$
$$5n = 30$$
$$n = 6.$$

The number to be found is 6.

Now let's make a few comments about our approach to Problem 1. Making a statement such as "Let n represent the number to be found" is often referred to as **declaring the variable**. It amounts to choosing a letter to use as a variable and indicating what the variable represents for a specific problem. This may seem like an insignificant idea, but as the problems become more complex, the process of declaring the variable becomes more important. It is also a good idea to choose a "meaningful" variable. For example, if the problem involves finding the width of a rectangle, then a choice of w for the variable is reasonable. Furthermore, it is true that some people can solve a problem such as Problem 1 without setting up an algebraic equation. However, as problems increase in difficulty, the translation from English to algebra becomes a key issue. Therefore, even with these relatively easy problems, we suggest that you concentrate on the translation process.

To *check* our answer for Problem 1, we must determine whether it satisfies the conditions stated in the original problem. Because 2 subtracted from 5(6) equals 28, we know that our answer of 6 is correct. Remember, when you are checking a potential answer for a word problem, it is *not* sufficient to check the result in the equation used to solve the problem, because the equation itself may be in error.

PROBLEM 2

Find three consecutive integers whose sum is -45.

Solution Let n represent the smallest integer; then $n + 1$ is the next integer and $n + 2$ is the largest of the three integers. Since the sum of the three consecutive integers is to be -45, we have the following equation.

$$n + (n + 1) + (n + 2) = -45$$
$$3n + 3 = -45$$
$$3n = -48$$
$$n = -16$$

If $n = -16$, then $n + 1$ is -15 and $n + 2$ is -14. Thus, the three consecutive integers are -16, -15, and -14.

PROBLEM 3

Tina is paid time-and-a-half for each hour worked over 40 hours in a week. Last week she worked 45 hours and earned \$380. What is her normal hourly rate?

Solution Let r represent her normal hourly rate. Then $\frac{3}{2}r$ represents $1\frac{1}{2}$ times her normal hourly rate (time-and-a-half). The following *guideline* can be used to help set up the equation.

regular wages for first 40 hours	+	wages for 5 hours of overtime	=	total wages
↓		↓		↓
$40r$		$5(\frac{3}{2}r)$		\$380

Solving this equation, we obtain

$$2[40r + 5(\tfrac{3}{2}r)] = 2(380)$$
$$2(40r) + 2[5(\tfrac{3}{2}r)] = 760$$
$$80r + 15r = 760$$
$$95r = 760$$
$$r = 8.$$

Her normal hourly rate is thus $8 per hour. (Check the answer back in the statement of the problem!)

PROBLEM 4

There are 51 students in a certain class. The number of females is 5 less than three times the number of males. Find the number of females and the number of males in the class.

Solution Let m represent the number of males; then $3m - 5$ represents the number of females. Since the total number of students is 51, we can set up and solve the following equation.

$$m + (3m - 5) = 51$$
$$4m - 5 = 51$$
$$4m = 56$$
$$m = 14$$

Therefore there are 14 males and $3(14) - 5 = 37$ females.

Problem Set 2.1

Solve each of the following equations.

1. $9x - 3 = -21$ **2.** $-5x + 4 = -11$ **3.** $13 - 2x = 14$

4. $17 = 6a + 5$ **5.** $3n - 2 = 2n + 5$ **6.** $4n + 3 = 5n - 9$

7. $-5a + 3 = -3a + 6$ **8.** $4x - 3 + 2x = 8x - 3 - x$

9. $-3(x + 1) = 7$ **10.** $5(2x - 1) = 13$

11. $4(2x - 1) = 3(3x + 2)$ **12.** $5x - 4(x - 6) = -11$

13. $4(n - 2) - 3(n - 1) = 2(n + 6)$ **14.** $-3(2t - 5) = 2(4t + 7)$

15. $3(2t - 1) - 2(5t + 1) = 4(3t + 4)$

16. $-(3x - 1) + (2x + 3) = -4 + 3(x - 1)$

17. $-2(y - 4) - (3y - 1) = -2 + 5(y + 1)$

18. $\dfrac{-3x}{4} = \dfrac{9}{2}$ **19.** $-\dfrac{6x}{7} = 12$ **20.** $\dfrac{n}{2} - \dfrac{1}{3} = \dfrac{13}{6}$

21. $\dfrac{3n}{4} - \dfrac{n}{12} = 6$ **22.** $\dfrac{2x}{3} - \dfrac{x}{5} = 7$ **23.** $\dfrac{h}{2} + \dfrac{h}{5} = 1$

24. $\dfrac{4y}{5} - 7 = \dfrac{y}{10}$ **25.** $\dfrac{y}{5} - 2 = \dfrac{y}{2} + 1$ **26.** $\dfrac{x+2}{3} + \dfrac{x-1}{4} = \dfrac{9}{2}$

27. $\dfrac{c+5}{7} + \dfrac{c-3}{4} = \dfrac{5}{14}$ **28.** $\dfrac{2x-5}{6} - \dfrac{3x-4}{8} = 0$ **29.** $\dfrac{n-3}{2} - \dfrac{4n-1}{6} = \dfrac{2}{3}$

30. $\dfrac{3x-1}{2} + \dfrac{x-3}{4} = \dfrac{1}{2}$ **31.** $\dfrac{2t+3}{6} - \dfrac{t-9}{4} = 5$ **32.** $\dfrac{2x+7}{9} - 4 = \dfrac{x-7}{12}$

33. $\dfrac{3n-1}{8} - 2 = \dfrac{2n+5}{7}$ **34.** $\dfrac{x+2}{3} + \dfrac{3x+1}{4} + \dfrac{2x-1}{6} = 2$

35. $\dfrac{2t-3}{6} + \dfrac{3t-2}{4} + \dfrac{5t+6}{12} = 4$ **36.** $\dfrac{3y-1}{8} + y - 2 = \dfrac{y+4}{4}$

37. $\dfrac{2x+1}{14} - \dfrac{3x+4}{7} = \dfrac{x-1}{2}$ **38.** $n + \dfrac{2n-3}{9} - 2 = \dfrac{2n+1}{3}$

39. $(x-3)(x-1) - x(x+2) = 7$ **40.** $(3n+4)(n-2) - 3n(n+3) = 3$

41. $(2y+1)(3y-2) - (6y-1)(y+4) = -20y$

42. $(4t-3)(t+2) - (2t+3)^2 = -1$

Solve each of the following problems by setting up and solving an algebraic equation.

43. Three is subtracted from a certain number and then that result is multiplied by 4 to produce 152. Find the number.

44. One is subtracted from twice a certain number and then that result is multiplied by 3 to produce -75. Find the number.

45. The sum of two numbers is 30. The larger number is 5 less than four times the smaller number. Find the two numbers.

46. One number is 5 less than another number. Find the numbers if five times the smaller number is 11 less than four times the larger number.

47. The sum of three consecutive integers is 21 larger than twice the smallest integer. Find the integers.

48. Find three consecutive even integers such that if the largest integer is subtracted from four times the smallest, the result is 6 more than twice the middle integer.

49. Find three consecutive odd integers such that three times the largest is 23 less than twice the sum of the two smallest integers.

50. Find two consecutive integers such that the difference of their squares is 37.

51. Find three consecutive integers such that the product of the two largest is 20 more than the square of the smallest integer.

52. Find four consecutive integers such that the product of the two largest is 46 more than the product of the two smallest integers.

53. The average of the salaries of Kelly, Renee, and Nina is $20,000 a year. If Kelly earns $4000 less than Renee and Nina's salary is two-thirds of Renee's salary, find the salary of each person.

54. Barry is paid double-time for each hour worked over 40 hours in a week. Last week he worked 47 hours and earned $378. What is his normal hourly rate?

55. Greg had 80 coins consisting of pennies, nickels, and dimes. The number of nickels was 5 more than one-third the number of pennies and the number of dimes was 1 less than one-fourth the number of pennies. How many coins of each kind did he have?

56. Rita has a collection of 105 coins consisting of nickels, dimes, and quarters. The number of dimes is 5 more than one-third the number of nickels, and the number of quarters is twice the number of dimes. How many coins of each kind does she have?

57. In a class of 43 students, the number of males is 8 less than twice the number of females. How many females and how many males are there in the class?

58. A precinct reported that 316 people had voted in an election. The number of Republican voters was 6 more than two-thirds the number of Democrats. How many Republicans and how many Democrats voted in that precinct?

59. Two years ago Janie was half as old as she will be 9 years from now. How old is she now?

60. The sum of the present ages of Eric and his father is 58 years. In 10 years his father will be twice as old as Eric will be at that time. Find their present ages.

61. Brad is presently 6 years older than Pedro. Five years ago Pedro's age was three-fourths of Brad's age at that time. Find the present ages of Brad and Pedro.

62. Tina is 4 years older than Sherry. In five years the sum of their ages will be 48. Find their present ages.

Miscellaneous Problems

63. Verify that for any three consecutive integers, the sum of the smallest and largest is equal to twice the middle integer.

64. Verify that no four consecutive integers can be found such that the product of the smallest and largest is equal to the product of the other two integers.

2.2

More Equations and Applications

In the previous section we considered linear equations, such as

$$\frac{x - 1}{3} + \frac{x + 2}{4} = \frac{1}{6},$$

having fractional coefficients with constants as denominators. Now let's consider equations that contain the variable in one or more of the denominators. Our approach to solving such equations remains essentially the same except **we must avoid any values of the variable that make a denominator zero.** Consider the following examples.

EXAMPLE 1

Solve $\dfrac{5}{3x} - \dfrac{1}{9} = \dfrac{1}{x}$.

Solution First we need to realize that *x cannot equal zero.* Let's indicate this restriction so that it is not forgotten; then we can proceed as follows:

$$\frac{5}{3x} - \frac{1}{9} = \frac{1}{x}, \qquad x \neq 0$$

$$9x\left(\frac{5}{3x} - \frac{1}{9}\right) = 9x\left(\frac{1}{x}\right) \qquad \text{Multiply both sides by the LCD.}$$

$$9x\left(\frac{5}{3x}\right) - 9x\left(\frac{1}{9}\right) = 9x\left(\frac{1}{x}\right)$$

$$15 - x = 9$$

$$-x = -6$$

$$x = 6$$

The solution set is $\{6\}$. (Check it!)

EXAMPLE 2

Solve $\dfrac{65 - n}{n} = 4 + \dfrac{5}{n}$.

Solution

$$\frac{65 - n}{n} = 4 + \frac{5}{n}, \qquad n \neq 0$$

$$n\left(\frac{65 - n}{n}\right) = n\left(4 + \frac{5}{n}\right)$$

$$65 - n = 4n + 5$$

$$60 = 5n$$

$$12 = n$$

The solution set is $\{12\}$.

EXAMPLE 3

Solve $\dfrac{a}{a - 2} + \dfrac{2}{3} = \dfrac{2}{a - 2}$.

Solution

$$\frac{a}{a - 2} + \frac{2}{3} = \frac{2}{a - 2}, \qquad a \neq 2$$

$$3(a - 2)\left(\frac{a}{a - 2} + \frac{2}{3}\right) = 3(a - 2)\left(\frac{2}{a - 2}\right)$$

$$3a + 2(a - 2) = 6$$

$$3a + 2a - 4 = 6$$

$$5a = 10$$

$$a = 2$$

Because our initial restriction was $a \neq 2$, we conclude that this equation *has no solution*. The solution set is $\emptyset$.

Example 3 illustrates the importance of recognizing the restrictions that must be made to exclude division by zero.

Ratio and Proportion

A **ratio** is the comparison of two numbers by division. The fractional form is frequently used to express ratios. For example, the ratio of a to b can be written a/b. A statement of equality between two ratios is called a **proportion**. Thus, if a/b and c/d are equal ratios, the proportion $a/b = c/d$ ($b \neq 0$ and $d \neq 0$) can be formed. There is a useful property of proportions:

$$\text{If } \frac{a}{b} = \frac{c}{d} \quad \text{then} \quad ad = bc.$$

This property can be deduced as follows:

$$\frac{a}{b} = \frac{c}{d}, \qquad b \neq 0 \text{ and } d \neq 0$$

$$bd\left(\frac{a}{b}\right) = bd\left(\frac{c}{d}\right) \qquad \text{Multiply both sides by } bd.$$

$$ad = bc$$

This is sometimes referred to as the **cross-multiplication property of proportions**.

Some equations can be treated as proportions and solved by using the cross-multiplication idea, as the next example illustrates.

EXAMPLE 4

Solve $\dfrac{3}{3x - 2} = \dfrac{4}{2x + 1}$.

Solution

$$\frac{3}{3x - 2} = \frac{4}{2x + 1}, \qquad x \neq \frac{2}{3}, \ x \neq -\frac{1}{2}$$

$$3(2x + 1) = 4(3x - 2) \qquad \text{Apply the cross-multiplication property.}$$

$$6x + 3 = 12x - 8$$

$$11 = 6x$$

$$\frac{11}{6} = x$$

The solution set is $\left\{\frac{11}{6}\right\}$.

Linear Equations Involving Decimals

To solve an equation such as $x + 2.4 = 0.36$ we can add -2.4 to both sides. However, as equations containing decimals become more complex, it is often easier to begin by *clearing the equation of all decimals*, which we accomplish by multiplying both sides by an appropriate power of 10. Let's consider two examples.

EXAMPLE 5

Solve $0.12t - 2.1 = 0.07t - 0.2$.

Solution

$$0.12t - 2.1 = 0.07t - 0.2$$
$$100(0.12t - 2.1) = 100(0.07t - 0.2) \qquad \text{Multiply both sides by 100.}$$
$$12t - 210 = 7t - 20$$
$$5t = 190$$
$$t = 38$$

The solution set is $\{38\}$.

EXAMPLE 6

Solve $0.8x + 0.9(850 - x) = 715$.

Solution

$$0.8x + 0.9(850 - x) = 715$$
$$10[0.8x + 0.9(850 - x)] = 10(715) \qquad \text{Multiply both sides by 10.}$$
$$10(0.8x) + 10[0.9(850 - x)] = 10(715)$$
$$8x + 9(850 - x) = 7150$$
$$8x + 7650 - 9x = 7150$$
$$-x = -500$$
$$x = 500$$

The solution set is $\{500\}$.

Changing Forms of Formulas

Many practical applications of mathematics involve the use of formulas. For example, to find the distance traveled in four hours at a rate of 55 miles per hour, we multiply the rate times the time; thus the distance is $55(4) = 220$ miles. The rule "distance equals rate times time" is commonly stated as a formula: $d = rt$. When using a formula, it is sometimes convenient first to change its form. For example, multiplying both sides of $d = rt$ by $1/t$ produces the equivalent form $r = d/t$. Multiplying both sides of $d = rt$ by $1/r$ produces another equivalent form, $t = d/r$. The following two examples further illustrate this process of obtaining equivalent forms of certain formulas.

EXAMPLE 7

If P dollars is invested at a simple rate of r percent, then the amount, A, accumulated after t years is given by the formula $A = P + Prt$. Solve this formula for P.

Solution

$$A = P + Prt$$

$$A = P(1 + rt) \qquad \text{Apply the distributive property to right side.}$$

$$\frac{A}{1 + rt} = P \qquad \text{Multiply both sides by } \frac{1}{1 + rt}.$$

$$P = \frac{A}{1 + rt} \qquad \text{Apply the symmetric property of equality.}$$

EXAMPLE 8

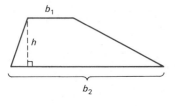

The area (A) of a trapezoid is given by the formula $A = \frac{1}{2}h(b_1 + b_2)$. Solve this equation for b_1.

Solution

$$A = \tfrac{1}{2}h(b_1 + b_2)$$

$$2A = h(b_1 + b_2) \qquad \text{Multiply both sides by 2.}$$

$$2A = hb_1 + hb_2 \qquad \text{Apply the distributive property to the right side.}$$

$$2A - hb_2 = hb_1 \qquad \text{Add } -hb_2 \text{ to both sides.}$$

$$\frac{2A - hb_2}{h} = b_1 \qquad \text{Multiply both sides by } \frac{1}{h}.$$

In Example 7, notice that the distributive property is used to change from the form $P + Prt$ to $P(1 + rt)$. However, in Example 8 the distributive property was used to change $h(b_1 + b_2)$ to $hb_1 + hb_2$. In both examples the goal is to *isolate the term* containing the variable being solved for so that an appropriate application of the multiplication property will produce the desired result. Also note the use of *subscripts* to identify the two bases of the trapezoid. Subscripts allow us to use the same letter b to identify the bases, but b_1 represents one base and b_2 the other.

More on Problem Solving

Volumes have been written on the topic of problem solving, but certainly one of the best-known sources is George Polya's book *How to Solve It*.* In this book, Polya suggests the following four-phase plan for solving problems:

  1. *Understanding the problem*;

  2. *Devising a plan* to solve the problem;

---

* George Polya, *How to Solve It* (Princeton: Princeton University Press), 1945.

3. *Carrying out the plan* to solve the problem;

4. *Looking back* at the completed solution to review and discuss it.

Let's comment briefly on each of the phases and offer some suggestions in terms of using an algebraic approach to solving problems.

Understanding the Problem

Read the problem carefully, making certain that you understand the meanings of all the words. Be especially alert for any technical terms used in the statement of the problem. Often it is helpful to sketch a figure, diagram, or chart to visualize and organize the conditions of the problem. Determine the known and unknown facts and if one of the previously mentioned pictorial devices is used, record these facts in the appropriate places of the diagram or chart.

Devising a Plan

This is the key part of the four-phase plan. It is sometimes referred to as the *analysis* of the problem. There are numerous strategies and techniques used to solve problems. We shall discuss some of these strategies at various places throughout this text; however, at this time we offer the following general suggestions.

1. Choose a meaningful variable to represent an unknown quantity in the problem (perhaps t if time is an unknown quantity) and represent any other unknowns in terms of that variable.

2. Look for a *guideline* that can be used to set up an equation. A guideline might be a formula such as $A = P + Prt$ from Example 7, or a statement of a relationship such as "the sum of the two numbers is 28." Sometimes a relationship suggested by a pictorial device can be used as a guideline for setting up the equation. Also, be alert to the possibility that this "new" problem might really be an "old" problem in a new setting, perhaps even stated in different vocabulary.

3. Form an equation containing the variable so that the conditions of the guideline are translated from English into algebra.

Carrying out the Plan

This phase is sometimes referred to as the *synthesis* of the plan. If phase two has been successfully completed, then carrying out the plan may simply be a matter of solving the equation and doing any further computations to answer all of the questions in the problem. Confidence in your plan creates a better working atmosphere for carrying it out. It is also in this phase that the calculator may become a valuable tool. The type of data and the amount of complexity involved in the computations are two factors that can influence your decision to use one.

Looking Back

This is an important but often overlooked part of problem solving. The following list of questions suggests some things for you to consider in this phase.

1. Is your answer to the problem a *reasonable* answer?

2. Have you *checked* your answer by substituting it back into the conditions stated in the problem?

3. Looking back over your solution, do you now see another plan that could be used to solve the problem?

4. Do you see a way of generalizing your procedure for this problem that could be used to solve other problems of this type?

5. Do you now see that this problem is closely related to another problem that you have previously solved?

6. Have you "tucked away for future reference" the technique used to solve this problem?

Looking back over the solution of a newly solved problem can lay important groundwork for solving problems in the future.

Keep the previous suggestions in mind as we tackle some more word problems. Perhaps it would also be helpful for you to attempt to solve these problems on your own before looking at our approach.

PROBLEM 1

One number is 65 larger than another number. If the larger number is divided by the smaller, the quotient is 6 and the remainder is 5. Find the numbers.

Solution Let n represent the smaller number. Then $n + 65$ represents the larger number. We can use the following relationship as a *guideline*.

$$\frac{\text{dividend}}{\text{divisor}} = \text{quotient} + \frac{\text{remainder}}{\text{divisor}}$$

$$\frac{n + 65}{n} = 6 + \frac{5}{n}$$

We solve this equation by multiplying both sides by n.

$$n\left(\frac{n + 65}{n}\right) = n\left(6 + \frac{5}{n}\right), \qquad n \neq 0$$

$$n + 65 = 6n + 5$$

$$60 = 5n$$

$$12 = n$$

If $n = 12$ then $n + 65$ equals 77. The two numbers are 12 and 77.

Sometimes the concepts of ratio and proportion can be used to set up an equation and solve a problem, as the next example illustrates.

PROBLEM 2

The ratio of male students to female students at a certain university is 5 to 7. If there is a total of 16,200 students, find the number of male and the number of female students.

Solution Let m represent the number of male students; then $16200 - m$ represents the number of female students. The following proportion can be set up and solved.

$$\frac{m}{16200 - m} = \frac{5}{7}$$

$$7m = 5(16200 - m)$$
$$7m = 81000 - 5m$$
$$12m = 81000$$
$$m = 6750$$

Therefore there are 6750 male students and $16200 - 6750 = 9450$ female students.

The next problem has a geometric setting. In such cases, the use of figures is very helpful.

PROBLEM 3

If two opposite sides of a square are each increased by 3 centimeters and the other two sides are each decreased by 2 centimeters, the area is increased by 8 square centimeters. Find the length of the side of the square.

Solution Let s represent the side of the square. Then the figures at left represent the square and the rectangle formed by increasing two opposite sides of the square by 3 centimeters and decreasing the other two sides by 2 centimeters. Since the area of the rectangle is 8 square centimeters more than the area of the square, the following equation can be set up and solved.

$$(s + 3)(s - 2) = s^2 + 8$$
$$s^2 + s - 6 = s^2 + 8$$
$$s = 14$$

Thus the length of a side of the original square is 14 centimeters.

Many consumer problems can be solved by using an algebraic approach. For example, let's consider a discount sale problem involving the relationship, "Original selling price minus discount equals discount sale price."

PROBLEM 4

Jim bought a pair of slacks at a 30% discount sale for $28. What was the original price of the slacks?

Solution Let p represent the original price of the slacks.

original price $-$ discount $=$ discount sale price

$$(100\%)(p) \quad - (30\%)(p) = \qquad \$28$$

We switch this equation to decimal form to solve it.

$$p - 0.3p = 28$$
$$0.7p = 28$$
$$p = 40$$

The original price of the slacks was $40.

Another basic relationship pertaining to consumer problems is "Selling price equals cost plus profit." Profit, also called markup, markon, and margin of profit, may be stated in different ways. It can be expressed as a percent of the cost, a percent of the selling price, or simply in terms of dollars and cents. Let's consider a problem for which the profit is stated as a percent of the selling price.

PROBLEM 5

A retailer of sporting goods bought a putter for $25. He wants to price the putter to make a profit of 20% of the selling price. What price should he mark on the putter?

Solution Let s represent the selling price.

$$\text{selling price} = \text{cost} + \text{profit}$$
$$\downarrow \qquad \downarrow \qquad \downarrow$$
$$s \qquad = \$25 + (20\%)(s)$$

Solving this equation involves using the methods we developed earlier for working with decimals.

$$s = 25 + (20\%)(s)$$
$$s = 25 + 0.2s$$
$$10s = 250 + 2s$$
$$8s = 250$$
$$s = 31.25$$

The selling price should be $31.25.

Certain types of investment problems can be solved by using an algebraic approach. As our final example of this section, let's consider one such problem.

PROBLEM 6

Cindy invested a certain amount of money at 10% interest and $1500 more than that amount at 11%. Her total yearly interest was $795. How much did she invest at each rate?

Solution Let d represent the amount invested at 10%; then $d + 1500$ represents the amount invested at 11%. The following guideline can be used to set up an equation.

$$\text{interest earned at } 10\% + \text{interest earned at } 11\% = \text{total interest}$$

$$(10\%)(d) \qquad + \qquad (11\%)(d + 1500) \qquad = \qquad \$795$$

We can solve this equation by multiplying both sides by 100.

$$0.1d + 0.11(d + 1500) = 795$$
$$10d + 11(d + 1500) = 79500$$
$$10d + 11d + 16500 = 79500$$
$$21d = 63000$$
$$d = 3000$$

Cindy invested $3000 at 10% and $3000 + $1500 = $4500 at 11%.

Don't forget phase four of Polya's problem-solving plan. We have not taken the space to look back over and discuss each of our examples. However, it would be beneficial for you to do so, keeping in mind the questions posed earlier under this phase.

Problem Set 2.2

Solve each of the following equations.

1. $\dfrac{x-2}{3} + \dfrac{x+1}{4} = \dfrac{1}{6}$

2. $\dfrac{5n-1}{4} - \dfrac{2n-3}{10} = \dfrac{3}{5}$

3. $\dfrac{5}{x} + \dfrac{1}{3} = \dfrac{8}{x}$

4. $\dfrac{5}{3n} - \dfrac{1}{9} = \dfrac{1}{n}$

5. $\dfrac{1}{3n} + \dfrac{1}{2n} = \dfrac{1}{4}$

6. $\dfrac{1}{x} - \dfrac{3}{2x} = \dfrac{1}{5}$

7. $\dfrac{35-x}{x} = 7 + \dfrac{3}{x}$

8. $\dfrac{n}{46-n} = 5 + \dfrac{4}{46-n}$

9. $\dfrac{n+67}{n} = 5 + \dfrac{11}{n}$

10. $\dfrac{n+52}{n} = 4 + \dfrac{1}{n}$

11. $\dfrac{5}{3x-2} = \dfrac{1}{x-4}$

12. $\dfrac{-2}{5x-3} = \dfrac{4}{4x-1}$

13. $\dfrac{4}{2y-3} - \dfrac{7}{3y-5} = 0$

14. $\dfrac{3}{2n+1} + \dfrac{5}{3n-4} = 0$

15. $\dfrac{n}{n+1} + 3 = \dfrac{4}{n+1}$

16. $\dfrac{a}{a+5} - 2 = \dfrac{3a}{a+5}$

17. $\dfrac{3x}{2x-1} - 4 = \dfrac{x}{2x-1}$

18. $\dfrac{x}{x-8} - 4 = \dfrac{8}{x-8}$

19. $\dfrac{3}{x+3} - \dfrac{1}{x-2} = \dfrac{5}{2x+6}$

20. $\dfrac{6}{x+3} + \dfrac{20}{x^2+x-6} = \dfrac{5}{x-2}$

21. $\dfrac{n}{n-3} - \dfrac{3}{2} = \dfrac{3}{n-3}$

22. $\dfrac{4}{x-2} + \dfrac{x}{x+1} = \dfrac{x^2-2}{x^2-x-2}$

23. $s = 9 + 0.25s$

24. $s = 1.95 + 0.35s$

25. $0.09x + 0.1(700 - x) = 67$

26. $0.08x + 0.09(950 - x) = 81$

27. $0.09x + 0.11(x + 125) = 68.75$

28. $0.08(x + 200) = 0.07x + 20$

29. $0.8(t - 2) = 0.5(9t + 10)$

30. $0.3(2n - 5) = 11 - 0.65n$

31. $0.92 + 0.9(x - 0.3) = 2x - 5.95$

32. $0.5(3x + 0.7) = 20.6$

Solve each of the following formulas for the indicated variable.

33. $P = 2l + 2w$ for w (Perimeter of a rectangle)

34. $V = \frac{1}{3}Bh$ for B (Volume of a pyramid)

35. $A = 2\pi r^2 + 2\pi rh$ for h (Surface area of a right circular cylinder)

36. $A = \frac{1}{2}h(b_1 + b_2)$ for h (Area of a trapezoid)

37. $C = \frac{5}{9}(F - 32)$ for F (Fahrenheit to Celsius)

38. $F = \frac{9}{5}C + 32$ for C (Celsius to Fahrenheit)

39. $V = C\left(1 - \dfrac{T}{N}\right)$ for T (Linear depreciation)

40. $V = C\left(1 - \dfrac{T}{N}\right)$ for N (Linear depreciation)

41. $I = kl(T - t)$ for T (Expansion allowance in highway construction)

42. $S = \dfrac{CRD}{12d}$ for d (Cutting speed of a circular saw)

43. $\dfrac{1}{R_n} = \dfrac{1}{R_1} + \dfrac{1}{R_2}$ for R_n (Resistance in parallel circuit design)

44. $f = \dfrac{1}{\dfrac{1}{a} + \dfrac{1}{b}}$ for b (Focal length of a camera lens)

Set up an equation and solve each of the following problems.

45. The sum of two numbers is 98. If the larger is divided by the smaller, the quotient is 4 and the remainder is 13. Find the numbers.

46. One number is 100 larger than another number. If the larger number is divided by the smaller, the quotient is 15 and the remainder is 2. Find the numbers.

47. What number must be added to both the numerator and denominator of $\frac{3}{5}$ to produce a rational number equivalent to $\frac{9}{7}$?

48. The denominator of a fraction is 4 more than its numerator. If 5 is added to the numerator and 11 is added to the denominator, a rational number equivalent to $\frac{1}{2}$ is produced. Find the original fraction.

49. A sum of $2250 is to be divided between two people in the ratio of 2 to 3. How much does each person receive?

50. One type of motor requires a mixture of oil and gasoline in a ratio of 1 to 15 (that is, 1 part of oil to 15 parts of gasoline). How many liters of each are contained in a 20-liter mixture?

51. The ratio of students to teaching faculty in a certain high school is 20 to 1. If the total number of students and faculty is 777, find the number of each.

52. The ratio of sodium to chlorine in common table salt is 5 to 3. Find the amount of each element in a salt compound weighing 200 pounds.

53. Gary bought a coat at a 20% discount sale for $52. What was the original price of the coat?

54. Roya bought a pair of slacks at a 30% discount sale for $33.60. What was the original price of the slacks?

55. After a 7% increase in salary, Laurie makes $1016.50 per month. How much did she earn per month before the increase?

56. Russ bought a car with 5% sales tax included for $11,025. What was the selling price of the car without the tax?

57. A retailer has some shoes that cost $28 per pair. At what price should they be sold to obtain a profit of 15% of the cost?

58. If a head of lettuce costs a retailer $.40, at what price should it be sold to make a profit of 45% of the cost?

59. Karla sold a bicycle for $97.50. This selling price respresented a 30% profit for her, based on what she had originally paid for the bike. Find Karla'a original cost for the bicycle.

60. If a ring costs a jeweler $250, at what price should it be sold to make a profit of 60% of the selling price?

61. A retailer has some skirts that cost $18 each. She wants to sell them at a profit of 40% of the selling price. What price should she charge for the skirts?

62. Suppose that an item costs a retailer $50. How much more profit could be gained by fixing a 50% profit based on selling price rather than a 50% profit based on cost?

63. Derek has some nickels and dimes worth $3.60. The number of dimes is one more than twice the number of nickels. How many nickels and dimes does he have?

64. Robin has a collection of nickels, dimes, and quarters worth $38.50. She has 10 more dimes than nickels and twice as many quarters as dimes. How many coins of each kind does she have?

65. A collection of 70 coins consisting of dimes, quarters, and half-dollars has a value of $17.75. There are three times as many quarters as dimes. Find the number of each kind of coin.

66. A certain amount of money is invested at 8% per year and $1500 more than that amount is invested at 9% per year. The annual interest from the 9% investment exceeds the annual interest from the 8% investment by $160. How much is invested at each rate?

67. A total of $5500 was invested, part of it at 9% per year and the remainder at 10% per year. If the total yearly interest amounted to $530, how much was invested at each rate?

68. A sum of $3500 is split between two investments, one paying 9% yearly interest and the other 11%. If the return on the 11% investment exceeds that on the 9% investment by $85 the first year, how much is invested at each rate?

69. Celia has invested $2500 at 11% yearly interest. How much must she invest at 12% so that the interest from both investments totals $695 after a year?

70. The length of a rectangle is 2 inches less than three times its width. If the perimeter of the rectangle is 108 inches, find its length and width.

71. The length of a rectangle is 4 centimeters more than its width. If the width is increased by 2 centimeters and the length increased by 3 centimeters, a new rectangle is formed having an area of 44 square centimeters more than the area of the original rectangle. Find the dimensions of the original rectangle.

72. The length of a picture without its border is 7 inches less than twice its width. If the border is 1 inch wide and its area is 62 square inches, what are the dimensions of the picture alone?

2.3

Quadratic Equations

A **quadratic equation** in the variable x is defined as any equation that can be written in the form

$$ax^2 + bx + c = 0,$$

where a, b, and c are real numbers and $a \neq 0$. The form $ax^2 + bx + c = 0$ is called the **standard form** of a quadratic equation. The choice of x for the variable is arbitrary. An equation such as $3t^2 + 5t - 4 = 0$ is a quadratic equation in the variable t.

Quadratic equations such as $x^2 + 2x - 15 = 0$, for which the polynomial is factorable, can be solved by applying the following property: **If $ab = 0$ then $a = 0$ or $b = 0$.** Our work might take on the following format.

$$x^2 + 2x - 15 = 0$$
$$(x + 5)(x - 3) = 0$$
$$x + 5 = 0 \qquad \text{or} \qquad x - 3 = 0$$
$$x = -5 \qquad \text{or} \qquad x = 3$$

The solution set for this equation is $\{-5, 3\}$.

Let's consider another example of this type.

EXAMPLE 1

Solve $6n^2 + n - 12 = 0$.

Solution

$$6n^2 + n - 12 = 0$$
$$(3n - 4)(2n + 3) = 0$$
$$3n - 4 = 0 \qquad \text{or} \qquad 2n + 3 = 0$$
$$3n = 4 \qquad \text{or} \qquad 2n = -3$$
$$n = \tfrac{4}{3} \qquad \text{or} \qquad n = -\tfrac{3}{2}$$

The solution set is $\{-\tfrac{3}{2}, \tfrac{4}{3}\}$.

Now suppose that we want to solve $x^2 = k$, where k is any real number. We can proceed as follows.

$$x^2 = k$$
$$x^2 - k = 0$$
$$(x + \sqrt{k})(x - \sqrt{k}) = 0$$
$$x + \sqrt{k} = 0 \qquad \text{or} \qquad x - \sqrt{k} = 0$$
$$x = -\sqrt{k} \qquad \text{or} \qquad x = \sqrt{k}$$

Thus, we can state the following property for any real number k.

PROPERTY 2.2

> The solution set of $x^2 = k$ is $\{-\sqrt{k}, \sqrt{k}\}$, which can also be written $\{\pm\sqrt{k}\}$.

Property 2.2, along with our knowledge of the square root, makes it very easy to solve quadratic equations of the form $x^2 = k$.

EXAMPLE 2

Solve each of the following:

(a) $x^2 = 72$ (b) $(3n - 1)^2 = 26$ (c) $(y + 2)^2 = -24$

Solutions

(a) $x^2 = 72$
$$x = \pm\sqrt{72}$$
$$x = \pm 6\sqrt{2}$$

The solution set is $\{\pm 6\sqrt{2}\}$.

(b)
$$(3n - 1)^2 = 26$$
$$3n - 1 = \pm\sqrt{26}$$

$$3n - 1 = \sqrt{26} \qquad \text{or} \qquad 3n - 1 = -\sqrt{26}$$
$$3n = 1 + \sqrt{26} \qquad \text{or} \qquad 3n = 1 - \sqrt{26}$$

$$n = \frac{1 + \sqrt{26}}{3} \qquad \text{or} \qquad n = \frac{1 - \sqrt{26}}{3}$$

The solution set is $\left\{\dfrac{1 \pm \sqrt{26}}{3}\right\}$.

(c) $(y + 2)^2 = -24$
$$y + 2 = \pm\sqrt{-24}$$
$$y + 2 = \pm 2i\sqrt{6} \qquad \text{Remember that } \sqrt{-24} = i\sqrt{24} = i\sqrt{4}\sqrt{6} = 2i\sqrt{6}.$$
$$y + 2 = 2i\sqrt{6} \qquad \text{or} \qquad y + 2 = -2i\sqrt{6}$$
$$y = -2 + 2i\sqrt{6} \qquad \text{or} \qquad y = -2 - 2i\sqrt{6}$$

The solution set is $\{-2 \pm 2i\sqrt{6}\}$.

Completing the Square

A factoring technique reviewed in Chapter 1 relied on recognizing *perfect-square trinomials*. In each of the following examples, the perfect-square trinomial on the right side of the identity is the result of squaring the binomial on the left side:

$$(x + 5)^2 = x^2 + 10x + 25, \qquad (x - 7)^2 = x^2 - 14x + 49,$$
$$(x + 9)^2 = x^2 + 18x + 81, \qquad (x - 12)^2 = x^2 - 24x + 144.$$

Notice that in each of the square trinomials, the constant term is equal to the square of one-half of the coefficient of the x-term. This relationship allows us to *form* a perfect-square trinomial by adding a proper constant term. For example, suppose that we want to form a perfect-square trinomial from $x^2 + 8x$. Since $\frac{1}{2}(8) = 4$ and $4^2 = 16$, the perfect-square trinomial is $x^2 + 8x + 16$. Now let's use this idea to solve a quadratic equation.

EXAMPLE 3

Solve $x^2 + 8x - 2 = 0$.

Solution

$$x^2 + 8x - 2 = 0$$
$$x^2 + 8x = 2$$
$$x^2 + 8x + 16 = 2 + 16 \qquad \text{We added 16 to the left side to form a}$$

We added 16 to the left side to form a perfect-square trinomial. Thus, 16 has to be added to the right side.

$$(x + 4)^2 = 18$$
$$x + 4 = \pm\sqrt{18}$$
$$x + 4 = \pm 3\sqrt{2}$$

$$x + 4 = 3\sqrt{2} \qquad \text{or} \qquad x + 4 = -3\sqrt{2}$$
$$x = -4 + 3\sqrt{2} \qquad \text{or} \qquad x = -4 - 3\sqrt{2}$$

The solution set is $\{-4 \pm 3\sqrt{2}\}$.

We have been using a relationship for a perfect-square trinomial that states, "The constant term is equal to the square of one-half of the coefficient of the *x*-term." This relationship holds only if the coefficient of x^2 is 1. Thus, a slight adjustment needs to be made when we are solving quadratic equations having a coefficient of x^2 other than 1. The next example shows how to make this adjustment.

EXAMPLE 4

Solve $2x^2 + 6x - 3 = 0$.

Solution

$$2x^2 + 6x - 3 = 0$$
$$2x^2 + 6x = 3$$

$$x^2 + 3x = \frac{3}{2} \qquad \text{Multiply both sides by } \frac{1}{2}.$$

$$x^2 + 3x + \frac{9}{4} = \frac{3}{2} + \frac{9}{4} \qquad \text{Add } \frac{9}{4} \text{ to both sides.}$$

$$\left(x + \frac{3}{2}\right)^2 = \frac{15}{4}$$

$$x + \frac{3}{2} = \pm\frac{\sqrt{15}}{2}$$

$$x + \frac{3}{2} = \frac{\sqrt{15}}{2} \qquad \text{or} \qquad x + \frac{3}{2} = -\frac{\sqrt{15}}{2}$$

$$x = -\frac{3}{2} + \frac{\sqrt{15}}{2} \qquad \text{or} \qquad x = -\frac{3}{2} - \frac{\sqrt{15}}{2}$$

$$x = \frac{-3 + \sqrt{15}}{2} \qquad \text{or} \qquad x = \frac{-3 - \sqrt{15}}{2}$$

The solution set is $\left\{ \dfrac{-3 \pm \sqrt{15}}{2} \right\}$.

The Quadratic Formula

The process used in Examples 3 and 4 is called **completing the square**. It can be used to solve *any* quadratic equation. If we use this process of completing the square to solve the general quadratic equation $ax^2 + bx + c = 0$, we obtain a formula known as the **quadratic formula**. The details are as follows:

$$ax^2 + bx + c = 0, \qquad a \neq 0$$

$$ax^2 + bx = -c$$

$$x^2 + \frac{b}{a}x = -\frac{c}{a} \qquad \text{Multiply both sides by } \frac{1}{a}.$$

$$x^2 + \frac{b}{a}x + \frac{b^2}{4a^2} = -\frac{c}{a} + \frac{b^2}{4a^2} \qquad \begin{array}{l}\text{Complete the square by} \\ \text{adding } \dfrac{b^2}{4a^2} \text{ to both sides.}\end{array}$$

$$\left(x + \frac{b}{2a}\right)^2 = \frac{b^2 - 4ac}{4a^2} \qquad \begin{array}{l}\text{Combine the right side} \\ \text{into a single fraction.}\end{array}$$

$$x + \frac{b}{2a} = \pm\sqrt{\frac{b^2 - 4ac}{4a^2}}$$

$$x + \frac{b}{2a} = \pm\frac{\sqrt{b^2 - 4ac}}{\sqrt{4a^2}}$$

$$x + \frac{b}{2a} = \pm\frac{\sqrt{b^2 - 4ac}}{2a}$$

$$x + \frac{b}{2a} = \frac{\sqrt{b^2 - 4ac}}{2a} \qquad \text{or} \qquad x + \frac{b}{2a} = -\frac{\sqrt{b^2 - 4ac}}{2a}$$

$$x = -\frac{b}{2a} + \frac{\sqrt{b^2 - 4ac}}{2a} \qquad \text{or} \qquad x = -\frac{b}{2a} - \frac{\sqrt{b^2 - 4ac}}{2a}$$

$$x = \frac{-b + \sqrt{b^2 - 4ac}}{2a} \qquad \text{or} \qquad x = \frac{-b - \sqrt{b^2 - 4ac}}{2a}$$

The quadratic formula can be stated as follows.

Quadratic Formula

If $a \neq 0$, then the solutions (roots) of the equation $ax^2 + bx + c = 0$ are given by

$$x = \frac{-b \pm \sqrt{b^2 - 4ac}}{2a}.$$

This formula can be used to solve any quadratic equation by expressing the equation in the standard form, $ax^2 + bx + c = 0$, and substituting the values for a, b, and c into the formula. Let's consider some examples.

EXAMPLE 5

Solve each of the following by using the quadratic formula.

(a) $3x^2 - x - 5 = 0$ **(b)** $25n^2 - 30n = -9$ **(c)** $t^2 - 2t + 4 = 0$

Solutions

(a) We need to think of $3x^2 - x - 5 = 0$ as $3x^2 + (-x) + (-5) = 0$; thus, $a = 3$, $b = -1$, and $c = -5$. We then substitute these values into the quadratic formula and simplify.

$$x = \frac{-b \pm \sqrt{b^2 - 4ac}}{2a}$$

$$x = \frac{-(-1) \pm \sqrt{(-1)^2 - 4(3)(-5)}}{2(3)}$$

$$x = \frac{1 \pm \sqrt{61}}{6}$$

The solution set is $\left\{ \dfrac{1 \pm \sqrt{61}}{6} \right\}$.

(b) The quadratic formula is usually stated in terms of the variable x, but again the choice of variable is arbitrary. The given equation, $25n^2 - 30n = -9$, needs to be changed to standard form: $25n^2 - 30n + 9 = 0$. From this we obtain $a = 25$, $b = -30$, and $c = 9$. Now we use the formula.

$$n = \frac{-(-30) \pm \sqrt{(-30)^2 - 4(25)(9)}}{2(25)}$$

$$n = \frac{30 \pm \sqrt{0}}{50}$$

$$n = \frac{3}{5}$$

The solution set is $\left\{ \frac{3}{5} \right\}$.

(c) We substitute $a = 1$, $b = -2$, and $c = 4$ into the quadratic formula.

$$t = \frac{-(-2) \pm \sqrt{(-2)^2 - 4(1)(4)}}{2(1)}$$

$$= \frac{2 \pm \sqrt{-12}}{2}$$

$$= \frac{2 \pm 2i\sqrt{3}}{2}$$

$$= \frac{2(1 \pm i\sqrt{3})}{2}$$

The solution set is $\{1 \pm i\sqrt{3}\}$.

From Example 5 we see that different kinds of solutions are obtained depending upon the radicand ($b^2 - 4ac$) inside the radical in the quadratic formula. For this reason, the number $b^2 - 4ac$ is called the **discriminant** of the quadratic equation. It can be used to determine the nature of the solutions as follows.

    1. If $b^2 - 4ac > 0$, the equation has two unequal real solutions.

    2. If $b^2 - 4ac = 0$, the equation has one real solution.

    3. If $b^2 - 4ac < 0$, the equation has two complex but nonreal solutions.

The following examples illustrate each of these situations. (You may want to solve the equations completely to verify our conclusions.)

EQUATION	DISCRIMINANT	NATURE OF SOLUTIONS
$4x^2 - 7x - 1 = 0$	$b^2 - 4ac = (-7)^2 - 4(4)(-1)$ $= 49 + 16$ $= 65$	two real solutions
$4x^2 + 12x + 9 = 0$	$b^2 - 4ac = (12)^2 - 4(4)(9)$ $= 144 - 144$ $= 0$	one real solution
$5x^2 + 2x + 1 = 0$	$b^2 - 4ac = (2)^2 - 4(5)(1)$ $= 4 - 20$ $= -16$	two complex solutions

There is another useful relationship involving the solutions of a quadratic equation of the form $ax^2 + bx + c = 0$ and the numbers a, b, and c. Suppose that we let x_1 and x_2 be the two roots of the equation. (If $b^2 - 4ac = 0$, then $x_1 = x_2$ and the "one-solution situation" can be thought of as two equal solutions.) By the quadratic formula we have

$$x_1 = \frac{-b + \sqrt{b^2 - 4ac}}{2a} \quad \text{and} \quad x_2 = \frac{-b - \sqrt{b^2 - 4ac}}{2a}.$$

Now let's consider both the sum and the product of the two roots.

Sum: $x_1 + x_2 = \dfrac{-b + \sqrt{b^2 - 4ac}}{2a} + \dfrac{-b - \sqrt{b^2 - 4ac}}{2a} = \dfrac{-2b}{2a} = -\dfrac{b}{a}$

Product: $(x_1)(x_2) = \left(\dfrac{-b + \sqrt{b^2 - 4ac}}{2a}\right)\left(\dfrac{-b - \sqrt{b^2 - 4ac}}{2a}\right)$

$$= \dfrac{b^2 - (b^2 - 4ac)}{4a^2}$$

$$= \dfrac{b^2 - b^2 + 4ac}{4a^2}$$

$$= \dfrac{4ac}{4a^2} = \dfrac{c}{a}$$

These relationships provide another way of checking potential solutions when solving quadratic equations. We will illustrate this point in a moment.

Solving Quadratic Equations: Which Method?

Which method should be used to solve a particular quadratic equation? There is no definite answer to that question; it depends upon the type of equation and perhaps your personal preference. However, it is to your advantage to be able to use all three techniques and to know the strengths and weaknesses of each technique. In the next two examples we will give our reasons for choosing a specific technique.

EXAMPLE 6

Solve $x^2 - 4x - 192 = 0$.

Solution The size of the constant term makes the factoring approach a little cumbersome for this problem. However, since the coefficient of the x^2-term is 1 and the coefficient of the x-term is even, the method of completing the square should work rather effectively.

$$x^2 - 4x - 192 = 0$$
$$x^2 - 4x = 192$$
$$x^2 - 4x + 4 = 192 + 4$$
$$(x - 2)^2 = 196$$
$$x - 2 = \pm\sqrt{196}$$
$$x - 2 = \pm 14$$

$x - 2 = 14$ or $x - 2 = -14$
$x = 16$ or $x = -12$

Check:

Sum of roots: $16 + (-12) = 4$ and $-\dfrac{b}{a} = -\left(\dfrac{-4}{1}\right) = 4$

Product of roots: $(16)(-12) = -192$ and $\dfrac{c}{a} = \dfrac{-192}{1} = -192$

The solution set is $\{-12, 16\}$.

EXAMPLE 7

Solve $2x^2 - x + 3 = 0$.

Solution It would be reasonable first to try factoring the polynomial $2x^2 - x + 3$. Unfortunately, it is not factorable using integers; thus we must solve the equation by completing the square or by using the quadratic formula. Since the coefficient of the x^2-term is not 1, let's avoid completing the square and use the formula instead.

$$
\begin{aligned}
x &= \frac{-b \pm \sqrt{b^2 - 4ac}}{2a} \\
&= \frac{-(-1) \pm \sqrt{(-1)^2 - 4(2)(3)}}{2(2)} \\
&= \frac{1 \pm \sqrt{-23}}{4} \\
&= \frac{1 \pm i\sqrt{23}}{4}
\end{aligned}
$$

Check:

Sum of roots: $\dfrac{1 + i\sqrt{23}}{4} + \dfrac{1 - i\sqrt{23}}{4} = \dfrac{2}{4} = \dfrac{1}{2}$ and $-\dfrac{b}{a} = -\dfrac{-1}{2} = \dfrac{1}{2}$

Product of roots: $\left(\dfrac{1 + i\sqrt{23}}{4}\right)\left(\dfrac{1 - i\sqrt{23}}{4}\right) = \dfrac{1 - 23i^2}{16} = \dfrac{1 + 23}{16} = \dfrac{24}{16} = \dfrac{3}{2}$

and $\dfrac{c}{a} = \dfrac{3}{2}$

The solution set is $\left\{\dfrac{1 \pm i\sqrt{23}}{4}\right\}$.

Being able to solve quadratic equations gives us more power for solving word problems. Let's conclude this section with a problem that can be solved using a quadratic equation.

PROBLEM

One leg of a right triangle is 7 meters longer than the other leg. If the length of the hypotenuse is 17 meters, find the length of each leg.

Solution Let l represent the length of one leg; then $l + 7$ represents the length of the other leg. Using the Pythagorean theorem as a guideline, we can set up and solve a quadratic equation.

$$l^2 + (l + 7)^2 = 17^2$$
$$l^2 + l^2 + 14l + 49 = 289$$
$$2l^2 + 14l - 240 = 0$$
$$l^2 + 7l - 120 = 0$$
$$(l + 15)(l - 8) = 0$$

$$l + 15 = 0 \qquad \text{or} \qquad l - 8 = 0$$
$$l = -15 \qquad \text{or} \qquad l = 8$$

The negative solution must be disregarded (since l is a length), so the length of one leg is 8 meters. The other leg, represented by $l + 7$, is $8 + 7 = 15$ meters long.

Problem Set 2.3

Solve each of the following equations by factoring or by using the property, "If $x^2 = k$ then $x = \pm\sqrt{k}$."

1. $x^2 - 3x - 28 = 0$ **2.** $x^2 - 4x - 12 = 0$ **3.** $3x^2 + 5x - 12 = 0$

4. $2x^2 - 13x + 6 = 0$ **5.** $2x^2 - 3x = 0$ **6.** $3n^2 = 3n$

7. $9y^2 = 12$ **8.** $(4n - 1)^2 = 16$ **9.** $(2n + 1)^2 = 20$

10. $3(4x - 1)^2 + 1 = 16$ **11.** $15n^2 + 19n - 10 = 0$ **12.** $6t^2 + 23t - 4 = 0$

13. $(x - 2)^2 = -4$ **14.** $24x^2 + 23x - 12 = 0$ **15.** $10y^2 + 33y - 7 = 0$

16. $(x - 3)^2 = -9$

Use the method of completing the square to solve each of the following equations. Check your solutions by using the sum-and-product-of-roots relationships.

17. $x^2 - 10x + 24 = 0$ **18.** $x^2 + x - 20 = 0$ **19.** $n^2 + 10n - 2 = 0$

20. $n^2 + 6n - 1 = 0$ **21.** $y^2 - 3y = -1$ **22.** $y^2 + 5y = -2$

23. $x^2 + 4x + 6 = 0$ **24.** $x^2 - 6x + 21 = 0$ **25.** $2t^2 + 12t - 5 = 0$

26. $3p^2 + 12p - 2 = 0$ **27.** $x^2 - 2x - 288 = 0$ **28.** $x^2 + 4x - 221 = 0$

29. $3n^2 + 5n - 1 = 0$ **30.** $2n^2 + n - 4 = 0$

Use the quadratic formula to solve each of the following equations. Check your solutions by using the sum-and-product-of-roots relationships.

31. $n^2 - 3n - 54 = 0$ **32.** $y^2 + 13y + 22 = 0$ **33.** $3x^2 + 16x = -5$

34. $10x^2 - 29x - 21 = 0$ **35.** $y^2 - 2y - 4 = 0$ **36.** $n^2 - 6n - 3 = 0$

37. $2a^2 - 6a + 1 = 0$ **38.** $2x^2 + 3x - 1 = 0$ **39.** $n^2 - 3n = -7$

40. $n^2 - 5n = -8$ **41.** $x^2 + 4 = 8x$ **42.** $x^2 + 31 = -14x$

43. $4x^2 - 4x + 1 = 0$ **44.** $x^2 + 24 = 0$

Solve each of the following quadratic equations by using the method that seems most appropriate to you.

45. $8x^2 + 10x - 3 = 0$ **46.** $18x^2 - 39x + 20 = 0$ **47.** $x^2 + 2x = 168$

48. $x^2 + 28x = -187$ **49.** $2t^2 - 3t + 7 = 0$ **50.** $3n^2 - 2n + 5 = 0$

51. $(3n - 1)^2 + 2 = 18$ **52.** $20y^2 + 17y - 10 = 0$ **53.** $4y^2 + 4y - 1 = 0$

54. $(5n + 2)^2 + 1 = -27$ **55.** $x^2 - 16x + 14 = 0$ **56.** $x^2 - 18x + 15 = 0$

57. $t^2 + 20t = 25$ **58.** $n^2 - 18n = 9$ **59.** $5x^2 - 2x - 1 = 0$

60. $7x^2 - 17x + 6 = 0$

61. Find the discriminant of each of the following quadratic equations and determine whether the equation has (1) two complex but nonreal solutions, (2) one real solution, or (3) two unequal real solutions.

(a) $4x^2 + 20x + 25 = 0$ (b) $x^2 + 4x + 7 = 0$

(c) $x^2 - 18x + 81 = 0$ (d) $36x^2 - 31x + 3 = 0$

(e) $2x^2 + 5x + 7 = 0$ (f) $16x^2 = 40x - 25$

(g) $6x^2 - 4x - 7 = 0$ (h) $5x^2 - 2x - 4 = 0$

Set up a quadratic equation and solve each of the following problems.

62. Find two consecutive even integers whose product is 528.

63. Find two consecutive whole numbers such that the sum of their squares is 265.

64. The sum of two integers is 4. The sum of the squares of the integers is 136. Find the integers.

65. Find two positive numbers having a sum of 22 and a product of 112.

66. One leg of a right triangle is 4 inches longer than the other leg. If the length of the hypotenuse is 20 inches, find the length of each leg.

67. The sum of the lengths of the two legs of a right triangle is 34 meters. If the length of the hypotenuse is 26 meters, find the length of each leg.

68. The lengths of the three sides of a right triangle are consecutive even integers. Find the length of each side.

69. The perimeter of a rectangle is 44 inches and its area is 112 square inches. Find the length and width of the rectangle.

70. A page for a magazine contains 70 square inches of type. The height of a page is twice the width. If the margin around the type is 2 inches uniformly, what are the dimensions of the page?

71. The length of a rectangle is 4 meters more than twice its width. If the area of the rectangle is 126 square meters, find its length and width.

72. The length of one side of a triangle is 3 centimeters less than twice the length of the altitude to that side. If the area of the triangle is 52 square centimeters, find the length of the side and the length of the altitude to that side.

73. A rectangular plot of ground measuring 12 meters by 20 meters is surrounded by a sidewalk of uniform width. The area of the sidewalk is 68 square meters. Find the width of the walk.

74. A piece of wire 60 inches long is cut into two pieces and then each piece is bent into the shape of a square. If the sum of the areas of the two squares is 117 square inches, find the length of each piece of wire.

75. A rectangular piece of cardboard is 4 inches longer than it is wide. From each of its corners a square piece 2 inches on a side is cut out. The flaps are then turned up to form an open box, which has a volume of 42 cubic inches. Find the length and width of the original piece of cardboard.

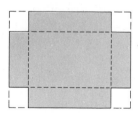

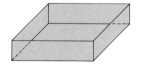

Figure for Exercise 75

Miscellaneous Problems

76. Solve each of the following equations for x.

(a) $x^2 - 7kx = 0$ (b) $x^2 = 25kx$

(c) $x^2 - 3kx - 10k^2 = 0$ (d) $6x^2 + kx - 2k^2 = 0$

(e) $9x^2 - 6kx + k^2 = 0$ **(f)** $k^2x^2 - kx - 6 = 0$

(g) $x^2 + \sqrt{2}x - 3 = 0$ **(h)** $x^2 - \sqrt{3}x + 5 = 0$

77. Solve each of the following for the indicated variable. (Assume that all letters represent positive numbers.)

(a) $A = \pi r^2$ for r **(b)** $E = c^2m - c^2m_0$ for c

(c) $s = \frac{1}{2}gt^2$ for t **(d)** $\dfrac{x^2}{a^2} + \dfrac{y^2}{b^2} = 1$ for x

(e) $\dfrac{x^2}{a^2} - \dfrac{y^2}{b^2} = 1$ for y **(f)** $s = \frac{1}{2}gt^2 + V_0t$ for t

78. Determine k so that $4x^2 - kx + 1 = 0$ has one real solution.

2.4

Applications of Linear and Quadratic Equations

Let's begin this section by considering two fractional equations, one that is equivalent to a linear equation and one that is equivalent to a quadratic equation.

EXAMPLE 1

Solve $\dfrac{3}{2x - 8} - \dfrac{x - 5}{x^2 - 2x - 8} = \dfrac{7}{x + 2}$.

Solution

$$\frac{3}{2x - 8} - \frac{x - 5}{x^2 - 2x - 8} = \frac{7}{x + 2}$$

$$\frac{3}{2(x - 4)} - \frac{x - 5}{(x - 4)(x + 2)} = \frac{7}{x + 2}, \quad x \neq 4, \quad x \neq -2$$

$$2(x - 4)(x + 2)\left(\frac{3}{2(x - 4)} - \frac{x - 5}{(x - 4)(x + 2)}\right) = 2(x - 4)(x + 2)\left(\frac{7}{x + 2}\right)$$

$$3(x + 2) - 2(x - 5) = 14(x - 4)$$

$$3x + 6 - 2x + 10 = 14x - 56$$

$$x + 16 = 14x - 56$$

$$72 = 13x$$

$$\frac{72}{13} = x$$

The solution set is $\{\frac{72}{13}\}$.

In Example 1, notice that we did not indicate the restrictions until the denominators were expressed in factored form. It is usually easier to determine the necessary restrictions at that step.

EXAMPLE 2

Solve $\dfrac{3n}{n^2 + n - 6} + \dfrac{2}{n^2 + 4n + 3} = \dfrac{n}{n^2 - n - 2}$.

Solution

$$\frac{3n}{n^2 + n - 6} + \frac{2}{n^2 + 4n + 3} = \frac{n}{n^2 - n - 2}$$

$$\frac{3n}{(n + 3)(n - 2)} + \frac{2}{(n + 3)(n + 1)} = \frac{n}{(n - 2)(n + 1)}, \qquad \begin{array}{l} n \neq -3, \\ n \neq 2, \\ n \neq -1 \end{array}$$

$$(n + 3)(n - 2)(n + 1)\left(\frac{3n}{(n + 3)(n - 2)} + \frac{2}{(n + 3)(n + 1)}\right) = (n + 3)(n - 2)(n + 1)\left(\frac{n}{(n - 2)(n + 1)}\right)$$

$$3n(n + 1) + 2(n - 2) = n(n + 3)$$
$$3n^2 + 3n + 2n - 4 = n^2 + 3n$$
$$3n^2 + 5n - 4 = n^2 + 3n$$
$$2n^2 + 2n - 4 = 0$$
$$n^2 + n - 2 = 0$$
$$(n + 2)(n - 1) = 0$$
$$n + 2 = 0 \qquad \text{or} \qquad n - 1 = 0$$
$$n = -2 \qquad \text{or} \qquad n = 1$$

The solution set is $\{-2, 1\}$.

More on Problem Solving

Before tackling a variety of applications of linear and quadratic equations, let's restate some suggestions made earlier in this chapter for solving word problems.

Suggestions for Solving Word Problems

1. Read the problem carefully, making certain that you understand the meanings of all the words. Be especially alert for any technical terms used in the statement of the problem.

2. Read the problem a second time (perhaps even a third time), to get an overview of the situation being described and to determine the known facts, as well as what is to be found.

3. Sketch any figure, diagram, or chart that might be helpful in analyzing the problem.

4. Choose a meaningful variable to represent an unknown quantity in the problem (for example, *l* for the length of a rectangle), and represent any other unknowns in terms of that variable.

5. Look for a *guideline* that can be used in setting up an equation. A guideline might be a formula, such as $A = lw$, or a relationship, such as "the fractional part of the job done by Bill plus the fractional part of the job done by Mary equals the total job."

6. Form an equation containing the variable to translate the conditions of the guideline from English to algebra.

7. Solve the equation and use the solution to determine all the facts requested in the problem.

8. Check all answers back into the *original statement of the problem.*

Suggestion 5 is a key part of the analysis of a problem. A formula to be used as a guideline may or may not be explicitly stated in the problem. Likewise, a relationship to be used as a guideline may not be actually stated in the problem but must be determined from what is stated. Let's consider some examples.

PROBLEM 1

A room contains 120 chairs. The number of chairs per row is one less than twice the number of rows. Find the number of rows and the number of chairs per row.

Solution Let r represent the number of rows. Then $2r - 1$ represents the number of chairs per row. The statement of the problem implies a formation of chairs such that the total number of chairs is equal to the number of rows times the number of chairs per row. This gives us an equation.

number of rows × number of chairs per row = total number of chairs

$$r \quad \times \quad (2r - 1) \quad = \quad 120$$

We solve this equation by the factorization method.

$$2r^2 - r = 120$$
$$2r^2 - r - 120 = 0$$
$$(2r + 15)(r - 8) = 0$$
$$2r + 15 = 0 \qquad \text{or} \qquad r - 8 = 0$$
$$2r = -15 \qquad \text{or} \qquad r = 8$$
$$r = -\tfrac{15}{2} \qquad \text{or} \qquad r = 8$$

The solution $-\tfrac{15}{2}$ must be disregarded, so there are 8 rows and $2(8) - 1 = 15$ chairs per row.

The basic relationship "distance equals rate times time" is used to help solve a variety of *uniform-motion problems.* This relationship may be expressed by any one of the following equations.

$$d = rt, \qquad r = \frac{d}{t}, \qquad t = \frac{d}{r}$$

PROBLEM 2

Domenica and Javier start from the same location at the same time and ride their bicycles in opposite directions for 4 hours, at which time they are 140 miles apart. If Domenica rides 3 miles per hour faster than Javier, find the rate of each rider.

Domenica riding at $r + 3$ mph for 4 hours

Javier riding at r mph for 4 hours

total of 140 miles

Solution Let r represent Javier's rate; then $r + 3$ represents Domenica's rate. Sketching a diagram may help in our analysis.

The fact that the total distance is 140 miles can be used as a guideline. We use the $d = rt$ equation.

distance Domenica rides + distance Javier rides = 140

$$4(r + 3) \quad + \quad 4r \quad = 140$$

Solving this equation yields Javier's speed.

$$4r + 12 + 4r = 140$$
$$8r = 128$$
$$r = 16$$

Thus Javier rides at 16 miles per hour and Domenica at $16 + 3 = 19$ miles per hour.

Some people find that it is helpful to use a chart or a table to organize the known and unknown facts in uniform-motion problems. Let's illustrate this approach.

PROBLEM 3

Riding on a moped, Sue takes 2 hours less to travel 60 miles than Ann takes to travel 50 miles on a bicycle. Sue travels 10 miles per hour faster than Ann. Find the times and rates of both girls.

	DISTANCE	TIME	$r = \dfrac{d}{t}$
ANN	50	t	$\dfrac{50}{t}$
SUE	60	$t - 2$	$\dfrac{60}{t-2}$

Solution Let t represent Ann's time; then $t - 2$ represents Sue's time. We can record the information in a table as shown at the left. The fact that Sue travels 10 miles per hour faster than Ann can be used as a guideline.

Sue's rate = Ann's rate + 10

$$\frac{60}{t - 2} = \frac{50}{t} + 10$$

Solving this equation yields Ann's time.

$$t(t - 2)\left(\frac{60}{t - 2}\right) = t(t - 2)\left(\frac{50}{t} + 10\right), \qquad t \neq 0, t \neq 2$$
$$60t = 50(t - 2) + 10t(t - 2)$$
$$60t = 50t - 100 + 10t^2 - 20t$$
$$0 = 10t^2 - 30t - 100$$
$$0 = t^2 - 3t - 10$$
$$0 = (t - 5)(t + 2)$$
$$t - 5 = 0 \quad \text{or} \quad t + 2 = 0$$
$$t = 5 \quad \text{or} \quad t = -2$$

The solution -2 must be disregarded, since we're solving for time. Therefore, Ann rides for 5 hours at $\frac{50}{5} = 10$ miles per hour and Sue rides for $5 - 2 = 3$ hours at $\frac{60}{3} = 20$ miles per hour.

There are various applications commonly classified as *mixture problems*. Even though these problems arise in many different areas, essentially the same mathematical approach can be used to solve them. A general suggestion for solving mixture-type problems is to *work in terms of a pure substance*. We will illustrate what we mean by that statement.

PROBLEM 4

How many milliliters of pure acid must be added to 50 milliliters of a 40% acid solution to obtain a 50% acid solution?

Solution Let a represent the number of milliliters of pure acid to be added. Thinking in terms of pure acid, we know that *the amount of pure acid to start with plus the amount of pure acid added equals the amount of pure acid in the final solution*. Let's use that as a guideline and set up an equation.

$$\underset{\displaystyle 40\%(50)}{\underset{\displaystyle \downarrow}{\genfrac{}{}{0pt}{}{\text{pure acid}}{\text{to start with}}}} + \underset{\displaystyle a}{\underset{\displaystyle \downarrow}{\genfrac{}{}{0pt}{}{\text{pure acid}}{\text{added}}}} = \underset{\displaystyle 50\%(50 + a)}{\underset{\displaystyle \downarrow}{\genfrac{}{}{0pt}{}{\text{pure acid in}}{\text{final solution}}}}$$

Solving this equation we obtain the amount of acid we must add.

$$0.4(50) + a = 0.5(50 + a)$$
$$4(50) + 10a = 5(50 + a)$$
$$200 + 10a = 250 + 5a$$
$$5a = 50$$
$$a = 10$$

We need to add 10 milliliters of pure acid.

There is another class of problems commonly referred to as *work problems*, or sometimes as *rate-time problems*. For example, if a certain machine produces 120 items in ten minutes, then we say that it is working at a rate of $\frac{120}{10} = 12$ items per minute. Likewise, if a person can do a certain job in five hours, then he is working at a rate of $\frac{1}{5}$ of the job per hour. In general, if Q is the quantity of something done in t units of time, then the rate, r, is given by $r = Q/t$. The rate is stated in terms of "*so much quantity per unit of time*." The uniform-motion problems discussed earlier are a special kind of rate-time problems for which the quantity is distance. Likewise, the use of tables to organize information, as we illustrated with the motion problems, is a convenient aid for rate-time problems in general. Let's consider some problems.

PROBLEM 5

Printing press A can produce 35 fliers per minute and press B can produce 50 fliers per minute. Suppose that 2225 fliers are produced by first using press A

alone for 15 minutes and then using presses A and B together until the job is done. How long does press B need to be used?

Solution Let m represent the number of minutes that press B is used. Then $m + 15$ represents the number of minutes press A is used. The information in the problem can be organized in a table as shown at the left. Since the total quantity (the total number of fliers) is 2225, we can set up and solve the following equation.

	QUANTITY	TIME	RATE
A	$35(m + 15)$	$m + 15$	35
B	$50\,m$	m	50

$$35(m + 15) + 50m = 2225$$
$$35m + 525 + 50m = 2225$$
$$85m = 1700$$
$$m = 20$$

Therefore press B must be used for 20 minutes.

PROBLEM 6

It takes Amy twice as long to deliver newspapers as it does Nancy. How long does it take each girl by herself if they can deliver the papers together in 40 minutes?

	QUANTITY	TIME	RATE
NANCY	1	m	$\dfrac{1}{m}$
AMY	1	$2m$	$\dfrac{1}{2m}$

Solution Let m represent the number of minutes that it takes Nancy by herself. Then $2m$ represents Amy's time by herself. Thus, the information can be organized as shown at the left. (Notice that the *quantity* is 1: there is one job to be done.) Since their combined rate is $\frac{1}{40}$, we can solve the following equation.

$$\frac{1}{m} + \frac{1}{2m} = \frac{1}{40}$$
$$40m\left(\frac{1}{m} + \frac{1}{2m}\right) = 40m\left(\frac{1}{40}\right)$$
$$40 + 20 = m$$
$$60 = m$$

Therefore, Nancy can deliver the papers by herself in 60 minutes and Amy can deliver them by herself in $2(60) = 120$ minutes.

Our final example of this section illustrates another approach that some people find works well for rate-time problems. The basic idea used in this approach involves representing the fractional parts of a job. For example, if a person can do a certain job in 7 hours, then at the end of 3 hours he has finished $\frac{3}{7}$ of the job. (Again a constant rate of work is being assumed.) At the end of 5 hours he has finished $\frac{5}{7}$ of the job and, in general, at the end of h hours he has finished $h/7$ of the job.

PROBLEM 7

Walt can mow a lawn in 50 minutes and his son, Mike, can mow the same lawn in 40 minutes. One day Mike started to mow the lawn by himself and worked for 10 minutes. Then Walt joined him with another mower and they finished the lawn. How long did it take them to finish mowing the lawn after Walt started to help?

Solution Let m represent the number of minutes that it takes them to finish the mowing after Walt starts to help. Since Mike has been mowing for 10 minutes, he has done $\frac{10}{40}$ or $\frac{1}{4}$ of the lawn when Walt starts. Thus there is $\frac{3}{4}$ of the lawn yet to mow. The following guideline can be used to set up an equation.

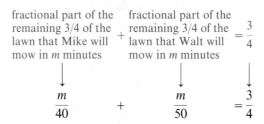

Solving this equation yields the time they mow the lawn together.

$$200\left(\frac{m}{40} + \frac{m}{50}\right) = 200\left(\frac{3}{4}\right)$$

$$5m + 4m = 150$$

$$9m = 150$$

$$m = \frac{150}{9} = \frac{50}{3}$$

They should finish the lawn in $16\frac{2}{3}$ minutes.

As you tackle word problems throughout this text, keep in mind that our primary objective is to expand your repertoire of problem-solving techniques. In the examples, we are sharing some of our ideas for solving problems, but don't hesitate to use your own ingenuity. Furthermore, don't become discouraged—all of us have difficulty with some problems. Give it your best shot.

Problem Set 2.4

Solve each of the following equations.

1. $\dfrac{x}{2x - 8} + \dfrac{16}{x^2 - 16} = \dfrac{1}{2}$

2. $\dfrac{3}{n - 5} - \dfrac{2}{2n + 1} = \dfrac{n + 3}{2n^2 - 9n - 5}$

3. $\dfrac{5t}{2t + 6} - \dfrac{4}{t^2 - 9} = \dfrac{5}{2}$

4. $\dfrac{x}{4x - 4} + \dfrac{5}{x^2 - 1} = \dfrac{1}{4}$

5. $2 + \dfrac{4}{n - 2} = \dfrac{8}{n^2 - 2n}$

6. $3 + \dfrac{6}{t - 3} = \dfrac{6}{t^2 - 3t}$

7. $\dfrac{a}{a + 2} + \dfrac{3}{a + 4} = \dfrac{14}{a^2 + 6a + 8}$

8. $\dfrac{3}{x + 1} + \dfrac{2}{x + 3} = 2$

9. $\dfrac{-2}{3x + 2} + \dfrac{x - 1}{9x^2 - 4} = \dfrac{3}{12x - 8}$

10. $\dfrac{-1}{2x - 5} + \dfrac{2x - 4}{4x^2 - 25} = \dfrac{5}{6x + 15}$

11. $\dfrac{n}{2n - 3} + \dfrac{1}{n - 3} = \dfrac{n^2 - n - 3}{2n^2 - 9n + 9}$

12. $\dfrac{3y}{y^2 + y - 6} + \dfrac{2}{y^2 + 4y + 3} = \dfrac{y}{y^2 - y - 2}$

13. $\dfrac{3y + 1}{3y^2 - 4y - 4} + \dfrac{9}{9y^2 - 4} = \dfrac{2y - 2}{3y^2 - 8y + 4}$

14. $\dfrac{4n + 10}{2n^2 - n - 6} - \dfrac{3n + 1}{2n^2 - 5n + 2} = \dfrac{2}{4n^2 + 4n - 3}$

15. $\dfrac{x + 1}{2x^2 + 7x - 4} - \dfrac{x}{2x^2 - 7x + 3} = \dfrac{1}{x^2 + x - 12}$

16. $\dfrac{3}{x - 2} + \dfrac{5}{x + 3} = \dfrac{8x - 1}{x^2 + x - 6}$

17. $\dfrac{7x + 2}{12x^2 + 11x - 15} - \dfrac{1}{3x + 5} = \dfrac{2}{4x - 3}$

18. $\dfrac{2n}{6n^2 + 7n - 3} - \dfrac{n - 3}{3n^2 + 11n - 4} = \dfrac{5}{2n^2 + 11n + 12}$

19. $\dfrac{3}{5x^2 + 18x - 8} + \dfrac{x + 1}{x^2 - 16} = \dfrac{5x}{5x^2 - 22x + 8}$

20. $\dfrac{2}{4x^2 + 11x - 3} - \dfrac{x + 1}{3x^2 + 8x - 3} = \dfrac{-4x}{12x^2 - 7x + 1}$

Solve each of the following problems.

21. An apple orchard contains 126 trees. The number of trees in each row is 4 less than twice the number of rows. Find the number of rows and the number of trees per row.

22. The sum of a number and its reciprocal is $\frac{10}{3}$. Find the number.

23. Jill starts at city A and travels toward city B at 50 miles per hour. At the same time, Russ starts at city B and travels on the same highway toward city A at 52 miles per hour. How long will it take before they meet if the two cities are 459 miles apart?

24. Two cars, which are 510 miles apart and whose speeds differ by 6 miles per hour, are moving toward each other. If they meet in 5 hours, find the speed of each car.

25. Rita rode her bicycle out into the country at a speed of 20 miles per hour and returned along the same route at 15 miles per hour. If the round trip took 5 hours and 50 minutes, how far out did she ride?

26. A jogger who can run an 8-minute mile starts a half mile ahead of a jogger who can run a 6-minute mile. How long will it take the faster jogger to catch the slower jogger?

27. It takes a freight train 2 hours more to travel 300 miles than it does an express train to travel 280 miles. The rate of the express is 20 miles per hour greater than the rate of the freight. Find the rates of both trains.

28. An airplane travels 2050 miles in the same time that a car travels 260 miles. If the rate of the plane is 358 miles per hour greater than the rate of the car, find the rate of the plane.

29. A container has 6 liters of a 40% alcohol solution in it. How much pure alcohol should be added to raise it to a 60% solution?

30. How many liters of a 60% acid solution must be added to 14 liters of a 10% acid solution to produce a 25% acid solution?

31. One solution contains 50% alcohol and another solution contains 80% alcohol. How many liters of each solution should be mixed to produce 10.5 liters of a 70% alcohol solution?

32. A contractor has a 24-pound mixture that is one-fourth cement and three-fourths sand. How much of a mixture that is half cement and half sand needs to be added to produce a mixture that is one-third cement?

33. A 10-quart radiator contains a 40% antifreeze solution. How much of the solution needs to be drained out and replaced with pure antifreeze in order to raise the solution to 70% antifreeze?

34. How much water must be evaporated from 20 gallons of a 10% salt solution in order to obtain a 20% salt solution?

35. One pipe can fill a tank in 4 hours and another pipe can fill the tank in 6 hours. How long will it take to fill the tank if both pipes are used?

36. Lolita and Doug working together can paint a shed in 3 hours and 20 minutes. If Doug can paint the shed by himself in 10 hours, how long would it take Lolita to paint the shed by herself?

37. An inlet pipe can fill a tank in 10 minutes. A drain can empty the tank in 12 minutes. If the tank is empty and both the pipe and drain are open, how long will it take before the tank overflows?

38. Mark can overhaul an engine in 20 hours and Phil can do the same job by himself in 30 hours. If they both work together for a time and then Mark finishes the job by himself in 5 hours, how long did they work together?

39. Pat and Mike working together can assemble a bookcase in 6 minutes. It takes Mike, working by himself, 9 minutes longer than it takes Pat working by himself to assemble the bookcase. How long does it take each, working alone, to do the job?

40. A computer company markets two card readers. Card reader *B* can read 8000 cards in 2 minutes less time than it takes card reader *A* to read 7500 cards. If the rate of card reader *A* is 250 cards per minute less than the rate of card reader *B*, find each rate.

41. Amelia can type 600 words in 5 minutes less time than it takes Paul to type 600 words. If Amelia types at a rate of 20 words per minute more than Paul types, find the rate of each.

42. A car that averages 16 miles per gallon of gasoline for city driving and 22 miles per gallon for highway driving uses 14 gallons in 296 miles of driving. How much of the driving was city driving?

43. Angie bought some golf balls for $14. If each ball had cost $.25 less, she could have purchased one more ball for the same amount of money. How many golf balls did Angie buy?

44. A new labor contract provides for a wage increase of $1 per hour and a reduction of 5 hours in the work week. A worker who received $320 per week under the old contract will receive $315 per week under the new contract. How long was the work week under the old contract?

45. Todd contracted to paint a house for $480. It took him 4 hours longer than he had anticipated, so he earned $.50 per hour less than he originally calculated. How long had he anticipated it would take him to paint the house?

2.5

Miscellaneous Equations

Our previous work with solving linear and quadratic equations provides us with a basis for solving a variety of other types of equations. For example, the technique of factoring and applying the property

If $ab = 0$ then $a = 0$ or $b = 0$

can sometimes be used for other than quadratic equations.

EXAMPLE 1

Solve $x^3 + 2x^2 - 9x - 18 = 0$.

Solution

$$x^3 + 2x^2 - 9x - 18 = 0$$
$$x^2(x + 2) - 9(x + 2) = 0$$
$$(x + 2)(x^2 - 9) = 0$$
$$(x + 2)(x + 3)(x - 3) = 0$$

$$x + 2 = 0 \quad \text{or} \quad x + 3 = 0 \quad \text{or} \quad x - 3 = 0$$
$$x = -2 \quad \text{or} \quad x = -3 \quad \text{or} \quad x = 3$$

The solution set is $\{-3, -2, 3\}$.

EXAMPLE 2

Solve $3x^5 + 5x^4 = 3x^3 + 5x^2$.

Solution

$$3x^5 + 5x^4 = 3x^3 + 5x^2$$
$$3x^5 + 5x^4 - 3x^3 - 5x^2 = 0$$
$$x^4(3x + 5) - x^2(3x + 5) = 0$$
$$(3x + 5)(x^4 - x^2) = 0$$
$$(3x + 5)(x^2)(x^2 - 1) = 0$$
$$(3x + 5)(x^2)(x + 1)(x - 1) = 0$$

$$3x + 5 = 0 \quad \text{or} \quad x^2 = 0 \quad \text{or} \quad x + 1 = 0 \quad \text{or} \quad x - 1 = 0$$
$$3x = -5$$
$$x = -\tfrac{5}{3} \quad \text{or} \quad x = 0 \quad \text{or} \quad x = -1 \quad \text{or} \quad x = 1$$

The solution set is $\{-\tfrac{5}{3}, 0, -1, 1\}$.

Be careful with an equation like the one in Example 2. Don't be tempted to divide both sides of the equation by x^2. In so doing, the solution of zero will be lost. In general, don't divide both sides of an equation by an expression containing the variable.

An equation such as

$$\sqrt{2x - 4} = x - 2,$$

which contains a radical with the variable in the radicand, is often referred to as a **radical equation**. To solve radical equations we need the following additional property of equality.

PROPERTY 2.3

> Let a and b be real numbers and n a positive integer.
>
> If $a = b$ then $a^n = b^n$.

Property 2.3 states that *we can raise both sides of an equation to a positive integral power. However,* when applying Property 2.3 we must be very careful. Raising both sides of an equation to a positive integral power sometimes produces results that do not satisfy the original equation. Consider the following example.

EXAMPLE 3

Solve $\sqrt{x} + 6 = x$.

Solution

$$\sqrt{x} + 6 = x$$
$$\sqrt{x} = x - 6$$
$$(\sqrt{x})^2 = (x - 6)^2 \qquad \text{Square both sides.}$$
$$x = x^2 - 12x + 36$$
$$0 = x^2 - 13x + 36$$
$$0 = (x - 4)(x - 9)$$
$$x - 4 = 0 \qquad \text{or} \qquad x - 9 = 0$$
$$x = 4 \qquad \text{or} \qquad x = 9$$

Check:

$$\sqrt{x} = x - 6 \qquad\qquad \sqrt{x} = x - 6$$
$$\sqrt{4} \overset{?}{=} 4 - 6 \qquad\qquad \sqrt{9} \overset{?}{=} 9 - 6$$
$$2 \neq -2 \qquad\qquad\quad 3 = 3$$

The only solution is 9, so the solution set is $\{9\}$.

Remark: Notice what happens if we square both sides of the original equation. We obtain $x + 12\sqrt{x} + 36 = x^2$, an equation more complex than the original one and still containing a radical. Therefore it is important first to isolate the term containing the radical on one side of the equation and then to square both sides of the equation.

In general, raising both sides of an equation to a positive integral power produces an equation that has all of the solutions of the original equation, *but* it may also have some extra solutions that will not satisfy the original equation. Such extra solutions are called **extraneous solutions**. Therefore, when using Property 2.3, we *must* check each potential solution in the original equation.

EXAMPLE 4

Solve $\sqrt[3]{2x + 3} = -3$.

Solution

$$\sqrt[3]{2x + 3} = -3$$
$$(\sqrt[3]{2x + 3})^3 = (-3)^3 \qquad \text{Cube both sides.}$$
$$2x + 3 = -27$$
$$2x = -30$$
$$x = -15$$

Check:

$$\sqrt[3]{2x + 3} = -3$$
$$\sqrt[3]{2(-15) + 3} \stackrel{?}{=} -3$$
$$\sqrt[3]{-27} \stackrel{?}{=} -3$$
$$-3 = -3$$

The solution set is $\{-15\}$.

EXAMPLE 5

Solve $\sqrt{x + 4} = \sqrt{x - 1} + 1$.

Solution

$$\sqrt{x + 4} = \sqrt{x - 1} + 1$$
$$(\sqrt{x + 4})^2 = (\sqrt{x - 1} + 1)^2 \qquad \text{Square both sides.}$$
$$x + 4 = x - 1 + 2\sqrt{x - 1} + 1 \qquad \begin{array}{l}\text{Remember the middle term} \\ \text{when squaring the binomial.}\end{array}$$

$$4 = 2\sqrt{x - 1}$$
$$2 = \sqrt{x - 1}$$
$$2^2 = (\sqrt{x - 1})^2 \qquad \text{Square both sides.}$$
$$4 = x - 1$$
$$5 = x$$

Check:

$$\sqrt{x + 4} = \sqrt{x - 1} + 1$$
$$\sqrt{5 + 4} \stackrel{?}{=} \sqrt{5 - 1} + 1$$
$$\sqrt{9} \stackrel{?}{=} \sqrt{4} + 1$$
$$3 = 3$$

The solution set is $\{5\}$.

Equations of Quadratic Form

An equation such as $x^4 + 5x^2 - 36 = 0$ is not a quadratic equation. However, if we let $u = x^2$, then $u^2 = x^4$. Substituting u for x^2 and u^2 for x^4 in $x^4 + 5x^2 - 36 = 0$ produces

$$u^2 + 5u - 36 = 0,$$

which is a quadratic equation. In general, an equation in the variable x is said to be of **quadratic form** if it can be written in the form

$$au^2 + bu + c = 0,$$

where $a \neq 0$ and u is some algebraic expression in x. We have two basic approaches to solving equations of quadratic form, as illustrated by the next two examples.

EXAMPLE 6

Solve $x^{2/3} + x^{1/3} - 6 = 0$.

Solution Let $u = x^{1/3}$; then $u^2 = x^{2/3}$ and the given equation can be rewritten $u^2 + u - 6 = 0$. Solving this equation yields two solutions.

$$u^2 + u - 6 = 0$$
$$(u + 3)(u - 2) = 0$$
$$u + 3 = 0 \quad \text{or} \quad u - 2 = 0$$
$$u = -3 \quad \text{or} \quad u = 2$$

Now, substituting $x^{1/3}$ for u, we have

$$x^{1/3} = -3 \quad \text{or} \quad x^{1/3} = -2,$$

from which we obtain

$$(x^{1/3})^3 = (-3)^3 \quad \text{or} \quad (x^{1/3})^3 = 2^3,$$
$$x = -27 \quad \text{or} \quad x = 8.$$

Check:

$$x^{2/3} + x^{1/3} - 6 = 0 \qquad\qquad x^{2/3} + x^{1/3} - 6 = 0$$
$$(-27)^{2/3} + (-27)^{1/3} - 6 \overset{?}{=} 0 \qquad (8)^{2/3} + (8)^{1/3} - 6 \overset{?}{=} 0$$
$$9 + (-3) - 6 \overset{?}{=} 0 \qquad\qquad 4 + 2 - 6 \overset{?}{=} 0$$
$$0 = 0 \qquad\qquad\qquad 0 = 0$$

The solution set is $\{-27, 8\}$.

EXAMPLE 7

Solve $x^4 + 5x^2 - 36 = 0$.

Solution

$$x^4 + 5x^2 - 36 = 0$$
$$(x^2 + 9)(x^2 - 4) = 0$$
$$x^2 + 9 = 0 \quad \text{or} \quad x^2 - 4 = 0$$
$$x^2 = -9 \quad \text{or} \quad x^2 = 4$$
$$x = \pm 3i \quad \text{or} \quad x = \pm 2$$

The solution set is $\{\pm 3i, \pm 2\}$.

Notice in Example 6 that we made a substitution (u for $x^{1/3}$) to change the original equation to a quadratic equation in terms of the variable u. Then, after solving for u, we substituted $x^{1/3}$ for u to obtain the solutions of the original equation. However, in Example 7 we factored the given polynomial and proceeded without changing to a quadratic equation. Which approach you use may depend upon the complexity of the given equation.

EXAMPLE 8

Solve $15x^{-2} - 11x^{-1} - 12 = 0$.

Solution Let $u = x^{-1}$; then $u^2 = x^{-2}$ and the given equation can be written and solved as follows.

$$15u^2 - 11u - 12 = 0$$
$$(5u + 3)(3u - 4) = 0$$

$$5u + 3 = 0 \qquad \text{or} \qquad 3u - 4 = 0$$
$$5u = -3 \qquad \text{or} \qquad 3u = 4$$
$$u = -\tfrac{3}{5} \qquad \text{or} \qquad u = \tfrac{4}{3}$$

Now substituting x^{-1} back for u we have

$$x^{-1} = -\tfrac{3}{5} \qquad \text{or} \qquad x^{-1} = \tfrac{4}{3},$$

from which we obtain

$$\frac{1}{x} = \frac{-3}{5} \qquad \text{or} \qquad \frac{1}{x} = \frac{4}{3}$$

$$-3x = 5 \qquad \text{or} \qquad 4x = 3$$

$$x = -\frac{5}{3} \qquad \text{or} \qquad x = \frac{3}{4}.$$

The solution set is $\left\{ -\tfrac{5}{3}, \tfrac{3}{4} \right\}$.

Problem Set 2.5

Solve each of the following equations. Don't forget that you *must* check potential solutions whenever Property 2.3 is applied.

1. $x^3 + x^2 - 4x - 4 = 0$

2. $x^3 - 5x^2 - x + 5 = 0$

3. $2x^3 - 3x^2 + 2x - 3 = 0$

4. $3x^3 + 5x^2 + 12x + 20 = 0$

5. $8x^5 + 10x^4 = 4x^3 + 5x^2$

6. $10x^5 + 15x^4 = 2x^3 + 3x^2$

7. $x^{3/2} = 4x$

8. $5x^4 = 6x^3$

9. $n^{-2} = n^{-3}$

10. $n^{4/3} = 4n$

11. $\sqrt{3x - 2} = 4$

12. $\sqrt{5x - 1} = -4$

13. $\sqrt{3x - 8} - \sqrt{x - 2} = 0$

14. $\sqrt{2x - 3} = 1$

15. $\sqrt{4x - 3} = -2$

16. $\sqrt{2x - 1} - \sqrt{x + 2} = 0$

17. $\sqrt[3]{2x + 3} + 3 = 0$

18. $\sqrt[3]{n^2 - 1} + 1 = 0$

19. $2\sqrt{n} + 3 = n$

20. $\sqrt{3t} - t = -6$

21. $\sqrt{3x - 2} = 3x - 2$

22. $5x - 4 = \sqrt{5x - 4}$

23 $\sqrt{2t-1}+2=t$

24. $p=\sqrt{-4p+17}+3$

25. $\sqrt{x+2}-1=\sqrt{x-3}$

26. $\sqrt{x+5}-2=\sqrt{x-7}$

27. $\sqrt{7n+23}-\sqrt{3n+7}=2$

28. $\sqrt{5t+31}-\sqrt{t+3}=4$

29. $\sqrt{3x+1}+\sqrt{2x+4}=3$

30. $\sqrt{2x-1}-\sqrt{x+3}=1$

31. $\sqrt{x-2}-\sqrt{2x-11}=\sqrt{x-5}$

32. $\sqrt{-2x-7}+\sqrt{x+9}=\sqrt{8-x}$

33. $\sqrt{1+2\sqrt{x}}=\sqrt{x+1}$

34. $\sqrt{7+3\sqrt{x}}=\sqrt{x}+1$

35. $x^4-5x^2+4=0$

36. $x^4-25x^2+144=0$

37. $2n^4-9n^2+4=0$

38. $3n^4-4n^2+1=0$

39. $x^4-2x^2-35=0$

40. $2x^4+5x^2-12=0$

41. $x^4-4x^2+1=0$

42. $x^4-8x^2+11=0$

43. $x^{2/3}+3x^{1/3}-10=0$

44. $x^{2/3}+x^{1/3}-2=0$

45. $6x^{2/3}-5x^{1/3}-6=0$

46. $3x^{2/3}-11x^{1/3}-4=0$

47. $x^{-2}+4x^{-1}-12=0$

48. $12t^{-2}-17t^{-1}-5=0$

49. $x-11\sqrt{x}+30=0$

50. $2x-11\sqrt{x}+12=0$

51. $x+3\sqrt{x}-10=0$

52. $6x-19\sqrt{x}-7=0$

Miscellaneous Problems

53. Verify that $x=a$ and $x^2=a^2$ are *not* equivalent equations.

54. Solve each of the following equations, expressing solutions to the nearest hundredth.

(a) $x^4-3x^2+1=0$

(b) $x^4-5x^2+2=0$

(c) $2x^4-7x^2+2=0$

(d) $3x^4-9x^2+1=0$

(e) $x^4-100x^2+2304=0$

(f) $4x^4-373x^2+3969=0$

2.6

Inequalities

Just as we use the symbol "$=$" to represent "is equal to," we also use the symbols "$<$" and "$>$" to represent "is less than" and "is greater than," respectively. Thus, various **statements of inequality** can be made as follows.

$a<b$ means a is less than b;

$a\le b$ means a is less than or equal to b;

$a>b$ means a is greater than b;

$a\ge b$ means a is greater than or equal to b.

The following are examples of **numerical statements of inequality**.

$$7+8>10 \qquad -4+(-6)\ge-10$$
$$-4>-6 \qquad 7-9\le-2$$
$$7-1<20 \qquad 3+4>12$$
$$8(-3)<5(-3) \qquad 7-1<0$$

Notice that only $3+4>12$ and $7-1<0$ are *false*; the other six are *true* numerical statements.

Algebraic inequalities contain one or more variables. The following are examples of algebraic inequalities.

$$x + 4 > 8 \qquad 3x + 2y \le 4$$
$$(x - 2)(x + 4) \ge 0 \qquad x^2 + y^2 + z^2 \le 16$$

An algebraic inequality such as $x + 4 > 8$ is neither true nor false as it stands and is called an **open sentence**. For each numerical value substituted for x, the algebraic inequality $x + 4 > 8$ becomes a numerical statement of inequality that is true or false. For example, if $x = -3$, then $x + 4 > 8$ becomes $-3 + 4 > 8$, which is false. If $x = 5$, then $x + 4 > 8$ becomes $5 + 4 > 8$, which is true. **Solving an algebraic inequality** refers to the process of finding the numbers that make it a true numerical statement. Such numbers are called the **solutions** of the inequality and are said to **satisfy** it.

The general process for solving inequalities closely parallels that for solving equations. We repeatedly replace the given inequality with equivalent but simpler inequalities until the solution set is obvious. The following properties provide the basis for producing equivalent inequalities.

PROPERTY 2.4

For all real numbers a, b, and c,

   1. If $a > b$ then $a + c > b + c$;
   2. If $a > b$ and $c > 0$ then $ac > bc$;
   3. If $a > b$ and $c < 0$ then $ac < bc$.

Similar properties exist if $>$ is replaced by $<$, $\le$, or $\ge$. Part 1 of Property 2.4 is commonly called the **addition property of inequality**. Parts 2 and 3 together make up the **multiplication property of inequality**. Pay special attention to part 3. **If both sides of an inequality are multiplied by a negative number, the inequality symbol must be reversed.** For example, if both sides of $3 < 5$ are multiplied by -2, the equivalent inequality $-6 > -10$ is produced.

Now let's consider using the addition and multiplication properties of inequality to help solve some inequalities.

EXAMPLE 1

Solve $3(2x - 1) < 8x - 7$.

Solution

$$3(2x - 1) < 8x - 7$$

$6x - 3 < 8x - 7$	Apply distributive property to left side.
$-2x - 3 < -7$	Add $-8x$ to both sides.
$-2x < -4$	Add 3 to both sides.
$-\frac{1}{2}(-2x) > -\frac{1}{2}(-4)$	Multiply both sides by $-\frac{1}{2}$, which reverses the inequality.
$x > 2$	

The solution set is $\{x \mid x > 2\}$.

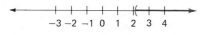

Figure 2.1

A graph of the solution set $\{x \mid x > 2\}$ in Example 1 is shown in Figure 2.1. The parenthesis indicates that 2 does not belong to the solution set.

Checking the solutions of an inequality presents a problem. Obviously, we cannot check all of the infinitely many solutions for a particular inequality. However, by checking at least one solution, especially when the multiplication property has been used, we might catch a mistake of forgetting to change the type of inequality. In Example 1 we are claiming that all numbers greater than 2 will satisfy the original inequality. Let's check one number, for example, 3.

$$3(2x - 1) < 8x - 7$$
$$3[2(3) - 1] \overset{?}{<} 8(3) - 7$$
$$3(5) \overset{?}{<} 17$$
$$15 < 17 \qquad \text{It checks!}$$

It is also convenient to express solution sets of inequalities using **interval notation**. For example, the symbol $(2, \infty)$ refers to the interval of all real numbers greater than 2. As on the graph in Figure 2.1, the left-hand parenthesis indicates that 2 is not to be included. The infinity symbol, ∞, along with the right-hand parenthesis, indicates that there is no right-hand endpoint. Following is a list of interval notations along with the sets and graphs that they represent. Notice the use of square brackets to *include* endpoints. Also recall that the notation $a < x < b$ is a compact way of expressing "x is greater than a *and* x is less than b." From now on, we will express solution sets of inequalities using interval notation.

TYPE OF INTERVAL	SET	INTERVAL NOTATION	GRAPH
open interval	$\{x \mid x > a\}$	(a, ∞)	
	$\{x \mid a < x < b\}$	(a, b)	
	$\{x \mid x < b\}$	$(-\infty, b)$	
half-open interval	$\{x \mid x \geq a\}$	$[a, \infty)$	
	$\{x \mid a < x \leq b\}$	$(a, b]$	
	$\{x \mid a \leq x < b\}$	$[a, b)$	
	$\{x \mid x \leq b\}$	$(-\infty, b]$	
closed interval	$\{x \mid a \leq x \leq b\}$	$[a, b]$	

EXAMPLE 2

Solve $\dfrac{x - 4}{6} - \dfrac{x - 2}{9} \leq \dfrac{5}{18}$.

Solution

$$\frac{x-4}{6} - \frac{x-2}{9} \le \frac{5}{18}$$

$$18\left(\frac{x-4}{6} - \frac{x-2}{9}\right) \le 18\left(\frac{5}{18}\right) \qquad \text{Multiply both sides by the LCD.}$$

$$18\left(\frac{x-4}{6}\right) - 18\left(\frac{x-2}{9}\right) \le 18\left(\frac{5}{18}\right)$$

$$3(x-4) - 2(x-2) \le 5$$

$$3x - 12 - 2x + 4 \le 5$$

$$x - 8 \le 5$$

$$x \le 13$$

The solution set is $(-\infty, 13]$.

The notation $a < x < b$ is a compact way of writing "$x > a$ *and* $x < b$." It is a convenient form for solving **compound inequalities** involving the connecting word "and," as the next example illustrates.

EXAMPLE 3

Solve $-2 < \dfrac{3x+2}{2} < 7$.

Solution

$$-2 < \quad \frac{3x+2}{2} \quad < 7$$

$$2(-2) < 2\left(\frac{3x+2}{2}\right) < 2(7) \qquad \text{Multiply through by 2.}$$

$$-4 < \quad 3x + 2 \quad < 14$$

$$-6 < \quad 3x \quad\quad < 12 \qquad \text{Add } -2 \text{ to all three quantities.}$$

$$-2 < \quad x \quad\quad < 4 \qquad \text{Multiply through by } \tfrac{1}{3}.$$

The solution set is the interval $(-2, 4)$.

Quadratic Inequalities

The equation $ax^2 + bx + c = 0$ has been referred to as the standard form of a quadratic equation in one variable. Similarly, the form $ax^2 + bx + c < 0$ is used to represent a **quadratic inequality**. (The symbol $<$ can be replaced by $>$, $\le$, or $\ge$ to produce other forms of quadratic inequalities.)

The number line becomes a very useful tool for helping to analyze quadratic inequalities. Let's consider some examples to illustrate the procedure.

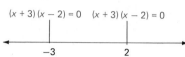

EXAMPLE 4

Solve $x^2 + x - 6 < 0$.

Solution First let's factor the polynomial.

$$x^2 + x - 6 < 0$$
$$(x + 3)(x - 2) < 0$$

Second, let's locate the values for which the product $(x + 3)(x - 2)$ is equal to zero. The numbers -3 and 2 divide the number line into three intervals:

The numbers less than -3,
The numbers between -3 and 2, and
The numbers greater than 2.

We can choose a **test number** from each of these intervals and see how it affects the signs of the factors $x + 3$ and $x - 2$ and, consequently, the sign of the product of these factors. For example, if $x < -3$ (try $x = -4$) then $x + 3$ is negative and $x - 2$ is negative; so their product is positive. If $-3 < x < 2$ (try $x = 0$) then $x + 3$ is positive and $x - 2$ is negative; so their product is negative. If $x > 2$ (try $x = 3$) then $x + 3$ is positive and $x - 2$ is positive; so their product is positive. This information can be conveniently arranged using a number line as follows.

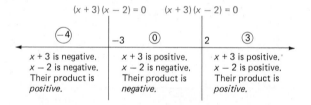

Therefore the given inequality, $x^2 + x - 6 < 0$, is satisfied by the numbers between -3 and 2. That is, the solution set is the open interval $(-3, 2)$.

Numbers such as -3 and 2 in the preceding example, for which the given polynomial or algebraic expression equals zero or is undefined, are referred to as **critical numbers**. Let's consider some additional examples making use of critical numbers and test numbers.

EXAMPLE 5

Solve $6x^2 + 17x - 14 \geq 0$.

Solution First, we factor the polynomial.

$$6x^2 + 17x - 14 \geq 0$$
$$(2x + 7)(3x - 2) \geq 0$$

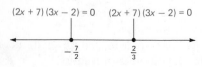

Second, we locate the values for which the product $(2x + 7)(3x - 2)$ equals zero. We suggest putting dots at $-\frac{7}{2}$ and $\frac{2}{3}$ to remind ourselves that these two numbers must be included in the solution set, since the given statement includes equality. Now let's choose a test number from each of the three intervals and observe the sign behavior of the factors.

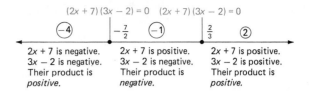

Using the concept of set union,* the solution set can be written $(-\infty, -\frac{7}{2}] \cup [\frac{2}{3}, \infty)$.

Let's conclude this section by considering a word problem that involves an inequality situation. All of the problem-solving techniques offered earlier continue to apply except that now we look for a guideline that can be used to generate an inequality rather than an equation.

PROBLEM 1

Lance has $500 to invest. If he invests $300 at 9%, at what rate must he invest the remaining $200 so that the total yearly interest from the two investments exceeds $47?

Solution Let r represent the unknown rate of interest. The following guideline can be used to set up an inequality.

interest from 9% investment + interest from r percent investment > $47

$$(9\%)(\$300) \quad + \quad r(\$200) \quad > \$47$$

We solve this inequality using methods we have already acquired.

$$(.09)(300) + 200r > 47$$
$$27 + 200r > 47$$
$$200r > 20$$
$$r > 0.1$$

The other $200 must be invested at a rate greater than 10%.

Problem Set 2.6

Express each of the following in interval notation and graph each of the intervals.

1. $x \le -2$ 2. $x > -1$ 3. $1 < x < 4$

4. $-1 < x \le 2$ 5. $2 > x > 0$ 6. $-3 \ge x$

7. $-2 \le x \le -1$ 8. $1 \le x$

---

* The **union** of sets A and B, written $A \cup B$, is defined as $A \cup B = \{x \mid x \in A \; or \; x \in B\}$.

Solve each of the following inequalities. Express the solution sets in interval notation.

9. $-2x + 1 > 5$

10. $6 - 3x < 12$

11. $-3n + 5n - 2 \geq 8n - 7 - 9n$

12. $3n - 5 > 8n + 5$

13. $6(2t - 5) - 2(4t - 1) \geq 0$

14. $3(2x + 1) - 2(2x + 5) < 5(3x - 2)$

15. $\frac{2}{3}x - \frac{3}{4} \leq \frac{1}{4}x + \frac{2}{3}$

16. $\frac{3}{5} - \frac{x}{2} \geq \frac{1}{2} + \frac{x}{5}$

17. $\frac{n + 2}{4} + \frac{n - 3}{8} < 1$

18. $\frac{2n + 1}{6} + \frac{3n - 1}{5} > \frac{2}{15}$

19. $\frac{x}{2} - \frac{x - 1}{5} \geq \frac{x + 2}{10} - 4$

20. $\frac{4x - 3}{6} - \frac{2x - 1}{12} < -2$

21. $0.09x + 0.1(x + 200) > 77$

22. $0.06x + 0.08(250 - x) \geq 19$

23. $0 < \frac{5x - 1}{3} < 2$

24. $-3 \leq \frac{4x + 3}{2} \leq 1$

25. $3 \geq \frac{7 - x}{2} \geq 1$

26. $-2 \leq \frac{5 - 3x}{4} \leq \frac{1}{2}$

27. $x^2 + 3x - 4 < 0$

28. $x^2 - 4 < 0$

29. $x^2 - 2x - 15 > 0$

30. $x^2 - 12x + 32 \geq 0$

31. $n^2 - n \leq 2$

32. $n^2 + 5n \leq 6$

33. $3t^2 + 11t - 4 > 0$

34. $2t^2 - 9t - 5 > 0$

35. $15x^2 - 26x + 8 \leq 0$

36. $6x^2 + 25x + 14 \leq 0$

37. $4x^2 - 4x + 1 > 0$

38. $9x^2 + 6x + 1 \leq 0$

39. $(x + 1)(x - 3) > (x + 1)(2x - 1)$

40. $(x - 2)(2x + 5) > (x - 2)(x - 3)$

41. $(x + 1)(x - 2) \geq (x - 4)(x + 6)$

42. $(2x - 1)(x + 4) \geq (2x + 1)(x - 3)$

43. $(x - 1)(x - 2)(x + 4) > 0$

44. $(x + 1)(x - 3)(x + 7) \geq 0$

45. $(x + 2)(2x - 1)(x - 5) \leq 0$

46. $(x - 3)(3x + 2)(x + 4) < 0$

47. $x^3 - 2x^2 - 24x \geq 0$

48. $x^3 + 2x^2 - 3x > 0$

49. $(x - 2)^2(x + 3) > 0$

50. $(x + 4)^2(x + 5) > 0$

Solve each of the following problems.

51. Felix has $1000 to invest. Suppose he invests $500 at 8% interest. At what rate must he invest the other $500 so that the two investments yield more than $100 of yearly interest?

52. Suppose that Annette invests $700 at 9%. How much must she invest at 11% so that the total yearly interest from the two investments exceeds $162?

53. Rhonda had scores of 94, 84, 86, and 88 on her first four history exams of the semester. What score must she obtain on the fifth exam to have an average of 90 or better for the five exams?

54. The average height of the two forwards and the center of a basketball team is 6 feet 8 inches. What must the average height of the two guards be so that the team average is at least 6 feet 4 inches?

55. If the temperature for a 24-hour period ranged between 41°F and 59°F, inclusive, what was the range in Celsius degrees? $(F = \frac{9}{5}C + 32)$

56. If the temperature for a 24-hour period ranged between $-20°C$ and $-5°C$, inclusive, what was the range in Fahrenheit degrees? $(C = \frac{5}{9}(F - 32))$

57. A person's intelligence quotient (IQ) is found by dividing mental age (M), as indicated by standard tests, by the chronological age (C), and then multiplying this ratio by

100. The formula $IQ = 100M/C$ can be used. If the IQ range of a group of 11-year-olds is given by $80 \le IQ \le 140$, find the mental-age range of this group.

58. Repeat Problem 57 for an IQ range of 70 to 125, inclusive, for a group of 9-year-olds.

59. A car can be rented from agency A at $75 per day plus $.10 a mile or from agency B at $50 a day plus $.20 a mile. If the car is driven m miles, for what values of m does it cost less to rent from agency A?

60. Suppose the distance that a projectile, fired from ground level, is above the ground is given by $30t - 16t^2$, where t is measured in seconds. During what time interval is the projectile at least 14 feet above ground level?

Miscellaneous Problems

61. The product $(x - 2)(x + 3)$ is positive if both factors are negative *or* if both factors are positive. Therefore, we can solve $(x - 2)(x + 3) > 0$ as follows:

$$(x - 2 < 0 \quad \text{and} \quad x + 3 < 0) \quad \text{or} \quad (x - 2 > 0 \quad \text{and} \quad x + 3 > 0)$$
$$(x < 2 \quad \text{and} \quad x < -3) \quad \text{or} \quad (x > 2 \quad \text{and} \quad x > -3)$$
$$x < -3 \quad \text{or} \quad x > 2$$

The solution set is $(-\infty, -3) \cup (2, \infty)$. Use this type of analysis to solve each of the following.

(a) $(x - 1)(x + 5) > 0$ **(b)** $(x + 2)(x - 4) \ge 0$

(c) $(x + 4)(x - 3) < 0$ **(d)** $(2x - 1)(x + 5) \le 0$

(e) $(x + 4)(x + 1)(x - 2) > 0$ **(f)** $(x + 2)(x - 1)(x - 3) < 0$

62. If $a > b > 0$, verify that $1/a < 1/b$.

63. If $a > b$, is it always true that $1/a < 1/b$? Defend your answer.

2.7

Inequalities Involving Quotients and Absolute Value

The same type of number-line analysis that we did in the previous section can be used for indicated quotients as well as for indicated products. In other words, inequalities such as

$$\frac{x - 2}{x + 3} > 0$$

can be solved very effectively using the same basic approach that we used with quadratic inequalities in the previous section. Let's illustrate this procedure.

EXAMPLE 1

Solve $\dfrac{x - 2}{x + 3} > 0$.

Solution First we find that at $x = 2$ the quotient $\dfrac{x - 2}{x + 3}$ equals zero and at $x = -3$ the quotient is undefined. The critical numbers -3 and 2 divide the number line into three intervals. Then, using a test number from each interval (such as -4, 1, and 3), we can observe the sign behavior of the quotient.

$\dfrac{x-2}{x+3}$ is undefined. $\dfrac{x-2}{x+3}=0$

-4	-3	$①$	2	$③$
$x-2$ is negative. $x+3$ is negative. The quotient $\dfrac{x-2}{x+3}$ is *positive*.		$x-2$ is negative. $x+3$ is positive. The quotient $\dfrac{x-2}{x+3}$ is *negative*.		$x-2$ is positive. $x+3$ is positive. The quotient $\dfrac{x-2}{x+3}$ is *positive*.

Therefore, the solution set for $\dfrac{x-2}{x+3}>0$ is $(-\infty,-3)\cup(2,\infty)$.

EXAMPLE 2

Solve $\dfrac{x+2}{x+4}\le 3$.

Solution First let's change the form of the given inequality.

$$\frac{x+2}{x+4}\le 3$$

$$\frac{x+2}{x+4}-3\le 0$$

$$\frac{x+2-3(x+4)}{x+4}\le 0$$

$$\frac{x+2-3x-12}{x+4}\le 0$$

$$\frac{-2x-10}{x+4}\le 0$$

Now we can proceed as before. If $x=-5$ then the quotient $\dfrac{-2x-10}{x+4}$ equals zero, and if $x=-4$, the quotient is undefined. Then, using test numbers such as -6, $-4\frac{1}{2}$, and -3, we can study the sign behavior of the quotient.

$\dfrac{-2x-10}{x+4}=0$ $\dfrac{-2x-10}{x+4}$ is undefined.

-6	-5	$-4\frac{1}{2}$	-4	-3
$-2x-10$ is positive. $x+4$ is negative. The quotient $\dfrac{-2x-10}{x+4}$ is *negative*.		$-2x-10$ is negative. $x+4$ is negative. The quotient $\dfrac{-2x-10}{x+4}$ is *positive*.		$-2x-10$ is negative. $x+4$ is positive. The quotient $\dfrac{-2x-10}{x+4}$ is *negative*.

Therefore, the solution set for $\dfrac{x+2}{x+4}\le 3$ is $(-\infty,-5]\cup(-4,\infty)$.

Absolute Value

In Section 1.1 we defined the **absolute value** of a real number by

$$|a| = \begin{cases} a, & \text{if } a \geq 0, \\ -a, & \text{if } a < 0. \end{cases}$$

We also interpreted the absolute value of any real number to be **the distance between the number and zero on the number line**. For example, $|6| = 6$ because the distance between 6 and 0 is six units. Likewise, $|-9| = 9$ because the distance between -9 and 0 is nine units.

Interpreting absolute value as distance on a number line provides a way of analyzing a variety of equations and inequalities involving absolute value. For example, suppose that we want to solve the equation $|x| = 4$. Thinking in terms of looking for a number x such that the distance between x and 0 is four units, we see that x must be 4 or -4. In other words, the equation $|x| = 4$ is equivalent to $x = 4$ or $x = -4$. The solution set is $\{-4, 4\}$.

The following general property should seem reasonable from the distance interpretation. It can be proved using the definition of absolute value.

PROPERTY 2.5

> For any real number $k > 0$,
>
> if $|x| = k$ then $x = k$ or $x = -k$.

EXAMPLE 3

Solve $|3x - 2| = 7$.

Solution

$$|3x - 2| = 7$$
$$3x - 2 = 7 \quad \text{or} \quad 3x - 2 = -7$$
$$3x = 9 \quad \text{or} \quad 3x = -5$$
$$x = 3 \quad \text{or} \quad x = -\tfrac{5}{3}$$

The solution set is $\{-\tfrac{5}{3}, 3\}$.

The distance interpretation for absolute value also provides a good basis for solving some inequalities involving absolute value. For example, to solve $|x| < 4$, we know that the distance between x and 0 must be less than four units. In other words, x is to be less than four units away from zero. Thus, $|x| < 4$ is equivalent to $-4 < x < 4$ and the solution set is the interval $(-4, 4)$. The following general property can be verified.

PROPERTY 2.6

> For any real number $k > 0$,
>
> if $|x| < k$, then $-k < x < k$.

EXAMPLE 4

Solve $|2x + 1| < 5$.

Solution

$$|2x + 1| < 5$$
$$-5 < 2x + 1 < 5$$
$$-6 < 2x < 4$$
$$-3 < x < 2$$

The solution set is the interval $(-3, 2)$.

Now suppose that we want to solve $|x| > 4$. The distance between x and 0 must be more than four units or, in other words, x is to be more than four units away from zero. Therefore, $|x| > 4$ is equivalent to $x < -4$ or $x > 4$, and the solution set is $(-\infty, -4) \cup (4, \infty)$. The following general property can be verified by using the definition of absolute value.

PROPERTY 2.7

> For any real number $k > 0$,
>
> $\quad$ if $|x| > k$ $\quad$ then $\quad$ $x < -k$ $\quad$ or $\quad$ $x > k$.

EXAMPLE 5

Solve $|4x - 3| > 9$.

Solution

$$|4x - 3| > 9$$

$4x - 3 < -9$	or	$4x - 3 > 9$
$4x < -6$	or	$4x > 12$
$x < -\frac{6}{4}$	or	$x > 3$
$x < -\frac{3}{2}$	or	$x > 3$

The solution set is $(-\infty, -\frac{3}{2}) \cup (3, \infty)$.

Properties 2.5 through 2.7 provide a sound basis for solving many equations and inequalities involving absolute value. However, if at any time you become doubtful as to which property applies, don't forget the distance interpretation for absolute value. In fact, it is the distance interpretation that allows us to "solve by inspection" such equations and inequalities as $|3x - 7| = -4$, $|2x - 1| < -3$, and $|5x + 2| > -4$. (Notice that in each of these examples, k is a negative number; therefore Properties 2.5–2.7 do not apply.)

$\quad |3x - 7| = -4 \quad$ has no solutions because the absolute value (distance) cannot be negative. The solution set is $\varnothing$.

$|2x - 1| < -3$ has no solutions because we cannot obtain an absolute value less than -3. The solution set is $\emptyset$.

$|5x + 2| > -4$ is satisfied by all real numbers because the absolute value of $5x + 2$, regardless of what number is substituted for x, will always be greater than -4. The solution set is $(-\infty, \infty)$.

The number-line approach used in Examples 1 and 2 of this section, along with Properties 2.6 and 2.7, provides a systematic way of solving absolute-value inequalities having the variable in the denominator of a fraction. Let's analyze one such problem.

EXAMPLE 6

Solve $\left|\dfrac{x - 2}{x + 3}\right| < 4$.

Solution By Property 2.6, $\left|\dfrac{x - 2}{x + 3}\right| < 4$ becomes $-4 < \dfrac{x - 2}{x + 3} < 4$, which can be written

$$\frac{x - 2}{x + 3} > -4 \quad \text{and} \quad \frac{x - 2}{x + 3} < 4.$$

Each part of this "and" statement can be solved as we handled Example 2 earlier.

(a)		(b)
$\dfrac{x - 2}{x + 3} > -4$	and	$\dfrac{x - 2}{x + 3} < 4$
$\dfrac{x - 2}{x + 3} + 4 > 0$	and	$\dfrac{x - 2}{x + 3} - 4 < 0$
$\dfrac{x - 2 + 4(x + 3)}{x + 3} > 0$	and	$\dfrac{x - 2 - 4(x + 3)}{x + 3} < 0$
$\dfrac{x - 2 + 4x + 12}{x + 3} > 0$	and	$\dfrac{x - 2 - 4x - 12}{x + 3} < 0$
$\dfrac{5x + 10}{x + 3} > 0$	and	$\dfrac{-3x - 14}{x + 3} < 0$

This solution set can be found as in Example 1 of this section.

This solution set can be found as in Example 1 of this section.

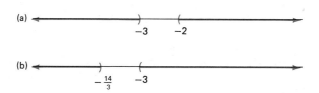

The intersection of the two pictured solution sets is the following set.

Therefore the solution set of $\left|\dfrac{x-2}{x+3}\right| < 4$ is $(-\infty, -\frac{14}{3}) \cup (-2, \infty)$.

Yes, Example 6 is a little messy, but it does illustrate the weaving together of previously used techniques to solve a more complicated problem. Don't be in a hurry when doing such problems. First, analyze the general approach to be taken and then carry out the details in a neatly organized format so as to minimize your chances of making careless errors.

Problem Set 2.7

Solve each of the following inequalities, expressing the solution sets in interval notation.

1. $\dfrac{x+1}{x-5} > 0$ **2.** $\dfrac{x+2}{x+4} \le 0$ **3.** $\dfrac{2x-1}{x+2} < 0$ **4.** $\dfrac{3x+2}{x-1} > 0$

5. $\dfrac{-x+3}{3x-1} \ge 0$ **6.** $\dfrac{-n-2}{n+4} < 0$ **7.** $\dfrac{n}{n+2} \ge 3$ **8.** $\dfrac{x}{x-1} > 2$

9. $\dfrac{x-1}{x+2} < 2$ **10.** $\dfrac{t-1}{t-5} \le 2$ **11.** $\dfrac{t-3}{t+5} > 1$ **12.** $\dfrac{x+2}{x+7} < 1$

13. $\dfrac{1}{x-2} < \dfrac{1}{x+3}$ **14.** $\dfrac{2}{x+1} > \dfrac{3}{x-4}$

Solve each of the following equations.

15. $|x-2| = 6$ **16.** $|x+3| = 4$ **17.** $|x+\frac{1}{4}| = \frac{2}{5}$ **18.** $|x-\frac{2}{3}| = \frac{3}{4}$

19. $|2n-1| = 7$ **20.** $|2n+1| = 11$ **21.** $|3x+4| = 5$ **22.** $|5x-3| = 10$

23. $|7x-1| = -4$ **24.** $|-2x-1| = 6$ **25.** $|-3x-2| = 8$

26. $|5x-4| = -3$ **27.** $\left|\dfrac{3}{k-1}\right| = 4$ **28.** $\left|\dfrac{-2}{n+3}\right| = 5$

29. $|3x-1| = |2x+3|$ **30.** $|2x+1| = |4x-3|$

31. $|-2n+1| = |-3n-1|$ **32.** $|-4n+5| = |-3n-5|$

33. $|x-2| = |x+4|$ **34.** $|2x-3| = |2x+5|$

Solve each of the following inequalities, expressing the solution sets in interval notation.

35. $|x| < 6$ **36.** $|x| \ge 4$ **37.** $|t-3| > 5$ **38.** $|n+2| < 1$

39. $|2x-1| \le 7$ **40.** $|2x+1| \ge 3$ **41.** $|3n+2| > 9$ **42.** $|5n-2| < 2$

43. $|4x-3| < -5$ **44.** $|2-x| > 1$ **45.** $|3-2x| < 4$

46. $|4x+5| > -3$ **47.** $|-1-x| \ge 8$ **48.** $|-2-x| \le 5$

49. $|x-1| + 2 < 4$ **50.** $|x+3| - 2 < 1$ **51.** $|x+4| - 1 > 1$

52. $|x-2| + 3 > 6$ **53.** $\left|\dfrac{x+1}{x-2}\right| < 3$ **54.** $\left|\dfrac{x-1}{x-4}\right| < 2$

55. $\left|\dfrac{x-1}{x+3}\right| > 1$ 56. $\left|\dfrac{x+4}{x-5}\right| \geq 3$ 57. $\left|\dfrac{n+2}{n}\right| \geq 4$

58. $\left|\dfrac{t+6}{t-2}\right| < 1$ 59. $\left|\dfrac{k}{2k-1}\right| \leq 2$ 60. $\left|\dfrac{k}{k+2}\right| > 4$

Miscellaneous Problems

61. Use the definition of absolute value and prove Property 2.5.
62. Prove Property 2.6.
63. Prove Property 2.7.
64. Solve each of the following inequalities by using the definition of absolute value. Do not use Properties 2.6 and 2.7.
    (a) $|x + 5| < 11$ (b) $|x - 4| \leq 10$ (c) $|2x - 1| > 7$
    (d) $|3x + 2| \geq 1$ (e) $|2 - x| < 5$ (f) $|3 - x| > 6$

Chapter 2 Summary

This chapter can be summarized in terms of three large topics, namely, (1) solving equations, (2) solving inequalities, and (3) problem solving.

Solving Equations

The following properties are used extensively in the equation-solving process.

1. $a = b$ if and only if $a + c = b + c$.   Addition property of equality
2. $a = b$ if and only if $ac = bc, c \neq 0$.   Multiplication property of equality
3. If $ab = 0$ then $a = 0$ or $b = 0$.
4. If $a = b$ then $a^n = b^n$, where n is a positive integer.

Remember that applying Property 4 may result in some extraneous solutions, so you *must* check all potential solutions.

The cross-multiplication property of proportions (if $a/b = c/d$ then $ad = bc$) can be used to solve some equations.

Quadratic equations can be solved by (1) factoring, (2) completing the square, or (3) using the quadratic formula, which can be stated as

$$x = \frac{-b \pm \sqrt{b^2 - 4ac}}{2a}.$$

The discriminant of a quadratic equation, $b^2 - 4ac$, indicates the nature of the solutions of the equation:

1. If $b^2 - 4ac > 0$, the equation has two unequal real solutions;
2. If $b^2 - 4ac = 0$, the equation has one real solution;
3. If $b^2 - 4ac < 0$, the equation has two complex but nonreal solutions.

If x_1 and x_2 are the solutions of a quadratic equation $ax^2 + bx + c = 0$, then (1) $x_1 + x_2 = -b/a$ and (2) $x_1x_2 = c/a$. These relationships can be used to check potential solutions.

The property "If $|x| = k$, then $x = k$ or $x = -k$ $(k > 0)$" is often helpful for solving equations involving absolute value.

Solving Inequalities

The following properties form a basis for solving inequalities.

    1. If $a > b$ then $a + c > b + c$.

    2. If $a > b$ and $c > 0$, then $ac > bc$.

    3. If $a > b$ and $c < 0$, then $ac < bc$.

Quadratic inequalities such as $(x + 3)(x - 7) > 0$ can be solved by considering the "sign behavior" of the individual factors.

The following properties play an important role in solving inequalities involving absolute value.

    1. If $|x| < k$, where $k > 0$, then $-k < x < k$.

    2. If $|x| > k$, where $k > 0$, then $x > k$ or $x < -k$.

Problem Solving

It would be helpful for you to reread pages 80–82. Some key problem-solving suggestions are given on those pages.

Chapter 2 Review Problem Set

Solve each of the following equations.

1. $2(3x - 1) - 3(x - 2) = 2(x - 5)$

2. $\dfrac{n - 1}{4} - \dfrac{2n + 3}{5} = 2$

3. $\dfrac{2}{x + 2} + \dfrac{5}{x - 4} = \dfrac{7}{2x - 8}$

4. $0.07x + 0.12(550 - x) = 56$

5. $(3x - 1)^2 = 16$

6. $4x^2 - 29x + 30 = 0$

7. $x^2 - 6x + 10 = 0$

8. $n^2 + 4n = 396$

9. $15x^3 + x^2 - 2x = 0$

10. $\dfrac{t + 3}{t - 1} - \dfrac{2t + 3}{t - 5} = \dfrac{3 - t^2}{t^2 - 6t + 5}$

11. $\dfrac{5 - x}{2 - x} - \dfrac{3 - 2x}{2x} = 1$

12. $x^4 + 4x^2 - 45 = 0$

13. $2n^{-4} - 11n^{-2} + 5 = 0$

14. $\left(x - \dfrac{2}{x}\right)^2 + 4\left(x - \dfrac{2}{x}\right) = 5$

15. $\sqrt{5 + 2x} = 1 + \sqrt{2x}$

16. $\sqrt{3 + 2n} + \sqrt{2 - 2n} = 3$

17. $\sqrt{3 - t} - \sqrt{3 + t} = \sqrt{t}$

18. $|5x - 1| = 7$

19. $|2x + 5| = |3x - 7|$

20. $\left|\dfrac{-3}{n - 1}\right| = 4$

Solve each of the following inequalities. Express the solution sets using interval notation.

21. $3(2 - x) + 2(x - 4) > -2(x + 5)$

22. $\frac{3}{5}x - \frac{1}{3} \leq \frac{2}{3}x + \frac{3}{4}$

23. $\dfrac{n - 1}{3} - \dfrac{2n + 1}{4} > \dfrac{1}{6}$

24. $0.08x + 0.09(700 - x) \geq 59$

25. $-16 \leq 7x - 2 \leq 5$

26. $5 > \dfrac{3y + 4}{2} > 1$

27. $x^2 - 3x - 18 < 0$

28. $n^2 - 5n \geq 14$

29. $(x - 1)(x - 4)(x + 2) < 0$

30. $\dfrac{x + 4}{2x - 3} \leq 0$

31. $\dfrac{5n - 1}{n - 2} > 0$

32. $\dfrac{x - 1}{x + 3} \geq 2$

33. $\dfrac{t + 5}{t - 4} < 1$

34. $|4x - 3| > 5$

35. $|3x + 5| \leq 14$

36. $|-3 - 2x| < 6$

37. $\left|\dfrac{x - 1}{x}\right| > 2$

38. $\left|\dfrac{n + 1}{n + 2}\right| < 1$

Solve each of the following problems.

39. The sum of three consecutive odd integers is 31 less than four times the largest integer. Find the integers.

40. The sum of two numbers is 74. If the larger is divided by the smaller, the quotient is 7 and the remainder is 2. Find the numbers.

41. The perimeter of a rectangle is 38 centimeters and its area is 84 square centimeters. Find the dimensions of the rectangle.

42. A sum of money amounting to $13.55 consists of nickels, dimes, and quarters. There are three times as many dimes as nickels and three less quarters than dimes. How many coins of each denomination are there?

43. A retailer has some shirts that cost him $14 each. He wants to sell them so as to make a profit of 30% of the selling price. What price should he charge for the shirts?

44. How many gallons of a solution of glycerine and water containing 55% glycerine should be added to 15 gallons of a 20% solution to give a 40% solution?

45. The sum of the present ages of Rosie and her mother is 47 years. In 5 years, Rosie will be one-half as old as her mother at that time. Find the present ages of both Rosie and her mother.

46. Kelly invested $800, part of it at 9% and the remainder at 12%. Her total yearly interest from the two investments was $85.50. How much did she invest at each rate?

47. Regina has scores of 93, 88, 89, and 95 on her first four math exams. What score must she get on the fifth exam to have an average of 92 or better for the five exams?

48. At how many minutes after 2 P.M. will the minute hand of a clock overtake the hour hand?

49. Russ started to mow the lawn, a task that usually takes him 40 minutes. After he had been working for 15 minutes, his friend Jay came along with his mower and began to help Russ. Working together, they finished the lawn in 10 minutes. How long would it have taken Jay to mow the lawn by himself?

50. Barry bought a number of shares of stock for $600. A week later the value of the stock increased $3 per share and he sold all but 10 shares and regained his original investment of $600. How many shares did he sell and at what price per share?

51. Larry drove 156 miles in one hour more than it took Mike to drive 108 miles. Mike drove at an average rate of 2 miles per hour faster than Larry. How fast did each one travel?

52. It takes Bill 2 hours longer to do a certain job than it takes Cindy. They worked together for 2 hours; then Cindy left and Bill finished the job in 1 hour. How long would it take each of them to do the job alone?

3

Coordinate Geometry and Graphing Techniques

3.1 Coordinate Geometry

3.2 Graphing Techniques: Linear Equations and Inequalities

3.3 Determining the Equation of a Line

3.4 More on Graphing

3.5 Circles, Ellipses, and Hyperbolas

René Descartes, a French mathematician of the seventeenth century, was able to transform geometric problems into an algebraic setting so that he could use the tools of algebra to solve the problems. This merging of algebraic and geometric ideas is the foundation of a branch of mathematics called **analytic geometry**, today more commonly called **coordinate geometry**. Basically, there are two kinds of problems in coordinate geometry: (1) Given an algebraic equation, find its geometric graph; (2) Given a set of conditions pertaining to a geometric graph, find its algebraic equation. We will discuss problems of both types in this chapter.

3.1

Coordinate Geometry

Recall that the real number line (Figure 3.1) exhibits a **one-to-one correspondence** between the set of real numbers and the points on a line. That is to say, to each real number there corresponds one and only one point on the line, and to each point on the line there corresponds one and only one real number. The number that corresponds to a particular point on the line is called the **coordinate** of that point.

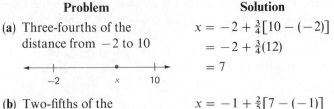

Figure 3.1

Suppose that on the number line we want to know the distance *from* -2 *to* 6. The "from–to" vocabulary implies a **directed distance**, which can be found to be $6 - (-2) = 8$ units. In other words, it is 8 units in a *positive direction* from -2 to 6. Likewise, the distance from 9 to -4 is $-4 - 9 = -13$, that is, 13 units in a *negative direction*. In general, if x_1 and x_2 are the coordinates of two points on the number line, then the distance **from x_1 to x_2** is given by $x_2 - x_1$, and the distance **from x_2 to x_1** is given by $x_1 - x_2$.

Now suppose that we want to find the distance *between* -2 and 6. The "between" vocabulary implies **distance without regard to direction**. Thus, the distance between -2 and 6 can be found by using either $|6 - (-2)| = 8$, or $|-2 - 6| = 8$. In general, if x_1 and x_2 are the coordinates of two points on the number line, the distance **between x_1 and x_2** can be found by using either $|x_2 - x_1|$ or $|x_1 - x_2|$.

Sometimes it is necessary to find the coordinate of a point located somewhere between two given points. For example, in Figure 3.2 suppose that we want to find the coordinate (x) of the point located two-thirds of the distance *from* 2 *to* 8. Since the total distance from 2 to 8 is $8 - 2 = 6$ units, we can start at 2 and move $\frac{2}{3}(6) = 4$ units toward 8. Thus,

$$x = 2 + \tfrac{2}{3}(6) = 2 + 4 = 6.$$

The following examples further illustrate the process of finding the coordinate of a point somewhere between two given points.

Figure 3.2

Problem	**Solution**
(a) Three-fourths of the distance from -2 to 10	$x = -2 + \frac{3}{4}[10 - (-2)]$ $= -2 + \frac{3}{4}(12)$ $= 7$
(b) Two-fifths of the distance from -1 to 7	$x = -1 + \frac{2}{5}[7 - (-1)]$ $= -1 + \frac{2}{5}(8)$ $= \frac{11}{5}$

(c) One-third of the distance from 9 to 1

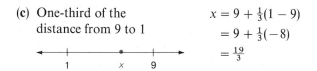

$$x = 9 + \tfrac{1}{3}(1 - 9)$$
$$= 9 + \tfrac{1}{3}(-8)$$
$$= \tfrac{19}{3}$$

(d) a/b of the distance from x_1 to x_2

$$x = x_1 + \frac{a}{b}(x_2 - x_1)$$

Problem (d) indicates that a general formula can be developed for this type of problem. However, it may be easier to remember the basic approach for doing such a problem than it is to memorize the formula.

As seen in Chapter 2, the real number line provides a geometric model for graphing solutions of algebraic equations and inequalities involving *one variable*. For example, the solutions of "$x > 2$ or $x \le -1$" are graphed in Figure 3.3.

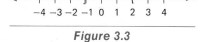

Figure 3.3

The Rectangular Coordinate System

To expand our work with coordinate geometry, we now consider two number lines, one vertical and one horizontal, perpendicular to each other at the point associated with zero on both lines (Figure 3.4). These number lines are referred to as the **horizontal** and **vertical axes**, or together as the **coordinate axes**. They partition the plane into four regions called **quadrants**. The quadrants are numbered counterclockwise from I through IV, as indicated in Figure 3.4. The point of intersection of the two axes is called the **origin**.

It is now possible to set up a one-to-one correspondence between *ordered pairs* of real numbers and the points in a plane. To each ordered pair of real numbers there corresponds a unique point in the plane and to each point in the plane there corresponds a unique ordered pair of real numbers. We have illustrated examples of this correspondence in Figure 3.5. The ordered pair $(3, 2)$ means that the point A is located three units to the right and two units up from the origin. The ordered pair $(-3, -5)$ means that point D is located three units to the left and five units down from the origin. The ordered pair $(0, 0)$ is associated with the origin.

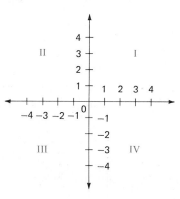

Figure 3.4

Remark: The notation $(-2, 4)$ was used in Chapter 2 to indicate an interval of the real number line. Now we are using the same notation to indicate an ordered pair of real numbers. This double meaning should not be confusing since the context of the material will definitely indicate the meaning being used at a particular time. Throughout this chapter we will be using the ordered-pair interpretation.

In general, the real numbers a and b in the ordered pair (a, b) are associated with a point; they are referred to as the **coordinates of the point**. The first number, a, is called the **abscissa**; it is the directed distance of the point from the vertical axis, measuring parallel to the horizontal axis. The second number, b, is called the **ordinate**; it is the directed distance from the horizontal axis, measuring

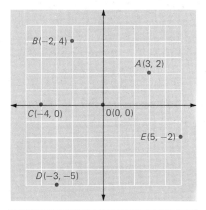

Figure 3.5

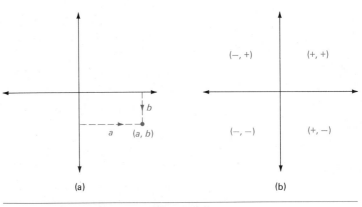

Figure 3.6

parallel to the vertical axis (Figure 3.6(a)). Thus, in the first quadrant, all points have a positive abscissa and a positive ordinate. In the second quadrant, all points have a negative abscissa and a positive ordinate. We have indicated the sign situations for all four quadrants in Figure 3.6(b). This system of associating points in a plane with pairs of real numbers is called the **rectangular coordinate system** or the **Cartesian coordinate system**.

The Distance between Two Points

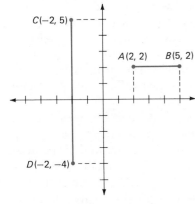

Figure 3.7

As we work with the rectangular coordinate system, it is sometimes necessary to express the length of certain line segments. In other words, we need to be able to find the *distance* between two points. Let's first consider two specific examples and then develop a general distance formula.

EXAMPLE 1

Find the distance between the points $A(2, 2)$ and $B(5, 2)$ and also between the points $C(-2, 5)$ and $D(-2, -4)$.

Solution Let's plot the points and draw $\overline{AB}$ and $\overline{CD}$ as in Figure 3.7. (The symbol $\overline{AB}$ denotes the line segment with endpoints A and B.) Since $\overline{AB}$ is parallel to the horizontal axis, its length can be expressed as $|5 - 2|$ or $|2 - 5|$. Thus, the length of $\overline{AB}$ (we shall use the notation AB to represent the length of $\overline{AB}$) is $AB = 3$ units. Likewise, since $\overline{CD}$ is parallel to the vertical axis, we obtain $CD = |5 - (-4)| = 9$ units.

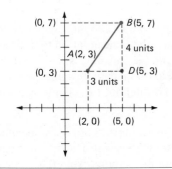

Figure 3.8

EXAMPLE 2

Find the distance between the points $A(2, 3)$ and $B(5, 7)$.

Solution Let's plot the points and form a right triangle using point D, as indicated in Figure 3.8. Notice that the coordinates of point D are $(5, 3)$. Because $\overline{AD}$ is parallel to the horizontal axis, as in Example 1, we have $AD = |5 - 2| = 3$ units. Likewise, $\overline{DB}$ is parallel to the vertical axis and therefore $DB = |7 - 3| = 4$

units. Applying the Pythagorean theorem, we obtain

$$(AB)^2 = (AD)^2 + (DB)^2$$
$$= 3^2 + 4^2$$
$$= 9 + 16$$
$$= 25.$$

Thus

$$AB = \sqrt{25} = 5 \text{ units.}$$

The approach used in Example 2 can be used to develop a general distance formula. However, before doing this let's make another notational agreement. For most problems in coordinate geometry it is customary to label the horizontal axis the **x-axis** and the vertical axis the **y-axis**. Then, ordered pairs representing points in the xy-plane are of the form (x, y); that is, x is the first coordinate and y is the second coordinate. Now let's develop a general distance formula.

Let $P_1(x_1, y_1)$ and $P_2(x_2, y_2)$ represent any two points in the xy-plane. Form a right triangle using point R, as indicated in Figure 3.9. The coordinates of the vertex of the right angle, point R, are (x_2, y_1). The length of $\overline{P_1 R}$ is $|x_2 - x_1|$ and the length of $\overline{RP_2}$ is $|y_2 - y_1|$. Letting d represent the length of $\overline{P_1 P_2}$ and applying the Pythagorean theorem, we obtain

$$d^2 = |x_2 - x_1|^2 + |y_2 - y_1|^2.$$

Since $|a|^2 = a^2$ for any real number a, the distance formula can be stated

$$d = \sqrt{(x_2 - x_1)^2 + (y_2 - y_1)^2}.$$

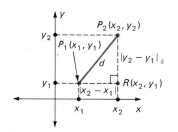

Figure 3.9

It makes no difference which point you call P_1 or P_2 when using the formula. Also, remember that if you forget the formula, there is no need to panic: merely form a right triangle and apply the Pythagorean theorem, as we did in Example 2.

Let's consider some examples illustrating the use of the distance formula.

EXAMPLE 3

Find the distance between $(-2, 5)$ and $(1, -1)$.

Solution Let $(-2, 5)$ be P_1 and $(1, -1)$ be P_2. Using the distance formula, we obtain

$$d = \sqrt{(x_2 - x_1)^2 + (y_2 - y_1)^2}$$
$$= \sqrt{[1 - (-2)]^2 + (-1 - 5)^2}$$
$$= \sqrt{3^2 + (-6)^2}$$
$$= \sqrt{9 + 36}$$
$$= \sqrt{45} = 3\sqrt{5}.$$

The distance between the two points is $3\sqrt{5}$ units.

In Example 3, notice the simplicity of the approach when we use the distance formula. No diagram was needed; we merely "plugged in" the values and did the computation. However, many times a figure *is* helpful in the analysis of the problem, as illustrated in the next example.

EXAMPLE 4

Verify that the points $(-3, 6)$, $(3, 4)$, and $(1, -2)$ are vertices of an isosceles triangle. (An isosceles triangle has two sides of the same length.)

Solution Let's plot the points and draw the triangle (Figure 3.10). The lengths d_1, d_2, and d_3 can all be found by using the distance formula.

$$d_1 = \sqrt{(3-1)^2 + [4-(-2)]^2}$$
$$= \sqrt{4 + 36}$$
$$= \sqrt{40} = 2\sqrt{10}$$

$$d_2 = \sqrt{(-3-3)^2 + (6-4)^2}$$
$$= \sqrt{36 + 4}$$
$$= \sqrt{40} = 2\sqrt{10}$$

$$d_3 = \sqrt{(-3-1)^2 + [6-(-2)]^2}$$
$$= \sqrt{16 + 64}$$
$$= \sqrt{80} = 4\sqrt{5}$$

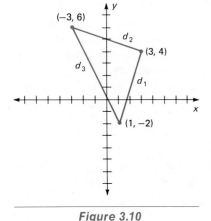

Figure 3.10

Since $d_1 = d_2$, it is an isosceles triangle.

Points of Division of a Line Segment

Earlier in this section we discussed the process of finding the coordinate of a point on a number line, given that it is located somewhere between two other points on the line. This same type of problem can occur in the *xy*-plane, and the approach used earlier can be extended to handle it. Let's consider some examples.

EXAMPLE 5

Find the coordinates of the point P that is two-thirds of the distance from $A(1, 2)$ to $B(7, 5)$.

Solution In Figure 3.11 we have plotted the given points A and B and completed a figure to help with the analysis of the problem. To find the coordinates of point P, we can proceed as follows. Point D is two-thirds of the distance from A to C because parallel lines cut off proportional segments on every transversal that intersects the lines. Therefore, since $\overline{AC}$ is parallel to the *x*-axis, it can be treated as a segment of the number line. Thus, we have

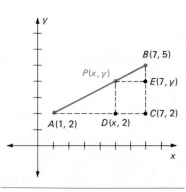

Figure 3.11

$$x = 1 + \tfrac{2}{3}(7 - 1) = 1 + \tfrac{2}{3}(6) = 5.$$

Similarly, $\overline{CB}$ is parallel to the y-axis, so it can also be treated as a segment of the number line. Thus we obtain

$$y = 2 + \tfrac{2}{3}(5 - 2)$$
$$= 2 + \tfrac{2}{3}(3)$$
$$= 4.$$

The point P has coordinates $(5, 4)$.

EXAMPLE 6

Find the coordinates of the midpoint of the line segment determined by the points $P_1(x_1, y_1)$ and $P_2(x_2, y_2)$.

Solution Figure 3.12 helps with the analysis of the problem. The line segment $\overline{P_1 R}$ is parallel to the x-axis and $S(x, y_1)$ is the midpoint of $\overline{P_1 R}$. Thus, we can determine the x-coordinate of S.

$$x = x_1 + \frac{1}{2}(x_2 - x_1)$$

$$= x_1 + \frac{1}{2}x_2 - \frac{1}{2}x_1$$

$$= \frac{1}{2}x_1 + \frac{1}{2}x_2$$

$$= \frac{x_1 + x_2}{2}.$$

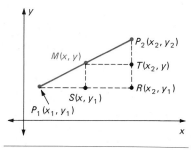

Figure 3.12

Similarly, $\overline{RP_2}$ is parallel to the y-axis and $T(x_2, y)$ is the midpoint of $\overline{RP_2}$. Therefore we can calculate the y-coordinate of T.

$$y = y_1 + \frac{1}{2}(y_2 - y_1)$$

$$= y_1 + \frac{1}{2}y_2 - \frac{1}{2}y_1$$

$$= \frac{1}{2}y_1 + \frac{1}{2}y_2$$

$$= \frac{y_1 + y_2}{2}.$$

Thus, the coordinates of the midpoint of a line segment determined by $P_1(x_1, y_1)$ and $P_2(x_2, y_2)$ are given by the following.

$$\left(\frac{x_1 + x_2}{2}, \frac{y_1 + y_2}{2}\right)$$

EXAMPLE 7

Find the coordinates of the midpoint of the line segment determined by the points $(-2, 4)$ and $(6, -1)$.

Solution Using the midpoint formula, we obtain

$$\left(\frac{x_1 + x_2}{2}, \frac{y_1 + y_2}{2}\right) = \left(\frac{-2 + 6}{2}, \frac{4 + (-1)}{2}\right)$$

$$= \left(\frac{4}{2}, \frac{3}{2}\right)$$

$$= \left(2, \frac{3}{2}\right).$$

We want to emphasize a couple of ideas from Examples 5, 6, and 7. If we want to find a point of division of a line segment, we will use the same approach as in Example 5. However, for the special case of the midpoint, the formula developed in Example 6 is convenient to use.

Problem Set 3.1

On a number line, find the following indicated distances.

1. From -4 to 6
2. From 5 to -14
3. From -6 to -11
4. From -7 to 10
5. Between -2 and 4
6. Between -4 and -12
7. Between 5 and -10
8. Between -2 and 13

For each of the following, find the coordinate of the indicated point on a number line.

9. Two-thirds of the distance from 1 to 10
10. Three-fourths of the distance from -2 to 14
11. One-third of the distance from -3 to 7
12. Two-fifths of the distance from -5 to 6
13. Three-fifths of the distance from -1 to -11
14. Five-sixths of the distance from 3 to -7

For each of the following, find the length of $\overline{AB}$ and the midpoint of $\overline{AB}$.

15. $A(2, 1)$, $B(10, 7)$ **16.** $A(-2, -1)$, $B(7, 11)$ **17.** $A(1, -1)$, $B(3, -4)$
18. $A(-5, 2)$, $B(-1, 6)$ **19.** $A(6, -4)$, $B(9, -7)$ **20.** $A(-3, 3)$, $B(0, -3)$
21. $A(\frac{1}{2}, \frac{1}{3})$, $B(-\frac{1}{3}, \frac{3}{2})$ **22.** $A(-\frac{3}{4}, 2)$, $B(-1, -\frac{5}{4})$

For Problems 23–28, find the coordinates of the indicated point in the xy-plane.

23. One-third of the distance from $(2, 3)$ to $(5, 9)$
24. Two-thirds of the distance from $(1, 4)$ to $(7, 13)$
25. Two-fifths of the distance from $(-2, 1)$ to $(8, 11)$
26. Three-fifths of the distance from $(2, -3)$ to $(-3, 8)$
27. Five-eighths of the distance from $(-1, -2)$ to $(4, -10)$
28. Seven-eighths of the distance from $(-2, 3)$ to $(-1, -9)$

Solve each of the following problems.

29. Find the coordinates of the point that is one-fourth of the distance from $(2, 4)$ to $(10, 13)$ by (a) using the midpoint formula twice, and (b) using the same approach as used for Problems 23–28.

30. If one endpoint of a line segment is $(-6, 4)$ and the midpoint of the segment is $(-2, 7)$, find the other endpoint.

31. Use the distance formula to verify that the points $(-2, 7)$, $(2, 1)$, and $(4, -2)$ lie on a straight line.

32. Use the distance formula to verify that the points $(-3, 8)$, $(7, 4)$, and $(5, -1)$ are vertices of a right triangle.

33. Verify that the points $(0, 3)$, $(2, -3)$, and $(-4, -5)$ are vertices of an isosceles triangle.

34. Verify that the points $(7, 12)$ and $(11, 18)$ divide the line segment joining $(3, 6)$ and $(15, 24)$ into three segments of equal length.

35. Find the perimeter of the triangle whose vertices are $(-6, -4)$, $(0, 8)$, and $(6, 5)$.

36. Verify that $(-4, 9)$, $(8, 4)$, $(3, -8)$, and $(-9, -3)$ are vertices of a square.

37. Verify that the points $(4, -5)$, $(6, 7)$, and $(-8, -3)$ lie on a circle having its center at $(-1, 2)$.

38. Suppose that $(-2, 5)$, $(6, 3)$, and $(-4, -1)$ are three vertices of a parallelogram. How many possibilities are there for the fourth vertex? Find the coordinates of each of those points. (Hint: The diagonals of a parallelogram bisect each other.)

39. Find x such that the line segment determined by $(x, -2)$ and $(-2, -14)$ is 13 units long.

40. Consider the triangle whose vertices are $(4, -6)$, $(2, 8)$, and $(-4, 2)$. Verify that the medians of this triangle intersect at a point that is two-thirds of the distance from a vertex to the midpoint of the opposite side. (**A median** of a triangle is the line segment determined by a vertex and the midpoint of the opposite side. Every triangle has three medians.)

41. Consider the line segment determined by $A(-1, 2)$ and $B(5, 11)$. Find the coordinates of a point P such that $AP/PB = 2/1$.

Miscellaneous Problems

42. The tools of coordinate geometry can be used to prove various geometric properties. For example, consider the following way of proving that the diagonals of a rectangle are equal in length.

First we draw a rectangle and coordinatize it by using a convenient position for the origin. Now we can use the distance formula to find the lengths of the diagonals $\overline{AC}$ and $\overline{BD}$.

$$AC = \sqrt{(l - 0)^2 + (w - 0)^2} = \sqrt{l^2 + w^2}$$
$$BD = \sqrt{(0 - l)^2 + (w - 0)^2} = \sqrt{l^2 + w^2}$$

Thus, $AC = BD$, and we have proven that the diagonals are equal in length.

Prove each of the following.

(a) The diagonals of an isosceles trapezoid are equal in length.

(b) The line segment joining the midpoints of two sides of a triangle is equal in length to one-half of the third side.

(c) The midpoint of the hypotenuse of a right triangle is equally distant from all three vertices.

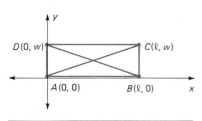

Figure for Exercise 42

(d) The diagonals of a parallelogram bisect each other.

(e) The line segments joining the midpoints of the opposite sides of a quadrilateral bisect each other.

(f) The medians of a triangle intersect at a point that is two-thirds of the distance from a vertex to the midpoint of the opposite side. (See Problem 40.)

3.2

Graphing Techniques: Linear Equations and Inequalities

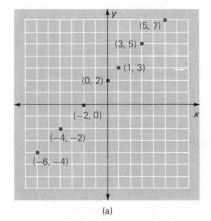

(a)

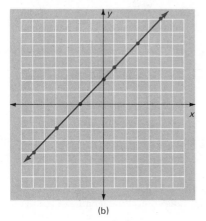

(b)

Figure 3.13

As you continue to study mathematics, you will find that the ability to sketch the graph of an equation quickly is an important skill. Thus, throughout precalculus and calculus courses, various curve-sketching techniques are discussed. We will use a good portion of this chapter to expand your repertoire of graphing techniques.

First, let's briefly review some basic ideas by considering the solutions for the equation $y = x + 2$. A **solution** of an equation in two variables is an ordered pair of real numbers that satisfy the equation. When using the variables x and y, the ordered pairs are of the form (x, y). We see that $(1, 3)$ is a solution for $y = x + 2$ because if x is replaced by 1 and y by 3, the true numerical statement $3 = 1 + 2$ is obtained. Likewise, $(-2, 0)$ is a solution because $0 = -2 + 2$ is a true statement. An infinite number of pairs of real numbers that satisfy $y = x + 2$ can be found by arbitrarily choosing values for x and, for each value of x chosen, by determining a corresponding value for y. Let's use a table to record some of the solutions for $y = x + 2$.

CHOOSE x	DETERMINE y From $y = x + 2$	SOLUTIONS FOR $y = x + 2$
0	2	(0, 2)
1	3	(1, 3)
3	5	(3, 5)
5	7	(5, 7)
−2	0	(−2, 0)
−4	−2	(−4, −2)
−6	−4	(−6, −4)

Plotting the points associated with the ordered pairs from the table produces Figure 3.13(a). The straight line containing the points (Figure 3.13(b)) is called the **graph of the equation** $y = x + 2$.

Graphing Linear Equations

Probably the most valuable graphing technique is the ability to recognize the kind of graph that is produced by a particular type of equation. For example, from previous mathematics courses you may remember that any equation of the form $Ax + By = C$, where A, B, and C are constants (A and B not both zero) and x and y are variables, is a **linear equation** and its graph is a **straight line**. Two comments about this description of a linear equation should be made.

First, the choice of x and y as variables is arbitrary; any two letters can be used to represent the variables. For example, an equation such as $3r + 2s = 9$ is also a linear equation in two variables. In order to avoid constantly changing the labeling of the coordinate axes when graphing equations, we will use the same two variables, x and y, in all equations. Second, the statement "any equation of the form $Ax + By = C$" technically means any equation of that form or equivalent to that form. For example, the equation $y = 2x - 1$ is equivalent to $-2x + y = -1$ and therefore is linear and produces a straight-line graph.

Before graphing some linear equations, let's give a general definition of the *intercepts* of a graph.

The x-coordinates of the points that a graph has in common with the x-axis are called the **x-intercepts** of the graph. (To compute the x-intercepts, let $y = 0$ and solve for x.)

The y-coordinates of the points that a graph has in common with the y-axis are called the **y-intercepts** of the graph. (To compute the y-intercepts, let $x = 0$ and solve for y.)

Once we know that any equation of the form $Ax + By = C$ produces a straight-line graph, along with the fact that two points determine a straight line, graphing linear equations becomes a simple process. We can find two points on the graph and draw the line determined by those two points. Usually the two points involving the intercepts are easy to find, and generally it's a good idea to plot a third point to serve as a check.

EXAMPLE 1

Graph $3x - 2y = 6$.

Solution First, let's find the intercepts. If $x = 0$, then

$$3(0) - 2y = 6$$
$$-2y = 6$$
$$y = -3.$$

Therefore, the point $(0, -3)$ is on the line. If $y = 0$, then

$$3x - 2(0) = 6$$
$$3x = 6$$
$$x = 2.$$

Thus, the point $(2, 0)$ is also on the line. Now let's find a check point. If $x = -2$, then

$$3(-2) - 2y = 6$$
$$-6 - 2y = 6$$
$$-2y = 12$$
$$y = -6.$$

So the point $(-2, -6)$ is also on the line.

In Figure 3.14, the three points are plotted and the graph of $3x - 2y = 6$ is drawn.

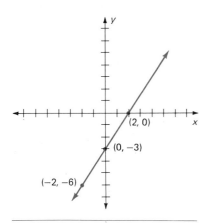

Figure 3.14

x	y
0	0
1	−2
−1	2

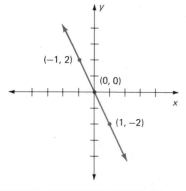

Figure 3.15

Notice in Example 1 that we did not solve the given equation for y in terms of x or for x in terms of y. Since we know it is a straight line, there is no need for an extensive table of values; thus, there is no need to change the form of the original equation. Furthermore, the point $(-2, -6)$ served as a check point. If it had not been on the line determined by the two intercepts, then we would have known that an error had been made in finding the intercepts.

EXAMPLE 2

Graph $y = -2x$.

Solution If $x = 0$, then $y = -2(0) = 0$; so the origin $(0, 0)$ is on the line. Since both intercepts are determined by the point $(0, 0)$, another point is necessary to determine the line. Then a third point should be found as a check point. The graph of $y = -2x$ is shown in Figure 3.15.

Example 2 illustrates the general concept that for the form $Ax + By = C$, if $C = 0$ then the line contains the origin. Stated another way, **the graph of any equation of the form $y = kx$, where k is any real number, is a straight line containing the origin**.

EXAMPLE 3

Graph $x = 2$.

Solution Since we are considering linear equations *in two variables*, the equation $x = 2$ is equivalent to $x + 0(y) = 2$. Any value of y can be used, but the x-value must always be 2. Therefore, some of the solutions are $(2, 0), (2, 1), (2, 2)$, $(2, -1)$, and $(2, -2)$. The graph of $x = 2$ is the vertical line shown in Figure 3.16.

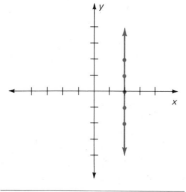

Figure 3.16

In general, the graph of any equation of the form $Ax + By = C$, where $A = 0$ or $B = 0$ (not both), is a line parallel to one of the axes. More specifically, **any equation of the form $x = a$, where a is any real number, is a line parallel to the y-axis** having an x-intercept of a. **Any equation of the form $y = b$, where b is a real number, is a line parallel to the x-axis** having a y-intercept of b.

Graphing Linear Inequalities

Linear inequalities in two variables are of the form $Ax + By > C$ or $Ax + By < C$, where A, B, and C are real numbers. (**Combined linear equality and inequality statements** are of the form $Ax + By \geq C$ or $Ax + By \leq C$.) Graphing linear inequalities is almost as easy as graphing linear equations. The following discussion will lead us to a simple, step-by-step process.

Let's consider the following equation and related inequalities.

$$x + y = 2, \qquad x + y > 2, \qquad x + y < 2$$

The straight line in Figure 3.17 is the graph of $x + y = 2$. The line divides the plane into two **half-planes**, one above the line and one below the line. For each point in the half-plane *above* the line, the ordered pair (x, y) associated

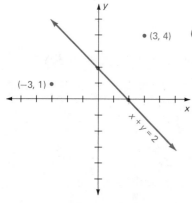

Figure 3.17

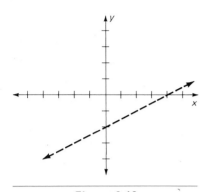

Figure 3.18

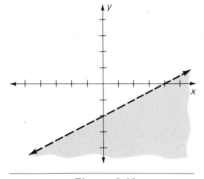

Figure 3.19

with the point satisfies the inequality $x + y > 2$. For example, the ordered pair $(3, 4)$ produces the true statement $3 + 4 > 2$. Likewise, for each point in the half-plane *below* the line, the ordered pair (x, y) associated with the point satisfies the inequality $x + y < 2$. For example, $(-3, 1)$ produces the true statement $-3 + 1 < 2$.

Now let's use these ideas from the previous discussion to help graph some inequalities.

EXAMPLE 4

Graph $x - 2y > 4$.

Solution First, graph $x - 2y = 4$ as a dashed line since equality is not included in $x - 2y > 4$ (Figure 3.18). Second, since *all* of the points in a specific half-plane satisfy either $x - 2y > 4$ or $x - 2y < 4$, let's try a *test point*. For example, try the origin.

$$x - 2y > 4 \quad \text{becomes} \quad 0 - 2(0) > 4, \quad \text{which is a false statement.}$$

Because the ordered pairs in the half-plane containing the origin do not satisfy $x - 2y > 4$, the ordered pairs in the other half-plane must satisfy it. Therefore, the graph of $x - 2y > 4$ is the half-plane *below* the line, as indicated by the shaded portion in Figure 3.19.

To graph a linear inequality, the following steps are suggested.

1. Graph the corresponding equality. Use a solid line if equality is included in the original statement and a dashed line if equality is not included.

2. Choose a *test point* not on the line and substitute its coordinates into the inequality. (The origin is a convenient point if it is not on the line.)

3. The graph of the original inequality is
   (a) the half-plane containing the test point if the inequality is satisfied by that point; or
   (b) the half-plane not containing the test point if the inequality is not satisfied by the point.

EXAMPLE 5

Graph $2x + 3y \geq -6$.

Solution

Step 1. Graph $2x + 3y = -6$ as a solid line (Figure 3.20).

Step 2. Choose the origin as a test point.

$$2x + 3y \geq -6 \quad \text{becomes} \quad 2(0) + 3(0) \geq -6, \quad \text{which is true.}$$

Step 3. Since the test point satisfies the given inequality, all points in the same half-plane as the test point satisfy it. Thus, the graph of $2x + 3y \geq -6$ is the line and the half-plane above the line (Figure 3.20).

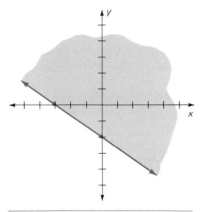

Figure 3.20

Problem Set 3.2

Graph each of the following linear equations.

1. $x - 2y = 4$ **2.** $2x + y = -4$ **3.** $3x + 2y = 6$ **4.** $2x - 3y = 6$

5. $4x - 5y = 20$ **6.** $5x + 4y = 20$ **7.** $x - y = 3$ **8.** $-x + y = 4$

9. $y = 3x - 1$ **10.** $y = -2x + 3$ **11.** $y = -x$ **12.** $y = 4x$

13. $x = 0$ **14.** $y = -1$ **15.** $y = \frac{2}{3}x$ **16.** $y = -\frac{1}{2}x$

Graph each of the following linear inequalities.

17. $x + 2y > 4$ **18.** $2x - y < -4$ **19.** $3x - 2y < 6$ **20.** $2x + 3y < 6$

21. $2x + 5y \leq 10$ **22.** $4x + 5y \leq 20$ **23.** $y > -x - 1$ **24.** $y < 3x - 2$

25. $y \leq -x$ **26.** $y \geq x$ **27.** $x + 2y < 0$ **28.** $3x - y > 0$

29. $x > -1$ **30.** $y < 3$

Miscellaneous Problems

From our work with absolute value, we know that $|x + y| = 4$ is equivalent to $x + y = 4$ or $x + y = -4$. Therefore, the graph of $|x + y| = 4$ is the two lines $x + y = 4$ and $x + y = -4$. Graph each of the following.

31. $|x - y| = 2$ **32.** $|2x + y| = 1$ **33.** $|x - 2y| \leq 4$

34. $|3x - 2y| \geq 6$ **35.** $|2x + 3y| > 6$ **36.** $|5x + 2y| < 10$

Using the definition of absolute value, the equation $y = |x| + 2$ becomes $y = x + 2$ for $x \geq 0$ and $y = -x + 2$ for $x < 0$. Therefore, the graph of $y = |x| + 2$ is as shown at left. Graph each of the following.

37. $y = |x| - 1$ **38.** $y = |x - 2|$ **39.** $|y| = x$

40. $|y| = |x|$ **41.** $y = 2|x|$ **42.** $|x| + |y| = 4$

3.3

Determining the Equation of a Line

As we stated earlier, there are basically two types of problems in coordinate geometry: (1) Given an algebraic equation, find its geometric graph; (2) Given a set of conditions pertaining to a geometric figure, find its algebraic equation. In the previous section, we considered some type 1 problems; that is, we did some graphing. Now we want to consider some type 2 problems that deal specifically with straight lines. In other words, given certain facts about a line, we need to be able to determine its algebraic equation.

As we work with straight lines, it is often helpful to be able to refer to the "steepness" or "slant" of a particular line. The concept of *slope* is used as a measure of the slant of a line. The **slope** of a line is the ratio of the vertical change of distance compared to the horizontal change of distance as we move from one point on a line to another. Consider the line in Figure 3.21. From point A to point B there is a vertical change of two units and a horizontal change of three units; therefore, the slope of the line is $\frac{2}{3}$.

A precise definition for slope can be given by considering the coordinates of the points P_1, P_2, and R in Figure 3.22. The horizontal change of distance as we

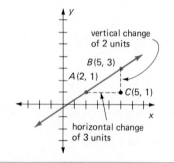

Figure 3.21

move from P_1 to P_2 is $x_2 - x_1$, and the vertical change is $y_2 - y_1$. Thus, the following definition is given.

DEFINITION 3.1

If P_1 and P_2 are any two different points on a line, P_1 with coordinates (x_1, y_1) and P_2 with coordinates (x_2, y_2), then the **slope** of the line (denoted by m) is

$$m = \frac{y_2 - y_1}{x_2 - x_1}, \qquad x_2 \neq x_1.$$

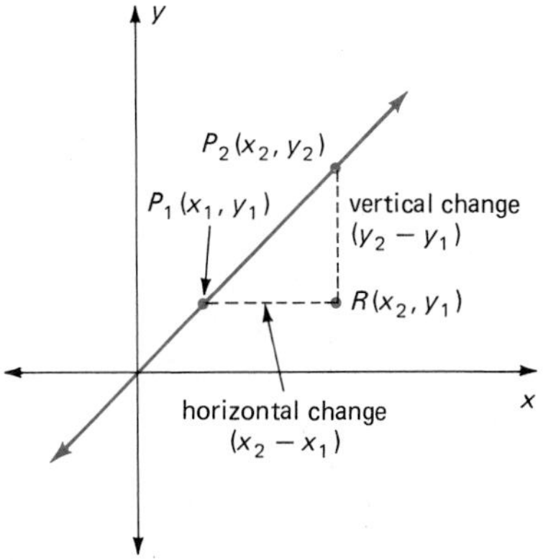

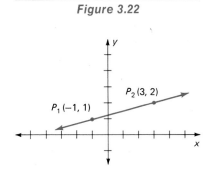

Figure 3.22

Since $\dfrac{y_2 - y_1}{x_2 - x_1} = \dfrac{y_1 - y_2}{x_1 - x_2}$, how we designate P_1 and P_2 is not important. Let's use Definition 3.1 to find the slopes of some lines.

EXAMPLE 1

Find the slope of the line determined by each of the following pairs of points and graph each line.

(a) $(-1, 1)$ and $(3, 2)$ (b) $(4, -2)$ and $(-1, 5)$ (c) $(2, -3)$ and $(-3, -3)$

Solutions

(a) Let $(-1, 1)$ be P_1 and $(3, 2)$ be P_2 (Figure 3.23).

$$m = \frac{y_2 - y_1}{x_2 - x_1} = \frac{2 - 1}{3 - (-1)} = \frac{1}{4}$$

(b) Let $(4, -2)$ be P_1 and $(-1, 5)$ be P_2 (Figure 3.24).

$$m = \frac{5 - (-2)}{-1 - 4} = \frac{7}{-5} = -\frac{7}{5}$$

(c) Let $(2, -3)$ be P_1 and $(-3, -3)$ be P_2 (Figure 3.25).

$$m = \frac{-3 - (-3)}{-3 - 2} = \frac{0}{-5} = 0$$

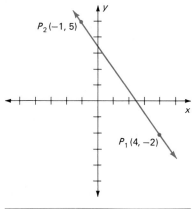

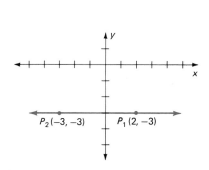

Figure 3.23

Figure 3.24

Figure 3.25

The three parts of Example 1 illustrate the three basic possibilities for slope; that is, the slope of a line can be positive, negative, or zero. A line having a positive slope rises as we move from left to right, as in Figure 3.23. A line having a negative slope falls as we move from left to right, as in Figure 3.24. A horizontal line, as in Figure 3.25, has a slope of zero. Finally, we need to realize that the concept of **slope is undefined for vertical lines**. This is due to the fact that for any vertical line, the horizontal change is zero as we move from one point on the line to another. Thus, the ratio $(y_2 - y_1)/(x_2 - x_1)$ will have a denominator of zero and be undefined. So in Definition 3.1, the restriction $x_2 \neq x_1$ is made.

Don't forget that **the slope of a line is a ratio**, the ratio of vertical change compared to horizontal change. For example, a slope of $\frac{2}{3}$ means that for every two units of vertical change there must be a corresponding three units of horizontal change.

Now let's consider some techniques for determining the equation of a line when given certain facts about the line.

EXAMPLE 2

Find the equation of the line having a slope of $\frac{2}{5}$ and containing the point $(3, 1)$.

Solution First, let's draw the line and record the given information (Figure 3.26). Then choose a point (x, y) that represents any point on the line other than the given point $(3, 1)$. The slope determined by $(3, 1)$ and (x, y) is to be $\frac{2}{5}$. Thus,

$$\frac{y - 1}{x - 3} = \frac{2}{5}$$

$$2(x - 3) = 5(y - 1)$$

$$2x - 6 = 5y - 5$$

$$2x - 5y = 1.$$

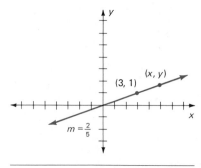

Figure 3.26

EXAMPLE 3

Find the equation of the line determined by $(1, -2)$ and $(-3, 4)$.

Solution First, let's draw the line determined by the two given points (Figure 3.27). These two points determine the slope of the line.

$$m = \frac{4 - (-2)}{-3 - 1} = \frac{6}{-4} = -\frac{3}{2}$$

Now we can use the same approach as in Example 2. Form an equation using one of the two given points, a point (x, y), and the slope of $-\frac{3}{2}$.

$$\frac{y + 2}{x - 1} = \frac{3}{-2}$$

$$3(x - 1) = -2(y + 2)$$

$$3x - 3 = -2y - 4$$

$$3x + 2y = -1$$

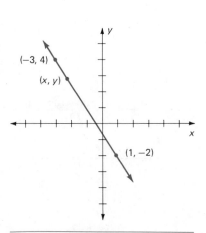

Figure 3.27

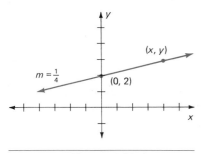

Figure 3.28

EXAMPLE 4

Find the equation of the line that has a slope of $\frac{1}{4}$ and a y-intercept of 2.

Solution A y-intercept of 2 means that the point $(0, 2)$ is on the line (Figure 3.28). Choosing a point (x, y), we can proceed as in the previous examples.

$$\frac{y - 2}{x - 0} = \frac{1}{4}$$

$$1(x - 0) = 4(y - 2)$$
$$x = 4y - 8$$
$$x - 4y = -8$$

At this point you might pause for a moment and look back over Examples 2, 3, and 4. Notice that the same basic approach is used in all three examples, that is, choosing a point (x, y) and using it to determine the equation that satisfies the conditions stated in the problem. This same approach will be used later with figures other than straight lines. Furthermore, you should realize that this approach can be used to develop some general forms of equations of straight lines.

Point–Slope Form

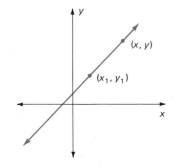

Figure 3.29

EXAMPLE 5

Find the equation of the line having a slope of m and containing the point (x_1, y_1).

Solution Choosing (x, y) to represent another point on the line (Figure 3.29), the slope of the line is given by

$$m = \frac{y - y_1}{x - x_1}, \qquad x \neq x_1,$$

from which we obtain

$$y - y_1 = m(x - x_1).$$

We refer to the equation

$$y - y_1 = m(x - x_1)$$

as the **point–slope form** of the equation of a straight line. Therefore, instead of using the approach of Example 2, we can substitute information into the point–slope form to write the equation of a line with a given slope and containing a given point. For example, the equation of the line having a slope of $\frac{3}{5}$ and containing the point $(2, 4)$ can be determined this way. We substitute $(2, 4)$ for (x_1, y_1) and $\frac{3}{5}$ for m in the point–slope equation.

$$y - 4 = \frac{3}{5}(x - 2)$$
$$5(y - 4) = 3(x - 2)$$
$$5y - 20 = 3x - 6$$
$$-14 = 3x - 5y$$

Slope–Intercept Form

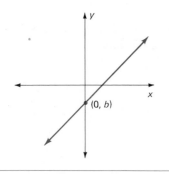

Figure 3.30

EXAMPLE 6

Find the equation of the line having a slope of m and a y-intercept of b.

Solution A y-intercept of b means that $(0, b)$ is on the line (Figure 3.30). Therefore, using the point–slope form with $(x_1, y_1) = (0, b)$, we obtain

$$y - y_1 = m(x - x_1)$$
$$y - b = m(x - 0)$$
$$y - b = mx$$
$$y = mx + b$$

We refer to the equation

$$y = mx + b$$

as the **slope–intercept form** of the equation of a straight line. It can be used for two primary purposes, as the next two examples illustrate.

EXAMPLE 7

Find the equation of the line that has a slope of $\frac{1}{4}$ and a y-intercept of 2.

Solution This is a restatement of Example 4, but this time we will use the slope–intercept form ($y = mx + b$) of the equation of a line to write its equation. Since $m = \frac{1}{4}$ and $b = 2$, we obtain

$$y = mx + b$$
$$y = \tfrac{1}{4}x + 2$$
$$4y = x + 8$$
$$-8 = x - 4y \qquad \text{Same result as in Example 4}$$

EXAMPLE 8

Find the slope and y-intercept of the line having an equation $2x - 3y = 7$.

Solution We can solve the equation for y in terms of x and then compare this result to the general slope–intercept form.

$$2x - 3y = 7$$
$$-3y = -2x + 7$$
$$y = \tfrac{2}{3}x - \tfrac{7}{3} \qquad y = mx + b$$

The slope of the line is $\frac{2}{3}$ and the y-intercept is $-\frac{7}{3}$.

In general, **if the equation of a nonvertical line is written in slope–intercept form, the coefficient of x is the slope of the line and the constant term is the y-intercept.**

Parallel and Perpendicular Lines

Since the concept of slope is used to indicate the slant of a line, it seems reasonable to expect slope to be related to the concepts of parallelism and perpendicularity. Such is the case, and the following two properties summarize this link.

PROPERTY 3.1

If two lines have slopes of m_1 and m_2, then

1. The two lines are parallel if and only if $m_1 = m_2$;
2. The two lines are perpendicular if and only if $m_1 m_2 = -1$.

We will test your ingenuity in devising proofs of these properties in the next problem set; however, for now we'll illustrate their use.

EXAMPLE 9

(a) Verify that the graphs of $3x + 2y = 9$ and $6x + 4y = 19$ are parallel lines.
(b) Verify that the graphs of $5x - 3y = 12$ and $3x + 5y = 27$ are perpendicular lines.

Solutions

(a) Let's change each equation to slope–intercept form.

$$3x + 2y = 9 \longrightarrow 2y = -3x + 9$$
$$y = -\tfrac{3}{2}x + \tfrac{9}{2}$$

$$6x + 4y = 19 \longrightarrow 4y = -6x + 19$$
$$= -\tfrac{6}{4}x + \tfrac{19}{4}$$
$$= -\tfrac{3}{2}x + \tfrac{19}{4}$$

Both lines have the same slope but different y-intercepts. Therefore, the two lines are parallel.

(b) Change each equation to slope–intercept form.

$$5x - 3y = 12 \longrightarrow -3y = -5x + 12$$
$$y = \tfrac{5}{3}x - 4$$

$$3x + 5y = 27 \longrightarrow 5y = -3x + 27$$
$$y = -\tfrac{3}{5}x + \tfrac{27}{5}$$

Because $(\tfrac{5}{3})(-\tfrac{3}{5}) = -1$, the product of the two slopes is -1 and the lines are perpendicular.

Remark: The statement, "The product of two slopes is -1" is equivalent to saying that the two slopes are *negative reciprocals* of each other, that is, $m_1 = -1/m_2$.

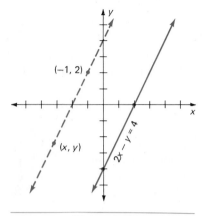

Figure 3.31

EXAMPLE 10

Find the equation of the line that contains the point $(-1, 2)$ and is parallel to the line with equation $2x - y = 4$.

Solution First, let's draw a figure to help in our analysis of the problem (Figure 3.31). Since the line through $(-1, 2)$ is to be parallel to the given line, it must have the same slope. So let's find the slope by changing $2x - y = 4$ to slope–intercept form.

$$2x - y = 4$$
$$-y = -2x + 4$$
$$y = 2x - 4$$

The slope of both lines is 2. Now, using the point–slope form with $(x_1, y_1) = (-1, 2)$, we obtain the equation of the line.

$$y - y_1 = m(x - x_1)$$
$$y - 2 = 2[x - (-1)]$$
$$y - 2 = 2(x + 1)$$
$$y - 2 = 2x + 2$$
$$-4 = 2x - y$$

EXAMPLE 11

Find the equation of the line that contains the point $(-1, -3)$ and is perpendicular to the line determined by $3x + 4y = 12$.

Solution Again let's start by drawing a figure to help with our analysis (Figure 3.32). Because the line through $(-1, -3)$ is to be perpendicular to the given line, its slope must be the negative reciprocal of the slope of the line with equation $3x + 4y = 12$. So let's find the slope of $3x + 4y = 12$ by changing to slope–intercept form.

$$3x + 4y = 12$$
$$4y = -3x + 12$$
$$y = -\tfrac{3}{4}x + 3$$

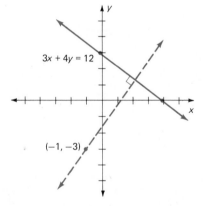

Figure 3.32

The slope of the desired line is $\tfrac{4}{3}$ (the negative reciprocal of $-\tfrac{3}{4}$) and we can proceed as before to obtain its equation.

$$y - y_1 = m(x - x_1)$$
$$y - (-3) = \tfrac{4}{3}[x - (-1)]$$
$$y + 3 = \tfrac{4}{3}(x + 1)$$
$$3y + 9 = 4x + 4$$
$$5 = 4x - 3y$$

Two forms of equations of straight lines are used extensively. They are the *standard form* and the *slope–intercept form* and can be described as follows:

Standard Form: $Ax + By = C$, where B and C are integers and A is a nonnegative integer (A and B are not both zero).

Slope–Intercept Form: $y = mx + b$, where m is a real number representing the slope of the line and b is a real number representing the y-intercept.

Problem Set 3.3

For Problems 1–8, find the slope of the line determined by each pair of points.

1. $(3, 1)$ and $(7, 4)$

2. $(-1, 2)$ and $(5, -3)$

3. $(-2, -1)$ and $(-1, -6)$

4. $(-2, -4)$ and $(3, 7)$

5. $(-4, 2)$ and $(-2, 2)$

6. $(4, -5)$ and $(-1, -5)$

7. $(a, 0)$ and $(0, b)$

8. (a, b) and (c, d)

9. Find x if the line through $(-2, 4)$ and $(x, 6)$ has a slope of $\frac{2}{9}$.

10. Find y if the line through $(1, y)$ and $(4, 2)$ has a slope of $\frac{5}{3}$.

11. Find x if the line through $(x, 4)$ and $(2, -5)$ has a slope of $-\frac{9}{4}$.

12. Find y if the line through $(5, 2)$ and $(-3, y)$ has a slope of $-\frac{7}{8}$.

For each of the following lines you are given one point and the slope of the line. Find the coordinates of three other points on the line.

13. $(3, 2)$; $m = \frac{2}{3}$

14. $(-1, 4)$; $m = \frac{5}{6}$

15. $(-1, -4)$; $m = 4$

16. $(-5, -3)$; $m = 2$

17. $(2, -1)$; $m = -\frac{3}{5}$

18. $(5, -1)$; $m = -\frac{2}{3}$

Write the equation of each of the following lines having the indicated slope and containing the indicated point. Express final equations in standard form.

19. $m = \frac{1}{3}$; $(2, 4)$

20. $m = \frac{3}{5}$; $(-1, 4)$

21. $m = 2$; $(-1, -2)$

22. $m = -3$; $(2, 5)$

23. $m = -\frac{2}{3}$; $(4, -3)$

24. $m = -\frac{1}{5}$; $(-3, 7)$

25. $m = 0$; $(5, -2)$

26. $m = \frac{4}{3}$; $(-4, -5)$

Write the equation of each of the following lines containing the indicated pair of points. Express final equations in standard form.

27. $(2, 3)$ and $(9, 8)$

28. $(1, -4)$ and $(4, 4)$

29. $(-1, 7)$ and $(5, 2)$

30. $(-3, 1)$ and $(6, -2)$

31. $(4, 2)$ and $(-1, 3)$

32. $(2, 7)$ and $(2, 5)$

33. $(4, -3)$ and $(-7, -3)$

34. $(-4, 2)$ and $(2, -3)$

Write the equation of each of the following lines having the indicated slope (m) and y-intercept (b). Express final equations in standard form.

35. $m = \frac{1}{2}$, $b = 3$

36. $m = \frac{5}{3}$, $b = -1$

37. $m = -\frac{3}{7}$, $b = 2$

38. $m = -3$, $b = -4$

39. $m = 4$, $b = \frac{3}{2}$

40. $m = \frac{2}{3}$, $b = \frac{3}{5}$

41. $m = -\frac{5}{6}$, $b = \frac{1}{4}$

42. $m = -\frac{4}{5}$, $b = 0$

Write the equation of each of the following lines satisfying the given conditions. Express final equations in standard form.

43. The x-intercept is 4 and the y-intercept is -5.

44. Contains the point $(3, -1)$ and is parallel to the x-axis

45. Contains the point $(-4, 3)$ and is parallel to the y-axis

46. Contains the point $(1, 2)$ and is parallel to the line $3x - y = 5$

47. Contains the point $(4, -3)$ and is parallel to the line $5x + 2y = 1$

48. Contains the origin and is parallel to the line $5x - 2y = 10$

49. Contains the point $(-2, 6)$ and is perpendicular to the line $x - 4y = 7$

50. Contains the point $(-3, -5)$ and is perpendicular to the line $3x + 7y = 4$

For each pair of lines in Problems 51–58, determine whether they are parallel, perpendicular, or intersecting lines that are not perpendicular.

51. $y = \frac{5}{6}x + 2$ **52.** $y = 5x - 1$ **53.** $5x - 7y = 14$ **54.** $2x - y = 4$
    $y = \frac{5}{6}x - 4$ $y = -\frac{1}{5}x + \frac{2}{3}$ $7x + 5y = 12$ $4x - 2y = 17$

55. $4x + 9y = 13$ **56.** $y = 5x$ **57.** $x + y = 0$ **58.** $2x - y = 14$
    $-4x + y = 11$ $y = -5x$ $x - y = 0$ $3x - y = 17$

For Problems 59–66, find the slope and the y-intercept of each of the given lines.

59. $2x - 3y = 4$ **60.** $3x + 4y = 7$ **61.** $x - 2y = 7$ **62.** $2x + y = 9$

63. $y = -3x$ **64.** $x - 5y = 0$

65. $7x - 5y = 12$ **66.** $-5x + 6y = 13$

67. The slope–intercept form of a line can also be used for graphing purposes. Suppose that we want to graph $y = \frac{2}{3}x + 1$. Since the y-intercept is 1, the point $(0, 1)$ is on the line. Furthermore, since the slope is $\frac{2}{3}$, another point can be found by moving two units *up* and three units to the *right*. Thus, the point $(3, 3)$ is also on the line. The two points $(0, 1)$ and $(3, 3)$ determine the line.

    Use the slope–intercept form to help graph each of the following lines.

  (a) $y = \frac{3}{4}x + 2$ **(b)** $y = \frac{1}{2}x - 4$ **(c)** $y = -\frac{4}{5}x + 1$

  (d) $y = -\frac{2}{3}x - 6$ **(e)** $y = -2x + \frac{5}{4}$ **(f)** $y = x - \frac{3}{2}$

68. Use the concept of slope to verify that $(-4, 6)$, $(6, 10)$, $(10, 0)$, and $(0, -4)$ are the vertices of a square.

69. Use the concept of slope to verify that $(6, 6)$, $(2, -2)$, $(-8, -5)$, and $(-4, 3)$ are vertices of a parallelogram.

70. Use the concept of slope to verify that the triangle determined by $(4, 3)$, $(5, 1)$, and $(3, 0)$ is a right triangle.

71. Use the concept of slope to verify that the quadrilateral whose vertices are $(0, 7)$, $(-2, -1)$, $(2, -2)$, and $(4, 6)$ is a rectangle.

72. Use the concept of slope to verify that the points $(8, -3)$, $(2, 1)$, and $(-4, 5)$ lie on a straight line.

73. The midpoints of the sides of a triangle are $(-3, 4)$, $(1, -4)$, and $(7, 2)$. Find the equations of the lines that contain the sides of the triangle.

74. The vertices of a triangle are $(2, 6)$, $(5, 1)$, and $(1, -4)$. Find the equations of the lines that contain the three altitudes of the triangle. (An altitude of a triangle is the perpendicular line segment from a vertex to the opposite side.)

75. The vertices of a triangle are $(1, -6)$, $(3, 1)$ and $(-2, 2)$. Find the equations of the lines containing the three medians of the triangle. (A median of a triangle is the line segment from a vertex to the midpoint of the opposite side.)

Miscellaneous Problems

76. The concept of slope is used for highway construction. The grade of a highway, expressed as a percent, means the number of feet that the highway changes in elevation for each 100 feet of horizontal change.

(a) A certain highway has a 2% grade. How many feet does it rise in a horizontal distance of 1 mile? (1 mile = 5280 feet)

(b) The grade of a highway up a hill is 30%. How much change in horizontal distance is there if the vertical height of the hill is 75 feet?

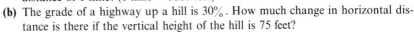

77. Slope is often expressed as the ratio **rise-to-run** in the construction of stairs.
   (a) If the ratio rise-to-run is to be $\frac{3}{5}$ for some stairs and the rise is 19 centimeters, find the measure of the run to the nearest centimeter.
   (b) If the ratio rise-to-run is to be $\frac{2}{3}$ for some stairs and the run is 28 centimeters, find the rise to the nearest centimeter.

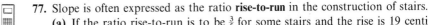

78. Suppose that a county ordinance requires a $2\frac{1}{4}$% fall for a sewage pipe from the house to the main pipe at the street. How much vertical drop must there be for a horizontal distance of 45 feet? Express the answer to the nearest tenth of a foot.

79. The form

$$\frac{y - y_1}{x - x_1} = \frac{y_2 - y_1}{x_2 - x_1}$$

is called the **two-point form** of the equation of a straight line. (1) Using points (x_1, y_1) and (x_2, y_2), develop the two-point form for the equation of a line. (2) Use the two-point form to write the equation of each of the following lines containing the indicated pair of points. Express the final equations in standard form.
   (a) (4, 3) and (5, 6)   **(b)** (−3, 5) and (2, −1)
   (c) (0, 0) and (−7, 2)   **(d)** (−3, −4) and (5, −1)

80. The form $(x/a) + (y/b) = 1$ is called the **intercept form** of the equation of a straight line. (1) Using a to represent the x-intercept and b the y-intercept, develop the intercept form. (2) Use the intercept form to write the equation of each of the following lines. Express the final equations in standard form.
   (a) $a = 2, b = 5$ **(b)** $a = -3, b = 1$
   (c) $a = 6, b = -4$ **(d)** $a = -1, b = -2$

81. Prove each of the following statements.
   (a) Two nonvertical parallel lines have the same slope.
   (b) Two lines with the same slope are parallel.
   (c) If two nonvertical lines are perpendicular, then their slopes are negative reciprocals of each other.
   (d) If the slopes of two lines are negative reciprocals of each other, then the lines are perpendicular.

82. Let $Ax + By = C$ and $A'x + B'y = C'$ represent two lines. Verify each of the following properties.
   (a) If $(A/A') = (B/B') \neq (C/C')$, then the lines are parallel.
   (b) If $AA' = -BB'$, then the lines are perpendicular.

83. The properties in Problem 82 give us another way to write the equation of a line parallel or perpendicular to a given line through a point not on the given line. For example, suppose that we want the equation of the line perpendicular to $3x + 4y = 6$ and containing the point $(1, 2)$. The form $4x - 3y = k$, where k is a constant, represents a family of lines perpendicular to $3x + 4y = 6$ because we have satisfied the condition $AA' = -BB'$. Therefore, to find the specific line of the family containing $(1, 2)$, we substitute 1 for x and 2 for y to determine k.

$$4x - 3y = k$$
$$4(1) - 3(2) = k$$
$$-2 = k$$

Thus, the equation of the desired line is $4x - 3y = -2$. Use the properties from Problem 82 to help write the equation of each of the following lines.

(a) Contains $(5, 6)$ and is parallel to the line $2x - y = 1$

(b) Contains $(-3, 4)$ and is parallel to the line $3x + 7y = 2$

(c) Contains $(2, -4)$ and is perpendicular to the line $2x - 5y = 9$

(d) Contains $(-3, -5)$ and is perpendicular to the line $4x + 6y = 7$

84. Some real-world situations can be described by the use of linear equations in two variables. If two pairs of values are known, then the equation can be determined by using the approach we used in Example 2 of this section. For each of the following, assume that the relationship can be expressed as a linear equation in two variables, and use the given information to determine the equation. Express the equation in standard form.

(a) A company produces 10 fiberglass shower stalls for $2015 and 15 stalls for $3015. Let y be the cost and x the number of stalls.

(b) A company can produce 6 boxes of candy for $8 and 10 boxes of candy for $13. Let y represent the cost and x the number of boxes of candy.

(c) Two banks on opposite corners of a town square have signs displaying the up-to-date temperature. One bank displays the temperature in Celsius degrees and the other in Fahrenheit. A temperature of $10°C$ was displayed at the same time as a temperature of $50°F$. On another day, a temperature of $-5°C$ was displayed at the same time as a temperature of $23°F$. Let y represent the temperature in Fahrenheit and x the temperature in Celsius.

85. The relationships that tie slope to parallelism and perpendicularity provide "firepower" for constructing coordinate-geometry proofs. (See Problem 81 on Page 151.) Prove each of the following using a coordinate-geometry approach.

(a) The diagonals of a square are perpendicular.

(b) The line segment joining the midpoints of two sides of a triangle is parallel to the third side.

(c) The line segments joining successive midpoints of the sides of a quadrilateral form a parallelogram.

(d) The line segments joining successive midpoints of the sides of a rectangle form a rhombus.

3.4

More On Graphing

As stated earlier, it is very helpful to recognize that a certain type of equation produces a particular kind of graph. However, we also need to develop some general graphing techniques to be used with equations for which we do not recognize the graph. Let's begin with the following suggestions and then add to the list throughout the remainder of the text.

1. Find the intercepts.
2. Solve the equation for y in terms of x or for x in terms of y if it is not already in such a form.
3. Set up a table of ordered pairs that satisfy the equation.
4. Plot the points associated with the ordered pairs and connect them with a smooth curve.

x	y	
0	−4	
2	0	intercepts
−2	0	
1	−3	
−1	−3	other
3	5	points
−3	5	

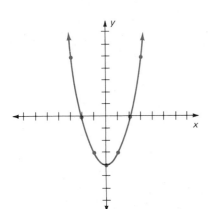

Figure 3.33

EXAMPLE 1

Graph $y = x^2 - 4$.

Solution First, let's find the intercepts. If $x = 0$, then

$$y = 0^2 - 4$$
$$y = -4.$$

This determines the point $(0, -4.)$ If $y = 0$, then

$$0 = x^2 - 4$$
$$4 = x^2$$
$$\pm 2 = x.$$

Thus, the points $(2, 0)$ and $(-2, 0)$ are determined.

Second, since the given equation expresses y in terms of x, the form is convenient for setting up a table of ordered pairs. Plotting these points and connecting them with a smooth curve produces Figure 3.33.

The curve in Figure 3.33 is said to be **symmetric with respect to the y-axis**. Stated another way, each half of the curve is a mirror image of the other half through the y-axis. Notice in the table of values that for each ordered pair (x, y), the ordered pair $(-x, y)$ is also a solution. Thus, a general test for y-axis symmetry can be stated as follows.

y-Axis Symmetry: The graph of an equation is symmetric with respect to the y-axis if replacing x with $-x$ results in an equivalent equation.

So the equation $y = x^2 - 4$ exhibits y-axis symmetry because replacing x with $-x$ produces $y = (-x)^2 - 4 = x^2 - 4$. Likewise, the equations $y = x^2 + 6$, $y = x^4$, and $y = x^4 + 2x^2$ exhibit y-axis symmetry.

EXAMPLE 2

Graph $x - 1 = y^2$.

Solution If $x = 0$, then

$$0 - 1 = y^2$$
$$-1 = y^2.$$

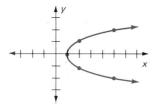

x	y	
1	0	intercept
2	1	
2	−1	other
5	2	points
5	−2	

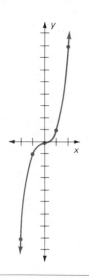

Figure 3.34

The equation $y^2 = -1$ has no real-number solutions. Therefore, this graph has no points on the y-axis. If $y = 0$, then

$$x - 1 = 0$$
$$x = 1.$$

So the point $(1, 0)$ is determined.

Solving the original equation for x produces $x = y^2 + 1$, from which the table of values at the left is easily determined. Plotting these points and connecting them with a smooth curve produces Figure 3.34.

The curve in Figure 3.34 is said to be **symmetric with respect to the x-axis**. That is to say, each half of the curve is a mirror image of the other half through the x-axis. Notice in the table of values that for each ordered pair (x, y), the ordered pair $(x, -y)$ is also a solution. The following general test for x-axis symmetry can be stated.

> **x-Axis Symmetry:** The graph of an equation is symmetric with respect to the x-axis if replacing y with $-y$ results in an equivalent equation.

Thus, the equation $x - 1 = y^2$ exhibits x-axis symmetry because replacing y with $-y$ produces $x - 1 = (-y)^2 = y^2$. Likewise, the equations $x = y^2$, $x = y^4 + 2$, and $x^3 = y^2$ exhibit x-axis symmetry.

EXAMPLE 3

Graph $y = x^3$.

Solution If $x = 0$, then

$$y = 0^3 = 0.$$

Thus, the origin $(0, 0)$ is on the graph.

The table of values at the left is easily determined from the equation. Plotting these points and connecting them with a smooth curve produces Figure 3.35.

The curve in Figure 3.35 is said to be **symmetric with respect to the origin**. Each half of the curve is a mirror image of the other half through the origin. In the table of values we see that for each ordered pair (x, y), the ordered pair $(-x, -y)$ is also a solution. The following general test for origin symmetry can be stated.

> **Origin Symmetry:** The graph of an equation is symmetric with respect to the origin if replacing x with $-x$ and y with $-y$ results in an equivalent equation.

The equation $y = x^3$ exhibits origin symmetry because replacing x with $-x$ and y with $-y$ produces $-y = -x^3$, which is equivalent to $y = x^3$. (Multiplying both sides of $-y = -x^3$ by -1 produces $y = x^3$.) Likewise, the equations $xy = 4$, $x^2 + y^2 = 10$, and $4x^2 - y^2 = 12$ exhibit origin symmetry.

x	y	
0	0	intercept
1	1	
2	8	other
−1	−1	points
−2	−8	

Figure 3.35

Remark: From the symmetry tests, we should observe that if a curve has both x-axis and y-axis symmetry, then it must have origin symmetry. However, it is possible for a curve to have origin symmetry and not be symmetrical to either axis. Figure 3.35 is an example of such a curve.

Another graphing consideration is that of **restricting a variable** to ensure real-number solutions. The following example illustrates this point.

EXAMPLE 4

Graph $y = \sqrt{x - 1}$.

Solution The radicand, $x - 1$, must be nonnegative. Therefore,

$$x - 1 \geq 0$$
$$x \geq 1.$$

The restriction $x \geq 1$ indicates that there is no y-intercept. The x-intercept can be found as follows: If $y = 0$, then

$$0 = \sqrt{x - 1}$$
$$0 = x - 1$$
$$1 = x.$$

The point $(1, 0)$ is on the graph.

Now, keeping the restriction in mind, the table of values at the left can be determined. Plotting these points and connecting them with a smooth curve produces Figure 3.36.

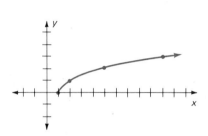

x	y	
1	0	intercept
2	1	other
5	2	points
10	3	

Figure 3.36

Now let's restate and add the concepts of symmetry and restrictions to the list of graphing suggestions. The order of the suggestions also indicates the order in which we usually attack a graphing problem if it is a "new" graph, that is, one that we do not recognize from its equation.

1. Determine the type of symmetry that the equation exhibits.
2. Find the intercepts.
3. Solve the equation for y in terms of x or for x in terms of y, if it is not already in such a form.
4. Determine the necessary restrictions so as to ensure real-number solutions.
5. Set up a table of ordered pairs that satisfy the equation. The type of symmetry and the restrictions will affect your choice of values in the table.
6. Plot the points associated with the ordered pairs and connect them with a smooth curve. Then, if appropriate, reflect this curve according to the symmetry possessed by the graph.

The final example of this section should help you pull all of these ideas together; it will illustrate the power of having these techniques at your fingertips.

x	y
2	0
3	$\sqrt{5}$
4	$2\sqrt{3}$
5	$\sqrt{21}$
6	$4\sqrt{2}$

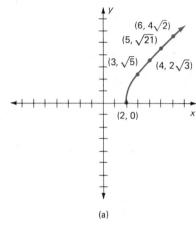

(a)

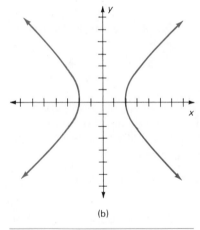

(b)

Figure 3.37

EXAMPLE 5

Graph $x^2 - y^2 = 4$.

Solution

Symmetry: The graph is symmetric with respect to both axes and the origin, because replacing x with $-x$ and y with $-y$ produces $(-x)^2 - (-y)^2 = 4$, which is equivalent to $x^2 - y^2 = 4$.

Intercepts: If $x = 0$, then

$$0^2 - y^2 = 4$$
$$-y^2 = 4$$
$$y^2 = -4.$$

Therefore, the graph contains no points on the y-axis. If $y = 0$, then

$$x^2 - 0^2 = 4$$
$$x^2 = 4$$
$$x = \pm 2.$$

Therefore, the points $(2, 0)$ and $(-2, 0)$ are on the graph.

Restrictions: Solving the given equation for y produces

$$x^2 - y^2 = 4$$
$$-y^2 = 4 - x^2$$
$$y^2 = x^2 - 4$$
$$y = \pm\sqrt{x^2 - 4}.$$

Therefore, $x^2 - 4 \geq 0$, which is equivalent to $x \geq 2$ or $x \leq -2$.

Table of values: Because of the restrictions and symmetries, we need only choose values corresponding to $x \geq 2$.

Plotting the graph: Plotting the points in the table of values and connecting them with a smooth curve produces Figure 3.37(a). Because of the symmetry with respect to both axes and the origin, the portion of the curve in Figure 3.37(a) can be reflected across both axes and through the origin to produce the complete curve in Figure 3.37(b).

Problem Set 3.4

For each of the following points, determine the points that are symmetric to the given point with respect to the x-axis, the y-axis, and the origin.

1. $(4, 3)$ **2.** $(-2, 5)$ **3.** $(-6, -1)$

4. $(3, -7)$ **5.** $(0, 4)$ **6.** $(-5, 0)$

Determine the type of symmetry (x-axis, y-axis, origin) possessed by each of the following graphs. Do not sketch the graph.

7. $y = x^2 - 6$ **8.** $x = y^2 + 1$ **9.** $x^3 = y^2$

10. $x^2y^2 = 4$ **11.** $x^2 + 2y^2 = 6$ **12.** $3x^2 - y^2 + 4x = 6$

13. $x^2 - 2x + y^2 - 3y - 4 = 0$ **14.** $xy = 4$ **15.** $y = x$

16. $2x - 3y = 15$ **17.** $y = x^3 + 2$ **18.** $y = x^4 + x^2$

19. $5x^2 - y^2 + 2y - 1 = 0$ **20.** $x^2 + y^2 - 2y - 4 = 0$

Use symmetry, intercepts, restrictions, and point-plotting to help graph each of the following.

21. $y = x^2$ **22.** $y = -x^2$ **23.** $y = x^2 + 2$ **24.** $y = -x^2 - 1$

25. $xy = 4$ **26.** $xy = -2$ **27.** $y = -x^3$ **28.** $y = x^3 + 2$

29. $y^2 = x^3$ **30.** $y^3 = x^2$ **31.** $y^2 - x^2 = 4$ **32.** $x^2 - 2y^2 = 8$

33. $y = -\sqrt{x}$ **34.** $y = \sqrt{x+1}$ **35.** $x^2 y = 4$ **36.** $xy^2 = 4$

37. $x^2 + 2y^2 = 8$ **38.** $2x^2 + y^2 = 4$ **39.** $y = \dfrac{4}{x^2 + 1}$ **40.** $y = \dfrac{-2}{x^2 + 1}$

3.5

Circles, Ellipses, and Hyperbolas

When we apply the distance formula

$$d = \sqrt{(x_2 - x_1)^2 + (y_2 - y_1)^2},$$

developed in Section 3.1, to the definition of a circle, we get what is known as the **standard form of the equation of a circle**. We start with a precise definition of a circle.

DEFINITION 3.2

> A **circle** is the set of all points in a plane equidistant from a given fixed point called the **center**. A line segment determined by the center and any point on the circle is called a **radius**.

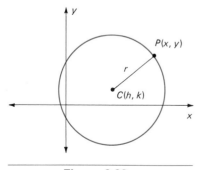

Figure 3.38

Now let's consider a circle having a radius of length r and a center at (h, k) on a coordinate system (Figure (3.38)). For any point P on the circle with coordinates (x, y), the length of a radius, denoted by r, can be expressed

$$r = \sqrt{(x - h)^2 + (y - k)^2}.$$

Squaring both sides of this equation, we obtain the standard form of the equation of a circle.

$$(x - h)^2 + (y - k)^2 = r^2$$

This form of the equation of a circle can be used to solve the two basic kinds of problems: (1) Given the coordinates of the center of a circle and the length of a radius of a circle, find its equation; (2) Given the equation of a circle, determine its graph. Let's illustrate each of these types of problems.

EXAMPLE 1

Find the equation of a circle having its center at $(-3, 5)$ and a radius of length four units.

Solution Substituting -3 for h, 5 for k, and 4 for r in the standard equation and simplifying gives us the equation of the circle.

$$(x - h)^2 + (y - k)^2 = r^2$$
$$[x - (-3)]^2 + (y - 5)^2 = 4^2$$
$$(x + 3)^2 + (y - 5)^2 = 4^2$$
$$x^2 + 6x + 9 + y^2 - 10y + 25 = 16$$
$$x^2 + y^2 + 6x - 10y + 18 = 0$$

Notice in Example 1 that we simplified the equation to the form $x^2 + y^2 + Dx + Ey + F = 0$, where D, E, and F are constants. This is another commonly used form when working with circles.

EXAMPLE 2

Graph $x^2 + y^2 - 6x + 4y + 9 = 0$.

Solution We can change the given equation into the standard form for a circle by completing the square on x and on y.

$$x^2 + y^2 - 6x + 4y + 9 = 0$$
$$(x^2 - 6x \quad) + (y^2 + 4y \quad) = -9$$
$$(x^2 - 6x + 9) + (y^2 + 4y + 4) = -9 + 9 + 4$$

| Add 9 to complete the square on x. | Add 4 to complete the square on y. | Add 9 and 4 to compensate for the 4 and 9 added on the left side. |

$$(x - 3)^2 + (y + 2)^2 = 2^2$$
$$(x - 3)^2 + [y - (-2)]^2 = 2^2$$

$$\qquad h \qquad\qquad k \qquad\qquad r$$

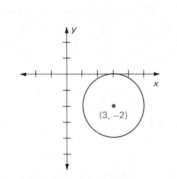

Figure 3.39

The center is at $(3, -2)$ and the length of the radius is two units. The circle is drawn in Figure 3.39.

Now suppose that we substitute 0 for h and 0 for k in the standard form of the equation of a circle.

$$(x - h)^2 + (y - k)^2 = r^2$$
$$(x - 0)^2 + (y - 0)^2 = r^2$$
$$x^2 + y^2 = r^2$$

The form $x^2 + y^2 = r^2$ is called the **standard form of the equation of a circle having its center at the origin**. For example, by inspection we can recognize that $x^2 + y^2 = 9$ is a circle with its center at the origin and a radius of length 3 units. Likewise, the equation $5x^2 + 5y^2 = 10$ is equivalent to $x^2 + y^2 = 2$; therefore its graph is a circle with its center at the origin and a radius of length $\sqrt{2}$ units. Furthermore, we can easily determine that the equation of the circle with its center at the origin and a radius of 8 units is $x^2 + y^2 = 64$.

Ellipses

Generally, it is true that any equation of the form $Ax^2 + By^2 = F$ (where $A = B$ and A, B, and F are nonzero constants having the same sign) is a circle with its center at the origin. The general equation $Ax^2 + By^2 = F$ can be used to describe other geometric figures by changing the restrictions on A and B. For example, if A, B, and F are of the same sign but $A \neq B$, then the graph of the equation $Ax^2 + By^2 = F$ is an **ellipse**. Let's consider two examples.

EXAMPLE 3

Graph $4x^2 + 9y^2 = 36$.

Solution Let's find the intercepts. If $x = 0$, then

$$4(0)^2 + 9y^2 = 36$$
$$9y^2 = 36$$
$$y^2 = 4$$
$$y = \pm 2.$$

Thus the points $(0, 2)$ and $(0, -2)$ are on the graph. If $y = 0$, then

$$4x^2 + 9(0)^2 = 36$$
$$4x^2 = 36$$
$$x^2 = 9$$
$$x = \pm 3.$$

The points $(3, 0)$ and $(-3, 0)$ are on the graph.

Plotting the four points that we have and knowing that it is an ellipse gives us a pretty good sketch of the figure (Figure 3.40).

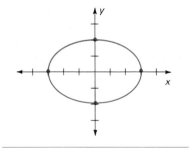

Figure 3.40

In Figure 3.40, the line segment with endpoints at $(-3, 0)$ and $(3, 0)$ is called the **major axis** of the ellipse. The shorter segment with endpoints at $(0, -2)$ and $(0, 2)$ is called the **minor axis**. Establishing the endpoints of the major and minor axes provides a basis for sketching an ellipse. Also notice that the equation $4x^2 + 9y^2 = 36$ exhibits symmetry with respect to both axes and the origin, as seen in Figure 3.40.

EXAMPLE 4

Graph $25x^2 + y^2 = 25$.

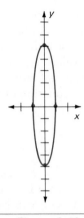

Figure 3.41

Solution The endpoints of the major and minor axes can be found by finding the intercepts. If $x = 0$, then

$$25(0)^2 + y^2 = 25$$
$$y^2 = 25$$
$$y = \pm 5.$$

The endpoints of the major axis are therefore at $(0, 5)$ and $(0, -5)$. If $y = 0$, then

$$25x^2 + (0)^2 = 25$$
$$25x^2 = 25$$
$$x^2 = 1$$
$$x = \pm 1.$$

The endpoints of the minor axis are at $(1, 0)$ and $(-1, 0)$. The ellipse is sketched in Figure 3.41.

Hyperbolas

The graph of an equation of the form $Ax^2 + By^2 = F$, where A and B are of *unlike* signs, is a **hyperbola**. The next two examples illustrate the graphing of hyperbolas.

EXAMPLE 5

Graph $x^2 - 4y^2 = 4$.

Solution If we let $y = 0$, then

$$x^2 - 4(0)^2 = 4$$
$$x^2 = 4$$
$$x = \pm 2.$$

Thus, the points $(2, 0)$ and $(-2, 0)$ are on the graph. If we let $x = 0$, then

$$0^2 - 4y^2 = 4$$
$$-4y^2 = 4$$
$$y^2 = -1.$$

Since $y^2 = -1$ has no real-number solutions, there are no points of the graph on the y-axis.

Notice that the equation $x^2 - 4y^2 = 4$ exhibits symmetry with respect to both axes and the origin. Now let's solve the given equation for y so as to have a more convenient form for finding other solutions.

$$x^2 - 4y^2 = 4$$
$$-4y^2 = 4 - x^2$$
$$4y^2 = x^2 - 4$$
$$y^2 = \frac{x^2 - 4}{4}$$
$$y = \frac{\pm\sqrt{x^2 - 4}}{2}$$

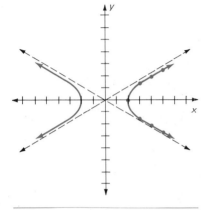

x	y	
2	0	intercepts
-2	0	
3	$\pm\frac{\sqrt{5}}{2}$	
4	$\pm\sqrt{3}$	other points
5	$\pm\frac{\sqrt{21}}{2}$	

Figure 3.42

Since the radicand, $x^2 - 4$, must be nonnegative, the values chosen for x must be such that $x \geq 2$ or $x \leq -2$. Symmetry and the points determined by the table at left provide the basis for sketching Figure 3.42.

Notice the dashed lines in Figure 3.42; they are called **asymptotes**. Each **branch** of the hyperbola approaches one of these lines but does not intersect it. Therefore, being able to sketch the asymptotes of a hyperbola is very helpful for graphing purposes. Fortunately, the equations of the asymptotes are easy to determine. They can be found by replacing the constant term in the given equation of the hyperbola with zero and then solving for y. (The reason this works will be discussed in a later chapter.) Thus, for the hyperbola in Example 5, we obtain

$$x^2 - 4y^2 = 0$$
$$-4y^2 = -x^2$$
$$y^2 = \tfrac{1}{4}x^2$$
$$y = \pm\tfrac{1}{2}x.$$

So the two lines $y = \tfrac{1}{2}x$ and $y = -\tfrac{1}{2}x$ are the asymptotes indicated by the dashed lines in Figure 3.42.

EXAMPLE 6

Graph $4y^2 - 9x^2 = 36$.

Solution If $x = 0$, then

$$4y^2 - 9(0)^2 = 36$$
$$4y^2 = 36$$
$$y^2 = 9$$
$$y = \pm 3.$$

The points $(0, 3)$ and $(0, -3)$ are on the graph. If $y = 0$, then

$$4(0)^2 - 9x^2 = 36$$
$$-9x^2 = 36$$
$$x^2 = -4$$

Since $x^2 = -4$ has no real-number solutions, we know that this hyperbola does not intersect the x-axis. Solving the equation for y yields

$$4y^2 - 9x^2 = 36$$
$$4y^2 = 9x^2 + 36$$
$$y^2 = \frac{9x^2 + 36}{4}$$
$$y = \frac{\pm\sqrt{9x^2 + 36}}{2} = \pm\frac{3\sqrt{x^2 + 4}}{2}.$$

x	y	
0	3	
0	−3	intercepts
1	$\pm\dfrac{3\sqrt{5}}{2}$	
2	$\pm 3\sqrt{2}$	other points
3	$\pm\dfrac{3\sqrt{13}}{2}$	

The table at the left shows some additional solutions. The equations of the asymptotes are determined as follows.

$$4y^2 - 9x^2 = 0$$
$$4y^2 = 9x^2$$
$$y^2 = \tfrac{9}{4}x^2$$
$$y = \pm\tfrac{3}{2}x$$

Sketching the asymptotes, plotting the points from the table, and using symmetry determines the hyperbola in Figure 3.43.

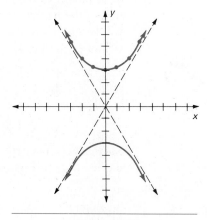

Figure 3.43

In summarizing this section, we do want you to be aware of the continuity pattern used. We started by using the definition of a circle to generate the standard form of the equation of a circle. Then ellipses and hyperbolas were discussed, not from a definition viewpoint, but by considering variations of the general equation $(Ax^2 + By^2 = F$, where A, B, and F are of the same sign and $A = B)$ of a circle with its center at the origin. In Chapter 9, parabolas, ellipses, and hyperbolas are developed from a definition viewpoint. In other words, we first define each of the concepts and then use those definitions to generate standard forms for their equations.

Problem Set 3.5

Write the equation of each of the following circles. Express the final equations in the form $x^2 + y^2 + Dx + Ey + F = 0$.

1. Center at $(2, 3)$ and $r = 5$. **2.** Center at $(-3, 4)$ and $r = 2$.

3. Center at $(-1, -5)$ and $r = 3$. **4.** Center at $(4, -2)$ and $r = 1$.

5. Center at $(3, 0)$ and $r = 3$. **6.** Center at $(0, -4)$ and $r = 6$.

7. Center at the origin and $r = 7$. **8.** Center at the origin and $r = 1$.

For Problems 9–18, find the center and the length of a radius of each of the following circles.

9. $x^2 + y^2 - 6x - 10y + 30 = 0$ **10.** $x^2 + y^2 + 8x - 12y + 43 = 0$

11. $x^2 + y^2 + 10x + 14y + 73 = 0$

12. $x^2 + y^2 + 6y - 7 = 0$

13. $x^2 + y^2 - 10x = 0$

14. $x^2 + y^2 - 4x + 2y = 0$

15. $x^2 + y^2 = 8$

16. $4x^2 + 4y^2 = 1$

17. $4x^2 + 4y^2 - 4x - 8y - 11 = 0$

18. $36x^2 + 36y^2 + 48x - 36y - 11 = 0$

19. Find the equation of the circle for which the line segment determined by $(-4, 9)$ and $(10, -3)$ is a diameter.

20. Find the equation of the circle that passes through the origin and has its center at $(-3, -4)$.

21. Find the equation of the circle that is tangent to both axes, has a radius of length seven units, and its center is in the fourth quadrant.

22. Find the equation of the circle that passes through the origin, has an x-intercept of -6, and a y-intercept of 12.

23. Find the equations of the circles that are tangent to the x-axis and have a radius of length five units. In each case the abscissa of the center is -3. (There is more than one circle that satisfies these conditions.)

Graph each of the following.

24. $4x^2 + 25y^2 = 100$

25. $9x^2 + 4y^2 = 36$

26. $x^2 - y^2 = 4$

27. $y^2 - x^2 = 9$

28. $x^2 + y^2 - 4x - 2y - 4 = 0$

29. $x^2 + y^2 - 4x = 0$

30. $4x^2 + y^2 = 4$

31. $x^2 + 9y^2 = 36$

32. $x^2 + y^2 + 2x - 6y - 6 = 0$

33. $y^2 - 3x^2 = 9$

34. $4x^2 - 9y^2 = 16$

35. $x^2 + y^2 + 4x + 6y - 12 = 0$

36. $2x^2 + 5y^2 = 50$

37. $4x^2 + 3y^2 = 12$

38. $x^2 + y^2 - 6x + 8y = 0$

39. $3x^2 - 2y^2 = 3$

40. $y^2 - 8x^2 = 9$

The graphs of equations of the form $xy = k$, where k is a nonzero constant, are also hyperbolas, sometimes referred to as **rectangular hyperbolas**. Graph each of the following.

41. $xy = 2$ **42.** $xy = 4$ **43.** $xy = -3$ **44.** $xy = -2$

45. What is the graph of $xy = 0$? Defend your answer.

46. We have graphed various equations of the form $Ax^2 + By^2 = F$, where F is a nonzero constant. Now describe the graph of each of the following.

(**a**) $x^2 + y^2 = 0$ (**b**) $2x^2 + 3y^2 = 0$

(**c**) $x^2 - y^2 = 0$ (**d**) $4x^2 - 9y^2 = 0$

Miscellaneous Problems

47. By expanding $(x - h)^2 + (y - k)^2 = r^2$, we obtain $x^2 - 2hx + h^2 + y^2 - 2ky + k^2 - r^2 = 0$. Comparing this result to the form $x^2 + y^2 + Dx + Ey + F = 0$, we see that $D = -2h$, $E = -2k$, and $F = h^2 + k^2 - r^2$. Therefore, the center and the length of a radius of a circle can be found by using $h = D/-2$, $k = E/-2$, and $r = \sqrt{h^2 + k^2 - F}$. Use these relationships to find the center and the length of a radius of each of the following circles.

(**a**) $x^2 + y^2 - 2x - 8y + 8 = 0$ (**b**) $x^2 + y^2 + 4x - 14y + 49 = 0$

(**c**) $x^2 + y^2 + 12x + 8y - 12 = 0$ (**d**) $x^2 + y^2 - 16x + 20y + 115 = 0$

(**e**) $x^2 + y^2 - 12y - 45 = 0$ (**f**) $x^2 + y^2 + 14x = 0$

48. Use a coordinate-geometry approach to prove that an angle inscribed in a semicircle is a right angle. (See accompanying figure.)

49. Use a coordinate-geometry approach to prove that a line segment from the center of a circle bisecting a chord is perpendicular to the chord. (Hint: let the ends of the chord be $(r, 0)$ and (a, b).)

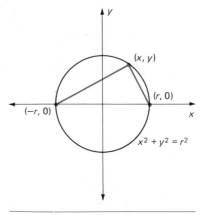

Figure for Exercise 48

Chapter 3 Summary

As emphasized throughout this chapter, coordinate geometry contains two basic kinds of problems:

1. Given an algebraic equation, determine its geometric graph;

2. Given a set of conditions pertaining to a geometric figure, determine its algebraic equation.

Let's review this chapter in terms of those two kinds of problems.

Graphing

The following graphing techniques were discussed in this chapter.

1. Recognize the type of graph that a certain kind of equation produces.
   (a) $Ax + By = C$ produces a straight line.
   (b) $x^2 + y^2 + Dx + Ey + F = 0$ produces a circle. The center and the length of a radius can be found by completing the square and comparing to the standard form of the equation of a circle: $(x - h)^2 + (y - k)^2 = r^2$.
   (c) $Ax^2 + By^2 = F$, where A, B and F have the same sign and $A = B$, produces a circle with the center at the origin.
   (d) $Ax^2 + By^2 = F$, where A, B, and F are of the same sign but $A \neq B$, produces an ellipse.
   (e) $Ax^2 + By^2 = F$, where A and B are of unlike signs, produces a hyperbola.

2. Determine the symmetry that a graph possesses.
   (a) The graph of an equation is symmetric with respect to the y-axis if replacing x with $-x$ results in an equivalent equation.

(b) The graph of an equation is symmetric with respect to the x-axis if replacing y with $-y$ results in an equivalent equation.

(c) The graph of an equation is symmetric with respect to the origin if replacing x with $-x$ and y with $-y$ results in an equivalent equation.

3. Find the intercepts. The x-intercept is found by letting $y = 0$ and solving for x. The y-intercept is found by letting $x = 0$ and solving for y.

4. Determine the necessary restrictions so as to ensure real-number solutions.

5. Set up a table of ordered pairs that satisfy the equation. The type of symmetry and the restrictions will affect your choice of values in the table. Furthermore, it may be convenient to change the form of the original equation by solving for y in terms of x or for x in terms of y.

6. Plot the points associated with the ordered pairs in the table and connect them with a smooth curve. Then, if appropriate, reflect the curve according to any symmetries possessed by the graph.

Determining Equations When Given Certain Conditions

You should review Examples 2, 3, and 4 of Section 3.3 to feel comfortable with the general approach of choosing a point (x, y) and using it to determine the equation that satisfies the conditions stated in the problem.

We developed some special forms that can be used to determine equations:

Point–slope form of a straight line:	$y - y_1 = m(x - x_1)$
Slope–intercept form of a straight line:	$y = mx + b$
Standard form of a circle:	$(x - h)^2 + (y - k)^2 = r^2$

The following formulas were used in different parts of the chapter.

Distance formula: $\qquad d = \sqrt{(x_2 - x_1)^2 + (y_2 - y_1)^2}$

Midpoint formula: The coordinates of the midpoint of a line segment determined by (x_1, y_1) and (x_2, y_2) are

$$\left(\frac{x_1 + x_2}{2}, \frac{y_1 + y_2}{2} \right).$$

Slope formula: $\qquad m = \dfrac{y_2 - y_1}{x_2 - x_1}$

Chapter 3 Review Problem Set

1. On a number line, find the coordinate of the point located three-fifths of the distance from -4 to 11.

2. On a number line, find the coordinate of the point located four-ninths of the distance from 3 to -15.

3. On the xy-plane, find the coordinates of the point located five-sixths of the distance from $(-1, -3)$ to $(11, 1)$.

4. If one endpoint of a line segment is at $(8, 14)$ and the midpoint of the segment is $(3, 10)$, find the coordinates of the other endpoint.

5. Verify that the points $(2, 2)$, $(6, 4)$, and $(5, 6)$ are vertices of a right triangle.

6. Verify that the points $(-3, 1)$, $(1, 3)$, and $(9, 7)$ lie in a straight line.

For Problems 7–12, identify any symmetries (x-axis, y-axis, origin) that each of the equations exhibits.

7. $x = y^2 + 4$ 　　　　　 **8.** $y = x^2 + 6x - 1$ 　　　　 **9.** $5x^2 - y^2 = 4$

10. $x^2 + y^2 - 2y - 4 = 0$ 　　 **11.** $y = -x$ 　　　　　 **12.** $y = \dfrac{6}{x^2 + 4}$

Graph each of the following (Problems 13–22).

13. $x^2 + y^2 - 6x + 4y - 3 = 0$ 　　　　　 **14.** $x^2 + 4y^2 = 16$

15. $x^2 - 4y^2 = 16$ 　　　　　　　　　　 **16.** $-2x + 3y = 6$

17. $2x - y < 4$ 　　　　　 **18.** $x^2 y^2 = 4$ 　　　　　 **19.** $4y^2 - 3x^2 = 8$

20. $x^2 + y^2 + 10y = 0$ 　　 **21.** $9x^2 + 2y^2 = 36$ 　　 **22.** $y \le -2x - 3$

23. Find the slope of the line determined by $(-3, -4)$ and $(-5, 6)$.

24. Find the slope of the line with equation $5x - 7y = 12$.

For Problems 25–28, write the equation of each of the lines satisfying the stated conditions. Express final equations in standard form $(Ax + By = C)$.

25. Contains the point $(7, 2)$ and has a slope of $-\frac{3}{4}$

26. Contains the points $(-3, -2)$ and $(1, 6)$

27. Contains the point $(2, -4)$ and is parallel to $4x + 3y = 17$

28. Contains the point $(-5, 4)$ and is perpendicular to $2x - y = 7$

For Problems 29–32, write the equation of each of the circles satisfying the stated conditions. Express final equations in the form $x^2 + y^2 + Dx + Ey + F = 0$.

29. Center at $(5, -6)$ and $r = 1$

30. The endpoints of a diameter are $(-2, 4)$ and $(6, 2)$

31. Center at $(-5, 12)$ and passes through the origin

32. Tangent to both axes, $r = 4$, and center is in the third quadrant

---

Chapters 1, 2, and 3 Cumulative Review Problem Set

Evaluate each of the following.

1. 3^{-3} 　　　　　　 **2.** -4^{-2} 　　　　　　 **3.** $\left(\frac{2}{3}\right)^{-2}$

4. $-\sqrt[3]{\frac{8}{27}}$ 　　　　 **5.** $\left(\frac{1}{27}\right)^{-2/3}$ 　　　　 **6.** $\dfrac{1}{\left(\frac{3}{4}\right)^{-2}}$

Perform the indicated operations and simplify. Express final answers using positive exponents only.

7. $(5x^{-3}y^{-2})(4xy^{-1})$ 　　　 **8.** $(-7a^{-3}b^2)(8a^4b^{-3})$ 　　 **9.** $(\frac{1}{2}x^{-2}y^{-1})^{-2}$

10. $\dfrac{80x^{-3}y^{-4}}{16xy^{-6}}$ 　　　 **11.** $\left(\dfrac{102x^{2/3}y^{3/4}}{6xy^{-1}}\right)^{-1}$ 　　 **12.** $\left(\dfrac{14a^3b^{-4}}{7a^{-1}b^3}\right)^2$

Express each of the following in simplest radical form. All variables represent positive real numbers.

13. $-5\sqrt{72}$

14. $2\sqrt{27x^3y^2}$

15 $\sqrt[3]{56x^4y^7}$

16. $\dfrac{3\sqrt{18}}{5\sqrt{12}}$

17. $\sqrt{\dfrac{3x}{7y}}$

18. $\dfrac{5}{\sqrt{2}-3}$

19. $\dfrac{3\sqrt{7}}{2\sqrt{2}-\sqrt{6}}$

20. $\dfrac{4\sqrt{x}}{\sqrt{x}+3\sqrt{y}}$

Perform the following indicated operations involving rational expressions. Express final answers in simplest form.

21. $\dfrac{12x^2y}{18x}\cdot\dfrac{9x^3y^3}{16xy^2}$

22. $\dfrac{-15ab^2}{14a^3b}\div\dfrac{20a}{7b^2}$

23. $\dfrac{3x^2+5x-2}{x^2-4}\cdot\dfrac{5x^2-9x-2}{3x^2-x}$

24. $\dfrac{2x-1}{4}+\dfrac{3x+2}{6}-\dfrac{x-1}{8}$

25. $\dfrac{5}{3n^2}-\dfrac{2}{n}+\dfrac{3}{2n}$

26. $\dfrac{5x}{x^2+6x-27}+\dfrac{3}{x^2-9}$

Solve each of the following equations.

27. $3(-2x-1)-2(3x+4)=-4(2x-3)$

28. $(2x-1)(3x+4)=(x+2)(6x-5)$

29. $\dfrac{3x-1}{4}-\dfrac{2x-1}{5}=\dfrac{1}{10}$

30. $9x^2-4=0$

31. $5x^3+10x^2-40x=0$

32. $7t^2-31t+12=0$

33. $x^4+15x^2-16=0$

34. $|5x-2|=3$

35. $2x^2-3x-1=0$

36. $(3x-2)(x+4)=(2x-1)(x-1)$

37. $\sqrt{5-t}+1=\sqrt{7+2t}$

38. $(2x-1)^2+4=0$

Solve each of the following inequalities. Express the solution sets using interval notation.

39. $-2(x-1)+(3-2x)>4(x+1)$

40. $2n+1+\dfrac{3n-1}{4}\geq\dfrac{n-1}{2}$

41. $0.09x+0.12(450-x)\geq 46.5$

42. $n^2+5n>24$

43. $6x^2+7x-3<0$

44. $(2x-1)(x+3)(x-4)>0$

45. $\dfrac{3x-2}{x+1}\leq 0$

46. $\dfrac{x+5}{x-1}\geq 2$

47. $|3x-1|>5$

48. $|5x-3|<12$

Graph each of the equations in Problems 49–54.

49. $x^2+4y^2=36$

50. $4x^2-y^2=4$

51. $y=-x^3-1$

52. $y=-x+3$

53. $y^2-5x^2=9$

54. $y=-\tfrac{3}{4}x-1$

Solve each of the following problems.

55. Find the center and the length of a radius of the circle with equation $x^2+y^2+14x-8y+56=0$.

56. Write the equation of the line that is parallel to $3x-4y=17$ and contains the point $(2,8)$.

57. Find the coordinates of the point located one-fifth of the distance from $(-3,4)$ to $(2,14)$.

58. Write the equation of the perpendicular bisector of the line segment determined by $(-3,4)$ and $(5,10)$.

4

Functions

4.1   The Concept of a Function

4.2   Linear and Quadratic Functions

4.3   More about Functions and Problem Solving

4.4   Combining Functions

4.5   Inverse Functions

4.6   Direct and Inverse Variations

One of the fundamental concepts of mathematics is that of a function. Functions are used to unify different areas of mathematics and they also serve as a meaningful way of applying mathematics to many real-world problems. They provide a means of studying quantities that vary with one another, that is, quantities such that a change in one produces a corresponding change in another.

In this chapter we will (1) introduce the basic ideas pertaining to the function concept, (2) use the idea of a function to unify some concepts from Chapter 3, and (3) discuss some applications using functions.

4.1

The Concept of a Function

The notion of correspondence is used in everyday situations and is central to the concept of a function. Consider the following correspondences.

1. To each person in a class, there corresponds an assigned seat.
2. To each day of a year, there corresponds an assigned integer that represents the average temperature for that day in a certain geographical location.
3. To each book in a library, there corresponds a whole number that represents the number of pages in the book.

Figure 4.1

Such correspondences can be represented pictorially, as in Figure 4.1. To each member in set A there corresponds *one and only one* member in set B. For example, in correspondence 1, set A would consist of the students in a class and set B would be the assigned seats. In the second example, set A would consist of the days of a year and set B would be a set of integers. Furthermore, the same integer might be assigned to more than one day of the year. (Different days might have the same average temperature.) The key idea is that *one and only one* integer is assigned *to each* day of the year. Likewise, in the third example, more than one book may have the same number of pages, but to each book there is assigned one and only one number of pages.

Mathematically, the general concept of a function can be defined as follows.

DEFINITION 4.1

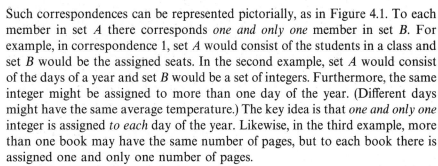

A **function** f is a correspondence between two sets X and Y that assigns to each element x of set X one and only one element y of set Y. The element y being assigned is called the **image** of x. The set X is called the **domain** of the function and the set of all images is called the **range** of the function.

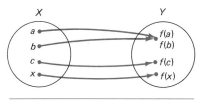

Figure 4.2

In Definition 4.1, the image y is usually denoted by $f(x)$. Thus, the symbol $f(x)$ (read "f of x" or "the **value** of f at x") represents the element in the range associated with the element x from the domain. Figure 4.2 pictorially represents this situation. Again, we emphasize that each member of the domain has precisely one image in the range; however, different members in the domain, such as a and b in Figure 4.2, may have the same image.

In Definition 4.1 we named the function f. It is common to name functions by means of a single letter and the letters f, g, and h are often used. We would suggest more meaningful choices when functions are used to portray real-world situations. For example, if a problem involves a profit function, then naming the function p or even P would seem natural. Be careful not to confuse f and $f(x)$. Remember that f is used to name a function, whereas $f(x)$ is an element of the range, namely, the element assigned to x by f.

The assignments made by a function are often expressed as ordered pairs. For example, referring back to Figure 4.2, the assignments could be expressed as $(a, f(a))$, $(b, f(b))$, $(c, f(c))$, and $(x, f(x))$, where the first components are from the domain and the second components from the range. Thus, a function can also be thought of as **a set of ordered pairs for which no two of the ordered pairs have the same first component**.

> ***Remark:*** In some texts, the concept of a **relation** is introduced first and then functions are defined as special kinds of relations. A relation is defined as "a set of ordered pairs" and then a function is defined as "a relation in which no two ordered pairs have the same first element."

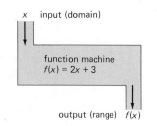

x input (domain)

function machine
$f(x) = 2x + 3$

output (range) $f(x)$

Figure 4.3

The ordered pairs representing a function can be generated by various means, such as a graph or a chart. However, one of the most common ways of generating ordered pairs is by use of equations. For example, the equation $f(x) = 2x + 3$ indicates that to each value of x in the domain, we assign $2x + 3$ from the range. For example,

$$f(1) = 2(1) + 3 = 5 \qquad \text{Produces the ordered pair } (1, 5);$$
$$f(4) = 2(4) + 3 = 11 \qquad \text{Produces the ordered pair } (4, 11);$$
$$f(-2) = 2(-2) + 3 = -1 \qquad \text{Produces the ordered pair } (-2, -1).$$

It may be helpful for you to picture mentally the concept of a function in terms of a "function machine," as illustrated in Figure 4.3. Each time that a value of x is put into the machine, the equation $f(x) = 2x + 3$ is used to generate one and only one value for $f(x)$ to be ejected from the machine.

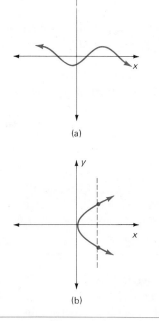

(a)

(b)

Figure 4.4

Using the ordered-pair interpretation of a function, we can define the **graph** of a function f to be the set of all points in a plane of the form $(x, f(x))$, where x is from the domain of f. In other words, the graph of f is the same as the graph of the equation $y = f(x)$. Furthermore, since $f(x)$, or y, takes on only one value for each value of x, we can easily tell whether or not a given graph represents a function. For example, in Figure 4.4(a), for any choice of x there is only one value for y. Geometrically, this means that no vertical line intersects the curve in more than one point. On the other hand, Figure 4.4(b) does not represent the graph of a function because certain values of x (all positive values) produce more than one value for y. In other words, some vertical lines intersect the curve in more than one point, as illustrated in Figure 4.4(b). A **vertical line test** for functions can be stated as follows.

> ***Vertical Line Test:*** If each vertical line intersects a graph in no more than one point, then the graph represents a function.

Let's consider some examples to help pull together some of these ideas about functions.

EXAMPLE 1

If $f(x) = x^2 - x + 4$ and $g(x) = x^3 - x^2$, find $f(3)$, $f(-2)$, $g(4)$, and $g(-3)$.

Solution

$$f(3) = 3^2 - 3 + 4 = 10, \qquad f(-2) = (-2)^2 - (-2) + 4 = 10,$$
$$g(4) = 4^3 - 4^2 = 48, \qquad g(-3) = (-3)^3 - (-3)^2 = -36$$

In Example 1, notice that we were working with two different functions in the same problem. That is why we used two different names, f and g.

EXAMPLE 2

If $f(x) = x^2 + 6$, find $f(a)$, $f(a + h)$, and $\dfrac{f(a + h) - f(a)}{h}$.

Solution

$$f(a) = a^2 + 6$$
$$f(a + h) = (a + h)^2 + 6 = a^2 + 2ah + h^2 + 6$$
$$\frac{f(a + h) - f(a)}{h} = \frac{(a^2 + 2ah + h^2 + 6) - (a^2 + 6)}{h}$$
$$= \frac{a^2 + 2ah + h^2 + 6 - a^2 - 6}{h}$$
$$= \frac{2ah + h^2}{h}$$
$$= \frac{h(2a + h)}{h} = 2a + h$$

The quotient $\dfrac{f(a + h) - f(a)}{h}$ is often called a **difference quotient** and is used extensively when studying the limit concept in calculus.

EXAMPLE 3

If $f(x) = 2x^2 + 3x - 4$, find $\dfrac{f(a + h) - f(a)}{h}$.

Solution

$$f(a) = 2a^2 + 3a - 4$$
$$f(a + h) = 2(a + h)^2 + 3(a + h) - 4$$
$$= 2(a^2 + 2ha + h^2) + 3a + 3h - 4$$
$$= 2a^2 + 4ha + 2h^2 + 3a + 3h - 4$$

Therefore,

$$f(a + h) - f(a) = (2a^2 + 4ha + 2h^2 + 3a + 3h - 4) - (2a^2 + 3a - 4)$$
$$= 2a^2 + 4ha + 2h^2 + 3a + 3h - 4 - 2a^2 - 3a + 4$$
$$= 4ha + 2h^2 + 3h$$

and

$$\frac{f(a + h) - f(a)}{h} = \frac{4ha + 2h^2 + 3h}{h}$$

$$= \frac{\cancel{h}(4a + 2h + 3)}{\cancel{h}}$$

$$= 4a + 2h + 3.$$

For our purposes in this text, if the domain of a function is not specifically indicated or determined by a real-world application, then we will assume the domain to be *all real-number* replacements for the variable, provided they represent elements in the domain and produce *real-number* functional values.

EXAMPLE 4

For the function $f(x) = \sqrt{x - 1}$, (a) specify the domain, (b) determine the range, and (c) evaluate $f(5)$, $f(50)$, and $f(25)$.

Solutions

(a) The radicand must be nonnegative, so $x - 1 \geq 0$ and thus $x \geq 1$. Therefore the domain (D) is

$$D = \{x \mid x \geq 1\}.$$

(b) The symbol $\sqrt{}$ indicates the nonnegative square root; thus, the range (R) is

$$R = \{f(x) \mid f(x) \geq 0\}.$$

(c) $f(5) = \sqrt{4} = 2$
$f(50) = \sqrt{49} = 7$
$f(25) = \sqrt{24} = 2\sqrt{6}$

As we will see later, the range of a function is often easier to determine after having graphed the function. However, our equation- and inequality-solving processes frequently are sufficient to determine the domain of a function. Let's consider some examples.

EXAMPLE 5

Determine the domain for each of the following functions.

(a) $f(x) = \dfrac{3}{2x - 5}$ (b) $g(x) = \dfrac{1}{x^2 - 9}$ (c) $f(x) = \sqrt{x^2 + 4x - 12}$

Solutions

(a) We can replace x with any real number except $\frac{5}{2}$, because $\frac{5}{2}$ makes the denominator zero. Thus, the domain is

$$D = \{x \mid x \neq \tfrac{5}{2}\}.$$

(b) We need to eliminate any values of x that will make the denominator zero. Therefore, let's solve the equation $x^2 - 9 = 0$.

$$x^2 - 9 = 0$$
$$x^2 = 9$$
$$x = \pm 3$$

The domain is thus the set

$$D = \{x \mid x \neq 3 \text{ and } x \neq -3\}.$$

(c) The radicand, $x^2 + 4x - 12$, must be nonnegative. Therefore, let's use a number-line approach, as we did in Chapter 2, to solve the inequality $x^2 + 4x - 12 \geq 0$.

$$x^2 + 4x - 12 \geq 0$$
$$(x + 6)(x - 2) \geq 0$$

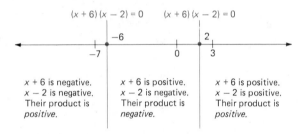

The product $(x + 6)(x - 2)$ is nonnegative if $x \leq -6$ or $x \geq 2$. Using interval notation, the domain can be expressed $(-\infty, -6] \cup [2, \infty)$.

Functions and function notation provide the basis for describing many real-world relationships. The next example illustrates this point.

EXAMPLE 6

Suppose a factory determines that the overhead for producing a quantity of a certain item is \$500 and the cost for each item is \$25. Express the total expenses as a function of the number of items produced and compute the expenses for producing 12, 25, 50, 75, and 100 items.

Solution Let n represent the number of items produced. Then $25n + 500$ represents the total expenses. Using E to represent the *expense function*, we have

$$E(n) = 25n + 500, \qquad \text{where } n \text{ is a whole number.}$$

Therefore, we obtain

$$E(12) = 25(12) + 500 = 800,$$
$$E(25) = 25(25) + 500 = 1125,$$
$$E(50) = 25(50) + 500 = 1750,$$
$$E(75) = 25(75) + 500 = 2375,$$
$$E(100) = 25(100) + 500 = 3000.$$

So the total expenses for producing 12, 25, 50, 75, and 100 items are $800, $1125, $1750, $2375, and $3000, respectively.

As stated before, an equation such as $f(x) = 5x - 7$ that is used to determine a function can also be written $y = 5x - 7$. In either form, x is referred to as the **independent variable** and y (or $f(x)$) as the **dependent variable**. Many formulas in mathematics and other related areas also determine functions. For example, the area formula for a circular region, $A = \pi r^2$, assigns to each positive real value for r a unique value for A. This formula determines a function f, where $f(r) = \pi r^2$. The variable r is the independent variable and A (or $f(r)$) is the dependent variable.

Problem Set 4.1

1. If $f(x) = -2x + 5$, find $f(3)$, $f(5)$, and $f(-2)$.
2. If $f(x) = x^2 - 3x - 4$, find $f(2)$, $f(4)$, and $f(-3)$.
3. If $g(x) = -2x^2 + x - 5$, find $g(3)$, $g(-1)$, and $g(-4)$.
4. If $g(x) = -x^2 - 4x + 6$, find $g(0)$, $g(5)$, and $g(-5)$.
5. If $h(x) = \frac{2}{3}x - \frac{3}{4}$, find $h(3)$, $h(4)$, and $h(-\frac{1}{2})$.
6. If $h(x) = -\frac{1}{2}x + \frac{2}{3}$, find $h(-2)$, $h(6)$, and $h(-\frac{2}{3})$.
7. If $f(x) = \sqrt{2x - 1}$, find $f(5)$, $f(\frac{1}{2})$, and $f(23)$.
8. If $f(x) = \sqrt{3x + 2}$, find $f(\frac{14}{3})$, $f(10)$, and $f(-\frac{1}{3})$.

For each of the given functions in Problems 9–16, find $\dfrac{f(a + h) - f(a)}{h}$.

9. $f(x) = 4x + 5$ 10. $f(x) = -7x - 2$ 11. $f(x) = x^2 - 3x$

12. $f(x) = -x^2 + 4x - 2$ 13. $f(x) = 2x^2 + 7x - 4$ 14. $f(x) = 3x^2 - x - 4$

15. $f(x) = x^3$ 16. $f(x) = x^3 - x^2 + 2x - 1$

For Problems 17–24, determine whether or not the indicated graph represents a function of x.

17.

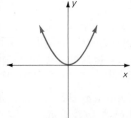

18.

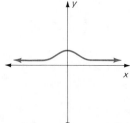

19.

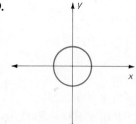

20.

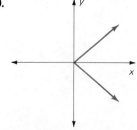

21.

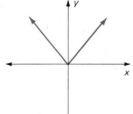

22.

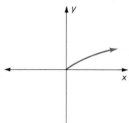

23.

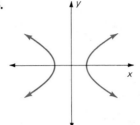

24.

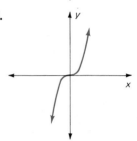

For Problems 25–32, determine the domain and the range of the given function.

25. $f(x) = \sqrt{x}$ **26.** $f(x) = \sqrt{3x - 4}$ **27.** $f(x) = x^2 + 1$

28. $f(x) = x^2 - 2$ **29.** $f(x) = x^3$ **30.** $f(x) = |x|$

31. $f(x) = x^4$ **32.** $f(x) = -\sqrt{x}$

For Problems 33–42, determine the domain of the given function.

33. $f(x) = \dfrac{3}{x - 4}$ **34.** $f(x) = \dfrac{-4}{x + 2}$ **35.** $f(x) = \dfrac{2x}{(x - 2)(x + 3)}$

36. $f(x) = \dfrac{5}{(2x - 1)(x + 4)}$ **37.** $f(x) = \sqrt{5x + 1}$ **38.** $f(x) = \dfrac{1}{x^2 - 4}$

39. $g(x) = \dfrac{3}{x^2 + 5x + 6}$ **40.** $f(x) = \dfrac{4x}{x^2 - x - 12}$ **41.** $g(x) = \dfrac{5}{x^2 + 4x}$

42. $g(x) = \dfrac{x}{6x^2 + 13x - 5}$

For Problems 43–50, express the domain of the given function using interval notation.

43. $f(x) = \sqrt{x^2 - 1}$ **44.** $f(x) = \sqrt{x^2 - 16}$

45. $f(x) = \sqrt{x^2 + 4}$ **46.** $f(x) = \sqrt{x^2 + 1} - 4$

47. $f(x) = \sqrt{x^2 - 2x - 24}$ **48.** $f(x) = \sqrt{x^2 - 3x - 40}$

49. $f(x) = \sqrt{12x^2 + x - 6}$ **50.** $f(x) = -\sqrt{8x^2 + 6x - 35}$

51. Suppose that the profit function for selling n items is given by

$$P(n) = -n^2 + 500n - 61500.$$

Evaluate $P(200)$, $P(230)$, $P(250)$, and $P(260)$.

52. The equation $A(r) = \pi r^2$ expresses the area of a circular region as a function of the length of a radius (r). Use 3.14 as an approximation for π and compute $A(2)$, $A(3)$, $A(12)$, and $A(17)$.

53. In a physics experiment, it is found that the equation $V(t) = 1667t - 6940t^2$ expresses the velocity of an object as a function of time (t). Compute $V(0.1)$, $V(0.15)$, and $V(0.2)$.

54. The height of a projectile fired vertically into the air (neglecting air resistance) at an initial velocity of 64 feet per second is a function of the time (t) and is given by the equation $h(t) = 64t - 16t^2$. Compute $h(1)$, $h(2)$, $h(3)$, and $h(4)$.

55. A car rental agency charges $50 per day plus $.32 a mile. Therefore, the daily charge for renting a car is a function of the number of miles traveled (m) and can be expressed as $C(m) = 50 + 0.32 m$. Compute $C(75)$, $C(150)$, $C(225)$, and $C(650)$.

56. The equation $I(r) = 500r$ expresses the amount of simple interest earned by an investment of $500 for one year as a function of the rate of interest (r). Compute $I(0.11)$, $I(0.12)$, $I(0.135)$, and $I(0.15)$.

57. Suppose that the height of a semielliptical archway is given by the function $h(x) = \sqrt{64 - 4x^2}$, where x is the distance from the center line of the arch. Compute $h(0)$, $h(2)$, and $h(4)$.

58. The equation $A(r) = 2\pi r^2 + 16\pi r$ expresses the total surface area of a right circular cylinder of height 8 centimeters as a function of the length of a radius (r). Use 3.14 as an approximation for π and compute $A(2)$, $A(4)$, and $A(8)$.

Miscellaneous Problems

A function f having the property that $f(-x) = f(x)$ for every x in its domain is called an **even function**. A function f having the property that $f(-x) = -f(x)$ for every x in its domain is called an **odd function**. For Problems 59–68, determine whether f is even, odd, or neither even nor odd.

59. $f(x) = x^2$ 60. $f(x) = x^3$ 61. $f(x) = x^2 + 1$

62. $f(x) = 3x - 1$ 63. $f(x) = x^2 + x$ 64. $f(x) = x^3 + 1$

65. $f(x) = x^5$ 66. $f(x) = x^4 + x^2 + 1$ 67. $f(x) = -x^3$

68. $f(x) = x^5 + x^3 + x$

Sometimes the rule of assignment for a function may consist of more than one part. For example, the function

$$f(x) = \begin{cases} 2x + 1 & \text{for } x \geq 0 \\ 3x - 1 & \text{for } x < 0 \end{cases}$$

assigns $2x + 1$ to x for nonnegative values of x and assigns $3x - 1$ to x for negative values. Thus, $f(4) = 2(4) + 1 = 9$ and $f(-2) = 3(-2) - 1 = -7$.

69. If $f(x) = \begin{cases} x & \text{for } x \geq 0 \\ x^2 & \text{for } x < 0 \end{cases}$, compute $f(4)$, $f(10)$, $f(-3)$, and $f(-5)$.

70. If $f(x) = \begin{cases} 3x + 2 & \text{for } x \geq 0 \\ 5x - 1 & \text{for } x < 0 \end{cases}$, compute $f(2)$, $f(6)$, $f(-1)$, and $f(-4)$.

71. If $f(x) = \begin{cases} 2x & \text{for } x \geq 0 \\ -2x & \text{for } x < 0 \end{cases}$, compute $f(3)$, $f(5)$, $f(-3)$, and $f(-5)$.

72. If $f(x) = \begin{cases} 2 & \text{for } x < 0 \\ x^2 + 1 & \text{for } 0 \leq x \leq 4 \\ -1 & \text{for } x > 4 \end{cases}$, compute $f(3)$, $f(6)$, $f(0)$, and $f(-3)$.

73. If $f(x) = \begin{cases} 1 & \text{for } x > 0 \\ 0 & \text{for } -1 < x \leq 0 \\ -1 & \text{for } x \leq -1 \end{cases}$, compute $f(2)$, $f(0)$, $f(-\frac{1}{2})$, and $f(-4)$.

4.2

Linear and Quadratic Functions

As we use the function concept in our study of mathematics, it is helpful to classify certain types of functions and become familiar with their equations, characteristics, and graphs. In this section we will discuss two special types of functions: *linear* and *quadratic* functions. These functions are a natural extension of our earlier study of linear and quadratic equations.

Linear Functions

Any function that can be written in the form

$$f(x) = ax + b,$$

where a and b are real numbers, is called a **linear function**. The following are examples of linear functions.

$$f(x) = -2x + 4, \qquad f(x) = 7x - 9, \qquad f(x) = \tfrac{2}{3}x + \tfrac{5}{6}$$

The equation $f(x) = ax + b$ can also be written $y = ax + b$. From our work with the slope–intercept form in Chapter 3, we know that $y = ax + b$ is the equation of a straight line having a slope of a and a y-intercept of b. This information can be used to graph linear functions, as illustrated by the following example.

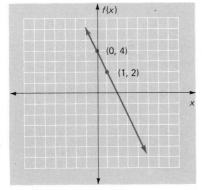

Figure 4.5

EXAMPLE 1

Graph $f(x) = -2x + 4$.

Solution Since the y-intercept is 4, the point $(0, 4)$ is on the line. Furthermore, because the slope is -2, we can move two units down and one unit to the right of $(0, 4)$ to determine the point $(1, 2)$. The line determined by $(0, 4)$ and $(1, 2)$ is drawn in Figure 4.5.

Note that in Figure 4.5 we labeled the vertical axis $f(x)$. It could also be labeled y, since $y = f(x)$. We will use the $f(x)$ labeling for most of our work with graphing functions.

Recall from Chapter 3 that we often graphed linear equations by finding the two intercepts. This same approach can be used with linear functions, as illustrated by the next example.

EXAMPLE 2

Graph $f(x) = 3x - 6$.

Solution First, we see that $f(0) = -6$; thus, the point $(0, -6)$ is on the graph. Second, by setting $3x - 6$ equal to zero and solving for x, we obtain

$$3x - 6 = 0$$
$$3x = 6$$
$$x = 2.$$

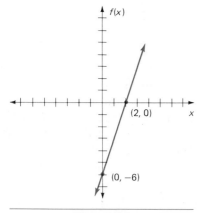

Figure 4.6

Therefore, $f(2) = 3(2) - 6 = 0$ and the point $(2, 0)$ is on the graph. The line determined by $(0, -6)$ and $(2, 0)$ is drawn in Figure 4.6.

As you graph functions by using function notation, it is often helpful to think of the ordinate of every point on the graph as the value of the function at a specific value of x. Geometrically, **the functional value is the directed distance of the point from the x-axis**. We have illustrated this idea in Figure 4.7 for the function $f(x) = x$ and in Figure 4.8 for the function $f(x) = 2$.

The linear function $f(x) = x$ is often called the **identity function**. Any linear function of the form $f(x) = ax + b$, where $a = 0$, is called a **constant function** and its graph is a horizontal line.

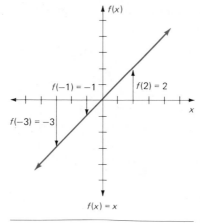

Figure 4.7

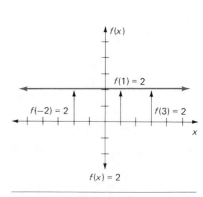

Figure 4.8

Quadratic Functions

Any function that can be written in the form

$$f(x) = ax^2 + bx + c,$$

where a, b, and c are real numbers and $a \neq 0$, is called a **quadratic function**. Furthermore, the graph of any quadratic function is a **parabola**. As we work with parabolas, the vocabulary indicated in Figure 4.9 will be used.

Graphing parabolas relies on being able to find the vertex, determining whether the parabola opens upward or downward, and locating two points on opposite sides of the axis of symmetry. It is also very helpful to compare the parabolas produced by various types of equations, such as $f(x) = x^2 + k$, $f(x) = ax^2$, $f(x) = (x - h)^2$, and $f(x) = a(x - h)^2 + k$. We are especially interested in how they compare to the **basic parabola** produced by the equation $f(x) = x^2$. The graph of $f(x) = x^2$ is shown in Figure 4.10. Notice that the graph of $f(x) = x^2$ is symmetric with respect to the y- or $f(x)$-axis. Remember that y-axis symmetry is exhibited by an equation if replacing x with $-x$ produces an equivalent equation. Therefore, since $f(-x) = (-x)^2 = x^2$, the equation $f(x) = x^2$ exhibits y-axis symmetry.

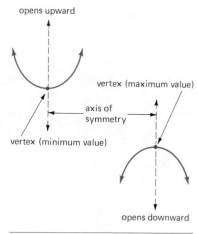

Figure 4.9

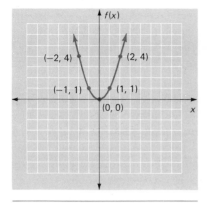

Figure 4.10

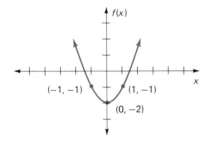

Figure 4.11

x	$f(x) = x^2$	$f(x) = 2x^2$
0	0	0
1	1	2
2	4	8
−1	1	2
−2	4	8

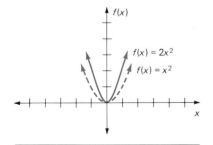

Figure 4.12

Now let's consider an equation of the form $f(x) = x^2 + k$, where k is a constant. (Keep in mind that all such equations exhibit y-axis symmetry.)

EXAMPLE 3

Graph $f(x) = x^2 - 2$.

Solution It should be observed that functional values for $f(x) = x^2 - 2$ are 2 less than corresponding functional values for $f(x) = x^2$. For example, $f(1) = -1$ for $f(x) = x^2 - 2$, but $f(1) = 1$ for $f(x) = x^2$. Thus, the graph of $f(x) = x^2 - 2$ is the same as the graph of $f(x) = x^2$ except *moved down 2 units* (Figure 4.11).

In general, **the graph of a quadratic function of the form $f(x) = x^2 + k$ is the same as the graph of $f(x) = x^2$ except moved up or down k units, depending on whether k is positive or negative.**

Now let's consider some quadratic functions of the form $f(x) = ax^2$, where a is a nonzero constant. (The graphs of these equations also have y-axis symmetry.)

EXAMPLE 4

Graph $f(x) = 2x^2$.

Solution Let's set up a table to make some comparisons of functional values. Notice in the table that the functional values for $f(x) = 2x^2$ are *twice* the corresponding functional values for $f(x) = x^2$. Thus, the parabola associated with $f(x) = 2x^2$ has the same vertex (the origin) as the graph of $f(x) = x^2$, but it is *narrower*, as shown in Figure 4.12.

EXAMPLE 5

Graph $f(x) = \frac{1}{2}x^2$.

Solution As we see from the table, the functional values for $f(x) = \frac{1}{2}x^2$ are *one-half* of the corresponding functional values for $f(x) = x^2$. Therefore, the parabola associated with $f(x) = \frac{1}{2}x^2$ is *wider* than the basic parabola, as shown in Figure 4.13.

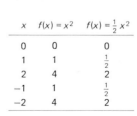

x	$f(x) = x^2$	$f(x) = \frac{1}{2}x^2$
0	0	0
1	1	$\frac{1}{2}$
2	4	2
−1	1	$\frac{1}{2}$
−2	4	2

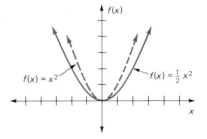

Figure 4.13

EXAMPLE 6

Graph $f(x) = -x^2$.

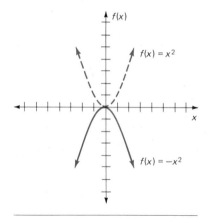

Figure 4.14

x	$f(x) = x^2$	$f(x) = (x-3)^2$
−1	1	16
0	0	9
1	1	4
2	4	1
3	9	0
4	16	1
5	25	4
6	36	9
7	49	16

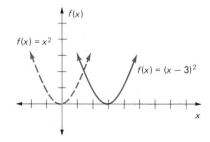

Figure 4.15

Solution It should be evident that the functional values for $f(x) = -x^2$ are the *opposites* of the corresponding functional values for $f(x) = x^2$. Therefore, the graph of $f(x) = -x^2$ is a reflection across the x-axis of the basic parabola (Figure 4.14).

In general, **the graph of a quadratic function of the form $f(x) = ax^2$ has its vertex at the origin and opens upward if a is positive and downward if a is negative. The parabola is "narrower" than the basic parabola if $|a| > 1$ and "wider" if $|a| < 1$.**

Let's continue our investigation of quadratic functions by considering those of the form $f(x) = (x - h)^2$, where h is a nonzero constant.

EXAMPLE 7

Graph $f(x) = (x - 3)^2$.

Solution A fairly extensive table of values illustrates a pattern. Notice that $f(x) = (x - 3)^2$ and $f(x) = x^2$ take on the same functional values, *but* for different values of x. More specifically, if $f(x) = x^2$ achieves a certain functional value at a specific value of x, then $f(x) = (x - 3)^2$ achieves that same functional value at x *plus three*. In other words, the graph of $f(x) = (x - 3)^2$ is the graph of $f(x) = x^2$ *moved three units to the right* (Figure 4.15).

In general, **the graph of a quadratic function of the form $f(x) = (x - h)^2$ is the same as the graph of $f(x) = x^2$ except moved to the right h units if h is positive or moved to the left h units if h is negative.**

The following diagram summarizes our work thus far for graphing quadratic functions.

$$f(x) = x^2 \nearrow \quad f(x) = x^2 + \boxed{k} \qquad \text{moves the parabola up or down}$$
$$f(x) = x^2 \rightarrow f(x) = \boxed{a}x^2 \qquad \text{affects the "width" and the way the parabola opens}$$
$$\text{basic parabola} \searrow \quad f(x) = (x - \boxed{h})^2 \qquad \text{moves the parabola right or left}$$

Now let's consider two examples that combine the previous ideas.

EXAMPLE 8

Graph $f(x) = 3(x - 2)^2 + 1$.

Solution

$$f(x) = 3(x - 2)^2 + 1$$

| narrows the parabola and opens it upward | moves the parabola 2 units to the right | moves the parabola 1 unit up |

The vertex is at $(2, 1)$ and the line $x = 2$ is the axis of symmetry. If $x = 1$, then $f(1) = 3(1 - 2)^2 + 1 = 4$. Thus, the point $(1, 4)$ is on the graph and so is its reflection, $(3, 4)$, across the line of symmetry. The parabola is drawn in Figure 4.16.

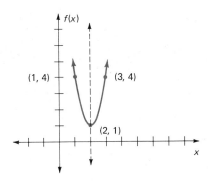

Figure 4.16

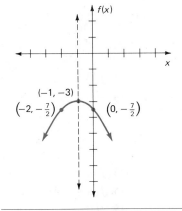

Figure 4.17

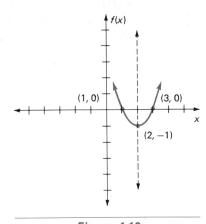

Figure 4.18

EXAMPLE 9

Graph $f(x) = -\frac{1}{2}(x + 1)^2 - 3$.

Solution

$$f(x) = -\frac{1}{2}[x - (-1)]^2 - 3$$

widens the parabola and opens it downward

moves the parabola 1 unit to the left

moves the parabola 3 units down

The vertex is at $(-1, -3)$ and the line $x = -1$ is the axis of symmetry. If $x = 0$, then $f(0) = -\frac{1}{2}(0 + 1)^2 - 3 = -\frac{7}{2}$. So the point $(0, -\frac{7}{2})$ is on the graph and so is its reflection, $(-2, -\frac{7}{2})$, across the line of symmetry. The parabola is drawn in Figure 4.17.

We are now ready to graph quadratic functions of the form $f(x) = ax^2 + bx + c$. The general approach is one of changing from the form $f(x) = ax^2 + bx + c$ to the form $f(x) = a(x - h)^2 + k$ and then proceeding as we did in Examples 8 and 9. The process of *completing the square* serves as the basis for making the change in form. Let's consider two examples to illustrate the details.

EXAMPLE 10

Graph $f(x) = x^2 - 4x + 3$.

Solution

$$f(x) = x^2 - 4x + 3$$
$$= (x^2 - 4x \quad) + 3$$
$$= (x^2 - 4x + 4) + 3 - 4$$
$$= (x - 2)^2 - 1$$

Add 4, which is the square of one-half of the coefficient of x.

Subtract 4 to compensate for the 4 that was added.

The graph of $f(x) = (x - 2)^2 - 1$ is the basic parabola moved 2 units to the right and 1 unit down (Figure 4.18).

EXAMPLE 11

Graph $f(x) = -2x^2 - 4x + 1$.

Solution

$$f(x) = -2x^2 - 4x + 1$$
$$= -2(x^2 + 2x \quad) + 1$$
$$= -2(x^2 + 2x + 1) + 1 + 2$$
$$= -2(x + 1)^2 + 3$$

Factor a -2 from the first two terms.

Add 1 inside the parentheses to complete the square.

Add 2 to compensate for the 1 inside the parentheses times the factor -2.

The graph of $f(x) = -2(x + 1)^2 + 3$ is drawn in Figure 4.19.

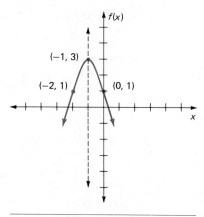

Figure 4.19

Problem Set 4.2

Graph each of the following linear functions.

1. $f(x) = 2x - 4$ **2.** $f(x) = 3x + 6$ **3.** $f(x) = -x + 1$

4. $f(x) = -2x - 4$ **5.** $f(x) = -2x$ **6.** $f(x) = 3x$

7. $f(x) = \frac{1}{2}x - \frac{3}{4}$ **8.** $f(x) = -\frac{2}{3}x + \frac{1}{2}$ **9.** $f(x) = -1$

10. $f(x) = -3$

Graph each of the following quadratic functions.

11. $f(x) = x^2 + 1$ **12.** $f(x) = x^2 - 3$ **13.** $f(x) = 3x^2$

14. $f(x) = -2x^2$ **15.** $f(x) = -x^2 + 2$ **16.** $f(x) = -3x^2 - 1$

17. $f(x) = (x + 2)^2$ **18.** $f(x) = (x - 1)^2$ **19.** $f(x) = -2(x + 1)^2$

20. $f(x) = 3(x - 2)^2$ **21.** $f(x) = (x - 1)^2 + 2$ **22.** $f(x) = -(x + 2)^2 + 3$

23. $f(x) = \frac{1}{2}(x - 2)^2 - 3$ **24.** $f(x) = 2(x - 3)^2 - 1$ **25.** $f(x) = x^2 + 2x + 4$

26. $f(x) = x^2 - 4x + 2$ **27.** $f(x) = x^2 - 3x + 1$ **28.** $f(x) = x^2 + 5x + 5$

29. $f(x) = 2x^2 + 12x + 17$ **30.** $f(x) = 3x^2 - 6x$

31. $f(x) = -x^2 - 2x + 1$ **32.** $f(x) = -2x^2 + 12x - 16$

33. $f(x) = 2x^2 - 2x + 3$ **34.** $f(x) = 2x^2 + 3x - 1$

35. $f(x) = -2x^2 - 5x + 1$ **36.** $f(x) = -3x^2 + x - 2$

Miscellaneous Problems

Graph each of the following functions.

37. $f(x) = \begin{cases} x & \text{for } x \geq 0 \\ 3x & \text{for } x < 0 \end{cases}$ **38.** $f(x) = \begin{cases} -x & \text{for } x \geq 0 \\ -4x & \text{for } x < 0 \end{cases}$

39. $f(x) = \begin{cases} 2x + 1 & \text{for } x \geq 0 \\ x^2 & \text{for } x < 0 \end{cases}$ **40.** $f(x) = \begin{cases} -x^2 & \text{for } x \geq 0 \\ 2x^2 & \text{for } x < 0 \end{cases}$

41. $f(x) = \begin{cases} 2 & \text{if } x \geq 0 \\ -1 & \text{if } x < 0 \end{cases}$ **42.** $f(x) = \begin{cases} 2 & \text{if } x > 2 \\ 1 & \text{if } 0 < x \leq 2 \\ -1 & \text{if } x \leq 0 \end{cases}$

43. $f(x) = \begin{cases} 1 & \text{if } 0 \le x < 1 \\ 2 & \text{if } 1 \le x < 2 \\ 3 & \text{if } 2 \le x < 3 \\ 4 & \text{if } 3 \le x < 4 \end{cases}$

44. $f(x) = \begin{cases} 2x + 3 & \text{if } x < 0 \\ x^2 & \text{if } 0 \le x < 2 \\ 1 & \text{if } x \ge 2 \end{cases}$

45. The **greatest integer function** is defined by the equation $f(x) = [x]$, where $[x]$ refers to the largest integer less than or equal to x. For example, $[2.6] = 2$, $[\sqrt{2}] = 1$, $[4] = 4$, and $[-1.4] = -2$. Graph $f(x) = [x]$ for $-4 \le x < 4$.

46. In the accompanying graph, the function f is said to be *increasing* in the intervals $(-\infty, x_1]$ and $[x_2, \infty)$, and f is said to be *decreasing* in the interval $[x_1, x_2]$.

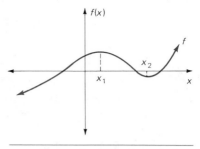

Figure for Exercise 46

More specifically, increasing and decreasing functions are defined as follows: Let f be a function, with the interval I a subset of the domain of f. Let x_1 and x_2 be in I. Then

1. f is **increasing** on I if $f(x_1) < f(x_2)$ whenever $x_1 < x_2$, and

2. f is **decreasing** on I if $f(x_1) > f(x_2)$ whenever $x_1 < x_2$.

Therefore, the function $f(x) = x^2$ is decreasing on $(-\infty, 0]$ and increasing on $[0, \infty)$. The function $f(x) = 2x + 1$ is increasing throughout its domain, so we say it is increasing over the interval $(-\infty, \infty)$.

For each of the following functions, indicate the intervals on which it is increasing and the intervals on which it is decreasing.

(a) $f(x) = x^2 + 1$ 　　　　　　　　**(b)** $f(x) = x^3$

(c) $f(x) = -3x + 1$ 　　　　　　　**(d)** $f(x) = (x - 3)^2 + 1$

(e) $f(x) = -(x + 2)^2 - 1$ 　　　　**(f)** $f(x) = x^2 - 2x + 6$

(g) $f(x) = -2x^2 - 16x - 35$ 　　　**(h)** $f(x) = x^2 + 3x - 1$

4.3

More about Functions and Problem Solving

In the previous section we used the process of completing the square to change a quadratic function such as $f(x) = x^2 - 4x + 3$ to the form $f(x) = (x - 2)^2 - 1$. From the form $f(x) = (x - 2)^2 - 1$, the vertex $(2, -1)$ and the axis of symmetry

$x = 2$ of the parabola are easily identified. In general, if we complete the square on

$$f(x) = ax^2 + bx + c,$$

we obtain

$$f(x) = a\left(x^2 + \frac{b}{a}x \quad\right) + c$$

$$= a\left(x^2 + \frac{b}{a}x + \frac{b^2}{4a^2}\right) + c - \frac{b^2}{4a}$$

$$= a\left(x + \frac{b}{2a}\right)^2 + \frac{4ac - b^2}{4a}.$$

Therefore, the parabola associated with the function $f(x) = ax^2 + bx + c$ has its vertex at

$$\left(-\frac{b}{2a}, \frac{4ac - b^2}{4a}\right),$$

and the equation of its axis of symmetry is $x = -b/2a$. These facts are illustrated in Figure 4.20.

By using the information from Figure 4.20, we now have another way of graphing quadratic functions of the form $f(x) = ax^2 + bx + c$, as indicated by the following steps.

1. Determine whether the parabola opens upward (if $a > 0$) or downward (if $a < 0$).

2. Find $-b/2a$, which is the x-coordinate of the vertex.

3. Find $f(-b/2a)$, which is the y-coordinate of the vertex, or find the y-coordinate by evaluating

$$\frac{4ac - b^2}{4a}.$$

4. Locate another point on the parabola and also locate its image across the axis of symmetry, which is the line with equation $x = -b/2a$.

The three points found in Steps 2, 3, and 4 should determine the general shape of the parabola. Let's illustrate this procedure with two examples.

EXAMPLE 1

Graph $f(x) = 3x^2 - 6x + 5$.

Solution

Step 1. Because $a = 3$, the parabola opens upward.

Step 2. $-\dfrac{b}{2a} = -\dfrac{-6}{6} = 1$.

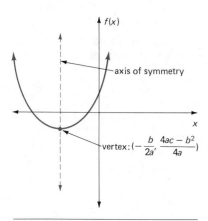

Figure 4.20

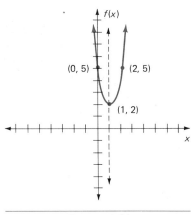

Figure 4.21

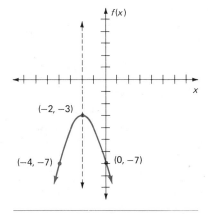

Figure 4.22

Step 3. $f\left(-\dfrac{b}{2a}\right) = f(1) = 3 - 6 + 5 = 2$. Thus, the vertex is at $(1, 2)$.

Step 4. Letting $x = 2$, we obtain $f(2) = 12 - 12 + 5 = 5$. Thus, $(2, 5)$ is on the graph and so is its reflection, $(0, 5)$, across the line of symmetry, $x = 1$.

The three points $(1, 2)$, $(2, 5)$, and $(0, 5)$ are used to graph the parabola in Figure 4.21.

EXAMPLE 2

Graph $f(x) = -x^2 - 4x - 7$.

Solution

Step 1. Since $a = -1$, the parabola opens downward.

Step 2. $-\dfrac{b}{2a} = -\dfrac{-4}{-2} = -2$.

Step 3. $f\left(-\dfrac{b}{2a}\right) = f(-2) = -(-2)^2 - 4(-2) - 7 = -3$. Thus, the vertex is at $(-2, -3)$.

Step 4. Letting $x = 0$, we obtain $f(0) = -7$. Thus, $(0, -7)$ is on the graph and so is its reflection, $(-4, -7)$, across the line of symmetry, $x = -2$.

The three points $(-2, -3)$, $(0, -7)$, and $(-4, -7)$ are used to draw the parabola in Figure 4.22.

In summary, to graph a quadratic function we basically have two methods.

1. We can express the function in the form $f(x) = a(x - h)^2 + k$ and use the values of a, h, and k to determine the parabola.

2. We can express the function in the form $f(x) = ax^2 + bx + c$ and use the approach demonstrated in Examples 1 and 2.

Back to Problem Solving

As we have seen, the vertex of the graph of a quadratic function is either the lowest or the highest point on the graph. Thus, the vocabulary **minimum value** or **maximum value** of a function is often used in applications of the parabola. The x-value of the vertex indicates where the minimum or maximum occurs and $f(x)$ yields the minimum or maximum value of the function. Let's consider some examples that use these ideas.

EXAMPLE 3

A farmer has 120 rods of fencing and wants to enclose a rectangular plot of land that requires fencing on only three sides, since it is bounded by a river on one side. Find the length and width of the plot that will maximize the area.

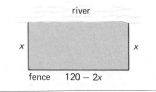

river

x x

fence $120 - 2x$

Figure 4.23

Solution Let x represent the width; then $120 - 2x$ represents the length, as indicated in Figure 4.23. The function $A(x) = x(120 - 2x)$ represents the area of the plot in terms of the width x. Since

$$A(x) = x(120 - 2x)$$
$$= 120x - 2x^2$$
$$= -2x^2 + 120x,$$

we have a quadratic function with $a = -2$, $b = 120$, and $c = 0$. Therefore, the *maximum* value ($a < 0$ so the parabola opens downward) of the function is obtained where the x-value is

$$-\frac{b}{2a} = -\frac{120}{2(-2)} = 30.$$

If $x = 30$, then $120 - 2x = 120 - 2(30) = 60$.

Thus, the farmer should make the plot 30 rods wide and 60 rods long to maximize the area at $(30)(60) = 1800$ square rods.

EXAMPLE 4

Find two numbers whose sum is 30, such that the sum of their squares is a minimum.

Solution Let x represent one of the numbers; then $30 - x$ represents the other number. By expressing the sum of their squares as a function of x, we obtain

$$f(x) = x^2 + (30 - x)^2,$$

which can be simplified to

$$f(x) = x^2 + 900 - 60x + x^2$$
$$= 2x^2 - 60x + 900.$$

This is a quadratic function with $a = 2$, $b = -60$, and $c = 900$. Therefore, the x-value where the *minimum* occurs is

$$-\frac{b}{2a} = -\frac{-60}{4} = 15.$$

If $x = 15$, then $30 - x = 30 - 15 = 15$. Thus, the two numbers should both be 15.

EXAMPLE 5

A golf pro-shop operator finds that she can sell 30 sets of golf clubs at $500 per set in a year. Furthermore, she predicts that for each $25 decrease in price, three extra sets of golf clubs could be sold. At what price should she sell the clubs to maximize gross income?

Solution Sometimes in analyzing such a problem it helps to start by setting up a table, as follows.

	NUMBER OF SETS	PRICE PER SET	INCOME
Three additional sets	30	$500	$15,000
can be sold for	33	$475	$15,675
a $25 decrease			
in price.	36	$450	$16,200

Let x represent the number of $25 decreases in price. Then the income can be expressed as a function of x:

$$f(x) = (30 + 3x)(500 - 25x).$$

number   price per
of sets   set

Simplifying this, we obtain

$$f(x) = 15,000 - 750x + 1500x - 75x^2$$
$$= -75x^2 + 750x + 15,000.$$

We complete the square in order to analyze the parabola.

$$f(x) = -75x^2 + 750x + 15,000$$
$$= -75(x^2 - 10x\ \ ) + 15,000$$
$$= -75(x^2 - 10x + 25) + 15,000 + 1875$$
$$= -75(x - 5)^2 + 16,875.$$

From this form we know that the vertex of the parabola is at $(5, 16875)$, and because $a = -75$, we know that a *maximum* occurs at the vertex. So five decreases of $25, that is, a $125 reduction in price, will give a maximum income of $16,875. The golf clubs should be sold at $375 per set.

More on Graphing Functions

Now suppose that we are faced with the problem of graphing a "new" function, one that is not familiar to us. We can use some of the graphing suggestions offered in Chapter 3. Let's restate those suggestions in terms of function vocabulary and notation.

1. Determine the domain of the function.
2. Find the y-intercept (we are labeling the y-axis with $f(x)$) by evaluating $f(0)$. Find the x-intercept (a) by finding the value(s) of x such that $f(x) = 0$.
3. Determine any types of symmetry that the equation possesses. If $f(-x) = f(x)$, then the function exhibits y-axis symmetry. If $f(-x) = -f(x)$, then the function exhibits origin symmetry. (Note that the definition of a function rules out the possibility that the graph of a function has x-axis symmetry.)
4. Set up a table of ordered pairs that satisfy the equation. The type of symmetry and the domain will affect your choice of values of x in the table.

x	$f(x) = x^4$
0	0
1	1
2	16
$\frac{1}{2}$	$\frac{1}{16}$

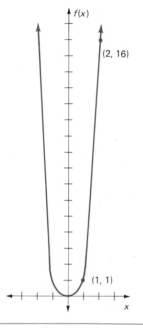

Figure 4.24

5. Plot the points associated with the ordered pairs and connect them with a smooth curve. Then, if appropriate, reflect this part of the curve according to any symmetries possessed by the graph.

EXAMPLE 6

Graph $f(x) = x^4$.

Solution The domain is the set of real numbers. Since $f(0) = 0$, the origin is on the graph. Because $f(-x) = (-x)^4 = x^4 = f(x)$, the graph has y-axis symmetry and we can concentrate our table of values on the positive values of x. If we connect the points associated with the ordered pairs from the table at the left with a smooth curve and then reflect across the vertical axis, we get the graph in Figure 4.24.

Remark: The curve in Figure 4.24 is not a parabola, even though it resembles one. This curve is flatter at the bottom and steeper.

EXAMPLE 7

Graph $f(x) = x^3$.

Solution The domain is the set of real numbers. Since $f(0) = 0$, the origin is on the graph. Because $f(-x) = (-x)^3 = -x^3 = -f(x)$, the graph is symmetrical with respect to the origin. Therefore, we can concentrate our table on the positive values of x. By connecting the points associated with the ordered pairs from the table below with a smooth curve and then reflecting it through the origin, we get the graph in Figure 4.25.

x	$f(x) = x^3$
0	0
1	1
2	8
$\frac{1}{2}$	$\frac{1}{8}$

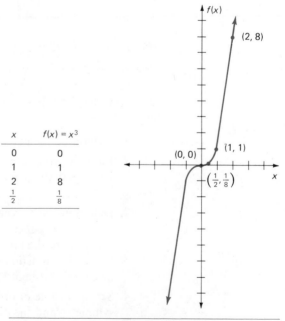

Figure 4.25

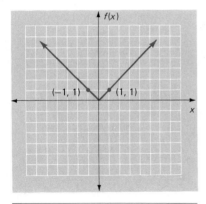

Figure 4.26

Sometimes a new function is defined in terms of old functions. In such cases, the definition plays an important role in the study of the new function. Consider the following example.

EXAMPLE 8

Graph $f(x) = |x|$.

Solution The concept of absolute value is defined for all real numbers by

$$|x| = \quad x \quad \text{if } x \geq 0$$
$$|x| = -x \quad \text{if } x < 0.$$

Therefore, the absolute value function can be expressed

$$f(x) = |x| = \begin{cases} x & \text{if } x \geq 0 \\ -x & \text{if } x < 0 \end{cases}.$$

The graph of $f(x) = x$ for $x \geq 0$ is the ray in the first quadrant and the graph of $f(x) = -x$ for $x < 0$ is the half-line (not including the origin) in the second quadrant, as indicated in Figure 4.26.

Recall that the graph of $f(x) = x^2 + k$ is the basic parabola moved vertically k units, and the graph of $f(x) = (x - h)^2$ is the basic parabola moved horizontally h units. Likewise, the graph of $f(x) = -x^2$ is the basic parabola reflected across the x-axis. These same variations apply to any basic function. In Figure 4.27 we have illustrated the graphs of some such variations of the basic absolute value function: (a) $f(x) = |x + 3|$, (b) $f(x) = |x| - 2$, and (c) $f(x) = -|x|$.

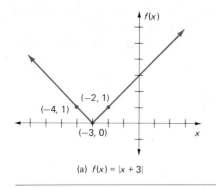

(a) $f(x) = |x + 3|$

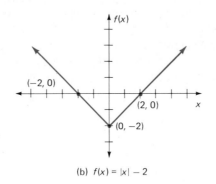

(b) $f(x) = |x| - 2$

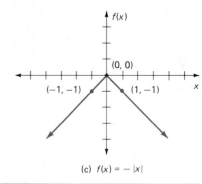

(c) $f(x) = -|x|$

Figure 4.27

Problem Set 4.3

Use the approach of Examples 1 and 2 of this section to graph each of the following quadratic functions.

1. $f(x) = x^2 - 8x + 15$ **2.** $f(x) = x^2 + 6x + 11$ **3.** $f(x) = 2x^2 + 20x + 52$

4. $f(x) = 3x^2 - 6x - 1$ **5.** $f(x) = -x^2 + 4x - 7$ **6.** $f(x) = -x^2 - 6x - 5$

7. $f(x) = -3x^2 + 6x - 5$ **8.** $f(x) = -2x^2 - 4x + 2$ **9.** $f(x) = x^2 + 3x - 1$

10. $f(x) = x^2 + 5x + 2$ **11.** $f(x) = -2x^2 + 5x + 1$ **12.** $f(x) = -3x^2 + 2x - 1$

For Problems 13–20, use the approach that you think is the most appropriate to graph each of the quadratic functions.

13. $f(x) = -x^2 + 3$ **14.** $f(x) = (x + 1)^2 + 1$ **15.** $f(x) = x^2 + x - 1$

16. $f(x) = -x^2 + 3x - 4$ **17.** $f(x) = -2x^2 + 4x + 1$ **18.** $f(x) = 4x^2 - 8x + 5$

19. $f(x) = -(x + \frac{5}{2})^2 + \frac{3}{2}$ **20.** $f(x) = x^2 - 4x$

For Problems 21–40, graph each of the functions.

21. $f(x) = x^4 + 2$ **22.** $f(x) = -x^4 - 1$ **23.** $f(x) = (x - 2)^4$

24. $f(x) = (x + 3)^4 + 1$ **25.** $f(x) = -x^3$ **26.** $f(x) = x^3 - 2$

27. $f(x) = (x + 2)^3$ **28.** $f(x) = (x - 3)^3 - 1$ **29.** $f(x) = |x - 1| + 2$

30. $f(x) = -|x + 2|$ **31.** $f(x) = |x + 1| - 3$ **32.** $f(x) = 2|x|$

33. $f(x) = x + |x|$ **34.** $f(x) = \dfrac{|x|}{x}$ **35.** $f(x) = \dfrac{1}{x}$

36. $f(x) = \dfrac{1}{x} + 3$ **37.** $f(x) = \dfrac{1}{x - 1}$ **38.** $f(x) = -\dfrac{1}{x}$

39. $f(x) = \dfrac{1}{x^2}$ **40.** $f(x) = -\dfrac{1}{x^2}$

41. Suppose that the equation $p(x) = -2x^2 + 280x - 1000$, where x represents the number of items sold, describes the profit function for a certain business. How many items should be sold to maximize the profit?

42. Suppose that the cost function for the production of a particular item is given by the equation $C(x) = 2x^2 - 320x + 12,920$, where x represents the number of items. How many items should be produced to minimize the cost?

43. The height of a projectile fired vertically into the air (neglecting air resistance) at an initial velocity of 96 feet per second is a function of time x and is given by the equation $f(x) = 96x - 16x^2$. Find the highest point reached by the projectile.

44. Find two numbers whose sum is 30, such that the sum of the square of one number plus ten times the other number is a minimum.

45. Find two numbers whose sum is 50 and whose product is a maximum.

46. Find two numbers whose difference is 40 and whose product is a minimum.

47. Two hundred and forty meters of fencing is available to enclose a rectangular playground. What should be the dimensions of the playground to maximize the area?

48. A motel advertises that they will provide dinner, dancing, and drinks for $50 per couple for a New Year's Eve party. They must have a guarantee of 30 couples. Furthermore, they will agree that for each couple in excess of 30, they will reduce the price per couple for all attending by $.50. How many couples will it take to maximize the motel's revenue?

49. A cable TV company has 1000 subscribers who each pay $15 per month. Based on a survey, they feel that for each decrease of $.25 on the monthly rate, they could obtain 20 additional subscribers. At what rate will maximum revenue be obtained and how many subscribers will it take at that rate?

50. A manufacturer finds that for the first 500 units of his product that are produced and sold, the profit is $50 per unit. The profit on each of the units beyond 500 is decreased by $.10 times the number of additional units sold. What level of output will maximize profit?

Miscellaneous Problems

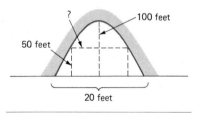

Figure for Exercise 51

51. Suppose that an arch is shaped like a parabola. It is 20 feet wide at the base and 100 feet high. How wide is the arch 50 feet above the ground?

52. A parabolic arch 27 feet high spans a parkway. If the center section of the parkway is 50 feet wide, how wide is the arch if it has a minimum clearance of 15 feet above the center section?

53. A parabolic arch spans a stream 200 feet wide. How high must the arch be above the stream to give a minimum clearance of 40 feet over a 120-foot-wide channel in the center?

4.4

Combining Functions

In subsequent mathematics courses, it is common to encounter functions that are defined in terms of sums, differences, products, and quotients of simpler functions. For example, if $h(x) = x^2 + \sqrt{x - 1}$, then we may consider the function h as the sum of f and g, where $f(x) = x^2$ and $g(x) = \sqrt{x - 1}$. In general, *if f and g are functions and D is the intersection of their domains*, then the following definitions can be made:

Sum: $\qquad (f + g)(x) = f(x) + g(x)$

Difference: $\quad (f - g)(x) = f(x) - g(x)$

Product: $\qquad (f \cdot g)(x) = f(x) \cdot g(x)$

Quotient: $\qquad \left(\dfrac{f}{g}\right)(x) = \dfrac{f(x)}{g(x)}, \qquad g(x) \neq 0$

EXAMPLE 1

If $f(x) = 3x - 1$ and $g(x) = x^2 - x - 2$, find **(a)** $(f + g)(x)$; **(b)** $(f - g)(x)$; **(c)** $(f \cdot g)(x)$; and **(d)** $(f/g)(x)$. Determine the domain of each.

Solutions

(a) $(f + g)(x) = f(x) + g(x) = (3x - 1) + (x^2 - x - 2) = x^2 + 2x - 3$

(b) $(f - g)(x) = f(x) - g(x)$
$$= (3x - 1) - (x^2 - x - 2)$$
$$= 3x - 1 - x^2 + x + 2$$
$$= -x^2 + 4x + 1$$

(c) $(f \cdot g)(x) = f(x) \cdot g(x)$
$$= (3x - 1)(x^2 - x - 2)$$
$$= 3x^3 - 3x^2 - 6x - x^2 + x + 2$$
$$= 3x^3 - 4x^2 - 5x + 2$$

(d) $\left(\dfrac{f}{g}\right)(x) = \dfrac{f(x)}{g(x)} = \dfrac{3x - 1}{x^2 - x - 2}$

The domain of both f and g is the set of all real numbers. Therefore, the domain of $f + g$, $f - g$, and $f \cdot g$ is the set of all real numbers. For f/g, the denominator $x^2 - x - 2$ cannot equal zero. Solving $x^2 - x - 2 = 0$ produces

$$(x - 2)(x + 1) = 0$$
$$x - 2 = 0 \quad \text{or} \quad x + 1 = 0$$
$$x = 2 \quad \text{or} \quad x = -1.$$

Therefore, the domain for f/g is the set of all real numbers except 2 and -1.

The Composition of Functions

Besides adding, subtracting, multiplying, and dividing functions, there is another important operation called *composition*. The composition of two functions can be defined as follows.

DEFINITION 4.2

The **composition** of functions f and g is defined by

$$(f \circ g)(x) = f(g(x))$$

for all x in the domain of g such that $g(x)$ is in the domain of f.

The left side, $(f \circ g)(x)$, of the equation in Definition 4.2 is read "the composition of f and g" and the right side is read "f of g of x." It may also be helpful for you to have a mental picture of Definition 4.2 as two function machines "hooked together" to produce another function (called the **composite function**), as illustrated in Figure 4.28. Notice that what comes out of the g function is substituted into the f function. Thus, composition is sometimes called the **substitution of functions**.

Figure 4.28 also illustrates the fact that $f \circ g$ is defined *for all x in the domain of g such that $g(x)$ is in the domain of f*. In other words, what comes out of g must be capable of being fed into f. Let's consider some examples.

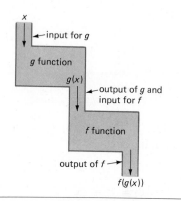

Figure 4.28

EXAMPLE 2

If $f(x) = x^2$ and $g(x) = 3x - 4$, find $(f \circ g)(x)$ and determine its domain.

Solution Applying Definition 4.2, we obtain

$$(f \circ g)(x) = f(g(x))$$
$$= f(3x - 4)$$
$$= (3x - 4)^2$$
$$= 9x^2 - 24x + 16.$$

Because g and f are both defined for all real numbers, so is $f \circ g$.

Definition 4.2, with f and g interchanged, defines the composition of g and f as $(g \circ f)(x) = g(f(x))$.

EXAMPLE 3

If $f(x) = x^2$ and $g(x) = 3x - 4$, find $(g \circ f)(x)$ and determine its domain.

Solution

$$
\begin{aligned}
(g \circ f)(x) &= g(f(x)) \\
&= g(x^2) \\
&= 3x^2 - 4
\end{aligned}
$$

Because f and g are defined for all real numbers, so is $g \circ f$.

The results of Examples 2 and 3 demonstrate an important idea, namely, that **the composition of functions is not a commutative operation**. In other words, $f \circ g \neq g \circ f$ for all functions f and g. However, as we will see in the next section, there is a special class of functions for which $f \circ g = g \circ f$.

EXAMPLE 4

If $f(x) = \sqrt{x}$ and $g(x) = 2x - 1$, find $(f \circ g)(x)$ and determine its domain.

Solution

$$
\begin{aligned}
(f \circ g)(x) &= f(g(x)) \\
&= f(2x - 1) \\
&= \sqrt{2x - 1}
\end{aligned}
$$

The domain and range of g is the set of all real numbers, but the domain of f is all *nonnegative* real numbers. Therefore $g(x)$, which is $2x - 1$, must be nonnegative:

$$
\begin{aligned}
2x - 1 &\geq 0 \\
2x &\geq 1 \\
x &\geq \tfrac{1}{2}
\end{aligned}
$$

Thus the domain of $f \circ g$ is $D = \{x \mid x \geq \tfrac{1}{2}\}$.

EXAMPLE 5

If $f(x) = \dfrac{2}{x - 1}$ and $g(x) = \dfrac{1}{x}$, find $(f \circ g)(x)$ and determine its domain.

Solution

$$
\begin{aligned}
(f \circ g)(x) &= f(g(x)) \\
&= f\left(\frac{1}{x}\right) \\
&= \frac{2}{\dfrac{1}{x} - 1} = \frac{2}{\dfrac{1 - x}{x}} \\
&= \frac{2x}{1 - x}
\end{aligned}
$$

The domain of g is all real numbers except zero, and the domain of f is all real numbers except one. Since $g(x)$, which is $1/x$, cannot equal 1,

$$\frac{1}{x} \neq 1$$

$$x \neq 1.$$

Therefore, the domain of $f \circ g$ is $D = \{x \,|\, x \neq 0 \text{ and } x \neq 1\}$.

Problem Set 4.4

For Problems 1–8, find $f + g$, $f - g$, $f \cdot g$, and $\dfrac{f}{g}$.

1. $f(x) = 3x - 4$, $g(x) = 5x + 2$
2. $f(x) = -6x - 1$, $g(x) = -8x + 7$
3. $f(x) = x^2 - 6x + 4$, $g(x) = -x - 1$
4. $f(x) = 2x^2 - 3x + 5$, $g(x) = x^2 - 4$
5. $f(x) = x^2 - x - 1$, $g(x) = x^2 + 4x - 5$
6. $f(x) = x^2 - 2x - 24$, $g(x) = x^2 - x - 30$
7. $f(x) = \sqrt{x - 1}$, $g(x) = \sqrt{x}$
8. $f(x) = \sqrt{x + 2}$, $g(x) = \sqrt{3x - 1}$

For Problems 9–22, find $(f \circ g)(x)$ and $(g \circ f)(x)$. Also specify the domain for each.

9. $f(x) = 2x$, $g(x) = 3x - 1$
10. $f(x) = 4x + 1$, $g(x) = 3x$
11. $f(x) = 5x - 3$, $g(x) = 2x + 1$
12. $f(x) = 3 - 2x$, $g(x) = -4x$
13. $f(x) = 3x + 4$, $g(x) = x^2 + 1$
14. $f(x) = 3$, $g(x) = -3x^2 - 1$
15. $f(x) = 3x - 4$, $g(x) = x^2 + 3x - 4$
16. $f(x) = 2x^2 - x - 1$, $g(x) = x + 4$
17. $f(x) = \dfrac{1}{x}$, $g(x) = 2x + 7$
18. $f(x) = \dfrac{1}{x^2}$, $g(x) = x$
19. $f(x) = \sqrt{x - 2}$, $g(x) = 3x - 1$
20. $f(x) = \dfrac{1}{x}$, $g(x) = \dfrac{1}{x^2}$
21. $f(x) = \dfrac{1}{x - 1}$, $g(x) = \dfrac{2}{x}$
22. $f(x) = \dfrac{4}{x + 2}$, $g(x) = \dfrac{3}{2x}$

Solve each of the following problems.

23. If $f(x) = 3x - 2$ and $g(x) = x^2 + 1$, find $(f \circ g)(-1)$ and $(g \circ f)(3)$.
24. If $f(x) = x^2 - 2$ and $g(x) = x + 4$, find $(f \circ g)(2)$ and $(g \circ f)(-4)$.
25. If $f(x) = 2x - 3$ and $g(x) = x^2 - 3x - 4$, find $(f \circ g)(-2)$ and $(g \circ f)(1)$.
26. If $f(x) = 1/x$ and $g(x) = 2x + 1$, find $(f \circ g)(1)$ and $(g \circ f)(2)$.
27. If $f(x) = \sqrt{x}$ and $g(x) = 3x - 1$, find $(f \circ g)(4)$ and $(g \circ f)(4)$.
28. If $f(x) = x + 5$ and $g(x) = |x|$, find $(f \circ g)(-4)$ and $(g \circ f)(-4)$.

For Problems 29–34, show that $(f \circ g)(x) = x$ and $(g \circ f)(x) = x$.

29. $f(x) = 2x$, $g(x) = \frac{1}{2}x$
30. $f(x) = \frac{3}{4}x$, $g(x) = \frac{4}{3}x$

31. $f(x) = x - 2$, $g(x) = x + 2$

32. $f(x) = 2x + 1$, $g(x) = \dfrac{x - 1}{2}$

33. $f(x) = 3x + 4$, $g(x) = \dfrac{x - 4}{3}$

34. $f(x) = 4x - 3$, $g(x) = \dfrac{x + 3}{4}$

Miscellaneous Problems

35. If $f(x) = 3x - 4$ and $g(x) = ax + b$, find conditions on a and b that will guarantee that $f \circ g = g \circ f$.

36. If $f(x) = x^2$ and $g(x) = \sqrt{x}$, with both having as domain the set of nonnegative real numbers, then show that $(f \circ g)(x) = x$ and $(g \circ f)(x) = x$.

37. If $f(x) = 3x^2 - 2x - 1$ and $g(x) = x$, find $f \circ g$ and $g \circ f$. (Recall that we have previously named $g(x) = x$ the *identity function*.)

38. In Problem Set 4.1, we defined an *even function* to be a function such that $f(-x) = f(x)$ and an *odd function* to be one such that $f(-x) = -f(x)$. Verify that (a) the sum of two even functions is an even function, and (b) the sum of two odd functions is an odd function.

4.5

Inverse Functions

Recall the *vertical line test*: "If each vertical line intersects a graph in no more than one point, then the graph represents a function." There is also a useful distinction made between two basic types of functions. Consider the graphs of the two functions in Figure 4.29: (a) $f(x) = 2x - 1$ and (b) $g(x) = x^2$. In part (a), any *horizontal line* will intersect the graph in no more than one point. Therefore, every value of $f(x)$ has only one value of x associated with it. Any function that has this property of having exactly one value of x associated with each value of $f(x)$ is called a **one-to-one function**. The function $g(x) = x^2$ is not a one-to-one

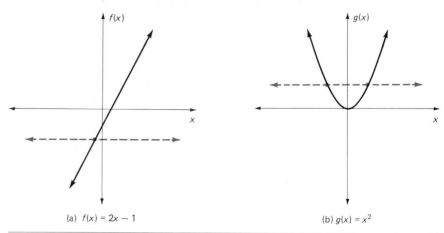

(a) $f(x) = 2x - 1$ (b) $g(x) = x^2$

Figure 4.29

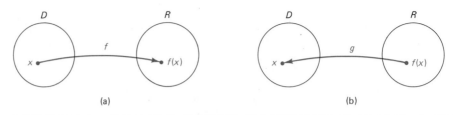

Figure 4.30

function because the horizontal line in Figure 4.29(b) intersects the parabola in two points.

Now let's consider a one-to-one function f that assigns the value $f(x)$ in its range R to each x in its domain D (Figure 4.30(a)). We can define a new function g that goes from R to D; it assigns $f(x)$ in R back to x in D, as indicated in Figure 4.30(b). The functions f and g are called *inverse functions* of one another. The following definition precisely states this concept.

DEFINITION 4.3

Let f be a one-to-one function with a domain of X and a range of Y. A function g with a domain of Y and a range of X is called the **inverse function** of f if

$$(f \circ g)(x) = x \quad \text{for every } x \text{ in } Y$$

and

$$(g \circ f)(x) = x \quad \text{for every } x \text{ in } X.$$

In Definition 4.3, note that for f and g to be inverses of each other, the domain of f must equal the range of g and the range of f must equal the domain of g. Furthermore, g must reverse the correspondences given by f, and f must reverse the correspondences given by g. In other words, inverse functions "undo" each other. Let's use Definition 4.3 to verify that two specific functions are inverses of each other.

EXAMPLE 1

Verify that $f(x) = 4x - 5$ and $g(x) = \dfrac{x + 5}{4}$ are inverse functions.

Solution Because the set of real numbers is the domain and range of both functions, we know that the domain of f equals the range of g and the range of f equals the domain of g. Furthermore,

$$(f \circ g)(x) = f(g(x))$$

$$= f\left(\frac{x + 5}{4}\right)$$

$$= 4\left(\frac{x + 5}{4}\right) - 5 = x$$

and

$$(g \circ f)(x) = g(f(x))$$
$$= g(4x - 5)$$
$$= \frac{4x - 5 + 5}{4} = x.$$

Therefore, f and g are inverses of each other.

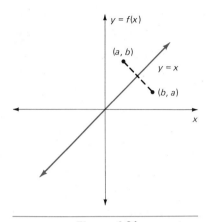

Figure 4.31

EXAMPLE 2

Verify that $f(x) = x^2 + 1$ for $x \geq 0$ and $g(x) = \sqrt{x - 1}$ for $x \geq 1$ are inverse functions.

Solution First, notice that the domain of f equals the range of g, namely, the set of nonnegative real numbers. Also, the range of f equals the domain of g, namely, the set of real numbers greater than or equal to 1. Furthermore,

$$(f \circ g)(x) = f(g(x))$$
$$= f(\sqrt{x - 1})$$
$$= (\sqrt{x - 1})^2 + 1$$
$$= x - 1 + 1 = x$$

and

$$(g \circ f)(x) = g(f(x))$$
$$= g(x^2 + 1)$$
$$= \sqrt{x^2 + 1 - 1} = \sqrt{x^2} = x. \qquad \sqrt{x^2} = x \text{ because } x \geq 1$$

Therefore, f and g are inverses of each other.

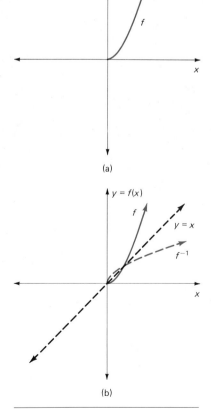

Figure 4.32

The inverse of a function f is commonly denoted by f^{-1} (read "f inverse" or "the inverse of f"). Do not confuse the -1 in f^{-1} with a negative exponent. The symbol f^{-1} *does not* mean $1/f^1$ but refers to the inverse function of function f.

Remember that a function can also be thought of as a set of ordered pairs no two of which have the same first element. Thinking along those lines, a one-to-one function further requires that no two of the ordered pairs have the same second element. Then, if the components of each ordered pair of a given one-to-one function are interchanged, the resulting function and the given function are inverses of each other. Thus, if

$$f = \{(1, 4), (2, 7), (5, 9)\}$$

then

$$f^{-1} = \{(4, 1), (7, 2), (9, 5)\}.$$

Graphically, two functions that are inverses of each other are **mirror images with reference to the line $y = x$**. This is due to the fact that ordered pairs (a, b) and (b, a) are reflections of each other with respect to the line $y = x$, as illustrated in Figure 4.31. (We will have you verify this in the next set of exercises.) Therefore, if the graph of a function f is known, as in Figure 4.32(a), then the graph of f^{-1} can be determined by reflecting f across the line $y = x$ (Figure 4.32(b)).

Finding Inverse Functions

The idea of inverse functions "undoing each other" provides the basis for a rather informal approach to finding the inverse of a function. Consider the function

$$f(x) = 2x + 1.$$

To each x this function assigns "twice x plus 1." To "undo" this function, we can "subtract 1 and divide by 2." So the inverse is

$$f^{-1}(x) = \frac{x - 1}{2}.$$

Now let's verify that f and f^{-1} are indeed inverses of each other.

$$(f \circ f^{-1})(x) = f(f^{-1}(x))$$

$$= f\left(\frac{x - 1}{2}\right)$$

$$= 2\left(\frac{x - 1}{2}\right) + 1$$

$$= x - 1 + 1 = x$$

$$(f^{-1} \circ f)(x) = f^{-1}(f(x))$$

$$= f^{-1}(2x + 1)$$

$$= \frac{2x + 1 - 1}{2}$$

$$= \frac{2x}{2} = x$$

Thus, the inverse of $f(x) = 2x + 1$ is $f^{-1}(x) = \frac{x - 1}{2}$.

This informal approach may not work very well with more complex functions, but it does emphasize how inverse functions are related to each other. A more formal and systematic technique for finding the inverse of a function can be described as follows.

1. Replace the symbol $f(x)$ by y.
2. Interchange x and y.
3. Solve the equation for y in terms of x.
4. Replace y by the symbol $f^{-1}(x)$.

The following examples illustrate this technique.

EXAMPLE 3

Find the inverse of $f(x) = \frac{2}{3}x + \frac{3}{5}$.

Solution Replacing $f(x)$ by y, the equation becomes

$$y = \tfrac{2}{3}x + \tfrac{3}{5}.$$

Interchanging x and y produces

$$x = \tfrac{2}{3}y + \tfrac{3}{5}.$$

Now, solving for y, we obtain

$$x = \frac{2}{3}y + \frac{3}{5}$$

$$15(x) = 15\left(\frac{2}{3}y + \frac{3}{5}\right)$$

$$15x = 10y + 9$$

$$15x - 9 = 10y$$

$$\frac{15x - 9}{10} = y.$$

Finally, replacing y by $f^{-1}(x)$ we can express the inverse function as

$$f^{-1}(x) = \frac{15x - 9}{10}.$$

The domain of f is equal to the range of f^{-1} (both are the set of real numbers) and the range of f equals the domain of f^{-1} (both are the set of real numbers). Furthermore, we could show that $(f \circ f^{-1})(x) = x$ and $(f^{-1} \circ f)(x) = x$. We leave this for you to complete.

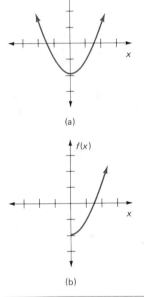

(a)

(b)

Figure 4.33

Does $f(x) = x^2 - 2$ have an inverse function? Sometimes a graph of the function helps to answer such a question. In Figure 4.33(a), it should be evident that f is not a one-to-one function and therefore cannot have an inverse. However, it should also be apparent from the graph that if we restrict the domain of f to be the nonnegative real numbers, then it is a one-to-one function and should have an inverse (Figure 4.33(b)). The next example illustrates how to find the inverse function.

EXAMPLE 4

Find the inverse of $f(x) = x^2 - 2$, where $x \geq 0$.

Solution Replacing $f(x)$ by y, the equation becomes

$$y = x^2 - 2, \qquad x \geq 0.$$

Interchanging x and y produces

$$x = y^2 - 2, \qquad y \geq 0.$$

Now let's solve for y, keeping in mind that y is to be nonnegative.

$$x = y^2 - 2$$
$$x + 2 = y^2$$
$$\sqrt{x + 2} = y, \qquad \text{where } x \geq -2$$

Finally, replacing y by $f^{-1}(x)$, the inverse function can be expressed

$$f^{-1}(x) = \sqrt{x + 2}, \qquad x \geq -2.$$

The domain of f equals the range of f^{-1} (both are the nonnegative real numbers), and the range of f equals the domain of f^{-1} (both are the real numbers greater than or equal to -2). It can be shown that $(f \circ f^{-1})(x) = x$ and $(f^{-1} \circ f)(x) = x$. We again leave this for you to complete.

Problem Set 4.5

For Problems 1–6, determine whether or not the graph represents a one-to-one function.

1.
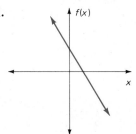

2.
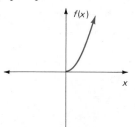

3.
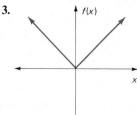

4.
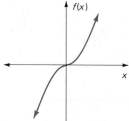

5.
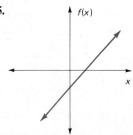

6.
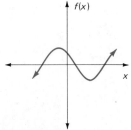

For each of the functions in Problems 7–10, (a) list the domain and range, (b) form the inverse function f^{-1}, and (c) list the domain and range of f^{-1}.

7. $f = \{(1, 5), (2, 9), (5, 21)\}$

8. $f = \{(1, 1), (4, 2), (9, 3), (16, 4)\}$

9. $f = \{(0, 0), (2, 8), (-1, -1), (-2, -8)\}$

10. $f = \{(-1, 1)(-2, 4), (-3, 9)(-4, 16)\}$

In each of Problems 11–18, verify that the two given functions are inverses of each other.

11. $f(x) = 5x - 9$ and $g(x) = \dfrac{x + 9}{5}$

12. $f(x) = -3x + 4$ and $g(x) = \dfrac{4 - x}{3}$

13. $f(x) = -\frac{1}{2}x + \frac{5}{6}$ and $g(x) = -2x + \frac{5}{3}$

14. $f(x) = x^3 + 1$ and $g(x) = \sqrt[3]{x - 1}$

15. $f(x) = \dfrac{1}{x-1}$ for $x > 1$ and $g(x) = \dfrac{x+1}{x}$ for $x > 0$

16. $f(x) = x^2 + 2$ for $x \geq 0$ and $g(x) = \sqrt{x-2}$ for $x \geq 2$

17. $f(x) = \sqrt{2x-4}$ for $x \geq 2$ and $g(x) = \dfrac{x^2+4}{2}$ for $x \geq 0$

18. $f(x) = x^2 - 4$ for $x \geq 0$ and $g(x) = \sqrt{x+4}$ for $x \geq -4$

For Problems 19–28, determine if f and g are inverse functions.

19. $f(x) = 3x$ and $g(x) = -\frac{1}{3}x$

20. $f(x) = \frac{3}{4}x - 2$ and $g(x) = \frac{4}{3}x + \frac{8}{3}$

21. $f(x) = x^3$ and $g(x) = \sqrt[3]{x}$

22. $f(x) = \dfrac{1}{x+1}$ and $g(x) = \dfrac{1-x}{x}$

23. $f(x) = x$ and $g(x) = \dfrac{1}{x}$

24. $f(x) = \frac{3}{5}x + \frac{1}{3}$ and $g(x) = \frac{5}{3}x - 3$

25. $f(x) = x^2 - 3$ for $x \geq 0$ and $g(x) = \sqrt{x+3}$ for $x \geq -3$

26. $f(x) = |x - 1|$ for $x \geq 1$ and $g(x) = |x + 1|$ for $x \geq 0$

27. $f(x) = \sqrt{x+1}$ and $g(x) = x^2 - 1$ for $x \geq 0$

28. $f(x) = \sqrt{2x-2}$ and $g(x) = \frac{1}{2}x^2 + 1$

For Problems 29–40, (a) find f^{-1} and (b) verify that $(f \circ f^{-1})(x) = x$ and $(f^{-1} \circ f)(x) = x$.

29. $f(x) = x - 4$

30. $f(x) = 2x - 1$

31. $f(x) = -3x - 4$

32. $f(x) = -5x + 6$

33. $f(x) = \frac{3}{4}x - \frac{5}{6}$

34. $f(x) = \frac{2}{3}x - \frac{1}{4}$

35. $f(x) = -\frac{2}{3}x$

36. $f(x) = \frac{4}{3}x$

37. $f(x) = \sqrt{x}$ for $x \geq 0$

38. $f(x) = \dfrac{1}{x}$ for $x \neq 0$

39. $f(x) = x^2 + 4$ for $x \geq 0$

40. $f(x) = x^2 + 1$ for $x \leq 0$

For Problems 41–48, (a) find f^{-1} and (b) graph f and f^{-1} on the same set of axes.

41. $f(x) = 3x$

42. $f(x) = -x$

43. $f(x) = 2x + 1$

44. $f(x) = -3x - 3$

45. $f(x) = \dfrac{2}{x-1}$ for $x > 1$

46. $f(x) = \dfrac{-1}{x-2}$ for $x > 2$

47. $f(x) = x^2 - 4$ for $x \geq 0$

48. $f(x) = \sqrt{x-3}$ for $x \geq 3$

Miscellaneous Problems

49. Explain why every nonconstant linear function has an inverse.

50. The function notation and the operation of composition can be used to find inverses as follows: To find the inverse of $f(x) = 5x + 3$ we know that $f(f^{-1}(x))$ must produce x. Therefore,

$$f(f^{-1}(x)) = 5[f^{-1}(x)] + 3 = x$$
$$5[f^{-1}(x)] = x - 3$$

$$f^{-1}(x) = \frac{x-3}{5}$$

Use this approach to find the inverse of each of the following functions.

(a) $f(x) = 3x - 9$ (b) $f(x) = -2x + 6$

(c) $f(x) = -x + 1$ (d) $f(x) = 2x$

(e) $f(x) = -5x$ (f) $f(x) = x^2 + 6$ for $x \geq 0$

51. If $f(x) = 2x + 3$ and $g(x) = 3x - 5$, find

(a) $(f \circ g)^{-1}(x)$ (b) $(f^{-1} \circ g^{-1})(x)$ (c) $(g^{-1} \circ f^{-1})(x)$

4.6

Direct and Inverse Variations

"The amount of simple interest earned by a fixed amount of money invested at a certain rate *varies directly* as the time."

"At a constant temperature, the volume of an enclosed gas *varies inversely* as the pressure."

Such statements illustrate two basic types of functional relationships, called **direct** and **inverse variation**, that are widely used, especially in the physical sciences. These relationships can be expressed by equations that determine functions. The purpose of this section is to investigate these special functions.

Direct Variation

The statement, "y varies directly as x" means

$$y = kx$$

where k is a nonzero constant, called the **constant of variation**. The phrase, "y is directly proportional to x" is also used to indicate direct variation; k is then referred to as the **constant of proportionality**.

> **Remark:** Notice that the equation $y = kx$ defines a function and can be written $f(x) = kx$. However, in this section it is more convenient not to use function notation but instead to use variables that are meaningful in terms of the physical entities involved in the particular problem.

Statements indicating direct variation may also involve powers of a variable. For example, "y varies directly as the square of x" can be written $y = kx^2$. In general, "y varies directly as the nth power of x $(n > 0)$" means

$$y = kx^n.$$

There are basically three types of problems when dealing with direct variation: (1) translating an English statement into an equation expressing the direct variation, (2) finding the constant of variation from given values of the variables, and (3) finding additional values of the variables once the constant of variation has been determined. Let's consider an example of each of these types of problems.

EXAMPLE 1

Translate the statement, "The tension on a spring varies directly as the distance it is stretched" into an equation using k as the constant of variation.

Solution Letting t represent the tension and d the distance, the equation is

$$t = kd.$$

EXAMPLE 2

If A varies directly as the square of e, and $A = 96$ when $e = 4$, find the constant of variation.

Solution Since A varies directly as the square of e, we have

$$A = ke^2.$$

Substituting 96 for A and 4 for e, we obtain

$$96 = k(4)^2$$
$$96 = 16k$$
$$6 = k.$$

The constant of variation is 6.

EXAMPLE 3

If y is directly proportional to x, and if $y = 6$ when $x = 8$, find the value of y when $x = 24$.

Solution The statement, "y is directly proportional to x" translates into

$$y = kx.$$

Letting $y = 6$ and $x = 8$, the constant of variation becomes

$$6 = k(8)$$
$$\frac{6}{8} = k$$
$$\frac{3}{4} = k$$

So, the specific equation is

$$y = \frac{3}{4}x.$$

Now, letting $x = 24$, we obtain

$$y = \frac{3}{4}(24) = 18.$$

Inverse Variation

The second basic type of variation, *inverse variation*, is defined as follows. The statement, "y varies inversely as x" means

$$y = \frac{k}{x},$$

where k is a nonzero constant and again is referred to as the constant of variation. The phrase, "y is inversely proportional to x" is also used to express inverse variation. As with direct variation, statements indicating variation may involve powers of x. For example, "y varies inversely as the square of x" can be written $y = k/x^2$. In general, "y varies inversely as the nth power of x $(n \to 0)$" means

$$y = \frac{k}{x^n}.$$

The following examples illustrate the three basic kinds of problems involving inverse variation.

EXAMPLE 4

Translate the statement, "The length of a rectangle of fixed area varies inversely as the width" into an equation using k as the constant of variation.

Solution Letting l represent the length and w the width, the equation is

$$l = \frac{k}{w}.$$

EXAMPLE 5

If y is inversely proportional to x and $y = 14$ when $x = 4$, find the constant of variation.

Solution Since y is inversely proportional to x, we have

$$y = \frac{k}{x}.$$

Substituting 4 for x and 14 for y, we obtain

$$14 = \frac{k}{4}.$$

Solving this equation yields

$$k = 56.$$

The constant of variation is 56.

EXAMPLE 6

The time required for a car to travel a certain distance varies inversely as the rate at which it travels. If it takes 4 hours at 50 miles per hour to travel the distance, how long will it take at 40 miles per hour?

Solution Let t represent time and r rate. The phrase, "time required . . . varies inversely as the rate" translates into

$$t = \frac{k}{r}.$$

Substituting 4 for t and 50 for r produces the constant of variation:

$$4 = \frac{k}{50}$$

$$k = 200.$$

So the specific equation is

$$t = \frac{200}{r}.$$

Now substituting 40 for r produces

$$t = \frac{200}{40} = 5.$$

It will take 5 hours at 40 miles per hour.

The terms *direct* and *inverse*, as applied to variation, refer to the relative behavior of the variables involved in the equation. That is to say, in *direct variation* $(y = kx)$, an assignment of **increasing absolute values for x** produces **increasing absolute values for y**. However, in *inverse variation* $(y = k/x)$, an assignment of **increasing absolute values for x** produces **decreasing absolute values for y**.

Joint Variation

Variation may involve more than two variables. The following table illustrates some different types of variation statements and their equivalent algebraic equations using k as the constant of variation.

VARIATION STATEMENT	ALGEBRAIC EQUATION
1. *y varies jointly* as *x* and *z*	$y = kxz$
2. *y varies jointly* as *x, z,* and *w*	$y = kxzw$
3. *V varies jointly* as *h* and the square of *r*	$V = khr^2$
4. *h* varies directly as *V* and inversely as *w*	$h = \dfrac{kV}{w}$
5. *y* is directly proportional to *x* and inversely proportional to the square of *z*	$y = \dfrac{kx}{z^2}$
6. *y varies jointly* as *w* and *z*, and inversely as *x*	$y = \dfrac{kwz}{x}$

Statements 1, 2, and 3 illustrate the concept of *joint variation*. Statements 4 and 5 show that both direct and inverse variation may occur in the same problem. Statement 6 combines joint variation with inverse variation.

The final two examples of this section illustrate different kinds of problems involving some of these variation situations.

EXAMPLE 7

The volume of a pyramid varies jointly as its altitude and the area of its base. If a pyramid having an altitude of 9 feet and a base with an area of 17 square feet has a volume of 51 cubic feet, find the volume of a pyramid with an altitude of 14 feet and a base with an area of 45 square feet.

Solution Let's use some variables as follows.

$$V = \text{volume}$$
$$B = \text{area of base}$$
$$h = \text{altitude}$$
$$k = \text{constant of variation}$$

The fact that the volume varies jointly as the altitude and the area of the base can be represented by the equation

$$V = kBh.$$

Substituting 51 for V, 17 for B, and 9 for h, we obtain

$$51 = k(17)(9)$$
$$51 = 153k$$
$$\tfrac{51}{153} = k$$
$$\tfrac{1}{3} = k.$$

Therefore, the specific equation is $V = \tfrac{1}{3}Bh$. Now, substituting 45 for B and 14 for h, we obtain

$$V = \tfrac{1}{3}(45)(14) = (15)(14) = 210.$$

The volume is 210 cubic feet.

EXAMPLE 8

Suppose that y varies jointly as x and z and inversely as w. If $y = 154$ when $x = 6$, $z = 11$, and $w = 3$, find y when $x = 8$, $z = 9$, and $w = 6$.

Solution The statement, "y varies jointly as x and z and inversely as w" translates into the equation

$$y = \frac{kxz}{w}.$$

Substituting 154 for y, 6 for x, 11 for z, and 3 for w produces

$$154 = \frac{(k)(6)(11)}{3}$$
$$154 = 22k$$
$$7 = k$$

So, the specific equation is

$$y = \frac{7xz}{w}.$$

Now, substituting 8 for x, 9 for z, and 6 for w, we obtain

$$y = \frac{7(8)(9)}{6} = 84.$$

Problem Set 4.6

For Problems 1–8, translate each of the statements of variation into an equation, using k as the constant of variation.

1. y varies directly as the cube of x.

2. a varies inversely as the square of b.

3. A varies jointly as l and w.

4. s varies jointly as g and the square of t.

5. At a constant temperature, the volume (V) of a gas varies inversely as the pressure (P).

6. y varies directly as the square of x and inversely as the cube of w.

7. The volume (V) of a cone varies jointly as its height (h) and the square of a radius (r).

8. I is directly proportional to r and t.

For Problems 9–18, find the constant of variation for each of the stated conditions.

9. y varies directly as x, and $y = 72$ when $x = 3$.

10. y varies inversely as the square of x, and $y = 4$ when $x = 2$.

11. A varies directly as the square of r, and $A = 154$ when $r = 7$.

12. V varies jointly as B and h, and $V = 104$ when $B = 24$ and $h = 13$.

13. A varies jointly as b and h, and $A = 81$ when $b = 9$ and $h = 18$.

14. s varies jointly as g and the square of t, and $s = -108$ when $g = 24$ and $t = 3$.

15. y varies jointly as x and z and inversely as w; $y = 154$ when $x = 6$, $z = 11$, and $w = 3$.

16. V varies jointly as h and the square of r, and $V = 1100$ when $h = 14$ and $r = 5$.

17. y is directly proportional to the square of x and inversely proportional to the cube of w, and $y = 18$ when $x = 9$ and $w = 3$.

18. y is directly proportional to x and inversely proportional to the square root of w, and $y = \frac{1}{5}$ when $x = 9$ and $w = 10$.

Solve each of the following problems.

19. If y is directly proportional to x, and $y = 5$ when $x = -15$, find the value of y when $x = -24$.

20. If y is inversely proportional to the square of x, and $y = \frac{1}{8}$ when $x = 4$, find y when $x = 8$.

21. If V varies jointly as B and h, and $V = 96$ when $B = 36$ and $h = 8$, find V when $B = 48$ and $h = 6$.

22. If A varies directly as the square of e, and $A = 150$ when $e = 5$, find A when $e = 10$.

23. The time required for a car to travel a certain distance varies inversely as the rate at which it travels. If it takes 3 hours at 50 miles per hour to travel the distance, how long will it take at 30 miles per hour?

24. The distance that a freely falling body falls varies directly as the square of the time it falls. If a body falls 144 feet in 3 seconds, how far will it fall in 5 seconds?

25. The period (the time required for one complete oscillation) of a simple pendulum varies directly as the square root of its length. If a pendulum 12 feet long has a period of 4 seconds, find the period of a pendulum of length 3 feet.

26. Suppose the number of days it takes to complete a construction job varies inversely as the number of people assigned to the job. If it takes 7 people 8 days to do the job, how long will it take 10 people to complete the job?

27. The number of days needed to assemble some machines varies directly as the number of machines and inversely as the number of people working. If it takes 4 people 32 days to assemble 16 machines, how many days will it take 8 people to assemble 24 machines?

28. The volume of a gas at a constant temperature varies inversely as the pressure. What is the volume of a gas under a pressure of 25 pounds if the gas occupies 15 cubic centimeters under a pressure of 20 pounds?

29. The volume (V) of a gas varies directly as the temperature (T) and inversely as the pressure (P). If $V = 48$ when $T = 320$ and $P = 20$, find V when $T = 280$ and $P = 30$.

30. The volume of a cylinder varies jointly as its altitude and the square of the radius of its base. If the volume of a cylinder is 1386 cubic centimeters when the radius of the base is 7 centimeters and its altitude is 9 centimeters, find the volume of a cylinder that has a base of radius 14 centimeters if the altitude of the cylinder is 5 centimeters.

31. The cost of labor varies jointly as the number of workers and the number of days that they work. If it costs $900 to have 15 people work for 5 days, how much will it cost to have 20 people work for 10 days?

32. The cost of publishing pamphlets varies directly as the number of pamphlets produced. If it costs $96 to publish 600 pamphlets, how much does it cost to publish 800 pamphlets?

Miscellaneous Problems

 In the previous problems, we have chosen numbers to make computations reasonable without the use of a calculator. However, often variation-type problems involve messy computations and the calculator becomes a very useful tool. Use your calculator to help solve the following problems.

33. The simple interest earned by a certain amount of money varies jointly as the rate of interest and the time (in years) that the money is invested.
   (a) If some money invested at 11% for 2 years earns $385, how much would the same amount earn at 12% for 1 year?
   (b) If some money invested at 12% for 3 years earns $819, how much would the same amount earn at 14% for 2 years?
   (c) If some money invested at 14% for 4 years earns $1960, how much would the same amount earn at 15% for 2 years?

34. The period (the time required for one complete oscillation) of a simple pendulum varies directly as the square root of its length. If a pendulum 9 inches long has a period of 2.4 seconds, find the period of a pendulum of length 12 inches. Express the answer to the nearest one-tenth of a second.

35. The volume of a cylinder varies jointly as its altitude and the square of the radius of its base. If the volume of a cylinder is 549.5 cubic meters when the radius of the base is 5 meters and its altitude is 7 meters, find the volume of a cylinder which has a base of radius 9 meters and an altitude of 14 meters.

36. If y is directly proportional to x and inversely proportional to the square of z, and if $y = 0.336$ when $x = 6$ and $z = 5$, find the constant of variation.

37. If y is inversely proportional to the square root of x, and $y = 0.08$ when $x = 225$, find y when $x = 625$.

Chapter 4 Summary

The function concept serves as a thread to tie this chapter together.

The Function Concept

DEFINITION 4.1

> A function f is a correspondence between two sets X and Y that assigns to each element x of set X one and only one element y of set Y. The element y being assigned is called the **image** of x. The set X is called the **domain** of the function and the set of all images is called the **range of the function**.

A function can also be thought of as a set of ordered pairs no two of which have the same first component. If each vertical line intersects a graph in no more than one point, then the graph represents a function.

If no member of the range is assigned to more than one member of the domain, then the function is a *one-to-one function*. If each horizontal line intersects the graph of a function in no more than one point, then the graph represents a one-to-one function.

Graphing Functions

Any function that can be written in the form

$$f(x) = ax + b,$$

where a and b are real numbers, is a **linear function**. The graph of a linear function is a straight line.

Any function that can be written in the form

$$f(x) = ax^2 + bx + c,$$

where a, b, and c are real numbers and $a \neq 0$, is a **quadratic function**. The graph of any quadratic function is a **parabola**, which can be drawn using either one of the following methods:

1. Express the function in the form $f(x) = a(x - h)^2 + k$ and use the values of a, h, and k to determine the parabola.

2. Express the function in the form $f(x) = ax^2 + bx + c$ and use the fact that the vertex is at

$$\left(-\frac{b}{2a}, f\left(-\frac{b}{2a}\right)\right)$$

and the axis of symmetry is

$$x = -\frac{b}{2a}.$$

The following suggestions are made for graphing functions that are unfamiliar.

1. Determine the domain of the function.

2. Find the intercepts.

3. Determine the type of symmetry that the equation exhibits.

4. Set up a table of values that satisfy the equation. The type of symmetry and the domain will affect your choice of values for x in the table.

5. Plot the points associated with the ordered pairs and connect them with a smooth curve. Then, if appropriate, reflect this part of the curve according to the symmetry possessed by the graph.

Operations on Functions

Sum of two functions: $\quad (f + g)(x) = f(x) + g(x)$

Difference of two functions: $\quad (f - g)(x) = f(x) - g(x)$

Product of two functions: $\quad (f \cdot g)(x) = f(x) \cdot g(x)$

Quotient of two functions: $\quad \left(\dfrac{f}{g}\right)(x) = \dfrac{f(x)}{g(x)}, \quad g(x) \neq 0$

DEFINITION 4.2

The **composition** of functions f and g is defined by

$$(f \circ g)(x) = f(g(x))$$

for all x in the domain of g such that $g(x)$ is in the domain of f.

Remember that the composition of functions is *not a commutative operation.*

Inverse Functions

DEFINITION 4.3

Let f be a one-to-one function with a domain of X and a range of Y. A function g, with a domain of Y and a range of X, is called the **inverse function of** f if

$$(f \circ g)(x) = x \quad \text{for every } x \text{ in } Y$$

and

$$(g \circ f)(x) = x \quad \text{for every } x \text{ in } X.$$

The inverse of a function f is denoted by f^{-1}. Graphically, two functions that are inverses of each other are mirror images with reference to the line $y = x$.

A systematic technique for finding the inverse of a function can be described as follows.

1. Let $y = f(x)$.
2. Interchange x and y.
3. Solve the equation for y in terms of x.
4. The inverse function $f^{-1}(x)$ is determined by the equation in Step 3.

Don't forget that the domain of f must equal the range of f^{-1}, and the domain of f^{-1} must equal the range of f.

Applications of Functions

Quadratic functions produce parabolas that have either a **minimum** or a **maximum value**. Therefore, a real-world minimum- or maximum-value problem that can be described by a quadratic function can be solved using the techniques of this chapter.

Relationships that involve **direct** and **inverse variation** can be expressed by equations that determine functions. The statement, "y varies directly as x" means

$$y = kx,$$

where k is the **constant of variation**. The statement, "y varies directly as the nth power of x $(n > 0)$" means

$$y = kx^n.$$

The statement, "y varies inversely as x" means

$$y = \frac{k}{x}.$$

The statement, "y varies inversely as the nth power of x $(n > 0)$" means

$$y = \frac{k}{x^n}.$$

The statement, "y varies jointly as x and w" means

$$y = kxw.$$

Chapter 4 Review Problem Set

1. If $f(x) = 3x^2 - 2x - 1$, find $f(2)$, $f(-1)$, and $f(-3)$.

2. For each of the following functions, find $\dfrac{f(a + h) - f(a)}{h}$.

(a) $f(x) = -5x + 4$ (b) $f(x) = 2x^2 - x + 4$

(c) $f(x) = -3x^2 + 2x - 5$

3. Determine the domain and range of the function $f(x) = x^2 + 5$.

4. Determine the domain of the function $f(x) = \dfrac{2}{2x^2 + 7x - 4}$.

5. Express the domain of $f(x) = \sqrt{x^2 - 7x + 10}$ using interval notation.

For Problems 6–15, graph each of the functions.

6. $f(x) = -2x + 2$ **7.** $f(x) = 2x^2 - 1$ **8.** $f(x) = -\sqrt{x - 2} + 1$

9. $f(x) = x^2 - 8x + 17$ **10.** $f(x) = -x^3 + 2$ **11.** $f(x) = 2|x - 1| + 3$

12. $f(x) = -2x^2 - 12x - 19$ **13.** $f(x) = -\frac{1}{3}x + 1$

14. $f(x) = -\dfrac{2}{x^2}$ **15.** $f(x) = 2|x| - x$

16. If $f(x) = 2x + 3$ and $g(x) = x^2 - 4x - 3$, find $f + g$, $f - g$, $f \cdot g$, and f/g.

For Problems 17–20, find $(f \circ g)(x)$ and $(g \circ f)(x)$. Also specify the domain for each.

17. $f(x) = 3x - 9$ and $g(x) = -2x + 7$ **18.** $f(x) = x^2 - 5$ and $g(x) = 5x - 4$

19. $f(x) = \sqrt{x - 5}$ and $g(x) = x + 2$ **20.** $f(x) = \dfrac{1}{x - 3}$ and $g(x) = \dfrac{1}{x + 2}$

Solve the following two problems.

21. If $f(x) = |x|$ and $g(x) = x^2 - x - 1$, find $(f \circ g)(1)$ and $(g \circ f)(-3)$.

22. Verify that $f(x) = x^2 + 8$ for $x \geq 0$ and $g(x) = \sqrt{x - 8}$ for $x \geq 8$ are inverse functions.

For Problems 23–26, determine whether the two functions are inverses of each other.

23. $f(x) = 7x - 1$ and $g(x) = \dfrac{x + 1}{7}$ **24.** $f(x) = -\frac{2}{3}x$ and $g(x) = \frac{3}{2}x$

25. $f(x) = x^2 - 6$ for $x \geq 0$ and $g(x) = \sqrt{x + 6}$ for $x \geq -6$

26. $f(x) = 2 - x^2$ for $x \geq 0$ and $g(x) = \sqrt{2 - x}$ for $x \leq 2$

For Problems 27–30, (a) find f^{-1} and (b) verify that $(f \circ f^{-1})(x) = x$ and $(f^{-1} \circ f)(x) = x$.

27. $f(x) = 4x + 5$ **28.** $f(x) = -3x - 7$

29. $f(x) = \frac{5}{6}x - \frac{1}{3}$ **30.** $f(x) = -2 - x^2$ for $x \geq 0$

31. A group of students is arranging a chartered flight to Europe. The charge per person is \$496 if 100 students go on the flight. If more than 100 students go, the charge per student is reduced by an amount equal to \$4 times the number of students above 100. How many students should the airline try to get in order to maximize their revenue?

32. If y varies directly as x and inversely as w, and if $y = 27$ when $x = 18$ and $w = 6$, find the constant of variation.

33. If y varies jointly as x and the square root of w, and if $y = 140$ when $x = 5$ and $w = 16$, find y when $x = 9$ and $w = 49$.

34. The weight of a body above the surface of the earth varies inversely as the square of its distance from the center of the earth. Assuming the radius of the earth to be 4000 miles, how much would a man weigh 1000 miles above the earth's surface if he weighs 200 pounds on the surface?

Exponential and Logarithmic Functions

5

5.1   Exponents and Exponential Functions

5.2   Applications of Exponential Functions

5.3   Logarithms

5.4   Logarithmic Functions

5.5   Exponential and Logarithmic Equations; Problem Solving

5.6   Computation with Common Logarithms (Optional)

In this chapter we will continue our study of exponents in several ways: (1) we will extend the meaning of an exponent, (2) we will work with some exponential functions, (3) we will introduce the concept of a logarithm, (4) we will work some logarithmic functions, and (5) we will use the concepts of exponent and logarithm to expand our problem-solving skills. Your calculator will be a valuable tool throughout this chapter.

5.1

Exponents and Exponential Functions

In Chapter 1, the expression b^n was defined to mean n factors of b, where n is any positive integer and b is any real number. For example,

$$2^3 = 2 \cdot 2 \cdot 2 = 8, \qquad (\tfrac{1}{3})^4 = (\tfrac{1}{3})(\tfrac{1}{3})(\tfrac{1}{3})(\tfrac{1}{3}) = \tfrac{1}{81},$$

$$(-4)^2 = (-4)(-4) = 16, \qquad -(0.5)^3 = -[(0.5)(0.5)(0.5)] = -0.125.$$

Also in Chapter 1, by defining $b^0 = 1$ and $b^{-n} = 1/b^n$, where n is any positive integer and b is any nonzero real number, we extended the concept of an exponent to include all integers. For example,

$$(0.76)^0 = 1, \qquad 2^{-3} = \frac{1}{2^3} = \frac{1}{8},$$

$$\left(\frac{2}{3}\right)^{-2} = \frac{1}{\left(\dfrac{2}{3}\right)^2} = \frac{1}{\dfrac{4}{9}} = \frac{9}{4}, \qquad (0.4)^{-1} = \frac{1}{(0.4)^1} = \frac{1}{0.4} = 2.5.$$

Finally, in Chapter 1 we provided for the use of any rational number as an exponent by defining

$$b^{m/n} = \sqrt[n]{b^m} = (\sqrt[n]{b})^m,$$

where n is a positive integer greater than 1 and b is a real number such that $\sqrt[n]{b}$ exists. For example,

$$27^{2/3} = (\sqrt[3]{27})^2 = 9, \qquad 16^{1/4} = \sqrt[4]{16^1} = 2,$$

$$\left(\frac{1}{9}\right)^{1/2} = \sqrt{\frac{1}{9}} = \frac{1}{3}, \qquad 32^{-1/5} = \frac{1}{32^{1/5}} = \frac{1}{\sqrt[5]{32}} = \frac{1}{2}.$$

If we were to make a formal extension of the concept of an exponent to include the use of irrational numbers, we would require some ideas from calculus; that is beyond the scope of this text. However, we can give you a brief glimpse at the general idea involved. Consider the number $2^{\sqrt{3}}$. By using the nonterminating and nonrepeating decimal representation 1.73205 . . . for $\sqrt{3}$, we can form the sequence of numbers 2^1, $2^{1.7}$, $2^{1.73}$, $2^{1.732}$, $2^{1.7320}$, $2^{1.73205}$, ... It is a reasonable idea that each successive power gets closer to $2^{\sqrt{3}}$. This is precisely what happens if n is irrational and b^n is properly defined by using the concept of a *limit*. Furthermore, this ensures that an expression such as 2^x will yield exactly one value for each value of x.

So from now on we can use any real number as an exponent and the basic properties stated in Chapter 1 can be extended to include all real numbers as exponents. Let's restate those properties with the restriction that the bases a and b are to be positive numbers to avoid expressions such as $(-4)^{1/2}$, which do not represent real numbers.

PROPERTY 5.1

> If a and b are positive real numbers and m and n are any real numbers, then the following properties hold.
>
> 1. $b^n \cdot b^m = b^{n+m}$ Product of two powers
> 2. $(b^n)^m = b^{mn}$ Power of a power
> 3. $(ab)^n = a^n b^n$ Power of a product
> 4. $\left(\dfrac{a}{b}\right)^n = \dfrac{a^n}{b^n}$ Power of a quotient
> 5. $\dfrac{b^n}{b^m} = b^{n-m}$ Quotient of two powers

Another property that can be used to solve certain types of equations involving exponents can be stated as follows.

PROPERTY 5.2

> If $b > 0$ but $b \neq 1$, and if m and n are real numbers, then
>
> $b^n = b^m$ if and only if $n = m$.

The following three examples illustrate the use of Property 5.2.

EXAMPLE 1

Solve $2^x = 32$.

Solution

$$2^x = 32$$
$$2^x = 2^5 \qquad 32 = 2^5$$
$$x = 5 \qquad \text{Apply Property 5.2.}$$

The solution set is $\{5\}$.

EXAMPLE 2

Solve $2^{3x} = \frac{1}{64}$.

Solution

$$2^{3x} = \frac{1}{64}$$

$$2^{3x} = \frac{1}{2^6}$$

$$2^{3x} = 2^{-6}$$
$$3x = -6 \qquad \text{Apply Property 5.2.}$$
$$x = -2$$

The solution set is $\{-2\}$.

EXAMPLE 3

Solve $(\frac{1}{5})^{x-4} = \frac{1}{125}$.

Solution

$$(\tfrac{1}{5})^{x-4} = \tfrac{1}{125}$$
$$(\tfrac{1}{5})^{x-4} = (\tfrac{1}{5})^3$$
$$x - 4 = 3 \qquad \text{Apply Property 5.2}$$
$$x = 7$$

The solution set is $\{7\}$.

If b is any positive number, then the expression b^x designates exactly one real number for every real value of x. Therefore, the equation $f(x) = b^x$ defines a function whose domain is the set of real numbers. Furthermore, if we place the additional restriction $b \neq 1$, then any equation of the form $f(x) = b^x$ describes a one-to-one function and is called an **exponential function**. This leads to the following definition.

DEFINITION 5.1

If $b > 0$ and $b \neq 1$, then the function f defined by

$$f(x) = b^x,$$

where x is any real number, is called the **exponential function with base b**.

Remark: The function $f(x) = 1^x$ is a constant function and therefore it is not a one-to-one function. Remember from Chapter 4 that one-to-one functions have inverses; this becomes a key issue in a later section.

Now let's consider graphing some exponential functions.

x	2^x
-2	$\frac{1}{4}$
-1	$\frac{1}{2}$
0	1
1	2
2	4
3	8

EXAMPLE 4

Graph the function $f(x) = 2^x$.

Solution Let's set up a table of values, keeping in mind that the domain is the set of real numbers and the equation $f(x) = 2^x$ exhibits no symmetry. Plotting these points and connecting them with a smooth curve produces Figure 5.1.

In the table for Example 4, we chose integral values for x to keep the computation simple. However, with the use of a calculator we could easily acquire functional values by using nonintegral exponents. Consider the following additional values for $f(x) = 2^x$.

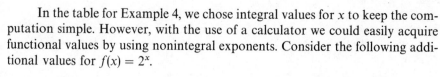

$$f(0.5) \approx 1.41, \qquad f(1.7) \approx 3.25, \qquad (\approx \text{means "is approximately equal to")}$$
$$f(-0.5) \approx 0.71, \qquad f(-2.6) \approx 0.16.$$

Use your calculator to check these results. Also notice that the points generated by these values do fit the graph in Figure 5.1.

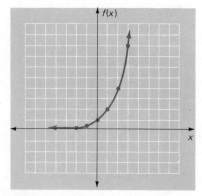

Figure 5.1

x	$\left(\frac{1}{2}\right)^x$
-2	4
-1	2
0	1
1	$\frac{1}{2}$
2	$\frac{1}{4}$
3	$\frac{1}{8}$

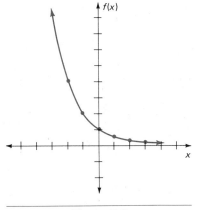

Figure 5.2

EXAMPLE 5

Graph $f(x) = \left(\frac{1}{2}\right)^x$.

Solution   Again, let's set up a table of values, plot the points, and connect them with a smooth curve. The graph is shown in Figure 5.2.

Remark: Since $\left(\frac{1}{2}\right)^x = 1/2^x = 2^{-x}$, the graphs of $f(x) = 2^x$ and $f(x) = \left(\frac{1}{2}\right)^x$ are reflections of each other across the y-axis. Therefore, Figure 5.2 could have been drawn by reflecting Figure 5.1 across the y-axis.

The graphs in Figures 5.1 and 5.2 illustrate a "general behavior pattern" of exponential functions. That is, if $b > 1$, then the graph of $f(x) = b^x$ goes up to the right, and the function is called an **increasing function**. If $0 < b < 1$, then the graph of $f(x) = b^x$ goes down to the right, and the function is called a **decreasing function**. These facts are illustrated in Figure 5.3. Notice that $b^0 = 1$ for any $b > 0$; thus, **all graphs of $f(x) = b^x$ contain the point $(0, 1)$**.

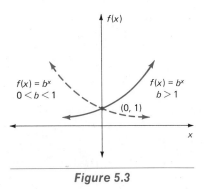

Figure 5.3

As you graph exponential functions, don't forget to use your previous graphing experience. For example, consider the following functions.

1. The graph of $f(x) = 2^x + 3$ is the graph of $f(x) = 2^x$ *moved up* 3 *units.*
2. The graph of $f(x) = 2^{x-4}$ is the graph of $f(x) = 2^x$ *moved to the right* 4 *units.*
3. The graph of $f(x) = -2^x$ is the graph of $f(x) = 2^x$ *reflected across the x-axis.*
4. The graph of $f(x) = 2^x + 2^{-x}$ is symmetrical with respect to the y-axis because $f(-x) = 2^{-x} + 2^x = f(x)$.

Furthermore, if you are faced with an exponential function that is not of the form $f(x) = b^x$ nor a variation thereof, don't forget the graphing suggestions offered in Chapter 3. Let's consider one such example.

EXAMPLE 6

Graph $f(x) = 2^{-x^2}$.

Solution Since $f(-x) = 2^{-(-x)^2} = 2^{-x^2} = f(x)$, we know that this curve is symmetrical with respect to the y-axis. Therefore, let's set up a table of values using nonnegative values for x. Plotting these points, connecting them with a smooth curve, and reflecting this portion of the curve across the y-axis produces the graph in Figure 5.4.

x	2^{-x^2}
0	1
$\frac{1}{2}$	0.84
1	0.5
$\frac{3}{2}$	0.21
2	0.06

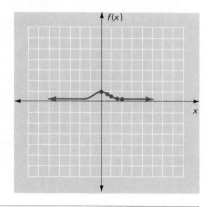

Figure 5.4

Problem Set 5.1

Solve each of the following equations.

1. $3^x = 27$ **2.** $2^x = 64$ **3.** $(\frac{1}{2})^x = \frac{1}{8}$ **4.** $(\frac{1}{2})^n = 4$

5. $3^{-x} = \frac{1}{81}$ **6.** $3^{x+1} = 9$ **7.** $5^{2n-1} = 125$ **8.** $2^{3-n} = 8$

9. $(\frac{2}{3})^t = \frac{9}{4}$ **10.** $(\frac{3}{4})^n = \frac{64}{27}$ **11.** $4^{3x-1} = 256$ **12.** $16^x = 64$

13. $4^n = 8$ **14.** $27^{4x} = 9^{x+1}$ **15.** $32^x = 16^{1-x}$ **16.** $(\frac{1}{8})^{-2t} = 2^{t+3}$

Graph each of the following exponential functions.

17. $f(x) = 3^x$ **18.** $f(x) = (\frac{1}{3})^x$ **19.** $f(x) = 4^x$

20. $f(x) = (\frac{1}{4})^x$ **21.** $f(x) = (\frac{2}{3})^x$ **22.** $f(x) = (\frac{3}{2})^x$

23. $f(x) = 2^x + 1$ **24.** $f(x) = 2^x - 3$ **25.** $f(x) = 2^{x-1}$

26. $f(x) = 2^{x+2}$ **27.** $f(x) = -3^x$ **28.** $f(x) = -2^x$

29. $f(x) = 2^{-x+1}$ **30.** $f(x) = 2^{-x-2}$ **31.** $f(x) = 2^x + 2^{-x}$

32. $f(x) = 2^{x^2}$ **33.** $f(x) = 3^{1-x^2}$ **34.** $f(x) = 2^{|x|}$

35. $f(x) = 2^{-|x|}$ **36.** $f(x) = 2^x - 2^{-x}$

5.2

Applications of Exponential Functions

Many real-world situations exhibiting growth or decay can be represented by equations that describe exponential functions. For example, suppose that an economist predicts an annual inflation rate of 5% per year for the next 10 years. This means that an item that presently costs $8 will cost $8(105%) =

$8(1.05) = \$8.40$ in a year from now. The same item will cost $[\$8(105\%)] \times (105\%) = \$8(1.05)^2 = \$8.82$ in 2 years. In general, the equation

$$P = P_0(1.05)^t$$

yields the predicted price P of an item in t years if the present cost is P_0 and the annual inflation rate is 5%. By using this equation, we can look at some future prices based on the prediction of a 5% inflation rate.

1. A \$3.27 container of hot cocoa mix will cost $\$3.27(1.05)^3 = \$3.79$ in 3 years.
2. A \$4.07 jar of coffee will cost $\$4.07(1.05)^5 = \$5.19$ in 5 years.
3. A \$9500 car will cost $\$9500(1.05)^7 = \$13,367$ (to the nearest dollar) in 7 years.

Suppose that it is estimated that the value of a car depreciates 15% per year for the first 5 years. Therefore, a car costing \$9500 will be worth $\$9500 \times (100\% - 15\%) = \$9500(85\%) = \$9500(0.85) = \$8075$ in 1 year. In 2 years the value of the car will have depreciated to $9500(0.85)^2 = \$6864$ (to the nearest dollar). The equation

$$V = V_0(0.85)^t$$

yields the value V of a car in t years if the initial cost is V_0 and it depreciates 15% per year. Therefore, we can estimate some car values to the nearest dollar:

1. A \$6900 car will be worth $\$6900(0.85)^3 = \$4237$ in 3 years.
2. A \$10,900 car will be worth $\$10,900(0.85)^4 = \$5690$ in 4 years.
3. A \$13,000 car will be worth $\$13,000(0.85)^5 = \$5768$ in 5 years.

Compound Interest

Compound interest provides another illustration of exponential growth. Suppose that \$500 (called the **principal**) is invested at an interest rate of 8% **compounded annually**. The interest earned the first year is $\$500(0.08) = \$40$, and this amount is added to the original \$500 to form a new principal of \$540 for the second year. The interest earned during the second year is $\$540(0.08) = \$43.20$, and this amount is added to \$540 to form a new principal of \$583.20 for the third year. Each year a new principal is formed by reinvesting the interest earned during that year.

In general, suppose that a sum of money P (called the principal) is invested at an interest rate of r percent compounded annually. The interest earned the first year is Pr, and the new principal for the second year is $P + Pr$ or $P(1 + r)$. Note that the new principal for the second year can be found by multiplying the original principal P by $(1 + r)$. In a like fashion, the new principal for the third year can be found by multiplying the previous principal, $P(1 + r)$, by $1 + r$, thus obtaining $P(1 + r)^2$. If this process is continued, then after t years the total amount of money accumulated, A, is given by

$$A = P(1 + r)^t.$$

Consider the following examples of investments made at a certain rate of interest compounded annually.

1. $750 invested for 5 years at 9% compounded annually produces

$$A = \$750(1.09)^5 = \$1153.97.$$

2. $1000 invested for 10 years at 11% compounded annually produces

$$A = \$1000(1.11)^{10} = \$2839.42.$$

3. $5000 invested for 20 years at 12% compounded annually produces

$$A = \$5000(1.12)^{20} = \$48{,}231.47.$$

If money invested at a certain rate of interest is to be compounded more than once a year, then the basic formula $A = P(1 + r)^t$ can be adjusted according to the number of compounding periods in a year. For example, for **compounding semiannually**, the formula becomes

$$A = P\left(1 + \frac{r}{2}\right)^{2t},$$

and for **compounding quarterly**, the formula becomes

$$A = P\left(1 + \frac{r}{4}\right)^{4t}.$$

In general, if n represents the number of **compounding periods** in a year, the formula becomes

$$A = P\left(1 + \frac{r}{n}\right)^{nt}.$$

The following examples illustrate the use of the formula.

1. $750 invested for 5 years at 9% compounded semiannually produces

$$A = \$750\left(1 + \frac{0.09}{2}\right)^{2(5)} = \$750(1.045)^{10} = \$1164.73.$$

2. $1000 invested for 10 years at 11% compounded quarterly produces

$$A = \$1000\left(1 + \frac{0.11}{4}\right)^{4(10)} = \$1000(1.0275)^{40} = \$2959.87.$$

3. $5000 invested for 20 years at 12% compounded monthly produces

$$A = \$5000\left(1 + \frac{0.12}{12}\right)^{12(20)} = \$5000(1.01)^{240} = \$54{,}462.77.$$

You may find it interesting to compare these results with those obtained earlier for compounding annually.

The Number e

An interesting situation occurs if we consider the compound-interest formula for $P = \$1$, $r = 100\%$, and $t = 1$ year. The formula becomes

$$A = 1\left(1 + \frac{1}{n}\right)^n.$$

n	$\left(1 + \dfrac{1}{n}\right)^n$
5	2.4883200
10	2.5937425
100	2.7048138
1000	2.7169236
10,000	2.7181459
100,000	2.7182818

The table at the left shows some values, rounded to seven decimals, for different values of n. The table suggests that as n increases, the value of $\left(1 + \dfrac{1}{n}\right)^n$ gets closer to some fixed number. This does in fact happen and the fixed number is called e. To five decimal places,

$$e = 2.71828.$$

Exponential expressions using e as a base are found in many real-world applications. Before considering some of those applications, let's take a look at the graph of the basic exponential function using e as a base, namely, $f(x) = e^x$.

EXAMPLE 1

Graph $f(x) = e^x$.

Solution For graphing purposes, let's use 2.72 as an approximation for e and express the functional values to the nearest tenth. The table on the left can be easily obtained by using a calculator. The graph of $f(x) = e^x$ is shown in Figure 5.5.

x	$(2.72)^x$
0	1.0
1	2.7
2	7.4
−1	0.4
−2	0.1

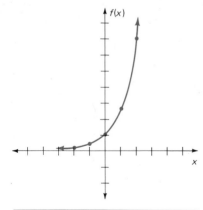

Figure 5.5

Let's return to the concept of compound interest. If the number of compounding periods in a year is increased indefinitely, we arrive at the concept of **compounding continuously**. Mathematically, this can be accomplished by applying the limit concept to the expression

$$P\left(1 + \frac{r}{n}\right)^{nt}.$$

We will not show the details here, but the following result is obtained. The formula

$$A = Pe^{rt}$$

yields the accumulated value (A) of a sum of money (P) that has been invested for t years at a rate of r percent compounded continuously. The following examples illustrate the use of this formula. (We are using 2.718 as an approximation for e in these calculations.)

1. $750 invested for 5 years at 9% compounded continuously produces

$$A = \$750(2.718)^{(0.09)(5)} = \$750(2.718)^{0.45} = \$1176.18.$$

2. $1000 invested for 10 years at 11% compounded continuously produces

$$A = \$1000(2.718)^{(0.11)(10)} = \$1000(2.718)^{1.1} = \$3003.82.$$

3. $5000 invested for 20 years at 12% compounded continuously produces

$$A = \$5000(2.718)^{(0.12)(20)} = \$5000(2.718)^{2.4} = \$55,102.17.$$

Again you may find it interesting to compare these results to those obtained earlier using different compounding periods.

The Law of Exponential Growth

The ideas behind compounding continuously carry over to other growth situations. The equation

$$Q(t) = Q_0 e^{kt} \qquad \text{Law of exponential growth}$$

is used as a mathematical model for numerous *growth* and *decay* applications. In this equation, $Q(t)$ represents the quantity of a given substance at any time t, the quantity Q_0 is the *initial amount* of the substance (when $t = 0$), and k is a constant that depends on the particular application. If $k < 0$, then $Q(t)$ decreases as t increases and the model is referred to as the **law of decay**.

Let's consider some growth and decay applications.

EXAMPLE 2

Suppose that in a certain bacterial culture, the equation $Q(t) = 15{,}000e^{0.3t}$ expresses the number of bacteria present as a function of the time, where t is expressed in hours. How many bacteria are present at the end of 3 hours?

Solution Using $Q(t) = 15{,}000e^{0.3t}$, we obtain

$$Q(3) = 15{,}000(2.718)^{0.3(3)}$$
$$= 15{,}000(2.718)^{0.9}$$
$$= 36{,}981 \qquad \text{To the nearest whole number}$$

EXAMPLE 3

Suppose the number of bacteria present in a certain culture after t minutes is given by the equation $Q(t) = Q_0 e^{0.05t}$, where Q_0 represents the initial number of bacteria. If 5000 bacteria are present after 20 minutes, how many bacteria were present initially?

Solution Since $Q(20) = 5000$, we obtain

$$5000 = Q_0(2.718)^{(0.05)(20)}$$
$$5000 = Q_0(2.718)^1$$

$$\frac{5000}{2.718} = Q_0$$

$$1840 = Q_0 \qquad \text{To the nearest whole number}$$

Therefore, there were approximately 1840 bacteria present initially.

EXAMPLE 4

The number of grams of a certain radioactive substance present after t seconds is given by the equation $Q(t) = 200e^{-0.3t}$. How many grams remain after 7 seconds?

Solution Using $Q(t) = 200e^{-0.3t}$, we obtain

$$Q(7) = 200e^{(-0.3)(7)}$$
$$= 200e^{-2.1}$$
$$= 200(2.718)^{-2.1}$$
$$= 24.5 \qquad \text{To the nearest tenth}$$

Thus, approximately 24.5 grams remain after 7 seconds.

The concept of *half-life* plays an important role in many decay-type problems. The **half-life** of a substance is the time required for one-half of the substance to disintegrate or disappear. For example, the half-life of radium is approximately 1600 years. Thus, if 2000 grams of radium are present now, 1600 years from now 1000 grams will be present.

EXAMPLE 5

Suppose that a certain substance has a half-life of 5 years. If there are presently 100 grams of the substance, then the equation $Q(t) = 100(2)^{-t/5}$ yields the amount remaining after t years. How much remains after 10 years? 18 years?

Solution

$$Q(10) = 100(2)^{-10/5}$$
$$= 100(2)^{-2}$$
$$= 100(\tfrac{1}{4})$$
$$= 25$$

Therefore, 25 grams remain after 10 years.

$$Q(18) = 100(2)^{-18/5}$$
$$= 100(2)^{-3.6}$$
$$= 8 \qquad \text{To the nearest whole number}$$

Therefore, approximately 8 grams remain after 18 years.

Problem Set 5.2

1. Assuming that the rate of inflation is 7% per year, the equation $P = P_0(1.07)^t$ yields the predicted price P of an item in t years if it presently costs P_0. Find the predicted price of each of the following items for the indicated years ahead.
   (a) $.55 can of soup in 3 years
   (b) $3.43 container of cocoa mix in 5 years
   (c) $1.76 jar of coffee creamer in 4 years
   (d) $.44 can of beans and bacon in 10 years
   (e) $9000 car in 5 years (to the nearest dollar)
   (f) $50,000 house in 8 years (to the nearest dollar)
   (g) $500 TV set in 7 years (to the nearest dollar)

2. Suppose that it is estimated that the value of a car depreciates 20% per year for the first 5 years. The equation $A = P_0(0.8)^t$ yields the value (A) of a car after t years if the original price is P_0. Find the value (to the nearest dollar) of each of the following cars after the indicated time.
   (a) $9000 car after 4 years
   (b) $5295 car after 2 years
   (c) $6395 car after 5 years
   (d) $15,595 car after 3 years

For Problems 3–12, use the formula

$$A = P\left[1 + \frac{r}{n}\right]^{nt}$$

to find the total amount of money accumulated at the end of the indicated time period for each of the following investments. Estimate to the nearest cent.

3. $250 for 5 years at 9% compounded annually

4. $350 for 7 years at 11% compounded annually

5. $300 for 6 years at 8% compounded semiannually

6. $450 for 10 years at 10% compounded semiannually

7. $600 for 12 years at 12% compounded quarterly

8. $750 for 15 years at 9% compounded quarterly

9. $1000 for 5 years at 12% compounded monthly

10. $1250 for 8 years at 9% compounded monthly

11. $600 for 10 years at $8\frac{1}{2}$% compounded annually

12. $1500 for 15 years at $9\frac{1}{4}$% compounded semiannually

For Problems 13–18, use the formula $A = Pe^{rt}$ to find the total amount of money accumulated at the end of the indicated time period by compounding continuously. Use 2.718 as an approximation for e.

13. $500 for 5 years at 8%

14. $750 for 7 years at 9%

15. $800 for 10 years at 10%

16. $1000 for 10 years at $9\frac{1}{2}$%

17. $1500 for 12 years at 9%

18. $2000 for 20 years at 11%

19. Complete the following chart, which will illustrate what happens to $1000 in 10 years based on different rates of interest and different compounding periods. Round your answers to the nearest dollar.

$1000 For 10 Years

	8%	10%	12%	14%
COMPOUNDED ANNUALLY				
COMPOUNDED SEMIANNUALLY				
COMPOUNDED QUARTERLY				
COMPOUNDED MONTHLY				
COMPOUNDED CONTINUOUSLY				

20. Complete the following chart, which will illustrate what happens to $1000 invested at 12% for different lengths of time and different compounding periods. Round all of your answers to the nearest dollar.

$1000 At 12%

	1 YEAR	5 YEARS	10 YEARS	20 YEARS
COMPOUNDED ANNUALLY				
COMPOUNDED-SEMIANNUALLY				
COMPOUNDED QUARTERLY				
COMPOUNDED MONTHLY				
COMPOUNDED CONTINUOUSLY				

21. Complete the following chart, which will illustrate what happens to $1000 invested at various rates of interest for different lengths of time, always compounded continuously.

$1000 Compounded Continuously

	8%	10%	12%	14%
5 YEARS				
10 YEARS				
15 YEARS				
20 YEARS				
25 YEARS				

For Problems 22–27, graph each of the exponential functions.

22. $f(x) = e^x + 1$ **23.** $f(x) = e^x - 2$ **24.** $f(x) = 2e^x$

25. $f(x) = -e^x$ **26.** $f(x) = e^{2x}$ **27.** $f(x) = e^{-x}$

28. Suppose that in a certain bacterial culture, the equation $Q(t) = 1000e^{0.4t}$ expresses the number of bacteria present as a function of the time t, where t is expressed in hours. How many bacteria are present at the end of 2 hours? 3 hours? 5 hours?

29. The number of bacteria present at a given time under certain conditions is given by the equation $Q(t) = 5000e^{0.05t}$, where t is expressed in minutes. How many bacteria are present at the end of 10 minutes? 30 minutes? 1 hour?

30. The number of bacteria present in a certain culture after t hours is given by the equation $Q(t) = Q_0e^{0.3t}$, where Q_0 represents the initial number of bacteria. If 6640 bacteria are present after 4 hours, how many bacteria were present initially?

31. The number of grams of a certain radioactive substance present after t seconds is given by the equation $Q(t) = 1500e^{-0.4t}$. How many grams remain after 5 seconds? 10 seconds? 20 seconds?

32. The atmospheric pressure, measured in pounds per square inch, is a function of the altitude above sea level. The equation $P(a) = 14.7e^{-0.21a}$, where a is the altitude measured in miles, can be used to approximate atmospheric pressure. Find the atmospheric pressure at each of the following locations.
(a) Mount McKinley in Alaska: altitude of 3.85 miles
(b) Denver: the "mile-high city"
(c) Asheville, North Carolina: altitude of 1985 feet
(d) Phoenix, Arizona: altitude of 1090 feet

33. Suppose that the present population of a city is 75,000 and the equation $P(t) = 75,000e^{0.01t}$ is used to estimate future growth. Estimate the population 10 years from now; 15 years from now; 25 years from now.

34. Suppose that a certain substance has a half-life of 20 years. If there are presently 2500 milligrams of the substance, then the equation $Q(t) = 2500(2)^{-t/20}$ yields the amount remaining after t years. How much remains after 40 years? 50 years?

35. Assume that the half-life of a certain radioactive substance is 1200 years and that the amount of the substance remaining after t years is given by the equation $Q(t) = 500(2)^{-t/1200}$. If the substance is measured in grams, how many grams are present now? How many grams will remain after 1500 years?

36. The half-life of radium is approximately 1600 years. If the initial amount is Q_0 milligrams, then the quantity $Q(t)$ remaining after t years is given by $Q(t) = Q_0 2^{kt}$. Find the value of k.

37. The half-life of a certain substance is 25 years. If the initial amount is Q_0 grams, then the quantity $Q(t)$ remaining after t years is given by $Q(t) = Q_0(2)^{kt}$. Find the value of k.

Miscellaneous Problems

Graph each of the following exponential functions.

38. $f(x) = x(2^x)$

39. $f(x) = \dfrac{e^x + e^{-x}}{2}$

40. $f(x) = \dfrac{2}{e^x + e^{-x}}$

41. $f(x) = \dfrac{e^x - e^{-x}}{2}$

42. $f(x) = \dfrac{2}{e^x - e^{-x}}$

5.3

Logarithms

In Sections 5.1 and 5.2, we gave meaning to exponential expressions of the form b^n, where b is any positive real number and n is any real number; we next used exponential expressions of the form b^n to define exponential functions; and then we used exponential functions to help solve problems. In the next three sections we will follow the same basic pattern with respect to a new concept, that of a *logarithm*. Let's begin with the following definition.

DEFINITION 5.2

> If r is any positive real number, then the unique exponent t such that $b^t = r$ is called the **logarithm of r with base b** and is denoted by $\log_b r$.

According to Definition 5.2, the logarithm of 16 base 2 is the exponent t such that $2^t = 16$; thus, we can write $\log_2 16 = 4$. Likewise, we can write $\log_{10} 1000 = 3$ because $10^3 = 1000$. In general, Definition 5.2 can be remembered in terms of the statement

$$\log_b r = t \quad \text{is equivalent to} \quad b^t = r.$$

Therefore, we can easily switch back and forth between exponential and logarithmic forms of equations, as the next examples illustrate.

$$\log_2 8 = 3 \quad \text{is equivalent to} \quad 2^3 = 8.$$
$$\log_{10} 100 = 2 \quad \text{is equivalent to} \quad 10^2 = 100.$$
$$\log_3 81 = 4 \quad \text{is equivalent to} \quad 3^4 = 81.$$
$$\log_{10} 0.001 = -3 \quad \text{is equivalent to} \quad 10^{-3} = 0.001.$$
$$2^7 = 128 \quad \text{is equivalent to} \quad \log_2 128 = 7.$$
$$5^3 = 125 \quad \text{is equivalent to} \quad \log_5 125 = 3.$$
$$(\tfrac{1}{2})^4 = \tfrac{1}{16} \quad \text{is equivalent to} \quad \log_{1/2} \tfrac{1}{16} = 4.$$
$$10^{-2} = 0.01 \quad \text{is equivalent to} \quad \log_{10} 0.01 = -2.$$

Some logarithms can be determined by changing to exponential form and using the properties of exponents, as the next two examples illustrate.

EXAMPLE 1

Evaluate $\log_{10} 0.0001$.

Solution Let $\log_{10} 0.0001 = x$. Then by changing to exponential form we have $10^x = 0.0001$, which can be solved as follows.

$$10^x = 0.0001$$
$$10^x = 10^{-4} \qquad 0.0001 = \frac{1}{10{,}000} = \frac{1}{10^4} = 10^{-4}$$
$$x = -4$$

Thus, we have $\log_{10} 0.0001 = -4$.

EXAMPLE 2

Evaluate $\log_9(\sqrt[5]{27}/3)$.

Solution Let $\log_9(\sqrt[5]{27}/3) = x$. Then by changing to exponential form we have $9^x = \sqrt[5]{27}/3$, which can be solved as follows.

$$9^x = \frac{(27)^{1/5}}{3}$$
$$(3^2)^x = \frac{(3^3)^{1/5}}{3}$$
$$3^{2x} = \frac{3^{3/5}}{3}$$
$$3^{2x} = 3^{-2/5}$$
$$2x = -\frac{2}{5}$$
$$x = -\frac{1}{5}$$

Therefore, we have $\log_9(\sqrt[5]{27}/3) = -\frac{1}{5}$.

Some equations involving logarithms can also be solved by changing to exponential form and using our knowledge of exponents.

EXAMPLE 3

Solve $\log_8 x = \frac{2}{3}$.

Solution Changing $\log_8 x = \frac{2}{3}$ to exponential form, we obtain

$$8^{2/3} = x.$$

Therefore,

$$x = (\sqrt[3]{8})^2 = 2^2 = 4.$$

The solution set is $\{4\}$.

EXAMPLE 4

Solve $\log_b(\frac{27}{64}) = 3$.

Solution Changing $\log_b(\frac{27}{64}) = 3$ to exponential form, we obtain

$$b^3 = \frac{27}{64}.$$

Therefore,

$$b = \sqrt[3]{\frac{27}{64}} = \frac{3}{4}.$$

The solution set is $\{\frac{3}{4}\}$.

Properties of Logarithms

There are some properties of logarithms that are a direct consequence of Definition 5.2 and the properties of exponents. For example, by writing the exponential equations $b^1 = b$ and $b^0 = 1$ in logarithmic form, the following property is obtained.

PROPERTY 5.3

> For $b > 0$ and $b \neq 1$,
>
> $$\log_b b = 1 \quad \text{and} \quad \log_b 1 = 0.$$

Therefore, according to Property 5.3, we can write

$$\log_{10} 10 = 1, \quad \log_4 4 = 1,$$
$$\log_{10} 1 = 0, \quad \log_5 1 = 0.$$

Also, from Definition 5.2 we know that $\log_b r$ is the exponent t such that $b^t = r$. Therefore, raising b to the $\log_b r$ power must produce r. This fact is stated in Property 5.4.

PROPERTY 5.4

> For $b > 0$, $b \neq 1$, and $r > 0$,
>
> $$b^{\log_b r} = r.$$

Therefore, according to Property 5.4, we can write

$$10^{\log_{10} 72} = 72, \qquad 3^{\log_3 85} = 85, \qquad e^{\log_e 7} = 7.$$

Because a logarithm is by definition an exponent, it is reasonable to predict that logarithms will have some properties that correspond to the basic exponential properties. This is an accurate prediction; these properties provide a basis for computational work with logarithms. Let's state the first of these properties and show how it can be verified by using our knowledge of exponents.

PROPERTY 5.5

For positive numbers b, r, and s, where $b \neq 1$,

$$\log_b rs = \log_b r + \log_b s.$$

To verify Property 5.5, we can proceed as follows. Let $m = \log_b r$ and $n = \log_b s$. Change each of these equations to exponential form:

$$m = \log_b r \quad \text{becomes} \quad r = b^m;$$

$$n = \log_b s \quad \text{becomes} \quad s = b^n.$$

Thus, the product rs becomes

$$rs = b^m \cdot b^n = b^{m+n}.$$

Now, by changing $rs = b^{m+n}$ back to logarithmic form, we obtain

$$\log_b rs = m + n.$$

Replacing m with $\log_b r$ and n with $\log_b s$ yields

$$\log_b rs = \log_b r + \log_b s.$$

The following two examples illustrate a use of Property 5.5.

EXAMPLE 5

If $\log_2 5 = 2.3222$ and $\log_2 3 = 1.5850$, evaluate $\log_2 15$.

Solution Because $15 = 5 \cdot 3$, we can apply Property 5.5 as follows:

$$\begin{aligned}
\log_2 15 &= \log_2(5 \cdot 3) \\
&= \log_2 5 + \log_2 3 \\
&= 2.3222 + 1.5850 = 3.9072.
\end{aligned}$$

EXAMPLE 6

If $\log_{10} 178 = 2.2504$ and $\log_{10} 89 = 1.9494$, evaluate $\log_{10}(178 \cdot 89)$.

Solution

$$\begin{aligned}
\log_{10}(178 \cdot 89) &= \log_{10} 178 + \log_{10} 89 \\
&= 2.2504 + 1.9494 = 4.1998.
\end{aligned}$$

Since $b^m/b^n = b^{m-n}$, we would expect a corresponding property pertaining to logarithms. Property 5.6 is that property. It can be verified by using an approach similar to the one used for Property 5.5. This verification is left for you to do as an exercise in the next problem set.

PROPERTY 5.6

For positive numbers b, r, and s, where $b \neq 1$,

$$\log_b\left(\frac{r}{s}\right) = \log_b r - \log_b s.$$

Property 5.6 can be used to change a division problem into an equivalent subtraction problem, as the next two examples illustrate.

EXAMPLE 7

If $\log_5 36 = 2.2265$ and $\log_5 4 = 0.8614$, evaluate $\log_5 9$.

Solution   Since $9 = \frac{36}{4}$, we can use Property 5.6 as follows:

$$
\begin{aligned}
\log_5 9 &= \log_5\left(\tfrac{36}{4}\right) \\
&= \log_5 36 - \log_5 4 \\
&= 2.2265 - 0.8614 = 1.3651.
\end{aligned}
$$

EXAMPLE 8

Evaluate $\log_{10}\left(\frac{379}{86}\right)$, given that $\log_{10} 379 = 2.5786$ and $\log_{10} 86 = 1.9345$.

Solution

$$
\begin{aligned}
\log_{10}\left(\tfrac{379}{86}\right) &= \log_{10} 379 - \log_{10} 86 \\
&= 2.5786 - 1.9345 = 0.6441.
\end{aligned}
$$

Another property of exponents states that $(b^n)^m = b^{mn}$. The corresponding property of logarithms is stated in Property 5.7. Again, we leave the verification of this property as an exercise for you to do in the next set of problems.

PROPERTY 5.7

If r is a positive real number, b is a positive real number other than 1, and p is any real number, then

$$\log_b r^p = p(\log_b r).$$

The next two examples illustrate a use of Property 5.7.

EXAMPLE 9

Evaluate $\log_2 22^{1/3}$, given that $\log_2 22 = 4.4598$.

Solution

$$
\begin{aligned}
\log_2 22^{1/3} &= \tfrac{1}{3}\log_2 22 \qquad \text{Property 5.7} \\
&= \tfrac{1}{3}(4.4598) = 1.4866.
\end{aligned}
$$

EXAMPLE 10

Evaluate $\log_{10}(8540)^{3/5}$, given that $\log_{10} 8540 = 3.9315$.

Solution

$$\log_{10}(8540)^{3/5} = \tfrac{3}{5}\log_{10} 8540$$
$$= \tfrac{3}{5}(3.9315) = 2.3589.$$

The properties of logarithms can be used together to allow us to change the forms of various logarithmic expressions. For example, an expression such as $\log_b \sqrt{xy/z}$ can be rewritten in terms of sums and differences of simpler logarithmic quantities:

$$\log_b \sqrt{\frac{xy}{z}} = \log_b \left(\frac{xy}{z}\right)^{1/2}$$

$$= \frac{1}{2}\log_b \left(\frac{xy}{z}\right) \qquad\qquad \text{Property 5.7}$$

$$= \frac{1}{2}(\log_b xy - \log_b z) \qquad\qquad \text{Property 5.6}$$

$$= \frac{1}{2}(\log_b x + \log_b y - \log_b z) \qquad \text{Property 5.5}$$

The properties of logarithms along with the link between logarithmic form and exponential form provide the basis for solving certain types of equations involving logarithms. Our final example of this section illustrates this idea.

EXAMPLE 11

Solve $\log_{10} x + \log_{10}(x + 9) = 1$.

Solution

$$\log_{10} x + \log_{10}(x + 9) = 1$$
$$\log_{10}[x(x + 9)] = 1 \qquad\qquad \text{Property 5.5}$$
$$x(x + 9) = 10^1 \qquad\qquad \text{Change to exponential form}$$
$$x^2 + 9x = 10$$
$$x^2 + 9x - 10 = 0$$
$$(x + 10)(x - 1) = 0$$
$$x + 10 = 0 \qquad \text{or} \qquad x - 1 = 0$$
$$x = -10 \qquad \text{or} \qquad x = 1$$

Since the left-hand side of the original equation is meaningful only if $x > 0$ and $x + 9 > 0$, the solution -10 must be discarded. Thus, the solution set is $\{1\}$.

Problem Set 5.3

Write each of the following in logarithmic form. For example, $2^4 = 16$ becomes $\log_2 16 = 4$.

1. $3^2 = 9$ **2.** $2^5 = 32$ **3.** $5^3 = 125$

4. $10^1 = 10$ **5.** $2^{-4} = \frac{1}{16}$ **6.** $(\frac{2}{3})^{-3} = \frac{27}{8}$

7. $10^{-2} = 0.01$ **8.** $10^5 = 100{,}000$

Write each of the following in exponential form. For example, $\log_2 8 = 3$ becomes $2^3 = 8$.

9. $\log_2 64 = 6$ **10.** $\log_3 27 = 3$ **11.** $\log_{10} 0.1 = -1$

12. $\log_5(\frac{1}{25}) = -2$ **13.** $\log_2(\frac{1}{16}) = -4$ **14.** $\log_{10} 0.00001 = -5$

Evaluate each of the following.

15. $\log_6 36$ **16.** $\log_3 243$ **17.** $\log_5(\frac{1}{5})$ **18.** $\log_4(\frac{1}{64})$

19. $\log_{10} 10$ **20.** $\log_{10} 1$ **21.** $\log_3 \sqrt{3}$ **22.** $\log_5 \sqrt[3]{25}$

23. $\log_3\left(\dfrac{\sqrt{27}}{3}\right)$ **24.** $\log_{1/2}\left(\dfrac{\sqrt[4]{8}}{2}\right)$ **25.** $\log_{1/4}\left(\dfrac{\sqrt[4]{32}}{2}\right)$ **26.** $\log_2\left(\dfrac{\sqrt[3]{16}}{4}\right)$

27. $10^{\log_{10} 7}$ **28.** $5^{\log_5 13}$ **29.** $\log_2(\log_5 5)$ **30.** $\log_6(\log_2 64)$

Solve each of the following equations.

31. $\log_5 x = 2$ **32.** $\log_{10} x = 3$ **33.** $\log_8 t = \frac{5}{3}$ **34.** $\log_4 m = \frac{3}{2}$

35. $\log_b 3 = \frac{1}{2}$ **36.** $\log_b 2 = \frac{1}{2}$ **37.** $\log_{10} x = 0$ **38.** $\log_{10} x = 1$

Given that $\log_{10} 2 = 0.3010$ and $\log_{10} 7 = 0.8451$, evaluate each of the following by using Properties 5.5–5.7.

39. $\log_{10} 14$ **40.** $\log_{10}(\frac{7}{2})$ **41.** $\log_{10} 4$ **42.** $\log_{10} 49$

43. $\log_{10} 343$ **44.** $\log_{10} 32$ **45.** $\log_{10} \sqrt{2}$ **46.** $\log_{10} \sqrt[3]{7}$

47. $\log_{10}(7)^{4/3}$ **48.** $\log_{10}(2)^{3/5}$ **49.** $\log_{10} 28$ **50.** $\log_{10} 56$

51. $\log_{10} 98$ **52.** $\log_{10} 20$ **53.** $\log_{10} 200$ **54.** $\log_{10} 70$

55. $\log_{10} 1400$ **56.** $\log_{10} 4900$

Express each of the following as the sum or difference of simpler logarithmic quantities. (Assume that all variables represent positive real numbers.) For example,

$$\log_b\left(\frac{x^3}{y^2}\right) = \log_b x^3 - \log_b y^2$$

$$= 3\log_b x - 2\log_b y.$$

57. $\log_b xyz$ **58.** $\log_b\left(\dfrac{x^2}{y}\right)$ **59.** $\log_b x^2 y^3$ **60.** $\log_b x^{2/3} y^{3/4}$

61. $\log_b \sqrt{xy}$ **62.** $\log_b \sqrt[3]{x^2 z}$ **63.** $\log_b \sqrt{\dfrac{x}{y}}$ **64.** $\log_b\left[x\left(\sqrt{\dfrac{x}{y}}\right)\right]$

Express each of the following as a simple logarithm. (Assume that all variables represent positive real numbers.) For example,

$$3\log_b x + 5\log_b y = \log_b x^3 y^5.$$

65. $\log_b x + \log_b y - \log_b z$ **66.** $2\log_b x - 4\log_b y$

67. $(\log_b x - \log_b y) - \log_b z$ **68.** $\log_b x - (\log_b y - \log_b z)$

69. $\log_b x + \frac{1}{2}\log_b y$

70. $2\log_b x + 4\log_b y - 3\log_b z$

71. $2\log_b x + \frac{1}{2}\log_b(x-1) - 4\log_b(2x+5)$

72. $\frac{1}{2}\log_b x - 3\log_b x + 4\log_b y$

Solve each of the following equations.

73. $\log_{10} 5 + \log_{10} x = 1$

74. $\log_{10} x + \log_{10} 25 = 2$

75. $\log_{10} 20 - \log_{10} x = 1$

76. $\log_{10} x + \log_{10}(x-3) = 1$

77. $\log_{10}(x+2) - \log_{10} x = 1$

78. $\log_{10} x + \log_{10}(x-21) = 2$

79. $\log_{10}(x-4) + \log_{10}(x-1) = 1$

80. $\log_{10}(x+2) + \log_{10}(x-1) = 1$

81. Verify Property 5.6.

82. Verify Property 5.7.

5.4

Logarithmic Functions

The concept of a logarithm can now be used to define a logarithmic function.

DEFINITION 5.3

If $b > 0$ and $b \neq 1$, then the function defined by

$$f(x) = \log_b x,$$

where x is any positive real number, is called the **logarithmic function with base b**.

x	$f(x)$
$\frac{1}{8}$	-3
$\frac{1}{4}$	-2
$\frac{1}{2}$	-1
1	0
2	1
4	2
8	3

We can obtain the graph of a specific logarithmic function in various ways. For example, we can change the equation $y = \log_2 x$ to the exponential equation $2^y = x$, from which we can determine a table of values. We will have you use this approach to graph some logarithmic functions in the next set of exercises.

We can also obtain the graph of a logarithmic function by setting up a table of values directly from the logarithmic equation. Let's illustrate this approach.

EXAMPLE 1

Graph $f(x) = \log_2 x$.

Solution Let's choose some values for x for which the corresponding values for $\log_2 x$ are easily determined. (Remember that logarithms are defined only for the positive real numbers.)

$$\log_2 \tfrac{1}{8} = -3 \quad \text{because} \quad 2^{-3} = \frac{1}{2^3} = \frac{1}{8}.$$

$$\log_2 1 = 0 \quad \text{because} \quad 2^0 = 1.$$

Plotting the points in the table (in the margin) and connecting them with a smooth curve produces Figure 5.6.

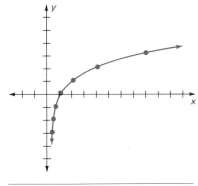

Figure 5.6

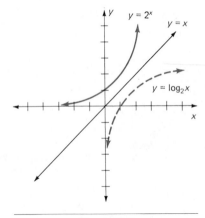

Figure 5.7

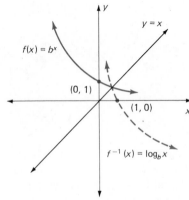

(a) $0 < b < 1$

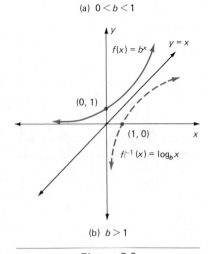

(b) $b > 1$

Figure 5.8

Now suppose that we consider two functions f and g as follows:

$f(x) = b^x$ Domain: all real numbers
Range: positive real numbers

$g(x) = \log_b x$ Domain: positive real numbers
Range: all real numbers

Furthermore, suppose that we consider the composition of f and g, and the composition of g and f.

$$(f \circ g)(x) = f(g(x)) = f(\log_b x) = b^{\log_b x} = x$$

$$(g \circ f)(x) = g(f(x)) = g(b^x) = \log_b b^x = x \log_b b = x(1) = x$$

Therefore, because the domain of f is the range of g and the range of f is the domain of g, and because $f(g(x)) = x$ and $g(f(x)) = x$, the two functions f and g are *inverses of each other.*

Remember from Chapter 4 that the graphs of a function and its inverse are reflections of each other through the line $y = x$. Thus, the graph of a logarithmic function can be determined by reflecting the graph of its inverse exponential function through the line $y = x$. This idea is illustrated in Figure 5.7, where the graph of $y = 2^x$ has been reflected across the line $y = x$ to produce the graph of $y = \log_2 x$.

Figure 5.3 illustrated the general behavior patterns of exponential functions with two graphs. We can now reflect each of these graphs through the line $y = x$ and observe the general behavior patterns of logarithmic functions, as shown in Figure 5.8.

Common Logarithms: Base 10

The properties of logarithms that we discussed in Section 5.3 are true for any valid base. However, since the Hindu-Arabic numeration system that we use is a base 10 system, logarithms to base 10 have historically been used for computational purposes. Base 10 logarithms are called **common logarithms**.

Originally, common logarithms were developed as an aid to numerical calculations. Today they are seldom used for that purpose because calculators and computers can handle messy computational problems more effectively. Thus, in this section we will restrict our discussion to evaluating common logarithms; then in an optional section at the end of this chapter we will illustrate a few of their computational characteristics.

As we know from earlier work, the definition of a logarithm allows us to evaluate $\log_{10} x$ for values of x that are integral powers of 10. Consider the following examples.

$$\log_{10} 1000 = 3 \quad \text{because} \quad 10^3 = 1000.$$

$$\log_{10} 100 = 2 \quad \text{because} \quad 10^2 = 100.$$

$$\log_{10} 10 = 1 \quad \text{because} \quad 10^1 = 10.$$

$$\log_{10} 1 = 0 \quad \text{because} \quad 10^0 = 1.$$

$$\log_{10} 0.1 = -1 \quad \text{because} \quad 10^{-1} = \frac{1}{10} = 0.1.$$

$$\log_{10} 0.01 = -2 \quad \text{because} \quad 10^{-2} = \frac{1}{10^2} = 0.01.$$

$$\log_{10} 0.001 = -3 \quad \text{because} \quad 10^{-3} = \frac{1}{10^3} = 0.001.$$

When we work exclusively with base 10 logarithms, it is customary to omit the numeral 10 that designates the base. Thus, the expression $\log_{10} x$ is simply written $\log x$, and a statement such as $\log_{10} 1000 = 3$ becomes $\log 1000 = 3$. We will follow this practice from now on in this chapter, but don't forget **the base is understood to be 10**.

To find the common logarithm of a positive number that is not an integral power of 10, we can use an appropriately equipped calculator or a table such as the one that appears inside the back cover of this text. Using a calculator equipped with a common logarithm function (ordinarily a key labeled log is used), we obtained the following results, rounded to four decimal places:

$$\log 1.75 = 0.2430$$

$$\log 23.8 = 1.3766$$

$$\log 134 = 2.1271$$

$$\log 0.192 = -0.7167$$

$$\log 0.0246 = -1.6091$$

Be sure that you can use a calculator and that you obtain these results.

Using Table 5.1 to find a common logarithm is relatively easy but it does require a little more effort than pushing a button, as you would with a calculator. Let's consider a small part of the table that appears in the back of the

Table 5.1 Common Logarithms

n	0	1	2	3	4	5	6	7	8	9
1.0	.0000	.0043	.0086	.0128	.0170	.0212	.0253	.0294	.0334	.0374
1.1	.0414	.0453	.0492	.0531	.0569	.0607	.0645	.0682	.0719	.0755
1.2	.0792	.0828	.0864	.0899	.0934	.0969	.1004	.1038	.1072	.1106
1.3	.1139	.1173	.1206	.1239	.1271	.1303	.1335	.1367	.1399	.1430
1.4	.1461	.1492	.1523	.1553	.1584	.1614	.1644	.1673	.1703	.1732
1.5	.1761	.1790	.1818	.1847	.1875	.1903	.1931	.1959	.1987	.2014
1.6	.2041	.2068	.2095	.2122	.2148	.2175	.2201	.2227	.2253	.2279
1.7	.2304	.2330	.2355	.2380	.2405	.2430	.2455	.2480	.2504	.2529
1.8	.2553	.2577	.2601	.2625	.2648	.2672	.2695	.2718	.2742	.2765
1.9	.2788	.2810	.2833	.2856	.2878	.2900	.2923	.2945	.2967	.2989
2.0	.3010	.3032	.3054	.3075	.3096	.3118	.3139	.3160	.3181	.3201
2.1	.3222	.3243	.3263	.3284	.3304	.3324	.3345	.3365	.3385	.3404
2.2	.3424	.3444	.3464	.3483	.3502	.3522	.3541	.3560	.3579	3598
2.3	.3617	.3636	.3655	.3674	.3692	.3711	.3729	.3747	.3766	.3784
2.4	.3802	.3820	.3838	.3856	.3874	.3892	.3909	.3927	.3945	.3962

(complete table on back endsheets of text)

book. Each number in the column headed n represents the first two significant digits of a number between 1 and 10, and each of the column headings 0 through 9 represents the third significant digit. To find the logarithm of a number such as 1.75, we look at the intersection of the row containing 1.7 and the column headed 5. Thus, we obtain

$$\log 1.75 = .2430.$$

Similarly, we can find that

$$\log 2.09 = .3201 \quad \text{and} \quad \log 2.40 = .3802.$$

Keep in mind that these values are rounded to four decimal places. Also, don't lose sight of the meaning of such logarithmic statements. In other words, $\log 1.75 = .2430$ means that 10 raised to the .2430 power is approximately 1.75. (Try it on your calculator!)

Now suppose that we want to use the table to find the logarithm of a positive number greater than 10 or less than 1. We can accomplish this by representing the number in scientific notation and applying the property $\log rs = \log r + \log s$. For example, to find $\log 134$ we can proceed as follows:

$$\begin{aligned}
\log 134 &= \log(1.34 \cdot 10^2) \\
&= \log 1.34 + \log 10^2 \\
&= \log 1.34 + 2\log 10 \\
&= .1271 + 2 = 2.1271.
\end{aligned}$$

from the   by inspection,
table   since $\log 10 = 1$

The decimal part (.1271) of the logarithm 2.1271 is called the **mantissa**, and the integral part (2) is called the **characteristic**. Thus, we can find the characteristic of a common logarithm by inspection (since it is the exponent of 10 when the number is written in scientific notation), and the mantissa we can get from a table. Let's consider two more examples.

$$\begin{aligned}
\log 23.8 &= \log(2.38 \cdot 10^1) \\
&= \log 2.38 + \log 10^1 \\
&= .3766 + 1 = 1.3766
\end{aligned}$$

from the   exponent
table   of 10

$$\begin{aligned}
\log 0.192 &= \log(1.92 \cdot 10^{-1}) \\
&= \log 1.92 + \log 10^{-1} \\
&= .2833 + (-1) = .2833 + (-1)
\end{aligned}$$

from the   exponent
table   of 10

Notice that in the last example, we expressed the logarithm of 0.192 as .2833 + (−1); we did not add .2833 and −1. This is the normal procedure when using a table of common logarithms because the mantissas given in the table are always positive numbers. However, adding .2833 and -1 produces -0.7167, which agrees with our earlier result obtained with a calculator.

If we read directly from the common logarithm table in this text, we are restricted to an approximation to *three* significant digits for the common logarithm of any number between 1.00 and 9.99. We can also use the table to obtain a reasonable approximation for the common logarithms of numbers with *four* significant digits by using a process called *linear interpolation*. This process is discussed in a later section of this chapter.

Antilogarithms

It is also necessary to be able to find the original number when given the common logarithm of that number; that is, given $\log x$ we need to be able to determine x. In this situation, x is referred to as the **antilogarithm** (abbreviated **antilog**) of $\log x$. Many calculators are equipped to find antilogarithms in two ways. Let's consider some examples.

EXAMPLE 2

Determine antilog .2430.

Solution A The phrase "determine the antilog of .2430" means to find a value for x such that $\log x = .2430$. Thus, changing $\log x = .2430$ to exponential form, we obtain $10^{.2430} = x$. Now on our calculator we can enter 10, press $\boxed{y^x}$, enter .2430, press $\boxed{=}$, and obtain 1.7498467, which becomes 1.75 when rounded to three significant digits.

Solution B Thinking in terms of functions, the *antilog function* is the inverse of the log function. Therefore on some calculators we can enter .2430, press $\boxed{\text{INV}}$, press $\boxed{\text{log}}$, and obtain 1.7498467, which rounds to 1.75.

EXAMPLE 3

Determine antilog -1.6091.

Solution Using the inverse routine, we can enter -1.6091, press $\boxed{\text{INV}}$, press $\boxed{\text{log}}$, and obtain 0.02459801, which becomes 0.0246 when rounded to three significant digits.

The common logarithm table at the back of the book can also be used to determine antilogarithms. We illustrate the procedure in the next three examples.

EXAMPLE 4

Determine antilog 1.3365.

Solution Finding an antilogarithm simply reverses the process we used to find a logarithm. Thus, antilog 1.3365 means that 1 is the characteristic and .3365 is the mantissa. We look for .3365 in the body of the common logarithm table: it is located at the intersection of the 2.1 row and the 7 column. Therefore, the antilogarithm is

$$2.17 \cdot 10^1 = 21.7.$$

EXAMPLE 5

Determine antilog(.1523 + (−2)).

Solution The mantissa (.1523) is located at the intersection of the 1.4 row and the 2 column in the log table. The characteristic is −2 and therefore the antilogarithm is

$$1.42 \cdot 10^{-2} = 0.0142.$$

EXAMPLE 6

Determine antilog −2.6038.

Solution The mantissas given in a log table are *positive* numbers. Thus we need to express −2.6038 in terms of a positive mantissa; this can be done by adding and subtracting 3 as follows:

$$(-2.6038 + 3) - 3 = .3962 + (-3).$$

Now we can look for .3962 in the log table; it is at the intersection of the 2.4 row and 9 column. Therefore the antilogarithm is

$$2.49 \cdot 10^{-3} = 0.00249.$$

Natural Logarithms

In many practical applications of logarithms, the number e (remember $e \approx 2.71828$) is used as a base. Logarithms with a base of e are called *natural logarithms*, and the symbol **ln x** is commonly used instead of $\log_e x$.

Natural logarithms can be found with an appropriately equipped calculator or with a table of natural logarithms. Using a calculator with a natural logarithm function (ordinarily a key labeled $\boxed{\ln x}$), we obtain the following results rounded to four decimal places.

$$\ln 3.21 = 1.1663$$
$$\ln 47.28 = 3.8561$$
$$\ln 842 = 6.7358$$
$$\ln 0.21 = -1.5606$$
$$\ln 0.0046 = -5.3817$$
$$\ln 10 = 2.3026$$

Be sure that you can use a calculator and that you obtain these results. Also, keep in mind the significance of a statement of the form $\ln 3.21 = 1.1663$. We are claiming that e raised to the 1.1663 power is approximately 3.21. Using 2.71828 as an approximation for e, we obtain

$$(2.71828)^{1.1663} = 3.2100908.$$

Furthermore, since $f(x) = \ln x$ and $g(x) = e^x$ are inverses of each other, we can enter 1.1663, press $\boxed{\text{INV}}$, press $\boxed{\ln x}$, and obtain 3.2100933 on a calculator. (The last two digits of this result differ from our previous result of 3.2100908 because a better approximation for e is used in this sequence of operations.)

Table 5.2 Natural Logarithms

n	$\ln n$	n	$\ln n$	n	$\ln n$	n	$\ln n$
0.1	-2.3026	2.6	0.9555	5.1	1.6292	7.6	2.0281
0.2	-1.6094	2.7	0.9933	5.2	1.6487	7.7	2.0412
0.3	-1.2040	2.8	1.0296	5.3	1.6677	7.8	2.0541
0.4	-0.9163	2.9	1.0647	5.4	1.6864	7.9	2.0669
0.5	-0.6931	3.0	1.0986	5.5	1.7047	8.0	2.0794
0.6	-0.5108	3.1	1.1314	5.6	1.7228	8.1	2.0919
0.7	-0.3567	3.2	1.1632	5.7	1.7405	8.2	2.1041
0.8	-0.2231	3.3	1.1939	5.8	1.7579	8.3	2.1163
0.9	-0.1054	3.4	1.2238	5.9	1.7750	8.4	2.1282
1.0	0.0000	3.5	1.2528	6.0	1.7918	8.5	2.1401
1.1	0.0953	3.6	1.2809	6.1	1.8083	8.6	2.1518
1.2	0.1823	3.7	1.3083	6.2	1.8245	8.7	2.1633
1.3	0.2624	3.8	1.3350	6.3	1.8405	8.8	2.1748
1.4	0.3365	3.9	1.3610	6.4	1.8563	8.9	2.1861
1.5	0.4055	4.0	1.3863	6.5	1.8718	9.0	2.1972
1.6	0.4700	4.1	1.4110	6.6	1.8871	9.1	2.2083
1.7	0.5306	4.2	1.4351	6.7	1.9021	9.2	2.2192
1.8	0.5878	4.3	1.4586	6.8	1.9169	9.3	2.2300
1.9	0.6419	4.4	1.4816	6.9	1.9315	9.4	2.2407
2.0	0.6931	4.5	1.5041	7.0	1.9459	9.5	2.2513
2.1	0.7419	4.6	1.5261	7.1	1.9601	9.6	2.2618
2.2	0.7885	4.7	1.5476	7.2	1.9741	9.7	2.2721
2.3	0.8329	4.8	1.5686	7.3	1.9879	9.8	2.2824
2.4	0.8755	4.9	1.5892	7.4	2.0015	9.9	2.2925
2.5	0.9163	5.0	1.6094	7.5	2.0149	10	2.3026

Table 5.2 contains the natural logarithms for numbers between 0.1 and 10, inclusive, at intervals of 0.1. Reading directly from the table, we obtain $\ln 1.6 = 0.4700$, $\ln 4.8 = 1.5686$, and $\ln 9.2 = 2.2192$.

We can use Table 5.2 to find the natural logarithm of a positive number less than 0.1 or greater than 10; we simply use the property $\ln rs = \ln r + \ln s$ and proceed as follows.

$$\begin{aligned} \ln 0.0084 &= \ln(8.4 \cdot 10^{-3}) \\ &= \ln 8.4 + \ln 10^{-3} \\ &= \ln 8.4 + (-3)(\ln 10) \\ &= 2.1282 - 3(2.3026) = 2.1282 - 6.9078 = -4.7796 \end{aligned}$$

from the table   from the table

$$\begin{aligned} \ln 190 &= \ln(1.9 \cdot 10^{2}) \\ &= \ln 1.9 + \ln 10^{2} \\ &= \ln 1.9 + 2\ln 10 \\ &= .6419 + 2(2.3026) = 5.2471 \end{aligned}$$

from the table   from the table

Problem Set 5.4

For Problems 1–6, graph each of the logarithmic functions.

1. $f(x) = \log_3 x$ **2.** $f(x) = \log_4 x$ **3.** $f(x) = \log_{10} x$

4. $f(x) = \log_5 x$ **5.** $f(x) = \log_{1/2} x$ **6.** $f(x) = \log_{1/3} x$

7. Graph $y = \log_3 x$ by changing the equation to exponential form.

8. Graph $y = \log_4 x$ by changing the equation to exponential form.

9. Graph $f(x) = \log_{1/2} x$ by reflecting the graph of $g(x) = (\frac{1}{2})^x$ across the line $y = x$.

10. Graph $f(x) = \log_{1/3} x$ by reflecting the graph of $g(x) = (\frac{1}{3})^x$ across the line $y = x$.

For Problems 11–14, graph each of the logarithmic functions. Recall that the graph of $f(x) = g(x) + 1$ is the graph of $g(x)$ moved up one unit.

11. $f(x) = 1 + \log_{10} x$ **12.** $f(x) = -2 + \log_{10} x$

13. $f(x) = \log_{10}(x - 1)$ **14.** $f(x) = \log_{10}(x + 2)$

For Problems 15–24, use a calculator to find each of the *common logarithms*. Express answers to four decimal places.

15. $\log 9.45$ **16.** $\log 1.07$ **17.** $\log 34.62$ **18.** $\log 578.1$

19. $\log 4721.4$ **20.** $\log 52,698$ **21.** $\log 0.612$ **22.** $\log 0.08134$

23. $\log 0.0047$ **24.** $\log 0.000076$

For Problems 25–36, find each *common logarithm* by using the table at the back of the book.

25. $\log 8.72$ **26.** $\log 6.04$ **27.** $\log 56.9$ **28.** $\log 48$

29. $\log 708$ **30.** $\log 14,100$ **31.** $\log 0.492$ **32.** $\log 0.023$

33. $\log 0.00528$ **34.** $\log 0.000415$ **35.** $\log 763,000$ **36.** $\log 9,180,000$

For Problems 37–46, use a calculator to find each antilogarithm. Express answers to five significant digits of accuracy. For example,

$$\text{antilog } 3.2147 = 1639.4569 = 1639.5. \qquad \text{to 5 significant digits}$$

37. antilog 1.5263 **38.** antilog 2.7185 **39.** antilog 3.9335

40. antilog 4.9547 **41.** antilog $.5517$ **42.** antilog 1.9006

43. antilog -0.1452 **44.** antilog -1.3148 **45.** antilog -2.6542

46. antilog -2.1928

For Problems 47–60, find each antilogarithm by using the table at the back of the book.

47. antilog 0.5502 **48.** antilog 0.9624 **49.** antilog 1.4829

50. antilog 1.9170 **51.** antilog 2.9926 **52.** antilog 3.4533

53. antilog 5.8062 **54.** antilog 4.6812 **55.** antilog$(-1 + .7340)$

56. antilog$(-2 + .7774)$ **57.** antilog$(-3 + .8639)$ **58.** antilog$(-4 + .0969)$

59. antilog $-.7471$ **60.** antilog -1.1232

For Problems 61–72, use a calculator to find each *natural logarithm*. Express answers to four decimal places.

61. $\ln 2$ **62.** $\ln 9$ **63.** $\ln 21.4$ **64.** $\ln 87.6$

65. $\ln 412$ **66.** $\ln 384.2$ **67.** $\ln 0.32$ **68.** $\ln 0.417$

69. $\ln 0.0715$ **70.** $\ln 0.006285$ **71.** $\ln 0.0008$ **72.** $\ln 52,173$

For Problems 73–82, find each *natural logarithm* by using the table on page 239.

73. ln 3.7   **74.** ln 2.8   **75.** ln 78   **76.** ln 140

77. ln 620   **78.** ln 9800   **79.** ln 0.42   **80.** ln 0.051

81. ln 0.0085   **82.** ln 0.00056

5.5

Exponential and Logarithmic Equations; Problem Solving

In Section 5.1 we solved exponential equations such as $3^x = 81$ by expressing both sides of the equation as a power of 3 and then applying the property "If $b^n = b^m$, then $n = m$." However, if we try this same approach with an equation such as $3^x = 5$, we face the difficulty of expressing 5 as a power of 3. We can solve this type of problem by using the properties of logarithms and the following property of equality.

PROPERTY 5.8

> If $x > 0$, $y > 0$, $b > 0$, and $b \neq 1$, then
>
> $x = y$ if and only if $\log_b x = \log_b y$.

Property 5.8 is stated in terms of any valid base b; however, for most applications we use either common logarithms or natural logarithms. Let's consider some examples.

EXAMPLE 1

Solve $3^x = 5$, to the nearest hundredth.

Solution By using common logarithms we can proceed as follows.

$$3^x = 5$$
$$\log 3^x = \log 5 \qquad \text{Property 5.8}$$
$$x \log 3 = \log 5 \qquad \log r^p = p \log r$$

$$x = \frac{\log 5}{\log 3}$$

$$x = \frac{.6990}{.4771} = 1.47 \qquad \text{To the nearest hundredth}$$

Check: When we use a calculator to raise 3 to the 1.47 power, we obtain $3^{1.47} = 5.0276871$. Thus, we say that to the nearest hundredth, the solution set for $3^x = 5$ is $\{1.47\}$.

A Word of Caution! The expression $\dfrac{\log 5}{\log 3}$ means that we must *divide*, not subtract, the logarithms. That is, $\dfrac{\log 5}{\log 3}$ *does not* mean $\log\left(\dfrac{5}{3}\right)$.

EXAMPLE 2

Solve $e^{x+1} = 19$, to the nearest hundredth.

Solution Since the base of e is used in the exponential expression, let's use natural logarithms to help solve this equation.

$$e^{x+1} = 5$$
$$\ln e^{x+1} = \ln 5 \qquad \text{Property 5.8}$$
$$(x+1)\ln e = \ln 5 \qquad \ln r^p = p\ln r$$
$$(x+1)(1) = \ln 5 \qquad \ln e = 1$$
$$x = \ln 5 - 1$$
$$x = 1.6094 - 1$$
$$x = .6094$$

Thus the solution set is $\{0.61\}$ to the nearest hundredth. (Check it!)

EXAMPLE 3

Solve $2^{3x-2} = 3^{2x+1}$, to the nearest hundredth.

Solution

$$2^{3x-2} = 3^{2x+1}$$
$$\log 2^{3x-2} = \log 3^{2x+1}$$
$$(3x-2)\log 2 = (2x+1)\log 3$$
$$3x\log 2 - 2\log 2 = 2x\log 3 + \log 3$$
$$3x\log 2 - 2x\log 3 = \log 3 + 2\log 2$$
$$x(3\log 2 - 2\log 3) = \log 3 + 2\log 2$$
$$x = \frac{\log 3 + 2\log 2}{3\log 2 - 2\log 3}$$

At this point we could evaluate x, but instead let's use the properties of logarithms to simplify the expression for x.

$$x = \frac{\log 3 + \log 2^2}{\log 2^3 - \log 3^2}$$
$$= \frac{\log 3 + \log 4}{\log 8 - \log 9}$$
$$= \frac{\log 12}{\log(\frac{8}{9})}$$

Now, evaluating x, we obtain

$$x \approx \frac{1.0792}{-.0512} = -21.08. \qquad \text{To the nearest hundredth}$$

The solution set is $\{-21.08\}$. (Check it!)

Logarithmic Equations

In Example 11 of Section 5.3, we solved the logarithmic equation

$$\log_{10} x + \log_{10}(x + 9) = 1$$

by simplifying the left side of the equation to $\log_{10}[x(x + 9)]$ and then changing the equation to exponential form to complete the solution. At this time, using Property 5.8, we can solve this type of logarithmic equation another way, and we can also expand our equation-solving capabilities. Let's consider some examples.

EXAMPLE 4

Solve $\log x + \log(x - 15) = 2$.

Solution Since $\log 100 = 2$, the given equation becomes

$$\log x + \log(x - 15) = \log 100.$$

Now, simplifying the left side and applying Property 5.8, we can proceed as follows.

$$\log[(x)(x - 15)] = \log 100$$
$$x(x - 15) = 100$$
$$x^2 - 15x - 100 = 0$$
$$(x - 20)(x + 5) = 0$$
$$x - 20 = 0 \quad \text{or} \quad x + 5 = 0$$
$$x = 20 \quad \text{or} \quad x = -5$$

The domain of a logarithmic function must contain only positive numbers, so x and $x - 15$ must be positive in this problem. Therefore, the solution of -5 is discarded and the solution set is $\{20\}$.

EXAMPLE 5

Solve $\ln(x + 2) = \ln(x - 4) + \ln 3$.

Solution

$$\ln(x + 2) = \ln(x - 4) + \ln 3$$
$$\ln(x + 2) = \ln[3(x - 4)]$$
$$x + 2 = 3(x - 4)$$
$$x + 2 = 3x - 12$$
$$14 = 2x$$
$$7 = x$$

The solution set is $\{7\}$.

EXAMPLE 6

Solve $\log_b(x + 2) + \log_b(2x - 1) = \log_b x$.

Solution

$$\log_b(x + 2) + \log_b(2x - 1) = \log_b x$$
$$\log_b[(x + 2)(2x - 1)] = \log_b x$$
$$(x + 2)(2x - 1) = x$$
$$2x^2 + 3x - 2 = x$$
$$2x^2 + 2x - 2 = 0$$
$$x^2 + x - 1 = 0$$

Using the quadratic formula, we obtain

$$x = \frac{-1 \pm \sqrt{1 + 4}}{2} = \frac{-1 \pm \sqrt{5}}{2}.$$

Since $x + 2$, $2x - 1$, and x all have to be positive, the solution of $(-1 - \sqrt{5})/2$ has to be discarded and the solution set is

$$\left\{ \frac{-1 + \sqrt{5}}{2} \right\}.$$

Problem Solving

In Section 5.2 we used the compound interest formula

$$A = P\left(1 + \frac{r}{n}\right)^{nt}$$

to determine the amount of money (A) accumulated at the end of t years if P dollars is invested at rate r of interest compounded n times per year. Now let's use this formula to solve other types of problems that deal with compound interest.

EXAMPLE 7

How long will it take $500 to double if it is invested at 12% compounded quarterly?

Solution To *double* $500 means that the $500 will grow into $1000. We want to find how long it will take; that is, what is t? Thus,

$$1000 = 500\left(1 + \frac{0.12}{4}\right)^{4t}$$
$$= 500(1 + 0.03)^{4t}$$
$$= 500(1.03)^{4t}$$

Multiplying both sides of $1000 = 500(1.03)^{4t}$ by $\frac{1}{500}$ yields

$$2 = (1.03)^{4t}.$$

Therefore,

$$\log 2 = \log(1.03)^{4t} \qquad \text{Property 5.8}$$
$$= 4t \log 1.03. \qquad \log r^p = p \log r$$

Solving for t, we obtain

$$\log 2 = 4t \log 1.03$$

$$\frac{\log 2}{\log 1.03} = 4t$$

$$\frac{\log 2}{4 \log 1.03} = t \qquad \text{Multiply both sides by } \tfrac{1}{4}.$$

$$\frac{.3010}{4(.0128)} = t$$

$$\frac{.3010}{.0512} = t$$

$$5.9 = t. \qquad \text{To the nearest tenth}$$

Therefore, we are claiming that $500 invested at 12% interest compounded quarterly will double itself in approximately 5.9 years.

Check: $500 invested at 12% compounded quarterly for 5.9 years will produce

$$A = \$500\left(1 + \frac{0.12}{4}\right)^{4(5.9)}$$

$$= \$500(1.03)^{23.6}$$

$$= \$1004.45.$$

EXAMPLE 8

At what rate of interest (to the nearest tenth of a percent) will an investment of $1000 yield $4000 in 10 years, if the money is compounded annually?

Solution Substituting the known facts into the compound interest formula, we obtain

$$4000 = 1000(1 + r)^{10}.$$

Multiplying both sides of this equation by $\frac{1}{1000}$ yields

$$4 = (1 + r)^{10}.$$

Therefore,

$$\log 4 = \log(1 + r)^{10} \qquad \text{Property 5.8}$$
$$= 10 \log(1 + r). \qquad \log r^p = p \log r$$

Multiplying both sides by $\frac{1}{10}$ produces

$$\frac{\log 4}{10} = \log(1 + r).$$

Using $\log 4 = .6021$, we obtain

$$\log(1 + r) = \frac{.6021}{10} = .0602.$$

Finding the antilog of .0602 and solving for r, we obtain

$$1.149 = 1 + r$$
$$r = 0.149 = 14.9\%.$$

Therefore, an interest rate of approximately 14.9% is needed.

Check: $1000 invested at 14.9% compounded annually for 10 years will produce

$$A = \$1000(1.149)^{10} = \$4010.52.$$

EXAMPLE 9

How long will it take $100 to triple itself if it is invested at 8% interest compounded continuously?

Solution To *triple itself* means that the $100 will grow into $300. Thus, using the formula for interest that is compounded continuously, we can proceed as follows.

$$A = Pe^{rt}$$
$$\$300 = \$100e^{0.08t}$$
$$3 = e^{0.08t}$$

$\ln 3 = \ln e^{0.08t}$	Property 5.8
$\ln 3 = 0.08t \ln e$	$\ln r^p = p \ln r$
$\ln 3 = 0.08t$	$\ln e = 1$

$$\frac{\ln 3}{0.08} = t$$

$$\frac{1.0986}{0.08} = t$$

$$13.7 = t \qquad \text{To the nearest tenth}$$

Therefore, in approximately 13.7 years, $100 will triple itself at 8% interest compounded continuously.

Check: $100 invested at 8% compounded continuously for 13.7 years produces

$$A = Pe^{rt}$$
$$= \$100e^{0.08(13.7)}$$
$$= \$100(2.718)^{1.096}$$
$$= \$299.18.$$

EXAMPLE 10

Suppose the number of bacteria present in a certain culture after t minutes is given by the equation $Q(t) = Q_0 e^{0.04t}$, where Q_0 represents the initial number of bacteria. How long will it take for the bacteria count to grow from 500 to 2000?

Solution Substituting into $Q(t) = Q_0 e^{0.04t}$ and solving for t, we obtain

$$2000 = 500 e^{0.04t}$$

$$4 = e^{0.04t}$$

$$\ln 4 = \ln e^{0.04t}$$

$$\ln 4 = 0.04t \ln e$$

$$\ln 4 = 0.04t$$

$$\frac{\ln 4}{0.04} = t$$

$$34.7 = t. \qquad \text{To the nearest tenth}$$

It should take approximately 34.7 minutes.

The basic approach of applying Property 5.8 and using either common or natural logarithms can also be used to evaluate a logarithm to some base other than 10 or e. The next example illustrates this idea.

EXAMPLE 11

Evaluate $\log_3 41$.

Solution Let $x = \log_3 41$. Changing to exponential form, we obtain

$$3^x = 41.$$

Now we can apply Property 5.8:

$$\log 3^x = \log 41$$

$$x \log 3 = \log 41$$

$$x = \frac{\log 41}{\log 3}$$

$$x = \frac{1.6128}{.4771}$$

$$x = 3.3804. \qquad \text{Rounded to four decimal places}$$

Therefore $\log_3 41 = 3.3804$. Thus, we are claiming that 3 raised to the 3.3804 power will produce approximately 41. (Check it!)

Using the method of Example 11 to evaluate $\log_a r$ produces the following formula, which is often referred to as the **change of base formula** for logarithms.

PROPERTY 5.9

If a, b, and r are positive numbers, with $a \neq 1$ and $b \neq 1$, then

$$\log_a r = \frac{\log_b r}{\log_b a}.$$

By using Property 5.9, we can easily determine a relationship between logarithms of different bases. For example, suppose that in Property 5.9 we let $a = 10$ and $b = e$. Then

$$\log_a r = \frac{\log_b r}{\log_b a}$$

becomes

$$\log_{10} r = \frac{\log_e r}{\log_e 10}$$

$$\log_e r = (\log_e 10)(\log_{10} r)$$
$$\log_e r = (2.3026)(\log_{10} r).$$

Thus, the natural logarithm of any positive number is approximately equal to 2.3026 times the common logarithm of the number.

Problem Set 5.5

Solve each of the following exponential equations and express approximate solutions to the nearest hundredth.

1. $2^x = 9$ **2.** $3^x = 20$ **3.** $5^t = 23$ **4.** $4^t = 12$

5. $2^{x+1} = 7$ **6.** $3^{x-2} = 11$ **7.** $7^{2t-1} = 35$ **8.** $5^{3t+1} = 9$

9. $e^x = 4.1$ **10.** $e^x = 30$ **11.** $e^{x-1} = 8.2$ **12.** $e^{x-2} = 13.1$

13. $2e^x = 12.4$ **14.** $3e^x - 1 = 17$ **15.** $3^{x-1} = 2^{x+3}$

16. $5^{2x+1} = 7^{x+3}$ **17.** $5^{x-1} = 2^{2x+1}$ **18.** $3^{2x+1} = 2^{3x+2}$

Solve each of the following logarithmic equations and express irrational solutions in simplest radical form.

19. $\log x + \log(x + 3) = 1$ **20.** $\log x + \log(x + 21) = 2$

21. $\log(2x - 1) - \log(x - 3) = 1$ **22.** $\log(3x - 1) = 1 + \log(5x - 2)$

23. $\log(x - 2) = 1 - \log(x + 3)$ **24.** $\log(x + 1) = \log 3 - \log(2x - 1)$

25. $\log(x + 1) - \log(x + 2) = \log \dfrac{1}{x}$ **26.** $\log(x + 2) - \log(2x + 1) = \log x$

27. $\ln(3t - 4) - \ln(t + 1) = \ln 2$ **28.** $\ln(2t + 5) = \ln 3 + \ln(t - 1)$

29. $\log(x^2) = (\log x)^2$ **30.** $\log \sqrt{x} = \sqrt{\log x}$

Evaluate each of the following logarithms to three decimal places. (Example 11 and/or Property 5.9 should be of some help.)

31. $\log_3 14$ **32.** $\log_4 94$ **33.** $\log_5 2.1$ **34.** $\log_6 0.345$

35. $\log_7 176$ **36.** $\log_8 296$ **37.** $\log_9 14.32$ **38.** $\log_7 0.024$

Solve each of the following problems.

39. How long will it take $1000 to double itself if it is invested at 9% interest compounded semiannually?

40. How long will it take $750 to be worth $1000 if it is invested at 12% interest compounded quarterly?

41. How long will it take $500 to triple itself if it is invested at 9% interest compounded continuously?

42. How long will it take $2000 to double itself if it is invested at 13% interest compounded continuously?

43. At what rate of interest (to the nearest tenth of a percent) compounded annually will an investment of $200 grow to $350 in 5 years?

44. At what rate of interest (to the nearest tenth of a percent) compounded continuously will an investment of $500 grow to $900 in 10 years?

45. A piece of machinery valued at $30,000 depreciates at a rate of 10% yearly. How long will it take until the machinery has a value of $15,000?

46. For a certain strain of bacteria, the number present after t hours is given by the equation $Q = Q_0 e^{0.34t}$, where Q_0 represents the initial number of bacteria. How long will it take 400 bacteria to increase to 4000 bacteria?

47. The number of grams of a certain radioactive substance present after t hours is given by the equation $Q = Q_0 e^{-0.45t}$, where Q_0 represents the initial number of grams. How long will it take 2500 grams to be reduced to 1250 grams?

48. The atmospheric pressure in pounds per square inch is expressed by the equation $P(a) = 14.7e^{-0.21a}$, where a is the altitude above sea level measured in miles. If the atmospheric pressure at Cheyenne, Wyoming, is approximately 11.53 pounds per square inch, find its altitude above sea level. Express your answer to the nearest hundred feet.

49. Suppose you are given the equation $P(t) = P_0 e^{0.02t}$ to predict population growth, where P_0 represents an initial population and t is the time in years. How long does this equation predict it will take a city of 50,000 to double its population?

50. In a certain bacterial culture, the equation $Q(t) = Q_0 e^{0.4t}$ yields the number of bacteria as a function of the time, where Q_0 is an initial number of bacteria and t is time measured in hours. How long will it take 500 bacteria to increase to 2000?

51. Radon is formed by the radioactive decay of radium. It has a half-life of approximately 4 days, and the equation $Q(t) = Q_0 e^{-4k}$ yields the amount of radon that remains after t days, if the initial amount is Q_0. Solve the equation $\frac{1}{2}Q_0 = Q_0 e^{-4k}$ for k. Express the answer to the nearest hundredth.

52. Polonium is also formed from the radioactive decay of radium and it has a half-life of approximately 140 days. The equation $Q(t) = Q_0 e^{-140k}$ yields the amount of polonium that remains after t days, if the initial amount is Q_0. Solve the equation $\frac{1}{2}Q_0 = Q_0 e^{-140k}$ for k and express the answer to the nearest thousandth.

Miscellaneous Problems

53. Use the approach of Example 11 and develop Property 5.9.

54. Let $r = b$ in Property 5.9 and verify that $\log_a b = 1/\log_b a$.

55. Solve the equation $\dfrac{5^x - 5^{-x}}{2} = 3$. Express your answer to the nearest hundredth.

56. Solve the equation $y = \dfrac{10^x + 10^{-x}}{2}$ for x in terms of y.

57. Solve the equation $y = \dfrac{e^x - e^{-x}}{2}$ for x in terms of y.

5.6

Computation with Common Logarithms (Optional)

As we mentioned earlier, the calculator has replaced the use of common logarithms for most computational purposes. Nevertheless, we feel that a brief look at how common logarithms are used will give you a better insight into the meaning of logarithms and their properties.

Although the computations in this section should be done by using logarithms and the table of common logarithms at the back of the book, we suggest that you check each problem by using a calculator. This will provide you with some additional practice with your calculator, and it will establish the validity of our work with logarithms.

Linear Interpolation

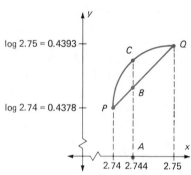

Figure 5.9

Let's begin by expanding our use of the common logarithm table at the back of the book. Suppose that we try to determine log 2.744 from the table. But the table contains only logarithms of numbers with, at most, three significant digits. However, by a process called **linear interpolation**, we can extend the capabilities of the table to include numbers with four significant digits.

First, let's consider a geometric basis of linear interpolation; then we will suggest a systematic procedure for carrying out the necessary calculations. Figure 5.9 shows a portion of the graph of $y = \log_{10} x$, with the curvature exaggerated to help illustrate the principle involved. The line segment $\overline{PQ}$ joining points P and Q is used to approximate the curve from P to Q. The actual value of log 2.744 is the ordinate of the point C, that is, the length of $\overline{AC}$. This cannot be determined from the table. Instead we will use the ordinate of point B (the length of AB) as an approximation for log 2.744.

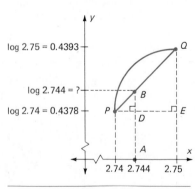

Figure 5.10

Now consider Figure 5.10, where line segments $\overline{DB}$ and $\overline{EQ}$ are drawn perpendicular to $\overline{PE}$. This forms similar right triangles $\triangle PDB$ and $\triangle PEQ$; therefore the lengths of their corresponding sides are proportional. Thus, we can write

$$\frac{PD}{PE} = \frac{DB}{EQ}.$$

Also from Figure 5.10 we see that

$$PD = 2.744 - 2.74 = .004,$$

$$PE = 2.75 - 2.74 = .01,$$

and

$$EQ = .4393 - .4378 = .0015.$$

Therefore, the preceding proportion becomes

$$\frac{.004}{.01} = \frac{DB}{.0015}.$$

Solving this proportion for *DB* yields

$$(.01)(DB) = (.004)(.0015)$$
$$DB = .0006.$$

Since $AB = AD + DB$, we have

$$AB = .4378 + .0006 = .4384.$$

Thus, we obtain the approximate value $\log 2.744 = .4384$.

Now we'll devise an abbreviated format for carrying out the calculations necessary to find $\log 2.744$.

$$
\begin{array}{cc}
x & \log x
\end{array}
$$

$$
4\left\{\begin{array}{l}2.740 \\ 2.744 \\ 2.750\end{array}\right\}10 \qquad k\left\{\begin{array}{l}.4378 \\ ? \\ .4393\end{array}\right\}.0015
$$

Notice that instead of .004 and .01, we have used 4 and 10 for the differences in the values of *x*, because the ratio .004/.01 equals 4/10. Setting up a proportion and solving for *k* yields

$$\frac{4}{10} = \frac{k}{.0015}$$

$$10k = 4(.0015) = .0060$$

$$k = .0006.$$

Thus, $\log 2.744 = .4378 + .0006 = .4384$.

The process of linear interpolation can also be used to approximate an antilogarithm when the mantissa is between two values in the table. The following example illustrates this procedure.

EXAMPLE 1

Find antilog 1.6157.

Solution From the table we see that the mantissa, .6157, is between .6149 and .6160. We can carry out the interpolation as follows.

$$
\begin{array}{cc}
\text{antilog } x & x
\end{array}
$$

$$
h\left\{\begin{array}{l}4.120 \\ ? \\ 4.130\end{array}\right\}.010 \qquad 8\left\{\begin{array}{l}.6149 \\ .6157 \\ .6160\end{array}\right\}11 \qquad \frac{.0008}{.0011} = \frac{8}{11}
$$

$$\frac{h}{.010} = \frac{8}{11}$$

$$11h = 8(.010) = .080$$

$$h = \frac{1}{11}(.080) = .007 \qquad \text{To the nearest thousandth}$$

Thus, antilog .6157 = 4.120 + .007 = 4.127. Therefore,

$$\text{antilog } 1.6157 = \text{antilog}(.6157 + 1)$$
$$= 4.127 \cdot 10^1$$
$$= 41.27.$$

Computation with Common Logarithms

Let's first restate the basic properties of logarithms in terms of common logarithms. Remember that we are writing $\log x$ instead of $\log_{10} x$.

If x and y are positive real numbers, then

1. $\log xy = \log x + \log y$,

2. $\log \dfrac{x}{y} = \log x - \log y$,

3. $\log x^p = p \log x$. p is any real number.

The following two properties of equality pertaining to logarithms will also be used. Both x and y are positive.

4. If $x = y$, then $\log x = \log y$.

5. If $\log x = \log y$, then $x = y$.

Now let's illustrate how we can use common logarithms for computational purposes.

EXAMPLE 2

Evaluate $\dfrac{(571.4)(8.236)}{71.68}$.

Solution Let $N = \dfrac{(571.4)(8.236)}{71.68}$. Then

$$\log N = \log \frac{(571.4)(8.236)}{71.68}$$
$$= \log 571.4 + \log 8.236 - \log 71.68$$
$$= 2.7569 + 0.9157 - 1.8554$$
$$= 1.8172.$$

Therefore,

$$N = \text{antilog } 1.8172$$
$$= \text{antilog}(.8172 + 1)$$
$$= 6.564 \cdot 10^1$$
$$= 65.64.$$

Check: By using a calculator, we obtain

$$N = \frac{(571.4)(8.236)}{71.68} = 65.653605.$$

When we use a table of logarithms, it is sometimes necessary to rewrite a logarithm so that the mantissa is positive. The next example illustrates this idea.

EXAMPLE 3

Find the quotient $\dfrac{1.73}{5.08}$.

Solution Let $N = \dfrac{1.73}{5.08}$. Then

$$\log N = \log \frac{1.73}{5.08}$$

$$= \log 1.73 - \log 5.08$$
$$= 0.2380 - 0.7059$$
$$= -.4679.$$

Now by adding 1 and subtracting 1, which changes the form but not the value, we obtain

$$\log N = -.4679 + 1 - 1$$
$$= .5321 - 1$$
$$= .5321 + (-1).$$

Therefore,

$$N = \text{antilog}(.5321 + (-1))$$
$$= 3.405 \cdot 10^{-1}$$
$$= 0.3405.$$

Check: By using a calculator we obtain

$$N = \frac{1.73}{5.08} = .34055118.$$

Sometimes it is necessary to change the form of a logarithm so that a subsequent calculation will produce an integer for the characteristic part of the logarithm. Let's consider an example to illustrate this idea.

EXAMPLE 4

Evaluate $\sqrt[4]{0.0767}$.

Solution Let $N = \sqrt[4]{0.0767} = (0.0767)^{1/4}$. Then

$$\log N = \log(0.0767)^{1/4}$$
$$= \tfrac{1}{4}\log 0.0767$$
$$= \tfrac{1}{4}(.8848 + (-2))$$
$$= \tfrac{1}{4}(-2 + .8848).$$

At this stage we recognize that applying the distributive property will produce a nonintegral characteristic, namely, $-\frac{1}{2}$. Therefore, let's add 4 and subtract 4 inside the parentheses, which will change the form but not the value:

$$\log N = \tfrac{1}{4}(-2 + .8848 + 4 - 4)$$
$$= \tfrac{1}{4}(4 - 2 + .8848 - 4)$$
$$= \tfrac{1}{4}(2.8848 - 4).$$

Now, applying the distributive property, we obtain

$$\log N = \tfrac{1}{4}(2.8848) - \tfrac{1}{4}(4)$$
$$= .7212 - 1$$
$$= .7212 + (-1).$$

Therefore,

$$N = \text{antilog}(.7212 + (-1))$$
$$= 5.262 \cdot 10^{-1}$$
$$= 0.5262.$$

Check: By using a calculator, we obtain

$$N = \sqrt[4]{0.0767} = 0.52625816.$$

Problem Set 5.6

Use the log table at the back of the book and linear interpolation to find each of the following common logarithms.

1. $\log 4.327$ **2.** $\log 27.43$ **3.** $\log 128.9$ **4.** $\log 3526$

5. $\log 0.8761$ **6.** $\log 0.07692$ **7.** $\log 0.005186$ **8.** $\log 0.0002558$

Use the log table at the back of the book and linear interpolation to find each of the following antilogarithms to four significant digits.

9. antilog .4690   **10.** antilog 1.7971   **11.** antilog 2.1925

12. antilog 3.7225   **13.** antilog$(.5026 + (-1))$ **14.** antilog$(.9397 + (-2))$

Use common logarithms and linear interpolation to help evaluate each of the following. Express your answers with four significant digits. Check your answers by using a calculator.

15. $(294)(71.2)$ **16.** $(192.6)(4.017)$ **17.** $\dfrac{23.4}{4.07}$ **18.** $\dfrac{718.5}{8.248}$

19. $(17.3)^5$ **20.** $(48.02)^3$ **21.** $\dfrac{(108)(76.2)}{13.4}$ **22.** $\dfrac{(126.3)(24.32)}{8.019}$

23. $\sqrt[5]{0.821}$ **24.** $\sqrt[4]{645.3}$ **25.** $(79.3)^{3/5}$ **26.** $(176.8)^{3/4}$

27. $\sqrt{\dfrac{(7.05)(18.7)}{0.521}}$ **28.** $\sqrt[3]{\dfrac{(41.3)(0.271)}{8.05}}$

Chapter 5 Summary

This chapter can be summarized around three main topics: (1) exponents and exponential functions, (2) logarithms and logarithmic functions, and (3) applications of exponential and logarithmic functions.

Exponents and Exponential Functions

If a and b are positive numbers, and m and n are real numbers, then the following properties hold.

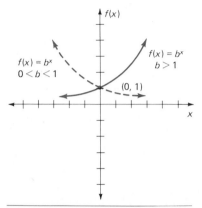

1. $b^n \cdot b^m = b^{n+m}$ (Product of two powers)
2. $(b^n)^m = b^{mn}$ (Power of a power)
3. $(ab)^n = a^n b^n$ (Power of a product)
4. $\left(\dfrac{a}{b}\right)^n = \dfrac{a^n}{b^n}$ (Power of a quotient)
5. $\dfrac{b^n}{b^m} = b^{n-m}$ (Quotient of two powers)

A function defined by an equation of the form

$$f(x) = b^x, \qquad b > 0 \text{ and } b \neq 1,$$

is called an **exponential function**. The figure at the left illustrates the general behavior of the graph of an exponential function of the form $f(x) = b^x$.

Graph of an exponential function

Logarithms and Logarithmic Functions

If r is any positive real number, then the unique exponent t such that $b^t = r$ is called the **logarithm of r with base b**; it is denoted by $\log_b r$.

The following properties of logarithms are used frequently.

1. $\log_b b = 1$
2. $\log_b 1 = 0$
3. $b^{\log_b r} = r$
4. $\log_b rs = \log_b r + \log_b s$
5. $\log_b \left(\dfrac{r}{s}\right) = \log_b r - \log_b s$
6. $\log_b(r^p) = p \log_b r$

Logarithms with a base of 10 are called **common logarithms**. The expression $\log_{10} x$ is commonly written $\log x$.

Many calculators are equipped with a common logarithm function. Often a key labeled  is used to find common logarithms.

The decimal part (.9425) of the logarithm 1.9425 is called the **mantissa**, and the integral part (1) is called the **characteristic**.

Calculators and tables can be used to find x when given $\log x$. The number x is referred to as an **antilogarithm**.

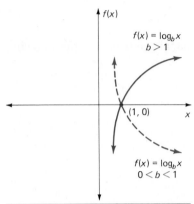

Graph of a logarithmic function

Natural logarithms are logarithms having a base of e, where e is an irrational number whose decimal approximation to eight digits is 2.7182818. Natural logarithms are denoted by $\log_e x$ or $\ln x$.

Many calculators are also equipped with a natural logarithm function. Often a key labeled $\boxed{\ln x}$ is used for this purpose.

A function defined by an equation of the form

$$f(x) = \log_b x, \qquad b > 0 \text{ and } b \neq 1,$$

is called a **logarithmic function**.

The graph of a logarithmic function (such as $y = \log_2 x$) can be determined by changing the equation to exponential form ($2^y = x$) and plotting points, or by reflecting the graph of the inverse function ($y = 2^x$) across the line $y = x$. This last approach is based on the fact that exponential and logarithmic functions are inverses of each other.

The figure at the left illustrates the general behavior of the graph of a logarithmic function of the form $f(x) = \log_b x$.

Applications

The following properties of equality are used frequently when solving exponential and logarithmic equations.

1. If $b > 0$, $b \neq 1$, and m and n are real numbers, then

$$b^n = b^m \quad \text{if and only if} \quad n = m.$$

2. If $x > 0$, $y > 0$, $b > 0$, and $b \neq 1$, then

$$x = y \quad \text{if and only if} \quad \log_b x = \log_b y.$$

A general formula for any principal (P) that is compounded n times per year for any number (t) of years at a given rate (r) is

$$A = P\left(1 + \frac{r}{n}\right)^{nt},$$

where A represents the total amount of money accumulated at the end of the t years.

As n gets infinitely large, the value of $\left(1 + \frac{1}{n}\right)^n$ approaches the number e, where e equals 2.71828 to five decimal places.

The formula

$$A = Pe^{rt}$$

yields the accumulated value (A) of a sum of money (P) that has been invested for t years at a rate of r percent *compounded continuously*.

The equation

$$Q(t) = Q_0 e^{kt}$$

is used as a mathematical model for exponential growth and decay problems. The formula

$$\log_a r = \frac{\log_b r}{\log_b a}$$

is often called the **change of base formula**.

Chapter 5 Review Problem Set

Evaluate each of the following.

1. $8^{5/3}$ **2.** $-25^{3/2}$ **3.** $(-27)^{4/3}$ **4.** $\log_6 216$

5. $\log_7(\frac{1}{49})$ **6.** $\log_2 \sqrt[3]{2}$ **7.** $\log_2\left(\dfrac{\sqrt[4]{32}}{2}\right)$ **8.** $\log_{10} 0.00001$

9. $\ln e$ **10.** $7^{\log_7 12}$

Solve each of the following equations. Express approximate solutions to the nearest hundredth.

11. $\log_{10} 2 + \log_{10} x = 1$ **12.** $\log_3 x = -2$

13. $4^x = 128$ **14.** $3^t = 42$

15. $\log_2 x = 3$ **16.** $(\frac{1}{27})^{3x} = 3^{2x-1}$

17. $2e^x = 14$ **18.** $2^{2x+1} = 3^{x+1}$

19. $\ln(x+4) - \ln(x+2) = \ln x$ **20.** $\log x + \log(x-15) = 2$

21. $\log(\log x) = 2$ **22.** $\log(7x-4) - \log(x-1) = 1$

23. $\ln(2t-1) = \ln 4 + \ln(t-3)$ **24.** $64^{2t+1} = 8^{-t+2}$

For Problems 25–28, if $\log 3 = .4771$ and $\log 7 = .8451$, evaluate each of the following.

25. $\log(\frac{7}{3})$ **26.** $\log 21$ **27.** $\log 27$ **28.** $\log 7^{2/3}$

29. Express each of the following as the sum or difference of simpler logarithmic quantities. Assume that all variables represent positive real numbers.

(a) $\log_b\left(\dfrac{x}{y^2}\right)$ **(b)** $\log_b \sqrt[4]{xy^2}$ **(c)** $\log_b\left(\dfrac{\sqrt{x}}{y^3}\right)$

30. Express each of the following as a single logarithm. Assume that all variables represent positive real numbers.

(a) $3\log_b x + 2\log_b y$ **(b)** $\frac{1}{2}\log_b y - 4\log_b x$

(c) $\frac{1}{2}(\log_b x + \log_b y) - 2\log_b z$

For Problems 31–34, approximate each of the logarithms to three significant decimal places.

31. $\log_2 3$ **32.** $\log_3 2$ **33.** $\log_4 191$ **34.** $\log_2 0.23$

For Problems 35–42, graph each of the functions.

35. $f(x) = (\frac{3}{4})^x$ **36.** $f(x) = 2^{x+2}$ **37.** $f(x) = e^{x-1}$

38. $f(x) = -1 + \log x$ **39.** $f(x) = 3^x - 3^{-x}$ **40.** $f(x) = e^{-x^2/2}$

41. $f(x) = \log_2(x-3)$ **42.** $f(x) = 3\log_3 x$

For Problems 43–45, use the compound interest formula

$$A = P\left(1 + \frac{r}{n}\right)^{nt}$$

to find the total amount of money accumulated at the end of the indicated time period for each of the investments.

43. $750 for 10 years at 11% compounded quarterly

44. $1250 for 15 years at 9% compounded monthly

45. $2500 for 20 years at 9.5% compounded semiannually

Solve the following problems.

46. How long will it take $100 to double itself if it is invested at 14% interest compounded annually?

47. How long will it take $1000 to be worth $3500 if it is invested at 10.5% interest compounded quarterly?

48. At what rate of interest (to the nearest tenth of a percent) compounded continuously will an investment of $500 grow to $1000 in 8 years?

49. Suppose that the present population of a city is 50,000 and, furthermore, suppose that the equation $P(t) = P_0 e^{0.02t}$ can be used to estimate future populations, where P_0 represents an initial population. Estimate the population of that city in 10 years, 15 years, and 20 years.

50. The number of bacteria present in a certain culture after t hours is given by the equation $Q = Q_0 e^{0.29t}$, where Q_0 represents the initial number of bacteria. How long will it take 500 bacteria to increase to 2000 bacteria?

Polynomial and Rational Functions

6

6.1   Dividing Polynomials

6.2   The Remainder and Factor Theorems

6.3   Polynomial Equations

6.4   Graphing Polynomial Functions

6.5   Graphing Rational Functions

6.6   More on Graphing Rational Functions

6.7   Partial Fractions

In earlier chapters we solved linear and quadratic equations and graphed linear and quadratic functions. In this chapter we will expand our equation-solving processes and graphing techniques to include more general polynomial equations and functions. Our knowledge of polynomial functions will then allow us to work with *rational functions*. The function concept will serve as a unifying thread throughout the chapter. To facilitate our study in this chapter, we will first review the concept of dividing polynomials, and then we will introduce a special division technique called *synthetic division*.

6.1

Dividing Polynomials

In Chapter 1 we used the properties

$$\frac{a+b}{c} = \frac{a}{c} + \frac{b}{c} \quad \text{and} \quad \frac{a-b}{c} = \frac{a}{c} - \frac{b}{c}$$

as a basis for dividing a polynomial by a monomial. For example,

$$\frac{18x^3 + 24x^2}{6x} = \frac{18x^3}{6x} + \frac{24x^2}{6x} = 3x^2 + 4x$$

and

$$\frac{35x^2y^3 - 55x^3y^4}{5xy^2} = \frac{35x^2y^3}{5xy^2} - \frac{55x^3y^4}{5xy^2} = 7xy - 11x^2y^2.$$

You may recall from a previous algebra course that the format used to divide a polynomial by a binomial resembles the long division format in arithmetic. Let's work through an example, describing this step-by-step process.

Step 1. Use the conventional long division format and arrange both the dividend and the divisor in descending powers of the variable.

$$3x + 1 \,\overline{)\, 3x^3 - 5x^2 + 10x + 1}$$

Step 2. Find the first term of the quotient by dividing the first term of the dividend by the first term of the divisor.

$$\begin{array}{r} x^2 \\ 3x + 1 \,\overline{)\, 3x^3 - 5x^2 + 10x + 1} \end{array}$$

Step 3. Multiply the entire divisor by the quotient term in Step 2 and place this product in position to be subtracted from the dividend.

$$\begin{array}{r} x^2 \\ 3x + 1 \,\overline{)\, 3x^3 - 5x^2 + 10x + 1} \\ 3x^3 + x^2 \end{array}$$

Step 4. Subtract.

$$\begin{array}{r} x^2 \\ 3x + 1 \,\overline{)\, 3x^3 - 5x^2 + 10x + 1} \\ 3x^3 + x^2 \\ \hline -6x^2 + 10x + 1 \end{array}$$

Step 5. Repeat Steps 2, 3, and 4 using $-6x^2 + 10x + 1$ as a new dividend.

$$\begin{array}{r} x^2 - 2x \\ 3x+1\overline{)3x^3 - 5x^2 + 10x + 1} \\ 3x^3 + x^2 \\ \hline -6x^2 + 10x + 1 \\ -6x^2 - 2x \\ \hline 12x + 1 \end{array}$$

Step 6. Repeat Steps 2, 3, and 4 using $12x + 1$ as a new dividend.

$$\begin{array}{r} x^2 - 2x + 4 \\ 3x+1\overline{)3x^3 - 5x^2 + 10x + 1} \\ 3x^3 + x^2 \\ \hline -6x^2 + 10x + 1 \\ -6x^2 - 2x \\ \hline 12x + 1 \\ 12x + 4 \\ \hline -3 \end{array}$$

Therefore $3x^3 - 5x^2 + 10x + 1 = (3x + 1)(x^2 - 2x + 4) + (-3)$, which is of the familiar form

dividend = (divisor)(quotient) + remainder.

This result is commonly called the **division algorithm for polynomials** and can be stated in general terms as follows.

Division Algorithm for Polynomials

If $f(x)$ and $g(x)$ are polynomials and $g(x) \neq 0$, then there exist unique polynomials $q(x)$ and $r(x)$ such that

$$f(x) = g(x)q(x) + r(x),$$

dividend   divisor   quotient   remainder

where $r(x) = 0$ or the degree of $r(x)$ is less than the degree of $g(x)$.

Let's consider one more example to illustrate this division process further.

EXAMPLE 1

Divide $t^2 - 3t + 2t^4 - 1$ by $t^2 + 4t$.

Solution Don't forget to arrange both the dividend and the divisor in descending powers of the variable.

$$\begin{array}{r} 2t^2 - 8t + 33 \\ t^2+4t\overline{)2t^4 + 0t^3 + t^2 - 3t - 1} \\ 2t^4 + 8t^3 \\ \hline -8t^3 + t^2 - 3t - 1 \\ -8t^3 - 32t^2 \\ \hline 33t^2 - 3t - 1 \\ 33t^2 + 132t \\ \hline -135t - 1 \end{array}$$

— Notice the inserting of a "t-cubed" term with a zero coefficient.

The division process is completed when the degree of the remainder is less than the degree of the divisor.

Synthetic Division

If the divisor is of the form $x - c$, where c is a constant, then the typical long division algorithm can be simplified into a process called **synthetic division**. First, let's consider another division problem using the regular division algorithm. Then, in a step-by-step fashion, we will demonstrate some shortcuts that will lead us into the synthetic division procedure. Consider the division problem $(2x^4 + x^3 - 17x^2 + 13x + 2) \div (x - 2)$.

$$
\begin{array}{r}
2x^3 + 5x^2 - 7x - 1 \\
x - 2 \overline{\smash{\big)}\ 2x^4 + x^3 - 17x^2 + 13x + 2} \\
\underline{2x^4 - 4x^3} \\
5x^3 - 17x^2 \\
\underline{5x^3 - 10x^2} \\
-7x^2 + 13x \\
\underline{-7x^2 + 14x} \\
-x + 2 \\
\underline{-x + 2}
\end{array}
$$

Because the dividend is written in descending powers of x, the quotient is produced in descending powers of x. In other words, the numerical coefficients are the *key issues*. So let's rewrite the above problem in terms of its coefficients.

$$
\begin{array}{r}
2 \quad 5 \quad -7 \quad -1 \\
1 - 2 \overline{\smash{\big)}\ 2 \quad 1 \quad -17 \quad 13 \quad 2} \\
②\quad -4 \\
\overline{5 \quad ⊝17} \\
⑤\quad -10 \\
\overline{-7 \quad ⑬} \\
⊝7 \quad 14 \\
\overline{-1 \quad ②} \\
⊝1 \quad 2
\end{array}
$$

Now observe that the circled numbers are simply repetitions of the numbers directly above them in the format. Thus, the circled numbers can be omitted and the format will be as follows (disregard the arrows for the moment):

$$
\begin{array}{r}
2 \quad 5 \quad -7 \quad -1 \\
1 - 2 \overline{\smash{\big)}\ 2 \quad 1 \quad -17 \quad 13 \quad 2} \\
-4 \\
5 \\
-10 \\
-7 \\
14 \\
-1 \\
2
\end{array}
$$

Next, by moving some numbers up, as indicated by the arrows, and by not writing the 1 that is the coefficient of x in the divisor, the following more compact form is obtained.

$$
\begin{array}{r}
\ \ 2 \ \ \ \ \ 5 \ \ \ -7 \ \ \ -1 \ \ \ \ \ \ \ \ \ (1) \\
-2\,\overline{|\ 2 \ \ \ \ 1 \ \ -17 \ \ \ \ 13 \ \ \ \ 2} \ \ \ (2) \\
\ \ -4 \ \ -10 \ \ \ \ 14 \ \ \ \ 2 \ \ \ (3) \\
\hline
\ \ \ \ 5 \ \ \ -7 \ \ \ -1 \ \ \ \ \ \ \ \ \ (4)
\end{array}
$$

Notice that line 4 reveals all of the coefficients of the quotient, (line 1), except for the first coefficient, 2. Thus, we can omit line 1, begin line 4 with the first coefficient, and then use the following form.

$$
\begin{array}{r}
-2\,\overline{|\ 2 \ \ \ \ 1 \ \ -17 \ \ \ \ 13 \ \ \ \ 2} \ \ \ (5) \\
\ -4 \ \ -10 \ \ \ \ 14 \ \ \ \ 2 \ \ \ (6) \\
\hline
\ 2 \ \ \ \ 5 \ \ \ -7 \ \ \ -1 \ \ \ \ 0 \ \ \ (7)
\end{array}
$$

Line 7 contains the coefficients of the quotient, where the zero indicates the remainder.

Finally, by changing the constant in the divisor to 2 (instead of -2), which changes the signs of the numbers in line 6, we can *add* the corresponding entries in lines 5 and 6 rather than subtract. Thus, the final synthetic division form for this problem is as follows.

$$
\begin{array}{r}
2\,\overline{|\ 2 \ \ \ \ 1 \ \ -17 \ \ \ \ 13 \ \ \ \ 2} \\
\ 4 \ \ \ \ 10 \ \ -14 \ \ -2 \\
\hline
\ 2 \ \ \ \ 5 \ \ \ -7 \ \ \ -1 \ \ \ \ 0
\end{array}
$$

Now we will consider another problem and indicate a step-by-step procedure for setting up and carrying out the synthetic division process. Suppose that we want to do the following division problem.

$$ x + 4\,\overline{|\ 2x^3 + 5x^2 - 13x - 2} $$

Step 1. Write the coefficients of the dividend as follows.

$$ \overline{|\,2 \ \ \ \ 5 \ \ -13 \ \ \ \ 2} $$

Step 2. In the divisor, use -4 instead of 4 so that later we can add rather than subtract.

$$ -4\,\overline{|\ 2 \ \ \ \ 5 \ \ -13 \ \ -2} $$

Step 3. Bring down the first coefficient of the dividend.

$$
\begin{array}{r}
-4\,\overline{|\ 2 \ \ \ \ 5 \ \ -13 \ \ -2} \\
\hline
\ 2
\end{array}
$$

Step 4. Multiply that first coefficient times the divisor, which yields $2(-4) = -8$. Add this result to the second coefficient of the dividend.

$$
\begin{array}{r}
-4\,\overline{|\ 2 \ \ \ \ 5 \ \ -13 \ \ -2} \\
\ -8 \\
\hline
\ 2 \ \ \ -3
\end{array}
$$

Step 5. Multiply $(-3)(-4)$, which yields 12; add this result to the third coefficient of the dividend.

$$
\begin{array}{r|rrrr}
-4 & 2 & 5 & -13 & -2 \\
   &   & -8 & 12 & \\
\hline
   & 2 & -3 & -1 &
\end{array}
$$

Step 6. Multiply $(-1)(-4)$, which yields 4; add this result to the last term of the dividend.

$$
\begin{array}{r|rrrr}
-4 & 2 & 5 & -13 & -2 \\
   &   & -8 & 12 & 4 \\
\hline
   & 2 & -3 & -1 & 2
\end{array}
$$

The last row indicates a quotient of $2x^2 - 3x - 1$ and a remainder of 2.

Let's consider two more examples showing only the final compact form of synthetic division.

EXAMPLE 2

Find the quotient and remainder for $(4x^4 - 2x^3 + 6x - 1) \div (x - 1)$.

Solution

$$
\begin{array}{r|rrrrr}
1 & 4 & -2 & 0 & 6 & -1 \\
  &   & 4 & 2 & 2 & 8 \\
\hline
  & 4 & 2 & 2 & 8 & 7
\end{array}
$$

Notice that a zero has been inserted as the coefficient of the missing x^2 term.

Thus, the quotient is $4x^3 + 2x^2 + 2x + 8$ and the remainder is 7.

EXAMPLE 3

Find the quotient and remainder for $(x^3 + 8x^2 + 13x - 6) \div (x + 3)$.

Solution

$$
\begin{array}{r|rrrr}
-3 & 1 & 8 & 13 & -6 \\
   &   & -3 & -15 & 6 \\
\hline
   & 1 & 5 & -2 & 0
\end{array}
$$

Thus, the quotient is $x^2 + 5x - 2$ and the remainder is zero.

Problem Set 6.1

Find the quotient and remainder for each of the following division problems.

1. $(12x^2 + 7x - 10) \div (3x - 2)$

2. $(20x^2 - 39x + 18) \div (5x - 6)$

3. $(3t^3 + 7t^2 - 10t - 4) \div (3t + 1)$

4. $(4t^3 - 17t^2 + 7t + 10) \div (4t - 5)$

5. $(6x^2 + 19x + 11) \div (3x + 2)$

6. $(20x^2 + 3x - 1) \div (5x + 2)$

7. $(3x^3 + 2x^2 - 5x - 1) \div (x^2 + 2x)$

8. $(4x^3 - 5x^2 + 2x - 6) \div (x^2 - 3x)$

9. $(5y^3 - 6y^2 - 7y - 2) \div (y^2 - y)$

10. $(8y^3 - y^2 - y + 5) \div (y^2 + y)$

11. $(4a^3 - 2a^2 + 7a - 1) \div (a^2 - 2a + 3)$ **12.** $(5a^3 + 7a^2 - 2a - 9) \div (a^2 + 3a - 4)$

13. $(3x^2 - 2xy - 8y^2) \div (x - 2y)$ **14.** $(4a^2 - 8ab + 4b^2) \div (a - b)$

Use *synthetic division* to determine the quotient and remainder for each of the following.

15. $(3x^2 + x - 4) \div (x - 1)$ **16.** $(2x^2 - 5x - 3) \div (x - 3)$

17. $(x^2 + 2x - 10) \div (x - 4)$ **18.** $(x^2 - 10x + 15) \div (x - 8)$

19. $(4x^2 + 5x - 4) \div (x + 2)$ **20.** $(5x^2 + 18x - 8) \div (x + 4)$

21. $(x^3 - 2x^2 - x + 2) \div (x - 2)$ **22.** $(x^3 - 5x^2 + 2x + 8) \div (x + 1)$

23. $(3x^4 - x^3 + 2x^2 - 7x - 1) \div (x + 1)$ **24.** $(2x^3 - 5x^2 - 4x + 6) \div (x - 2)$

25. $(x^3 - 7x - 6) \div (x + 2)$ **26.** $(x^3 + 6x^2 - 5x - 1) \div (x - 1)$

27. $(x^4 + 4x^3 - 7x - 1) \div (x - 3)$ **28.** $(2x^4 + 3x^2 + 3) \div (x + 2)$

29. $(x^3 + 6x^2 + 11x + 6) \div (x + 3)$ **30.** $(x^3 - 4x^2 - 11x + 30) \div (x - 5)$

31. $(x^5 - 1) \div (x - 1)$ **32.** $(x^5 - 1) \div (x + 1)$

33. $(x^5 + 1) \div (x - 1)$ **34.** $(x^5 + 1) \div (x + 1)$

35. $(2x^3 + 3x^2 - 2x + 3) \div (x + \frac{1}{2})$ **36.** $(9x^3 - 6x^2 + 3x - 4) \div (x - \frac{1}{3})$

37. $(4x^4 - 5x^2 + 1) \div (x - \frac{1}{2})$ **38.** $(3x^4 - 2x^3 + 5x^2 - x - 1) \div (x + \frac{1}{3})$

6.2

The Remainder and Factor Theorems

Let's consider the division algorithm (stated in the previous section) when the dividend, $f(x)$, is divided by a *linear polynomial* of the form $x - c$. Then the division algorithm,

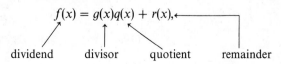

$$f(x) = g(x)q(x) + r(x),$$

dividend   divisor   quotient   remainder

becomes

$$f(x) = (x - c)q(x) + r(x).$$

Because the degree of the remainder, $r(x)$, must be less than the degree of the divisor, $x - c$, the remainder is a constant. Therefore, letting R represent the remainder, we have

$$f(x) = (x - c)q(x) + R.$$

If we evaluate f at c, we obtain

$$f(c) = (c - c)q(c) + R$$
$$= 0 \cdot q(c) + R$$
$$= R.$$

In other words, if a polynomial is divided by a linear polynomial of the form $x - c$, then the remainder is the value of the polynomial at c. Let's state this more formally as the **remainder theorem**.

PROPERTY 6.1
Remainder Theorem   If a polynomial $f(x)$ is divided by $x - c$, then the remainder is equal to $f(c)$.

EXAMPLE 1

If $f(x) = x^3 + 2x^2 - 5x - 1$, find $f(2)$ (a) by using synthetic division and the remainder theorem, and (b) by evaluating $f(2)$ directly.

Solutions

(a)

$$
\begin{array}{r|rrrr}
2 & 1 & 2 & -5 & -1 \\
  &   & 2 & 8 & 6 \\
\hline
  & 1 & 4 & 3 & ⑤ \\
\end{array}
\quad\longleftarrow R = f(2)
$$

(b) $f(2) = 2^3 + 2(2)^2 - 5(2) - 1 = 8 + 8 - 10 - 1 = 5$

EXAMPLE 2

If $f(x) = x^4 + 7x^3 + 8x^2 + 11x + 5$, find $f(-6)$ **(a)** by using synthetic division and the remainder theorem, and **(b)** by evaluating $f(-6)$ directly.

Solutions

(a)

$$
\begin{array}{r|rrrrr}
-6 & 1 & 7 & 8 & 11 & 5 \\
   &   & -6 & -6 & -12 & 6 \\
\hline
   & 1 & 1 & 2 & -1 & ⑪ \\
\end{array}
\quad\longleftarrow R = f(-6)
$$

(b) $f(-6) = (-6)^4 + 7(-6)^3 + 8(-6)^2 + 11(-6) + 5$
$$= 1296 - 1512 + 288 - 66 + 5$$
$$= 11$$

In Example 2, notice that finding $f(-6)$ using synthetic division and the remainder theorem involves much easier computation than evaluating $f(-6)$ directly. This is often the case.

EXAMPLE 3

Find the remainder when $x^3 + 3x^2 - 13x - 15$ is divided by $x + 1$.

Solution Letting $f(x) = x^3 + 3x^2 - 13x - 15$ and writing $x + 1$ as $x - (-1)$, we can apply the remainder theorem:

$$f(-1) = (-1)^3 + 3(-1)^2 - 13(-1) - 15 = 0.$$

Thus the remainder is zero.

Example 3 illustrates an important special case of the remainder theorem: the situation in which the remainder is *zero*. In this case we say that $x + 1$ is a **factor** of $x^3 + 3x^2 - 13x - 15$.

The Factor Theorem

A general *factor theorem* can be formulated by considering the equation

$$f(x) = (x - c)q(x) + R.$$

If $x - c$ is a factor of $f(x)$, then the remainder R, which is also $f(c)$, must be zero. Conversely, if $R = f(c) = 0$, then $f(x) = (x - c)q(x)$; in other words, $x - c$ is a factor of $f(x)$. The **factor theorem** can be stated as follows.

PROPERTY 6.2
Factor Theorem

A polynomial $f(x)$ has a factor $x - c$ if and only if $f(c) = 0$.

EXAMPLE 4

Is $x - 1$ a factor of $x^3 + 5x^2 + 2x - 8$?

Solution Letting $f(x) = x^3 + 5x^2 + 2x - 8$ and computing $f(1)$, we obtain

$$f(1) = 1^3 + 5(1)^2 + 2(1) - 8 = 0.$$

Therefore, by the factor theorem, $x - 1$ is a factor of $f(x)$.

EXAMPLE 5

Is $x + 3$ a factor of $2x^3 + 5x^2 - 6x - 7$?

Solution Using synthetic division, we obtain the following.

$$
\begin{array}{r|rrrr}
-3 & 2 & 5 & -6 & -7 \\
   &   & -6 & 3 & 9 \\
\hline
   & 2 & -1 & -3 & ②
\end{array}
\quad \longleftarrow R = f(-3)
$$

Since $f(-3) \neq 0$, we know that $x + 3$ is not a factor of the given polynomial.

In Examples 4 and 5 we were concerned only with determining whether a linear polynomial of the form $x - c$ was a factor of another polynomial. For such problems, it is reasonable to compute $f(c)$ either directly or by synthetic division, whichever way seems easier. However, if more information is required, such as the complete factorization of the given polynomial, then the use of synthetic division becomes appropriate, as the next two examples illustrate.

EXAMPLE 6

Show that $x - 1$ is a factor of $x^3 - 2x^2 - 11x + 12$ and find the other linear factors of the polynomial.

Solution Let's use synthetic division to divide $x^3 - 2x^2 - 11x + 12$ by $x - 1$.

$$
\begin{array}{r|rrrr}
1 & 1 & -2 & -11 & 12 \\
  &   & 1 & -1 & -12 \\
\hline
  & 1 & -1 & -12 & 0
\end{array}
$$

The last line indicates a quotient of $x^2 - x - 12$ and a remainder of zero. The zero remainder means that $x - 1$ is a factor. Furthermore, we can write

$$x^3 - 2x^2 - 11x + 12 = (x - 1)(x^2 - x - 12).$$

The quadratic polynomial $x^2 - x - 12$ can be factored as $(x - 4)(x + 3)$ by using our conventional factoring techniques. Thus we obtain

$$x^3 - 2x^2 - 11x + 12 = (x - 1)(x - 4)(x + 3).$$

EXAMPLE 7

Show that $x + 4$ is a factor of $f(x) = x^3 - 5x^2 - 22x + 56$ and complete the factorization of $f(x)$.

Solution We use synthetic division to divide $x^3 - 5x^2 - 22x + 56$ by $x + 4$.

$$
\begin{array}{r|rrrr}
-4 & 1 & -5 & -22 & 56 \\
   &   & -4 & 36 & -56 \\
\hline
   & 1 & -9 & 14 & 0
\end{array}
$$

The last line indicates a quotient of $x^2 - 9x + 14$ and a remainder of zero. The zero remainder means that $x + 4$ is a factor. Furthermore, we can write

$$x^3 - 5x^2 - 22x + 56 = (x + 4)(x^2 - 9x + 14)$$

and then complete the factoring to obtain

$$f(x) = x^3 - 5x^2 - 22x + 56 = (x + 4)(x - 7)(x - 2).$$

The factor theorem also plays a significant role in determining some general factorization ideas, as the last example of this section illustrates.

EXAMPLE 8

Verify that $x + 1$ is a factor of $x^n + 1$ whenever n is an odd positive integer.

Solution Letting $f(x) = x^n + 1$ and computing $f(-1)$, we obtain

$$
\begin{aligned}
f(-1) &= (-1)^n + 1 \\
      &= -1 + 1 \qquad \text{Any odd power of } -1 \text{ is } -1. \\
      &= 0.
\end{aligned}
$$

Since $f(-1) = 0$, we know that $x + 1$ is a factor of $f(x)$.

Problem Set 6.2

For Problems 1–10, find $f(c)$ **(a)** by using synthetic division and the remainder theorem, and **(b)** by evaluating $f(c)$ directly.

1. $f(x) = x^2 + x - 8$ and $c = 2$
2. $f(x) = x^3 + x^2 - 2x - 4$ and $c = -1$
3. $f(x) = 3x^3 + 4x^2 - 5x + 3$ and $c = -4$
4. $f(x) = 2x^4 + x^2 + 6$ and $c = 1$

5. $f(x) = x^4 - 2x^3 - 3x^2 + 5x - 1$ and $c = -2$

6. $f(x) = 2x^4 + x^3 - 4x^2 - x + 1$ and $c = 2$

7. $f(t) = 6t^3 - 35t^2 + 8t - 10$ and $c = 6$

8. $f(t) = 2t^5 - 1$ and $c = -2$

9. $f(n) = 3n^4 - 2n^3 + 4n - 1$ and $c = 3$

10. $f(n) = -2n^4 + 4n - 5$ and $c = -3$

For Problems 11–18, find $f(c)$ *either* by using synthetic division and the remainder theorem or by evaluating $f(c)$ directly.

11. $f(x) = 5x^6 - x^3 - 1$ and $c = -1$

12. $f(x) = 2x^3 - 3x^2 - 5x + 4$ and $c = 4$

13. $f(x) = x^4 - 8x^3 + 9x^2 - 15x + 2$ and $c = 7$

14. $f(t) = 5t^3 - 8t^2 + 9t - 4$ and $c = -5$

15. $f(n) = -2n^4 + 2n^2 - n - 5$ and $c = -2$

16. $f(x) = 4x^7 + 3$ and $c = 3$

17. $f(x) = 2x^3 - 5x^2 + 4x - 3$ and $c = \frac{1}{2}$

18. $f(x) = 3x^3 + 4x^2 - 5x - 7$ and $c = -\frac{1}{3}$

For Problems 19–28, use the factor theorem to help answer some questions about factors.

19. Is $x - 2$ a factor of $3x^2 - 4x - 4$?

20. Is $x + 3$ a factor of $6x^2 + 13x - 15$?

21. Is $x + 2$ a factor of $x^3 + x^2 - 7x - 10$?

22. Is $x - 3$ a factor of $2x^3 - 3x^2 - 10x + 3$?

23. Is $x - 1$ a factor of $3x^3 + 5x^2 - x - 2$?

24. Is $x + 4$ a factor of $x^3 - 4x^2 + 2x - 8$?

25. Is $x - 2$ a factor of $x^3 - 8$?   **26.** Is $x + 2$ a factor of $x^3 + 8$?

27. Is $x - 3$ a factor of $x^4 - 81$?   **28.** Is $x + 3$ a factor of $x^4 - 81$?

For Problems 29–34, use synthetic division to show that $g(x)$ is a factor of $f(x)$ and complete the factorization of $f(x)$.

29. $g(x) = x + 2$; $f(x) = x^3 + 7x^2 + 4x - 12$

30. $g(x) = x - 1$; $f(x) = 3x^3 + 19x^2 - 38x + 16$

31. $g(x) = x - 3$; $f(x) = 6x^3 - 17x^2 - 5x + 6$

32. $g(x) = x + 2$; $f(x) = 12x^3 + 29x^2 + 8x - 4$

33. $g(x) = x + 1$; $f(x) = x^3 - 2x^2 - 7x - 4$

34. $g(x) = x - 5$; $f(x) = 2x^3 + x^2 - 61x + 30$

For Problems 35–38, find the value(s) of k that make(s) the second polynomial a factor of the first.

35. $x^3 - kx^2 + 5x + k$; $x - 2$ **36.** $k^2x^4 + 3kx^2 - 4$; $x - 1$

37. $x^3 + 4x^2 - 11x + k$; $x + 2$ **38.** $kx^3 + 19x^2 + x - 6$; $x + 3$

39. Show that $x + 2$ is a factor of $x^{12} - 4096$.

40. Argue that $f(x) = 2x^4 + x^2 + 3$ has no factor of the form $x - c$, where c is a real number.

41. Verify that $x - 1$ is a factor of $x^n - 1$ for all positive integral values of n.

42. Verify that $x + 1$ is a factor of $x^n - 1$ for all even positive integral values of n.

43. **(a)** Verify that $x - y$ is a factor of $x^n - y^n$ whenever n is a positive integer.
  (b) Verify that $x + y$ is a factor of $x^n - y^n$ whenever n is an even positive integer.
  (c) Verify that $x + y$ is a factor of $x^n + y^n$ whenever n is an odd positive integer.

Miscellaneous Problems

The remainder and factor theorems are true for any complex value of c. Therefore, for Problems 44–46, find $f(c)$ **(a)** by using synthetic division and the remainder theorem, and **(b)** by evaluating $f(c)$ directly.

44. $f(x) = x^3 - 5x^2 + 2x + 1$ and $c = i$

45. $f(x) = x^2 + 4x - 2$ and $c = 1 + i$

46. $f(x) = x^3 + 2x^2 + x - 2$ and $c = 2 - 3i$

Solve the following problems.

47. Show that $x - 2i$ is a factor of $f(x) = x^4 + 6x^2 + 8$.

48. Show that $x + 3i$ is a factor of $f(x) = x^4 + 14x^2 + 45$.

49. Consider changing the form of the polynomial $f(x) = x^3 + 4x^2 - 3x + 2$ as follows:

$$f(x) = x^3 + 4x^2 - 3x + 2$$
$$= x(x^2 + 4x - 3) + 2$$
$$= x[x(x + 4) - 3] + 2$$

The final form, $f(x) = x[x(x + 4) - 3] + 2$, is called the **nested form** of the polynomial. It is particularly well suited for evaluating functional values of f either by hand or with a calculator.

For each of the following, find the indicated functional values, using the nested form of the given polynomial.

  (a) $f(4)$, $f(-5)$, and $f(7)$ for $f(x) = x^3 + 5x^2 - 2x + 1$
  (b) $f(3)$, $f(6)$, and $f(-7)$ for $f(x) = 2x^3 - 4x^2 - 3x + 2$
  (c) $f(4)$, $f(5)$, and $f(-3)$ for $f(x) = -2x^3 + 5x^2 - 6x - 7$
  (d) $f(5)$, $f(6)$, and $f(-3)$ for $f(x) = x^4 + 3x^3 - 2x^2 + 5x - 1$

6.3

Polynomial Equations

In Chapter 2 we solved a large variety of *linear equations* of the form $ax + b = 0$ and *quadratic equations* of the form $ax^2 + bx + c = 0$. Linear and quadratic equations are special cases of a general class of equations referred to as **polynomial equations**. The equation

$$a_n x^n + a_{n-1} x^{n-1} + \cdots + a_1 x + a_0 = 0,$$

where the coefficients $a_0, a_1, \ldots, a_n$ are real numbers and n is a positive integer, is called a **polynomial equation of degree n**. The following are examples of polynomial equations.

$$\sqrt{2}x - 6 = 0 \qquad \text{Degree 1}$$

$$\tfrac{3}{4}x^2 - \tfrac{2}{3}x + 5 = 0 \qquad \text{Degree 2}$$

$$4x^3 - 3x^2 - 7x - 9 = 0 \qquad \text{Degree 3}$$

$$5x^4 - x + 6 = 0 \qquad \text{Degree 4}$$

Remark: The most general polynomial equation allows complex numbers as coefficients. However, for our purposes in this text, we will restrict the coefficients to real numbers. Such equations are often referred to as **polynomial equations over the reals**.

In general, solving polynomial equations of degree greater than 2 can be very difficult and often requires mathematics beyond the scope of this text. However, there are some general methods for solving polynomial equations that you should know, since there are certain types of polynomial equations that we *can* solve using the techniques available to us at this time.

Let's begin by listing some previously encountered polynomial equations and their solution sets.

Equation	Solution set
$3x + 4 = 7$	$\{1\}$
$x^2 + x - 6 = 0$	$\{-3, 2\}$
$2x^3 - 3x^2 - 2x + 3 = 0$	$\{-1, 1, \tfrac{3}{2}\}$
$x^4 - 16 = 0$	$\{-2, 2, -2i, 2i\}$

Notice that in each of these examples, the number of solutions corresponds to the degree of the equation. The first-degree equation has one solution, the second-degree equation has two solutions, the third-degree equation has three solutions, and the fourth-degree equation has four solutions.

Now consider the equation

$$(x - 4)^2(x + 5)^3 = 0.$$

It can be written

$$(x - 4)(x - 4)(x + 5)(x + 5)(x + 5) = 0,$$

which implies that

$$x - 4 = 0 \quad \text{or} \quad x - 4 = 0 \quad \text{or} \quad x + 5 = 0 \quad \text{or} \quad x + 5 = 0 \quad \text{or} \quad x + 5 = 0.$$

Therefore

$$x = 4 \quad \text{or} \quad x = 4 \quad \text{or} \quad x = -5 \quad \text{or} \quad x = -5 \quad \text{or} \quad x = -5.$$

We say that the solution set of the original equation is $\{-5, 4\}$, but we also say that the equation has a solution of 4 with a **multiplicity of two**, and a solution of -5 with a **multiplicity of three**. Furthermore, notice that the sum of the multiplicities is 5, which agrees with the degree of the equation.

The following general property can be stated.

PROPERTY 6.3

A polynomial equation of degree n has n solutions, where any solution of multiplicity p is counted p times.

Finding Rational Solutions

As stated earlier, solving polynomial equations of degree greater than two can be very difficult. However, *rational* solutions of polynomial equations with integral coefficients can be found using techniques of this chapter. The following property restricts the possible rational solutions of such an equation.

PROPERTY 6.4
Rational Root Theorem

Consider the polynomial equation

$$a_n x^n + a_{n-1} x^{n-1} + \cdots + a_1 x + a_0 = 0,$$

where the coefficients $a_0, a_1, \ldots, a_n$ are integers. If the rational number c/d, reduced to lowest terms, is a solution of the equation, then c is a factor of the constant term a_0, and d is a factor of the leading coefficient a_n.

The "why" behind the rational root theorem is based on some simple factoring ideas, as indicated by the following outline of a proof for the theorem.

Outline of Proof If c/d is to be a solution, then

$$a_n \left(\frac{c}{d}\right)^n + a_{n-1} \left(\frac{c}{d}\right)^{n-1} + \cdots + a_1 \left(\frac{c}{d}\right) + a_0 = 0.$$

Multiplying both sides of this equation by d^n and then adding $-a_0 d^n$ to both sides yields

$$a_n c^n + a_{n-1} c^{n-1} d + \cdots + a_1 c d^{n-1} = -a_0 d^n.$$

Because c is a factor of the left side of this equation, c must also be a factor of $-a_0 d^n$. Furthermore, because c/d is in reduced form, c and d have no common factors other than -1 or 1. Thus c must be a factor of a_0. In the same way, from the equation

$$a_{n-1} c^{n-1} d + \cdots + a_1 c d^{n-1} + a_0 d^n = -a_n c^n,$$

we can conclude that d is a factor of the left side and therefore d is also a factor of a_n.

The rational root theorem, synthetic division, the factor theorem, and some previous knowledge about solving linear and quadratic equations all merge to form a basis for finding rational solutions. Let's consider some examples.

EXAMPLE 1

Find all rational solutions of $3x^3 + 8x^2 - 15x + 4 = 0$.

Solution If c/d is a rational solution, then c must be a factor of 4 and d must be a factor of 3. Therefore, the possible values for c and d are as follows:

For c: ± 1, ± 2, ± 4
For d: ± 1, ± 3

Thus, the possible values for c/d are

$$\pm 1, \ \pm \tfrac{1}{3}, \ \pm 2, \ \pm \tfrac{2}{3}, \ \pm 4, \ \pm \tfrac{4}{3}.$$

By using synthetic division, we can test $x - 1$.

$$
\begin{array}{r|rrrr}
1 & 3 & 8 & -15 & 4 \\
  &   & 3 & 11 & -4 \\
\hline
  & 3 & 11 & -4 & 0
\end{array}
$$

This shows that $x - 1$ is a factor of the given polynomial; therefore, 1 is a rational solution of the equation. Furthermore, the synthetic division result also indicates how to factor the given polynomial.

$$3x^3 + 8x^2 - 15x + 4 = 0$$
$$(x - 1)(3x^2 + 11x - 4) = 0$$

The quadratic factor can be further factored by using our previous techniques.

$$(x - 1)(3x^2 + 11x - 4) = 0$$
$$(x - 1)(3x - 1)(x + 4) = 0$$

$x - 1 = 0$ or $3x - 1 = 0$ or $x + 4 = 0$
    $x = 1$ or $x = \tfrac{1}{3}$ or $x = -4$

Thus, the entire solution set consists of rational numbers and can be listed as $\{-4, \tfrac{1}{3}, 1\}$.

In Example 1, we were fortunate that the first time we used synthetic division, we got a rational solution. But this often does not happen; then we need to conduct a little organized search, as the next example illustrates.

EXAMPLE 2

Find all rational solutions of $3x^3 + 7x^2 - 22x - 8 = 0$.

Solution If c/d is a rational solution, then c must be a factor of -8 and d must be a factor of 3. Therefore, the possible values for c and d are as follows:

For c: ± 1, ± 2, ± 4, ± 8
For d: ± 1, ± 3

Thus, the possible values for c/d are

$$\pm 1, \ \pm \tfrac{1}{3}, \ \pm 2, \ \pm \tfrac{2}{3}, \ \pm 4, \ \pm \tfrac{4}{3}, \ \pm 8, \ \pm \tfrac{8}{3}.$$

Let's begin our search for rational solutions by trying the integers first.

$$
\begin{array}{r|rrrr}
1 & 3 & 7 & -22 & -8 \\
  &   & 3 & 10 & -12 \\
\hline
  & 3 & 10 & -12 & \boxed{-20}
\end{array}
$$

⟵ This indicates that $x - 1$ is not a factor and thus 1 is not a solution.

$$
\begin{array}{r|rrrr}
-1 & 3 & 7 & -22 & -8 \\
   &   & -3 & -4 & 26 \\
\hline
   & 3 & 4 & -26 & \boxed{18}
\end{array}
$$

⟵ This indicates that -1 is not a solution.

$$
\begin{array}{r|rrrr}
2 & 3 & 7 & -22 & -8 \\
  &   & 6 & 26 & 8 \\
\hline
  & 3 & 13 & 4 & 0
\end{array}
$$

Now we know that $x - 2$ is a factor and we can proceed as follows.

$$3x^3 + 7x^2 - 22x - 8 = 0$$
$$(x - 2)(3x^2 + 13x + 4) = 0$$
$$(x - 2)(3x + 1)(x + 4) = 0$$

$$x - 2 = 0 \quad \text{or} \quad 3x + 1 = 0 \quad \text{or} \quad x + 4 = 0$$
$$x = 2 \quad \text{or} \quad 3x = -1 \quad \text{or} \quad x = -4$$
$$x = 2 \quad \text{or} \quad x = -\tfrac{1}{3} \quad \text{or} \quad x = -4$$

The solution set is $\{-4, -\tfrac{1}{3}, 2\}$.

In Examples 1 and 2, we were solving third-degree equations. Therefore, once we found one linear factor by synthetic division, we were able to factor the remaining quadratic factor in the usual way. However, if the given equation is of degree four or more, we may need to find more than one linear factor by synthetic division, as the next example illustrates.

EXAMPLE 3

Solve $x^4 - 6x^3 + 22x^2 - 30x + 13 = 0$.

Solution The possible values for c/d are ± 1 and ± 13. By synthetic division we test 1.

$$
\begin{array}{r|rrrrr}
1 & 1 & -6 & 22 & -30 & 13 \\
  &   & 1 & -5 & 17 & -13 \\
\hline
  & 1 & -5 & 17 & -13 & 0
\end{array}
$$

This indicates that $x - 1$ is a factor of the given polynomial. The bottom line of the synthetic division indicates that the given polynomial can now be factored as follows.

$$x^4 - 6x^3 + 22x^2 - 30x + 13 = 0$$
$$(x - 1)(x^3 - 5x^2 + 17x - 13) = 0$$

Therefore,

$$x - 1 = 0 \quad \text{or} \quad x^3 - 5x^2 + 17x - 13 = 0.$$

Now we can use the same approach to look for rational solutions of $x^3 - 5x^2 + 17x - 13 = 0$. The possible values for c/d are again ± 1, and ± 13. By synthetic division we again test 1.

$$
\begin{array}{r|rrrr}
1 & 1 & -5 & 17 & -13 \\
  &   & 1 & -4 & 13 \\
\hline
  & 1 & -4 & 13 & 0
\end{array}
$$

This indicates that $x - 1$ is also a factor of $x^3 - 5x^2 + 17x - 13$, and the other factor is $x^2 - 4x + 13$. Now we can solve the original equation:

$$
\begin{aligned}
x^4 - 6x^3 + 22x^2 - 30x + 13 &= 0 \\
(x - 1)(x^3 - 5x^2 + 17x - 13) &= 0 \\
(x - 1)(x - 1)(x^2 - 4x + 13) &= 0
\end{aligned}
$$

$$x - 1 = 0 \quad \text{or} \quad x - 1 = 0 \quad \text{or} \quad x^2 - 4x + 13 = 0$$
$$x = 1 \quad \text{or} \quad x = 1 \quad \text{or} \quad x^2 - 4x + 13 = 0$$

Using the quadratic formula on $x^2 - 4x + 13 = 0$ produces

$$x = \frac{4 \pm \sqrt{16 - 52}}{2} = \frac{4 \pm \sqrt{-36}}{2} = \frac{4 \pm 6i}{2} = 2 \pm 3i$$

Thus, the original equation has a rational solution of 1 with a multiplicity of two and two complex solutions, $2 + 3i$ and $2 - 3i$. We list the solution set as $\{1, 2 \pm 3i\}$.

Example 3 illustrates two general properties. First, notice that the coefficient of x^4 is 1; this forces the possible rational solutions to be integers. In general, **the possible rational solutions of $x^n + a_{n-1}x^{n-1} + \cdots + a_1 x + a_0 = 0$ are the integral factors of a_0.**

Second, notice that the complex solutions of Example 4 are conjugates of each other. The following general property can be stated.

PROPERTY 6.5

> If a polynomial equation with real coefficients has any nonreal complex solutions, they must occur in conjugate pairs.

Remark: The justification for Property 6.5 is based on some properties of conjugates that were presented in Problem 79 of Problem Set 1.8. We will not show the details of such a proof at this time.

Each of Properties 6.3, 6.4, and 6.5 yields some information about the solutions of a polynomial equation. Before stating one more property, which will give us some additional information, we need to illustrate two ideas.

In a polynomial that is arranged in descending powers of x, if two successive terms differ in sign, we say there is a **variation in sign**. Terms with zero coefficients are disregarded when counting sign variations. For example, the polynomial

$$+3x^3 - 2x^2 + 4x + 7$$

has *two* sign variations, whereas the polynomial

$$+x^5 - 4x^3 + x - 5$$

has *three* variations.

Another idea that we need to understand is the fact that the solutions of

$$a_n(-x)^n + a_{n-1}(-x)^{n-1} + \cdots + a_1(-x) + a_0 = 0$$

are the opposites of the solutions of

$$a_n x^n + a_{n-1} x^{n-1} + \cdots + a_1 x + a_0 = 0.$$

In other words, if a new equation is formed by replacing x with $-x$ in a given equation, then the solutions of the new equation are the opposites of the solutions of the original equation. For example, the solution set of $x^2 + 7x + 12 = 0$ is $\{-4, -3\}$; the solution set of $(-x)^2 + 7(-x) + 12 = 0$, which simplifies to $x^2 - 7x + 12 = 0$, is $\{3, 4\}$.

Now we can state a property that often helps us to determine the nature of the solutions of a polynomial equation without actually solving the equation.

PROPERTY 6.6
Descartes's Rule
of Signs

Let $a_n x^n + a_{n-1} x^{n-1} + \cdots + a_1 x + a_0 = 0$ be a polynomial equation with real coefficients.

1. The number of **positive real solutions** of the given equation is either equal to the number of variations in sign of the polynomial or less than the number of variations by a positive even integer.

2. The number of **negative real solutions** of the given equation is either equal to the number of variations in sign of the polynomial $a_n(-x)^n + a_{n-1}(-x)^{n-1} + \cdots + a_1(-x) + a_0$ or less than that number of variations by a positive even integer.

Property 6.6, along with Properties 6.3 and 6.5, allows us to acquire some information about the solutions of a polynomial equation without actually solving the equation. Let's consider some equations and indicate how much we know about their solutions without solving them.

$$x^3 + 3x^2 + 5x + 4 = 0$$

1. No variations of sign in $x^3 + 3x^2 + 5x + 4$ means that there are *no positive solutions.*

2. Replacing x with $-x$ in the given polynomial produces $(-x)^3 + 3(-x)^2 + 5(-x) + 4$, which simplifies to $-x^3 + 3x^2 - 5x + 4$. This

polynomial contains 3 variations of sign; thus there are 3 or 1 *negative solutions.*

Conclusion: The given equation has either 3 negative real solutions or 1 negative real solution and 2 nonreal complex solutions.

$2x^4 + 3x^2 - x - 1 = 0$

1. There is 1 variation of the sign in the given polynomial; thus, the equation has 1 *positive solution.*

2. Replacing x with $-x$ produces $2(-x)^4 + 3(-x)^2 - (-x) - 1$, which simplifies to $2x^4 + 3x^2 + x - 1$ and contains 1 variation of sign. Thus, the given equation has 1 *negative solution.*

Conclusion: The given equation has 1 positive, 1 negative, and 2 nonreal complex solutions.

$3x^4 + 2x^2 + 5 = 0$

1. No variations of sign in the given polynomial means that there are *no positive solutions.*

2. Replacing x with $-x$ produces $3(-x)^4 + 2(-x)^2 + 5$, which simplifies to $3x^4 + 2x^2 + 5$ and still contains no variations of sign. Thus, there are *no negative solutions.*

Conclusion: The given equation contains 4 nonreal complex solutions. We also know that these solutions will appear in conjugate pairs.

$2x^5 - 4x^3 + 2x - 5 = 0$

1. Three variations of sign in the given polynomial implies that *the number of positive solutions is* 3 *or* 1.

2. Replacing x with $-x$ produces $2(-x)^5 - 4(-x)^3 + 2(-x) - 5 = -2x^5 + 4x^3 - 2x - 5$, which contains 2 variations of sign. Thus *the number of negative solutions is* 2 *or* 0.

Conclusion: The given equation has

3 positive and 2 negative solutions, or

3 positive and 2 nonreal complex solutions, or

1 positive, 2 negative, and 2 nonreal complex solutions, or

1 positive and 4 nonreal complex solutions.

It should be evident from the previous discussions that sometimes we can truly pinpoint the nature of the solutions of a polynomial equation. However, for some equations (such as the last example), if we use the properties discussed in this section, the best that we can do is to restrict the nature of the solutions to a few possibilities.

As a final thought, perhaps you would find it helpful to look back over Examples 1, 2, and 3 of this section and show that the solution sets do satisfy Properties 6.3, 6.5, and 6.6.

Problem Set 6.3

Use the rational root theorem and the factor theorem to help solve each of the following equations. Be sure that the number of solutions for each equation agrees with Property 6.3, taking into account multiplicity of solutions.

1. $x^3 + x^2 - 4x - 4 = 0$
2. $x^3 - 2x^2 - 11x + 12 = 0$
3. $6x^3 + x^2 - 10x + 3 = 0$
4. $8x^3 - 2x^2 - 41x - 10 = 0$
5. $3x^3 + 13x^2 - 52x + 28 = 0$
6. $15x^3 + 14x^2 - 3x - 2 = 0$
7. $x^3 - 2x^2 - 7x - 4 = 0$
8. $x^3 - x^2 - 8x + 12 = 0$
9. $x^4 - 4x^3 - 7x^2 + 34x - 24 = 0$
10. $x^4 + 4x^3 - x^2 - 16x - 12 = 0$
11. $x^3 - 10x - 12 = 0$
12. $x^3 - 4x^2 + 8 = 0$
13. $3x^4 - x^3 - 8x^2 + 2x + 4 = 0$
14. $2x^4 + 3x^3 - 11x^2 - 9x + 15 = 0$
15. $6x^4 - 13x^3 - 19x^2 + 12x = 0$
16. $x^3 - x^2 + x - 1 = 0$
17. $x^4 - 3x^3 + 2x^2 + 2x - 4 = 0$
18. $x^4 + x^3 - 3x^2 - 17x - 30 = 0$
19. $2x^5 - 5x^4 + x^3 + x^2 - x + 6 = 0$
20. $4x^4 + 12x^3 + x^2 - 12x + 4 = 0$

Verify that the following equations have no rational solutions.

21. $x^4 - x^3 - 8x^2 - 3x + 1 = 0$
22. $x^4 + 3x - 2 = 0$
23. $2x^4 - 3x^3 + 6x^2 - 24x + 5 = 0$
24. $3x^4 - 4x^3 - 10x^2 + 3x - 4 = 0$
25. $x^5 - 2x^4 + 3x^3 + 4x^2 + 7x - 1 = 0$
26. $x^5 + 2x^4 - 2x^3 + 5x^2 - 2x - 3 = 0$

27. The rational root theorem pertains to polynomial equations with integral coefficients. However, if the coefficients are nonintegral rational numbers, we can first apply the *multiplication property of equality* to produce an equivalent equation with integral coefficients. Solve each of the following equations using this method.

    (a) $\frac{1}{10}x^3 + \frac{1}{2}x^2 + \frac{1}{5}x - \frac{4}{5} = 0$
    (b) $\frac{1}{10}x^3 + \frac{1}{5}x^2 - \frac{1}{2}x - \frac{3}{5} = 0$
    (c) $x^3 + \frac{9}{2}x^2 - x - 12 = 0$
    (d) $x^3 - \frac{5}{6}x^2 - \frac{22}{3}x + \frac{5}{2} = 0$

For Problems 28–37, use Descartes's rule of signs (Property 6.6) to determine the possibilities for the nature of the solutions for each of the equations. *Do not solve the equations.*

28. $6x^2 + 7x - 20 = 0$
29. $8x^2 - 14x + 3 = 0$
30. $2x^3 + x - 3 = 0$
31. $4x^3 + 3x + 7 = 0$
32. $3x^3 - 2x^2 + 6x + 5 = 0$
33. $4x^3 + 5x^2 - 6x - 2 = 0$
34. $x^5 - 3x^4 + 5x^3 - x^2 + 2x - 1 = 0$
35. $2x^5 + 3x^3 - x + 1 = 0$
36. $x^5 + 32 = 0$
37. $2x^6 + 3x^4 - 2x^2 - 1 = 0$

Miscellaneous Problems

38. Use the rational root theorem to argue that $\sqrt{2}$ is not a rational number. (Hint: The solutions of $x^2 - 2 = 0$ are $\pm\sqrt{2}$.)

39. Use the rational root theorem to argue that $\sqrt{12}$ is not a rational number.

40. Defend the statement, "Every polynomial equation of odd degree with real coefficients has at least one real-number solution."

41. The following synthetic division shows that 2 is a solution of $x^4 + x^3 + x^2 - 9x - 10 = 0$.

$$
\begin{array}{r|rrrrr}
2 & 1 & 1 & 1 & -9 & -10 \\
  &   & 2 & 6 & 14 & 10 \\
\hline
  & 1 & 3 & 7 & 5 & 0 \longleftarrow
\end{array}
$$

Notice that the new quotient row (indicated by the arrow) consists entirely of non-negative numbers. This indicates that searching for solutions greater than 2 would be a waste of time, since larger divisors would continue to increase each of the numbers (except the 1 on the far left) in the new quotient row. (Try 3 as a divisor!) Thus, we say that 2 is an **upper bound** for the real-number solutions of the given equation.

Now consider the following synthetic division, which shows that -1 is also a solution of $x^4 + x^3 + x^2 - 9x - 10 = 0$.

$$
\begin{array}{r|rrrrr}
-1 & 1 & 1 & 1 & -9 & -10 \\
   &   & -1 & 0 & -1 & 10 \\
\hline
   & 1 & 0 & 1 & -10 & 0 \longleftarrow
\end{array}
$$

The new quotient row (indicated by the arrow) shows that there is no need to look for solutions less than -1 because any divisor less than -1 would increase the size (in absolute value) of each number in the new quotient row (except the 1 on the far left). (Try -2 as a divisor!) Thus, we say that -1 is a **lower bound** for the real-number solutions of the given equation.

The following general property can be stated: If $a_n x^n + a_{n-1} x^{n-1} + \cdots + a_1 x + a_0 = 0$ is a polynomial equation with real coefficients, and $a_n > 0$, and if the polynomial is divided synthetically by $x - c$, then:

1. If $c > 0$ and all numbers in the new quotient row of the synthetic division are nonnegative, then c is an upper bound for the real-number solutions of the given equation.

2. If $c < 0$ and the numbers in the new quotient row alternate in sign (with 0 considered either positive or negative, as needed), then c is a lower bound for the real-number solutions of the given equation.

Find the smallest positive integer and the largest negative integer that are upper and lower bounds, respectively, for the real-number solutions of each of the following equations. Keep in mind that the integers that serve as bounds do not necessarily have to be solutions of the equation.

(a) $x^3 - 3x^2 + 25x - 75 = 0$ (b) $x^3 + x^2 - 4x - 4 = 0$

(c) $x^4 + 4x^3 - 7x^2 - 22x + 24 = 0$ (d) $3x^3 + 7x^2 - 22x - 8 = 0$

(e) $x^4 - 2x^3 - 9x^2 + 2x + 8 = 0$

6.4

Graphing Polynomial Functions

Just as we have a vocabulary to deal with linear, quadratic, and polynomial equations, we also have terms that classify functions. In Chapter 4 we defined a **linear function** by means of the equation

$$f(x) = ax + b$$

and a **quadratic function** by means of the equation

$$f(x) = ax^2 + bx + c.$$

Both of these are special cases of a general class of functions called **polynomial functions**. Any function of the form

$$f(x) = a_n x^n + a_{n-1} x^{n-1} + \cdots + a_1 x + a_0$$

is called a **polynomial function of degree *n***, where a_n is a nonzero real number, $a_{n-1}, \ldots, a_1, a_0$ are real numbers, and *n* is a nonnegative integer. The following are examples of polynomial functions.

$f(x) = 5x^3 - 2x^2 + x - 4$	Degree 3
$f(x) = -2x^4 - 5x^3 + 3x^2 + 4x - 1$	Degree 4
$f(x) = 3x^5 + 2x^2 - 3$	Degree 5

Remark: Our previous work with polynomial equations is sometimes presented as "finding zeros of polynomial functions." The *solutions*, or *roots*, of a polynomial equation are also called the **zeros** of the polynomial function. For example, -2 and 2 are solutions of $x^2 - 4 = 0$ and they are zeros of $f(x) = x^2 - 4$. That is, $f(-2) = 0$ and $f(2) = 0$.

For a complete discussion of graphing polynomial functions, we would need some tools from calculus. However, the graphing techniques that we have discussed so far will allow us to graph certain kinds of polynomial functions. For example, polynomial functions of the form

$$f(x) = ax^n$$

are quite easy to graph. We know from our previous work that if $n = 1$, then functions such as $f(x) = 2x$, $f(x) = -3x$, and $f(x) = \frac{1}{2}x$ are lines through the origin having slopes of 2, -3, and $\frac{1}{2}$, respectively.

Furthermore, if $n = 2$ we know that the graphs of functions of the form $f(x) = ax^2$ are parabolas that are symmetrical with respect to the *y*-axis and that have their vertices at the origin.

We have also previously graphed the special case of $f(x) = ax^n$ where $a = 1$ and $n = 3$, namely, the function $f(x) = x^3$. This graph is shown in Figure 6.1.

The graphs of functions of the form $f(x) = ax^3$ for which $a \neq 1$ are slight variations of $f(x) = x^3$ and can be easily determined by plotting a few points. We have indicated the graphs of $f(x) = \frac{1}{2}x^3$ and $f(x) = -x^3$ in Figure 6.2(a) and (b).

There are two general patterns that emerge from studying functions of the form $f(x) = x^n$. If *n* is odd and greater than 3, then the graph of $f(x) = x^n$ closely resembles Figure 6.1. For example, the graph of $f(x) = x^5$ is shown in Figure 6.3. Notice that it "flattens out" a little more around the origin than the graph of $f(x) = x^3$, and it increases and decreases more rapidly because of the larger exponent. If *n* is even and greater than 2, then the graphs of $f(x) = x^n$ are not parabolas, but they do resemble the basic parabola except they're flatter at the bottom and steeper. Figure 6.4 shows the graph of $f(x) = x^4$.

Graphs of functions of the form $f(x) = ax^n$, where *n* is an integer greater than 2 and $a \neq 1$, are variations of those shown in Figures 6.1 and 6.4. If *n* is

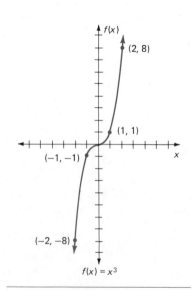

Figure 6.1

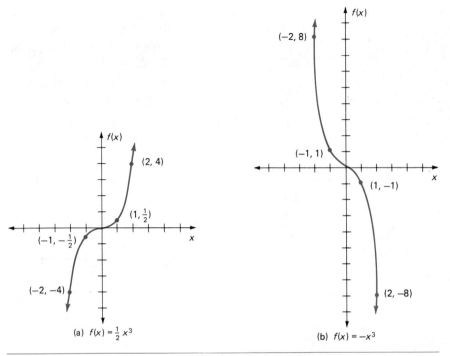

(a) $f(x) = \frac{1}{2} x^3$

(b) $f(x) = -x^3$

Figure 6.2

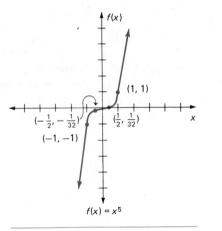

$f(x) = x^5$

Figure 6.3

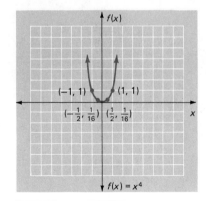

$f(x) = x^4$

Figure 6.4

odd, the curve is symmetrical about the origin; if n is even, the graph is symmetrical about the y-axis.

In Chapter 4 we found that functions such as $f(x) = x^2 + 2$, $f(x) = \sqrt{x - 1}$, and $f(x) = -|x|$ could easily be graphed by comparing them to a basic function as follows.

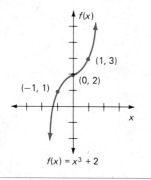

Figure 6.5

1. The graph of $f(x) = x^2 + 2$ is the same as the graph of the basic parabola $f(x) = x^2$ except *moved up two units.*
2. The graph of $f(x) = \sqrt{x - 1}$ is the same as the graph of the basic square root function $f(x) = \sqrt{x}$ except *moved to the right one unit.*
3. The graph of $f(x) = -|x|$ is the graph of the basic absolute value function $f(x) = |x|$ *reflected across the x-axis.*

In a similar fashion, we can easily sketch the graphs of such functions as $f(x) = x^3 + 2$, $f(x) = (x - 1)^5$, and $f(x) = -x^4$, as indicated in Figures 6.5, 6.6, and 6.7, respectively.

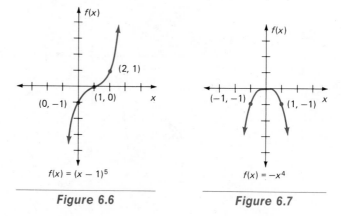

Figure 6.6   **Figure 6.7**

Graphing Polynomial Functions in Factored Form

As we mentioned earlier, a complete discussion of graphing polynomials of degree greater than 2 requires some tools from calculus. In fact, as the degree increases, the graphs often become more complicated. We do know that polynomial functions produce smooth continuous curves with a number of turning points, as illustrated in Figures 6.8 and 6.9. Some typical graphs of polynomial functions of odd degree are shown in Figure 6.8. As suggested by the graphs, every polynomial function of odd degree has at least *one real zero,* that is, at least one real number c such that $f(c) = 0$. Geometrically, the zeros of the function are the x-intercepts of the graph. In Figure 6.9 we have illustrated some possible graphs of polynomial functions of even degree.

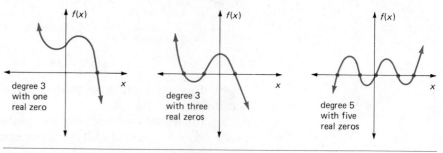

Figure 6.8

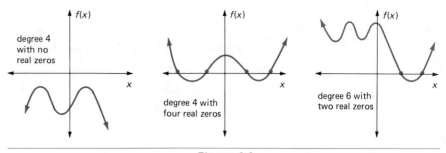

Figure 6.9

As indicated by the graphs in Figures 6.8 and 6.9, polynomial functions usually have **turning points** where the function either changes from increasing to decreasing or from decreasing to increasing. In calculus we are able to verify that *a polynomial function of degree n has at most n − 1 turning points*. Now let's illustrate how this information, along with some other techniques, can be used to graph polynomial functions that are expressed in factored form.

EXAMPLE 1

Graph $f(x) = (x + 2)(x - 1)(x - 3)$.

Solution First, let's find the x-intercepts (zeros of the function) by setting each factor equal to zero and solving for x.

$$x + 2 = 0 \qquad \text{or} \qquad x - 1 = 0 \qquad \text{or} \qquad x - 3 = 0$$
$$x = -2 \qquad \text{or} \qquad x = 1 \qquad \text{or} \qquad x = 3$$

Thus, the points $(-2, 0)$, $(1, 0)$, and $(3, 0)$ are on the graph. Second, the points associated with the x-intercepts divide the x-axis into four intervals as follows.

In each of these intervals, $f(x)$ is either always positive or always negative. That is to say, the graph is either completely above or completely below the x-axis. The sign can be determined by selecting a *test value* for x in each of the intervals. Any additional points that are easily obtained improve the accuracy of the graph. The following table summarizes these results.

INTERVAL	TEST VALUE	SIGN OF $f(x)$	LOCATION OF GRAPH
$x < -2$	$f(-3) = -24$	negative	below x-axis
$-2 < x < 1$	$f(0) = 6$	positive	above x-axis
$1 < x < 3$	$f(2) = -4$	negative	below x-axis
$x > 3$	$f(4) = 18$	positive	above x-axis

Additional values: $f(-1) = 8$

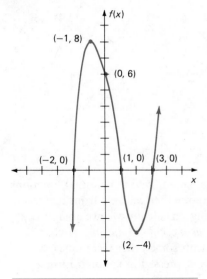

(−1, 8)

(0, 6)

(−2, 0) (1, 0) (3, 0)

(2, −4)

Figure 6.10

Making use of the x-intercepts and the information in the table, we have sketched the graph in Figure 6.10. (The points $(-3, -24)$ and $(4, 18)$ are not shown but they are used to indicate a rapid decrease and increase of the curve in those regions.)

    Remark: In Figure 6.10 we have indicated turning points of the graph at $(2, -4)$ and $(-1, 8)$. Keep in mind that these are only approximations: again the tools of calculus are needed to find the exact turning points.

EXAMPLE 2

Graph $f(x) = -x^4 + 3x^3 - 2x^2$.

Solution The polynomial can be factored as follows:

$$\begin{aligned} f(x) &= -x^4 + 3x^3 - 2x^2 \\ &= -x^2(x^2 - 3x + 2) \\ &= -x^2(x - 1)(x - 2). \end{aligned}$$

Now we can find the x-intercepts.

$$-x^2 = 0 \quad \text{or} \quad x - 1 = 0 \quad \text{or} \quad x - 2 = 0$$
$$x = 0 \quad \text{or} \quad x = 1 \quad \text{or} \quad x = 2$$

Thus the points $(0,0)$, $(1,0)$, and $(2,0)$ are on the graph and divide the x-axis into four intervals.

The following table determines some points and summarizes the sign behavior of $f(x)$.

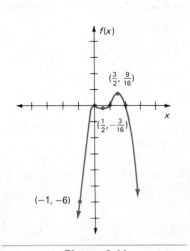

$\left(\frac{3}{2}, \frac{9}{16}\right)$

$\left(\frac{1}{2}, -\frac{3}{16}\right)$

$(-1, -6)$

Figure 6.11

INTERVAL	TEST VALUE	SIGN OF $f(x)$	LOCATION OF GRAPH
$x < 0$	$f(-1) = -6$	negative	below x-axis
$0 < x < 1$	$f(\frac{1}{2}) = -\frac{3}{16}$	negative	below x-axis
$1 < x < 2$	$f(\frac{3}{2}) = \frac{9}{16}$	positive	above x-axis
$x > 2$	$f(3) = -18$	negative	below x-axis

Making use of the table and the x-intercepts, the graph is illustrated in Figure 6.11.

EXAMPLE 3

Graph $f(x) = x^3 + 3x^2 - 4$.

Solution By using the rational root theorem, synthetic division, and the factor theorem, we can factor the given polynomial as follows.

$$f(x) = x^3 + 3x^2 - 4$$
$$= (x - 1)(x^2 + 4x + 4)$$
$$= (x - 1)(x + 2)^2$$

Now we can find the x-intercepts.

$$x - 1 = 0 \quad \text{or} \quad (x + 2)^2 = 0$$
$$x = 1 \quad \text{or} \quad x = -2$$

Thus the points $(-2, 0)$ and $(1, 0)$ are on the graph and divide the x-axis into three intervals.

The following table determines some points and summarizes the sign behavior of $f(x)$.

INTERVAL	TEST VALUE	SIGN OF $f(x)$	LOCATION OF GRAPH
$x < -2$	$f(-3) = -4$	negative	below x-axis
$-2 < x < 1$	$f(0) = -4$	negative	below x-axis
$x > 1$	$f(2) = 16$	positive	above x-axis

Additional values: $f(-1) = -2$; $f(-4) = -20$

With the results of the table and the x-intercepts, we have sketched the graph in Figure 6.12.

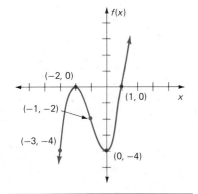

Figure 6.12

Problem Set 6.4

Graph each of the following polynomial functions.

1. $f(x) = x^3 - 3$

2. $f(x) = (x + 1)^3$

3. $f(x) = (x - 2)^3 + 1$

4. $f(x) = -(x - 3)^3$

5. $f(x) = x^4 - 2$

6. $f(x) = (x + 3)^4$

7. $f(x) = (x + 1)^4 + 3$

8. $f(x) = -x^5$

9. $f(x) = (x - 1)^5 + 2$

10. $f(x) = -(x - 2)^4$

11. $f(x) = (x - 1)(x + 1)(x - 3)$

12. $f(x) = (x - 2)(x + 1)(x + 3)$

13. $f(x) = (x + 4)(x + 1)(1 - x)$

14. $f(x) = x(x + 2)(2 - x)$

15. $f(x) = -x(x + 3)(x - 2)$

16. $f(x) = -x^2(x - 1)(x + 1)$

17. $f(x) = (x + 3)(x + 1)(x - 1)(x - 2)$

18. $f(x) = (2x - 1)(x - 2)(x - 3)$

19. $f(x) = (x - 1)^2(x + 2)$ **20.** $f(x) = (x + 2)^3(x - 4)$

21. $f(x) = (x + 1)^2(x - 1)^2$ **22.** $f(x) = x(x - 2)^2(x + 1)$

Graph each of the following polynomial functions by first factoring the given polynomial. You may need to use some factoring techniques from Chapter 1, as well as the rational root theorem and the factor theorem.

23. $f(x) = x^3 + x^2 - 2x$ **24.** $f(x) = -x^3 - x^2 + 6x$

25. $f(x) = -x^4 - 3x^3 - 2x^2$ **26.** $f(x) = x^4 - 6x^3 + 8x^2$

27. $f(x) = x^3 - x^2 - 4x + 4$ **28.** $f(x) = x^3 + 2x^2 - x - 2$

29. $f(x) = x^3 - 13x + 12$ **30.** $f(x) = x^3 - x^2 - 9x + 9$

31. $f(x) = x^3 - 2x^2 - 11x + 12$ **32.** $f(x) = 2x^3 - 3x^2 - 3x + 2$

33. $f(x) = -x^3 + 6x^2 - 11x + 6$ **34.** $f(x) = x^4 - 5x^2 + 4$

For each of the following, find (**a**) the y-intercepts, (**b**) the x-intercepts, and (**c**) the intervals of x for which $f(x) > 0$ and for which $f(x) < 0$. *Do not* sketch the graph.

35. $f(x) = (x - 5)(x + 4)(x - 3)$ **36.** $f(x) = (x + 3)(x - 6)(8 - x)$

37. $f(x) = (x - 4)^2(x + 3)^3$ **38.** $f(x) = (x + 3)^4(x - 1)^3$

39. $f(x) = (x + 2)^2(x - 1)^3(x - 2)$ **40.** $f(x) = x(x - 6)^2(x + 4)$

41. $f(x) = (x + 2)^5(x - 4)^2$

Miscellaneous Problems

42. A polynomial function with real coefficients is **continuous everywhere**; that is, its graph has no holes or breaks. This is the basis for the following property: **If $f(x)$ is a polynomial with real coefficients, and if $f(a)$ and $f(b)$ are of opposite sign, then there is at least one real zero between a and b.** This property, along with our previous knowledge of polynomial functions, provides the basis for locating and approximating irrational solutions of a polynomial equation.

Consider the equation $x^3 + 2x - 4 = 0$. Applying Descartes's rule of signs, we can determine that this equation has 1 positive real solution and 2 nonreal complex solutions. (You may want to confirm this!) The rational root theorem indicates that the only possible positive *rational* solutions are 1, 2, and 4. Using a little more compact format for synthetic division, we obtain the following results when testing for 1 and 2 as possible solutions.

	1	0	2	-4
1	1	1	3	-1
2	1	2	6	8

Since $f(1) = -1$ (negative) and $f(2) = 8$ (positive), there must be an *irrational* solution between 1 and 2. Furthermore, since -1 is closer to 0 than 8, our "guess" is that the solution is closer to 1 than to 2. Let's start looking at 1.0, 1.1, 1.2, etc., until we can clamp the solution between two numbers.

	1	0	2	-4
1.0	1	1	3	-1
1.1	1	1.1	3.21	-0.469
1.2	1	1.2	3.44	0.128

A calculator is very helpful at this time

Since $f(1.1) = -0.469$ and $f(1.2) = 0.128$, the irrational solution must be between 1.1 and 1.2. Furthermore, since 0.128 is closer to 0 than -0.469, our guess is that the solution is closer to 1.2 than to 1.1. Let's start looking at 1.15, 1.16, and so on.

	1	0	2	-4
1.15	1	1.15	3.3225	-0.179
1.16	1	1.16	3.3456	-0.119
1.17	1	1.17	3.3689	-0.058
1.18	1	1.18	3.3924	0.003

Since $f(1.17) = -0.058$ and $f(1.18) = 0.003$, the irrational solution must be between 1.17 and 1.18. Therefore, we can use 1.2 as a rational approximation to the nearest tenth.

For each of the following equations, verify that the equation has exactly one irrational solution, and find an approximation, to the nearest tenth, of that solution.

(a) $x^3 + x - 6 = 0$ (b) $x^3 - 6x - 4 = 0$

(c) $x^3 - 27x + 18 = 0$ (d) $x^3 - x^2 - x - 1 = 0$

(e) $x^3 - 24x - 32 = 0$ (f) $x^3 - 5x^2 + 3 = 0$

6.5

Graphing Rational Functions

A function of the form

$$f(x) = \frac{p(x)}{q(x)}, \qquad q(x) \neq 0,$$

where $p(x)$ and $q(x)$ are both polynomial functions, is called a **rational function**. The following are examples of rational functions:

$$f(x) = \frac{2}{x - 1}, \qquad f(x) = \frac{x}{x - 2},$$

$$f(x) = \frac{x^2}{x^2 - x - 6}, \qquad f(x) = \frac{x^3 - 8}{x + 4}.$$

In each example, the domain of the rational function is the set of all real numbers except those that make the denominator zero. For example, the domain of $f(x) = \dfrac{2}{x - 1}$ is the set of all real numbers except 1. As you will see, these exclusions from the domain are important numbers from a graphing standpoint. They represent breaks in an otherwise continuous curve.

Let's set the stage for graphing rational functions by considering in detail the function $f(x) = 1/x$. First, note that at $x = 0$, the function is undefined. Second, let's consider a rather extensive table of values to show some number trends and to build a basis for defining the concept of an *asymptote*.

x	$f(x) = \dfrac{1}{x}$	
1	1	
2	0.5	These values indicate that the value of $f(x)$ is positive and approaches zero from above as x gets larger and larger.
10	0.1	
100	0.01	
1000	0.001	
0.5	2	
0.1	10	These values indicate that $f(x)$ is positive and is getting larger and larger as x approaches zero from the right.
0.01	100	
0.001	1000	
0.0001	10000	
−0.5	−2	
−0.1	−10	These values indicate that $f(x)$ is negative and is getting smaller and smaller as x approaches zero from the left.
−0.01	−100	
−0.001	−1000	
−0.0001	−10000	
−1	−1	
−2	−0.5	These values indicate that $f(x)$ is negative and approaches zero from below as x gets smaller and smaller.
−10	−0.1	
−100	−0.01	
−1000	−0.001	

Using a few points from this table and the patterns we discussed, we have sketched $f(x) = 1/x$ in Figure 6.13. Notice that the graph approaches, but does not touch, either axis. We say that the y-axis (or $f(x)$-axis) is a *vertical asymptote* and the x-axis is a *horizontal asymptote*. In general, the following definitions can be given.

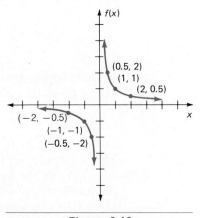

Figure 6.13

Vertical Asymptote

A line $x = a$ is a **vertical asymptote** for the graph of a function f if it satisfies either of the following two properties:

1. $f(x)$ either increases or decreases without bound as x approaches the number a from the right, *or*

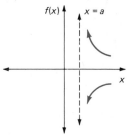

2. $f(x)$ either increases or decreases without bound as x approaches the number a from the left.

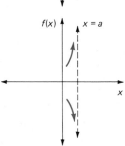

Horizontal Asymptote

A line $y = b$ (or $f(x) = b$) is a **horizontal asymptote** for the graph of a function f if it satisfies either of the following two properties:

1. $f(x)$ approaches the number b from above or below as x gets infinitely small, *or*

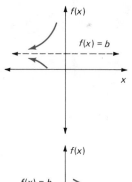

2. $f(x)$ approaches the number b from above or below as x gets infinitely large.

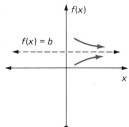

Remark: Observe that the equation $f(x) = 1/x$ exhibits origin symmetry because $f(-x) = -f(x)$. Thus, the graph in Figure 6.13 could have been drawn by first determining the part of the curve in the first quadrant and then reflecting that through the origin.

The following suggestions will help you graph rational functions of the type being considered in this section.

1. Check for y-axis and origin symmetry.
2. Find any vertical asymptote(s) by setting the denominator equal to zero and solving it for x.
3. Find any horizontal asymptote(s) by studying the behavior of $f(x)$ as x gets infinitely large or as x gets infinitely small.
4. Study the behavior of the graph when it is close to the asymptotes.
5. Plot as many points as necessary to determine the shape of the graph. This may be affected by whether the graph has any symmetry.

Keep these suggestions in mind as you study the following examples.

EXAMPLE 1

Graph $f(x) = \dfrac{-2}{x-1}$.

Solution Since $x = 1$ makes the denominator zero, the line $x = 1$ is a vertical asymptote; we have indicated this with a dashed line in Figure 6.14. Now let's look for a horizontal asymptote by checking some large and small values of x.

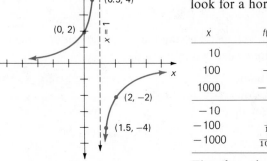

x	$f(x)$	
10	$-\frac{2}{9}$	This portion of the table shows that as x gets very large, the value of $f(x)$ approaches 0 from below.
100	$-\frac{2}{99}$	
1000	$-\frac{2}{999}$	
-10	$\frac{2}{11}$	This portion shows that as x gets very small, the value of $f(x)$ approaches 0 from above.
-100	$\frac{2}{101}$	
-1000	$\frac{2}{1001}$	

Therefore, the x-axis is a horizontal asymptote.

Finally, let's check the behavior of the graph near the vertical asymptote.

x	$f(x)$	
2	-2	As x approaches 1 from the right side, the value of $f(x)$ gets smaller and smaller.
1.5	-4	
1.1	-20	
1.01	-200	
1.001	-2000	
0	2	As x approaches 1 from the left side, the value of $f(x)$ gets larger and larger.
0.5	4	
0.9	20	
0.99	200	
0.999	2000	

The graph of $f(x) = \dfrac{-2}{x-1}$ is shown in Figure 6.14.

Figure 6.14

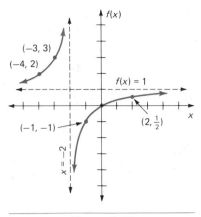

Figure 6.15

EXAMPLE 2

Graph $f(x) = \dfrac{x}{x + 2}$.

Solution Since $x = -2$ makes the denominator zero, the line $x = -2$ is a vertical asymptote. To study the behavior of $f(x)$ as x gets very large or very small, let's change the form of the rational expression by dividing both the numerator and the denominator by x:

$$f(x) = \frac{x}{x + 2} = \frac{\dfrac{x}{x}}{\dfrac{x + 2}{x}} = \frac{1}{\dfrac{x}{x} + \dfrac{2}{x}} = \frac{1}{1 + \dfrac{2}{x}}.$$

Now we can see that (1) as x gets larger and larger, the value of $f(x)$ approaches 1 from below, and (2) as x gets smaller and smaller, the value of $f(x)$ approaches 1 from above. (Perhaps you should check these claims by "plugging in" some values for x.) Thus, the line $f(x) = 1$ is a horizontal asymptote. Drawing the asymptotes (dashed lines) and plotting a few points allows us to complete the graph in Figure 6.15.

In the next two examples, pay special attention to the role of symmetry. It will allow us to direct our efforts on quadrants I and IV and then to reflect that portion of the curve across the vertical axis to complete the graph.

EXAMPLE 3

Graph $f(x) = \dfrac{2x^2}{x^2 + 4}$.

Solution First, notice that $f(-x) = f(x)$; therefore, this graph is symmetrical with respect to the y-axis. Second, the denominator $x^2 + 4$ cannot equal zero for any real number x. Thus, there is no vertical asymptote. Third, dividing both numerator and denominator of the rational expression by x^2 produces

$$f(x) = \frac{2x^2}{x^2 + 4} = \frac{\dfrac{2x^2}{x^2}}{\dfrac{x^2 + 4}{x^2}} = \frac{2}{\dfrac{x^2}{x^2} + \dfrac{4}{x^2}} = \frac{2}{1 + \dfrac{4}{x^2}}.$$

Now we can see that as x gets larger and larger, the value of $f(x)$ approaches 2 from below. Therefore, the line $f(x) = 2$ is a horizontal asymptote. So we can plot a few points using positive values for x, sketch this part of the curve, and then reflect across the $f(x)$-axis to obtain the complete graph in Figure 6.16.

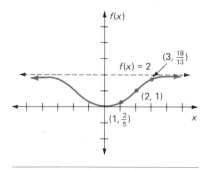

Figure 6.16

EXAMPLE 4

Graph $f(x) = \dfrac{3^i}{x^2 - 4}$.

Solution First, notice that $f(-x) = f(x)$; therefore, this graph is symmetrical about the $f(x)$-axis. Second, by setting the denominator equal to zero and solving for x, we obtain

$$x^2 - 4 = 0$$
$$x^2 = 4$$
$$x = \pm 2.$$

The lines $x = 2$ and $x = -2$ are vertical asymptotes. Next, we can see that $\dfrac{3}{x^2 - 4}$ approaches zero from above as x gets larger and larger. Finally, we can plot a few points using positive values for x (not 2), sketch this part of the curve, and then reflect it across the $f(x)$-axis to obtain the complete graph in Figure 6.17.

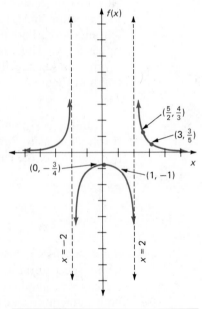

Figure 6.17

Problem Set 6.5

Graph each of the following rational functions.

1. $f(x) = \dfrac{-1}{x}$

2. $f(x) = \dfrac{1}{x^2}$

3. $f(x) = \dfrac{3}{x + 1}$

4. $f(x) = \dfrac{-1}{x - 3}$

5. $f(x) = \dfrac{2}{(x - 1)^2}$

6. $f(x) = \dfrac{-3}{(x + 2)^2}$

7. $f(x) = \dfrac{x}{x - 3}$

8. $f(x) = \dfrac{2x}{x - 1}$

9. $f(x) = \dfrac{-3x}{x + 2}$

10. $f(x) = \dfrac{-x}{x + 1}$

11. $f(x) = \dfrac{1}{x^2 - 1}$

12. $f(x) = \dfrac{-2}{x^2 - 4}$

13. $f(x) = \dfrac{-2}{(x+1)(x-2)}$ **14.** $f(x) = \dfrac{3}{(x+2)(x-4)}$ **15.** $f(x) = \dfrac{2}{x^2+x-2}$

16. $f(x) = \dfrac{-1}{x^2+x-6}$ **17.** $f(x) = \dfrac{x+2}{x}$ **18.** $f(x) = \dfrac{2x-1}{x}$

19. $f(x) = \dfrac{4}{x^2+2}$ **20.** $f(x) = \dfrac{4x^2}{x^2+1}$ **21.** $f(x) = \dfrac{2x^4}{x^4+1}$

22. $f(x) = \dfrac{x^2-4}{x^2}$

Miscellaneous Problems

23. The rational function $f(x) = \dfrac{(x-2)(x+3)}{x-2}$ has a domain of all the real numbers except 2 and can be simplified to $f(x) = x + 3$. Thus, its graph is a straight line with a hole at $(2, 5)$. Graph each of the following functions.

(a) $f(x) = \dfrac{(x+4)(x-1)}{x+4}$ **(b)** $f(x) = \dfrac{x^2-5x+6}{x-2}$

(c) $f(x) = \dfrac{x-1}{x^2-1}$ **(d)** $f(x) = \dfrac{x+2}{x^2+6x+8}$

6.6

More on Graphing Rational Functions

The rational functions that we studied in the previous section "behaved rather well." In fact, once we established the vertical and horizontal asymptotes, a little bit of point-plotting usually determined the graph rather easily. Such is not always the case with rational functions. In this section we want to investigate some rational functions that behave a little differently.

Since vertical asymptotes occur at values of x for which the denominator is zero, there can be no points of a graph on a vertical asymptote. However, recall that horizontal asymptotes are created by the behavior of $f(x)$ as x gets infinitely large or infinitely small. This does not restrict the possibility that for some values of x, there will be points of the graph on the horizontal asymptote. Let's consider some examples.

EXAMPLE 1

Graph $f(x) = \dfrac{x^2}{x^2-x-2}$.

Solution First, let's identify the vertical asymptotes by setting the denominator equal to zero and solving for x.

$$x^2 - x - 2 = 0$$
$$(x-2)(x+1) = 0$$
$$x - 2 = 0 \quad \text{or} \quad x + 1 = 0$$
$$x = 2 \quad \text{or} \quad x = -1$$

Thus, the lines $x = 2$ and $x = -1$ are vertical asymptotes. Next, we can divide both the numerator and the denominator of the rational expression by x^2.

$$f(x) = \frac{x^2}{x^2 - x - 2} = \frac{\dfrac{x^2}{x^2}}{\dfrac{x^2 - x - 2}{x^2}} = \frac{1}{1 - \dfrac{1}{x} - \dfrac{2}{x^2}}$$

Now we can see that as x gets larger and larger, the value of $f(x)$ approaches 1 from above. Thus, the line $f(x) = 1$ is a horizontal asymptote. To determine if any points of the graph are *on* the horizontal asymptote, we can see if the equation

$$\frac{x^2}{x^2 - x - 2} = 1$$

has any solutions.

$$\frac{x^2}{x^2 - x - 2} = 1$$
$$x^2 = x^2 - x - 2$$
$$0 = -x - 2$$
$$x = -2$$

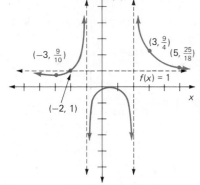

Figure 6.18

Therefore, the point $(-2, 1)$ is on the graph. Now by drawing the asymptotes, plotting a few points including $(-2, 1)$, and studying the behavior of the function close to the asymptotes, we can sketch the curve in Figure 6.18.

EXAMPLE 2

Graph $f(x) = \dfrac{x}{x^2 - 4}$.

Solution First, notice that $f(-x) = -f(x)$; therefore, this graph has origin symmetry. Second, let's identify the vertical asymptotes.

$$x^2 - 4 = 0$$
$$x^2 = 4$$
$$x = \pm 2$$

Thus, the lines $x = -2$ and $x = 2$ are vertical asymptotes. Next, by dividing the numerator and denominator of the rational expression by x^2, we obtain

$$f(x) = \frac{x}{x^2 - 4} = \frac{\dfrac{x}{x^2}}{\dfrac{x^2 - 4}{x^2}} = \frac{\dfrac{1}{x}}{1 - \dfrac{4}{x^2}}.$$

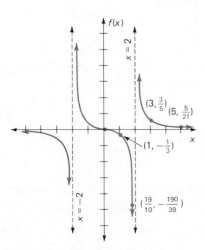

Figure 6.19

From this form we can see that as x gets larger and larger, the value of $f(x)$ approaches zero from above. Therefore, the x-axis is a horizontal asymptote.

Since $f(0) = 0$, we know that the origin is a point of the graph. Finally, by concentrating our point-plotting on positive values of x, we can sketch the portion of the curve to the right of the vertical axis and then use the fact that the graph is symmetric with respect to the origin to complete the graph. Figure 6.19 shows the completed graph.

EXAMPLE 3

Graph $f(x) = \dfrac{3x}{x^2 + 1}$.

Solution First, observe that $f(-x) = -f(x)$; therefore, this graph is symmetrical with respect to the origin. Second, since $x^2 + 1$ is a positive number for all real-number values of x, there are no vertical asymptotes for this graph. Next, by dividing the numerator and denominator of the rational expression by x^2, we obtain

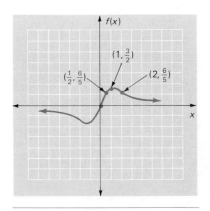

Figure 6.20

$$f(x) = \frac{3x}{x^2 + 1} = \frac{\dfrac{3x}{x^2}}{\dfrac{x^2 + 1}{x^2}} = \frac{\dfrac{3}{x}}{1 + \dfrac{1}{x^2}}.$$

From this form we see that as x gets larger and larger, the value of $f(x)$ approaches zero from above. Thus, the x-axis is a horizontal asymptote. Since $f(0) = 0$, the origin is a point of the graph. Finally, by concentrating our point-plotting on positive values of x, we can sketch the portion of the curve to the right of the vertical axis and then use origin symmetry to complete the graph, as shown in Figure 6.20.

Oblique Asymptotes

Thus far we have restricted our study of rational functions to those for which the degree of the numerator is less than or equal to the degree of the denominator. As our final example of graphing rational functions, we will consider a function for which the degree of the numerator is one greater than the degree of the denominator.

EXAMPLE 4

Graph $f(x) = \dfrac{x^2 - 1}{x - 2}$.

Solution First, let's observe that $x = 2$ is a vertical asymptote. Second, since the degree of the numerator is greater than the degree of the denominator, we can change the form of the rational expression by division. We use synthetic division.

$$
\begin{array}{r|rrr}
2 & 1 & 0 & -1 \\
  &   & 2 & 4 \\
\hline
  & 1 & 2 & 3
\end{array}
$$

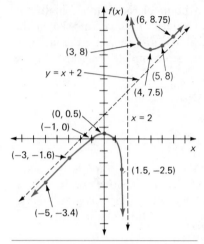

Figure 6.21

Therefore, the original function can be rewritten

$$f(x) = \frac{x^2 - 1}{x - 2} = x + 2 + \frac{3}{x - 2}.$$

Now, for very large values of $|x|$, the fraction $\frac{3}{x - 2}$ is close to zero. Therefore, as $|x|$ gets larger and larger, the graph of $f(x) = x + 2 + \frac{3}{x - 2}$ gets closer and closer to the line $f(x) = x + 2$. We call this line an **oblique asymptote** and indicate it with a dashed line in Figure 6.21. Finally, since this is a new situation, it may be necessary to plot a rather large number of points on both sides of the vertical asymptote, so let's make a rather extensive table of values. The graph of the function is shown in Figure 6.21.

x	$f(x) = \dfrac{x^2 - 1}{x - 2}$
2.1	34.1
2.5	10.5
3	8
4	7.5
5	8
6	8.75
10	12.375

These values indicate the behavior of $f(x)$ to the right of the vertical asymptote $x = 2$.

x	$f(x)$
1.9	−26.1
1.5	−2.5
1	0
0	0.5
−1	0
−3	−1.6
−5	−3.4
−10	−8.25

These values indicate the behavior of $f(x)$ to the left of the vertical asymptote $x = 2$.

Problem Set 6.6

Graph each of the following rational functions. Be sure first to check for symmetry and identify the asymptotes.

1. $f(x) = \dfrac{x^2}{x^2 + x - 2}$

2. $f(x) = \dfrac{x^2}{x^2 + 2x - 3}$

3. $f(x) = \dfrac{2x^2}{x^2 - 2x - 8}$

4. $f(x) = \dfrac{-x^2}{x^2 + 3x - 4}$

5. $f(x) = \dfrac{-x}{x^2 - 1}$

6. $f(x) = \dfrac{2x}{x^2 - 9}$

7. $f(x) = \dfrac{x}{x^2 + x - 6}$

8. $f(x) = \dfrac{-x}{x^2 - 2x - 8}$

9. $f(x) = \dfrac{x^2}{x^2 - 4x + 3}$

10. $f(x) = \dfrac{1}{x^3 + x^2 - 6x}$

11. $f(x) = \dfrac{x}{x^2 + 2}$

12. $f(x) = \dfrac{6x}{x^2 + 1}$

13. $f(x) = \dfrac{-4x}{x^2 + 1}$ **14.** $f(x) = \dfrac{-5x}{x^2 + 2}$ **15.** $f(x) = \dfrac{x^2 + 2}{x - 1}$

16. $f(x) = \dfrac{x^2 - 3}{x + 1}$ **17.** $f(x) = \dfrac{x^2 - x - 6}{x + 1}$ **18.** $f(x) = \dfrac{x^2 + 4}{x + 2}$

19. $f(x) = \dfrac{x^2 + 1}{1 - x}$ **20.** $f(x) = \dfrac{x^3 + 8}{x^2}$

6.7

Partial Fractions

In Chapter 1 we reviewed the process of adding rational expressions. For example,

$$\frac{3}{x - 2} + \frac{2}{x + 3} = \frac{3(x + 3) + 2(x - 2)}{(x - 2)(x + 3)} = \frac{3x + 9 + 2x - 4}{(x - 2)(x + 3)} = \frac{5x + 5}{(x - 2)(x + 3)}.$$

Now suppose that we want to reverse the process. That is, suppose we are given the rational expression

$$\frac{5x + 5}{(x - 2)(x + 3)}$$

and want to express it as the sum of two simpler rational expressions, called **partial fractions**. This process, called **partial fraction decomposition**, has several applications in calculus and differential equations. The following property provides the basis for partial fraction decomposition.

PROPERTY 6.7

Let $f(x)$ and $g(x)$ be polynomials with real coefficients, such that the degree of $f(x)$ is less than the degree of $g(x)$. The indicated quotient $f(x)/g(x)$ can be decomposed into partial fractions as follows:

1. If $g(x)$ has a linear factor of the form $ax + b$, then the partial fraction decomposition will contain a term of the form

$$\frac{A}{ax + b},\qquad \text{where } A \text{ is a constant.}$$

2. If $g(x)$ has a linear factor of the form $ax + b$ raised to the kth power, then the partial fraction decomposition will contain terms of the form

$$\frac{A_1}{ax + b} + \frac{A_2}{(ax + b)^2} + \cdots + \frac{A_k}{(ax + b)^k},$$

where $A_1, A_2, \ldots, A_k$ are constants.

3. If $g(x)$ has a quadratic factor of the form $ax^2 + bx + c$, where $b^2 - 4ac < 0$, then the partial fraction decomposition will contain a term of the form

$$\frac{Ax + B}{ax^2 + bx + c}, \qquad \text{where } A \text{ and } B \text{ are constants.}$$

4. If $g(x)$ has a quadratic factor of the form $ax^2 + bx + c$ raised to the kth power, where $b^2 - 4ac < 0$, then the partial fraction decomposition will contain terms of the form

$$\frac{A_1 x + B_1}{ax^2 + bx + c} + \frac{A_2 x + B_2}{(ax^2 + bx + c)^2} + \cdots + \frac{A_k x + B_k x}{(ax^2 + bx + c)^k},$$

where $A_1, A_2, \ldots, A_k, B_1, B_2, \ldots, B_k$ are constants.

Notice that Property 6.7 applies only to **proper fractions**, that is, fractions for which the degree of the numerator is less than the degree of the denominator. If the numerator is not of lower degree, we can divide and then apply Property 6.7 to the remainder, which will be a proper fraction. For example,

$$\frac{x^3 - 3x^2 - 3x - 5}{x^2 - 4} = x - 3 + \frac{x - 17}{x^2 - 4}$$

and the proper fraction $\dfrac{x - 17}{x^2 - 4}$ can be decomposed into partial fractions by applying Property 6.7.

Now let's consider some examples to illustrate the four cases in Property 6.7.

EXAMPLE 1

Find the partial fraction decomposition of $\dfrac{11x + 2}{2x^2 + x - 1}$.

Solution The denominator can be expressed as $(x + 1)(2x - 1)$. Therefore, according to Part 1 of Property 6.7, each of the linear factors produces a partial fraction of the form *constant over linear factor*. In other words, we can write

$$\frac{11x + 2}{(x + 1)(2x - 1)} = \frac{A}{x + 1} + \frac{B}{2x - 1} \tag{1}$$

for some constants A and B. To find A and B, we multiply both sides of equation (1) by the least common denominator $(x + 1)(2x - 1)$:

$$11x + 2 = A(2x - 1) + B(x + 1). \tag{2}$$

Equation (2) is an **identity**: *it is true for all values of x.*

Therefore, let's choose some convenient values for x that will determine the values for A and B. If we let $x = -1$, then equation (2) becomes an equation only in A.

$$11(-1) + 2 = A[2(-1) - 1] + B(-1 + 1)$$
$$-9 = -3A$$
$$3 = A$$

If we let $x = \frac{1}{2}$, then equation (2) becomes an equation only in B.

$$11(\tfrac{1}{2}) + 2 = A[2(\tfrac{1}{2}) - 1] + B(\tfrac{1}{2} + 1)$$
$$\tfrac{15}{2} = \tfrac{3}{2}B$$
$$5 = B$$

Therefore, the given rational expression can now be written

$$\frac{11x + 2}{2x^2 + x - 1} = \frac{3}{x + 1} + \frac{5}{2x - 1}.$$

In Example 1, after Property 6.7 has been applied, the key idea is the statement that equation (2) is true for all values of x. If we had chosen *any* two values for x, we still would have been able to determine the values for A and B. For example, letting $x = 1$ and then $x = 2$ produces the equations $13 = A + 2B$ and $24 = 3A + 3B$. Solving this system of two equations in two unknowns produces $A = 3$ and $B = 5$. In Example 1, our choices of letting $x = -1$ and then $x = \frac{1}{2}$ simply eliminated the need for solving a system of equations to find A and B.

EXAMPLE 2

Find the partial fraction decomposition of

$$\frac{-2x^2 + 7x + 2}{x(x - 1)^2}.$$

Solution By Part 1 of Property 6.7, there is a partial fraction of the form A/x corresponding to the factor of x. Next, by applying Part 2 of Property 6.7, the squared factor $(x - 1)^2$ gives rise to a sum of partial fractions of the form

$$\frac{B}{x - 1} + \frac{C}{(x - 1)^2}.$$

Therefore, the complete partial fraction decomposition is of the form

$$\frac{-2x^2 + 7x + 2}{x(x - 1)^2} = \frac{A}{x} + \frac{B}{x - 1} + \frac{C}{(x - 1)^2}. \tag{1}$$

Multiplying both sides of equation (1) by $x(x - 1)^2$ produces

$$-2x^2 + 7x + 2 = A(x - 1)^2 + Bx(x - 1) + Cx, \tag{2}$$

which is true for all values of x. If we let $x = 1$, then equation (2) becomes an equation only in C.

$$-2(1)^2 + 7(1) + 2 = A(1 - 1)^2 + B(1)(1 - 1) + C(1)$$
$$7 = C$$

If we let $x = 0$, then equation (2) becomes an equation just in A.

$$-2(0)^2 + 7(0) + 2 = A(0 - 1)^2 + B(0)(0 - 1) + C(0)$$
$$2 = A$$

If we let $x = 2$, then equation (2) becomes an equation in A, B, and C.

$$-2(2)^2 + 7(2) + 2 = A(2 - 1)^2 + B(2)(2 - 1) + C(2)$$
$$8 = A + 2B + 2C.$$

But since we already know that $A = 2$ and $C = 7$, we can easily determine B.

$$8 = 2 + 2B + 14$$
$$-8 = 2B$$
$$-4 = B$$

Therefore the original rational expression can be written

$$\frac{-2x^2 + 7x + 2}{x(x - 1)^2} = \frac{2}{x} - \frac{4}{x - 1} + \frac{7}{(x - 1)^2}.$$

EXAMPLE 3

Find the partial fraction decomposition of

$$\frac{4x^2 + 6x - 10}{(x + 3)(x^2 + x + 2)}.$$

Solution By Part 1 of Property 6.7, there is a partial fraction of the form $A/(x + 3)$ corresponding to the factor $x + 3$. By Part 3 of Property 6.7, there is also a partial fraction of the form

$$\frac{Bx + C}{x^2 + x + 2}.$$

Thus the complete partial fraction decomposition is of the form

$$\frac{4x^2 + 6x - 10}{(x + 3)(x^2 + x + 2)} = \frac{A}{x + 3} + \frac{Bx + C}{x^2 + x + 2}. \tag{1}$$

Multiplying both sides of equation (1) by $(x + 3)(x^2 + x + 2)$ produces

$$4x^2 + 6x - 10 = A(x^2 + x + 2) + (Bx + C)(x + 3), \tag{2}$$

which is true for all values of x. If we let $x = -3$, then equation (2) becomes an equation in A alone.

$$4(-3)^2 + 6(-3) - 10 = A[(-3)^2 + (-3) + 2] + [B(-3) + C][(-3) + 3]$$
$$8 = 8A$$
$$1 = A$$

If we let $x = 0$, then equation (2) becomes an equation in A and C.

$$4(0)^2 + 6(0) - 10 = A(0^2 + 0 + 2) + [B(0) + C](0 + 3)$$
$$-10 = 2A + 3C$$

Since $A = 1$, we obtain the value of C.

$$-10 = 2 + 3C$$
$$-12 = 3C$$
$$-4 = C$$

If we let $x = 1$, then equation (2) becomes an equation in A, B, and C.

$$4(1)^2 + 6(1) - 10 = A(1^2 + 1 + 2) + [B(1) + C](1 + 3)$$
$$0 = 4A + 4B + 4C$$

But since $A = 1$ and $C = -4$, we obtain the value of B.

$$0 = 4 + 4B - 16$$
$$12 = 4B$$
$$3 = B$$

Therefore the original rational expression can now be written

$$\frac{4x^2 + 6x - 10}{(x + 3)(x^2 + x + 2)} = \frac{1}{x + 3} + \frac{3x - 4}{x^2 + x + 2}.$$

EXAMPLE 4

Find the partial fraction decomposition of

$$\frac{x^3 + x^2 + x + 3}{(x^2 + 1)^2}.$$

Solution By Part 4 of Property 6.7, the partial fraction decomposition of this fraction is of the form

$$\frac{x^3 + x^2 + x + 3}{(x^2 + 1)^2} = \frac{Ax + B}{x^2 + 1} + \frac{Cx + D}{(x^2 + 1)^2}. \qquad (1)$$

Multiplying both sides of equation (1) by $(x^2 + 1)^2$ produces

$$x^3 + x^2 + x + 3 = (Ax + B)(x^2 + 1) + Cx + D, \qquad (2)$$

which is true for all values of x. Since equation (2) is an identity, we know that the coefficients of similar terms on both sides of the equation must be equal. Therefore, let's collect similar terms on the right side of equation (2).

$$x^3 + x^2 + x + 3 = Ax^3 + Ax + Bx^2 + B + Cx + D$$
$$= Ax^3 + Bx^2 + (A + C)x + B + D$$

Now we can equate coefficients from both sides:

$$1 = A, \qquad 1 = B, \qquad 1 = A + C, \qquad \text{and} \qquad 3 = B + D.$$

From these equations we can determine that $A = 1$, $B = 1$, $C = 0$, and $D = 2$. Therefore the original rational expression can be written

$$\frac{x^3 + x^2 + x + 3}{(x^2 + 1)^2} = \frac{x + 1}{x^2 + 1} + \frac{2}{(x^2 + 1)^2}.$$

Problem Set 6.7

Find the partial fraction decomposition for each of the following rational expressions.

1. $\dfrac{11x - 10}{(x - 2)(x + 1)}$

2. $\dfrac{11x - 2}{(x + 3)(x - 4)}$

3. $\dfrac{-2x - 8}{x^2 - 1}$

4. $\dfrac{-2x + 32}{x^2 - 4}$

5. $\dfrac{20x - 3}{6x^2 + 7x - 3}$

6. $\dfrac{-2x - 8}{10x^2 - x - 2}$

7. $\dfrac{x^2 - 18x + 5}{(x - 1)(x + 2)(x - 3)}$

8. $\dfrac{-9x^2 + 7x - 4}{x^3 - 3x^2 - 4x}$

9. $\dfrac{-6x^2 + 7x + 1}{x(2x - 1)(4x + 1)}$

10. $\dfrac{15x^2 + 20x + 30}{(x + 3)(3x + 2)(2x + 3)}$

11. $\dfrac{2x + 1}{(x - 2)^2}$

12. $\dfrac{-3x + 1}{(x + 1)^2}$

13. $\dfrac{-6x^2 + 19x + 21}{x^2(x + 3)}$

14. $\dfrac{10x^2 - 73x + 144}{x(x - 4)^2}$

15. $\dfrac{-2x^2 - 3x + 10}{(x^2 + 1)(x - 4)}$

16. $\dfrac{8x^2 + 15x + 12}{(x^2 + 4)(3x - 4)}$

17. $\dfrac{3x^2 + 10x + 9}{(x + 2)^3}$

18. $\dfrac{2x^3 + 8x^2 + 2x + 4}{(x + 1)^2(x^2 + 3)}$

19. $\dfrac{5x^2 + 3x + 6}{x(x^2 - x + 3)}$

20. $\dfrac{x^3 + x^2 + 2}{(x^2 + 2)^2}$

21. $\dfrac{2x^3 + x + 3}{(x^2 + 1)^2}$

22. $\dfrac{4x^2 + 3x + 14}{x^3 - 8}$

Chapter 6 Summary

Two themes unify this chapter: (1) solving polynomial equations and (2) graphing polynomial and rational functions.

Solving Polynomial Equations

The following concepts and properties provide the basis for solving polynomial equations.

1. Synthetic division
2. The factor theorem: A polynomial $f(x)$ has a factor $x - c$ if and only if $f(c) = 0$.
3. Property 6.3: A polynomial equation of degree n has n solutions, where any solution of multiplicity p is counted p times.
4. The rational root theorem: Consider the polynomial equation

$$a_n x^n + a_{n-1} x^{n-1} + \cdots + a_1 x + a_0 = 0,$$

   where *the coefficients are integers*. If the rational number c/d, reduced to lowest terms, is a solution of the equation, then c is a factor of the constant term, a_0, and d is a factor of the leading coefficient, a_n.
5. Property 6.5: If a polynomial equation with real coefficients has any nonreal complex solutions, they must occur in conjugate pairs.
6. Descartes's rule of signs: Let $a_n x^n + a_{n-1} x^{n-1} + \cdots + a_1 x + a_0 = 0$ be a polynomial equation with real coefficients.

(a) The number of *positive real solutions* either is equal to the number of sign variations in the given polynomial or is less than that number of sign variations by a positive even integer.

(b) The number of *negative real solutions* either is equal to the number of sign variations in

$$a_n(-x)^n + a_{n-1}(-x)^{n-1} + \cdots + a_1(-x) + a_0,$$

or is less than that number of sign variations by a positive even integer.

Graphing Polynomial and Rational Functions

Graphs of polynomial functions of the form $f(x) = ax^n$, where n is an integer greater than 2 and $a \neq 1$, are variations of the graphs shown in Figures 6.1 and 6.4. If n is odd, the curve is symmetrical about the origin, and if n is even, the graph is symmetrical about the vertical axis.

Graphs of polynomial functions of the form $f(x) = ax^n$ can be translated horizontally and vertically and reflected across the x-axis. For example:

1. The graph of $f(x) = 2(x - 4)^3$ is the graph of $f(x) = 2x^3$ moved 4 units to the right.

2. The graph of $f(x) = 3x^4 + 4$ is the graph of $f(x) = 3x^4$ moved up 4 units.

3. The graph of $f(x) = -x^5$ is the graph of $f(x) = x^5$ reflected across the x-axis.

To graph a polynomial function that is expressed in factored form, the following steps are helpful.

1. Find the x-intercepts, which are also called the *zeros* of the polynomial.

2. Use a test value in each of the intervals determined by the x-intercepts to find out whether the function is positive or negative over that interval.

3. Plot any additional points that are needed to determine the graph.

To graph a rational function, the following steps are useful.

1. Check for vertical-axis and origin symmetry.

2. Find any vertical asymptotes by setting the denominator equal to zero and solving it for x.

3. Find any horizontal asymptotes by studying the behavior of $f(x)$ as x gets very large or very small. This may require changing the form of the original rational expression.

4. Study the behavior of the graph when it is close to the asymptotic lines.

5. Plot as many points as necessary to determine the graph. This may be affected by whether the graph has any symmetries.

Be sure that you understand the process of partial fraction decomposition outlined in Property 6.7.

Chapter 6 Review Problem Set

For Problems 1 and 2, find the quotient and remainder of the division problems.

1. $(6x^3 + 11x^2 - 27x + 32) \div (2x + 7)$ **2.** $(2a^3 - 3a^2 + 13a - 1) \div (a^2 - a + 6)$

For Problems 3–6, use synthetic division to determine the quotient and remainder.

3. $(3x^3 - 4x^2 + 6x - 2) \div (x - 1)$ **4.** $(5x^3 + 7x^2 - 9x + 10) \div (x + 2)$

5. $(-2x^4 + x^3 - 2x^2 - x - 1) \div (x + 4)$ **6.** $(-3x^4 - 5x^2 + 9) \div (x + 3)$

For Problems 7–10, find $f(c)$ either by using synthetic division and the remainder theorem or by evaluating $f(c)$ directly.

7. $f(x) = 4x^5 - 3x^3 + x^2 - 1$ and $c = 1$

8. $f(x) = 4x^3 - 7x^2 + 6x - 8$ and $c = -3$

9. $f(x) = -x^4 + 9x^2 - x - 2$ and $c = -2$

10. $f(x) = x^4 - 9x^3 + 9x^2 - 10x + 16$ and $c = 8$

For Problems 11–14, use the factor theorem to help answer some questions about factors.

11. Is $x + 2$ a factor of $2x^3 + x^2 - 7x - 2$?

12. Is $x - 3$ a factor of $x^4 + 5x^3 - 7x^2 - x + 3$?

13. Is $x - 4$ a factor of $x^5 - 1024$?

14. Is $x + 1$ a factor of $x^5 + 1$?

For Problems 15–18, use the rational root theorem and the factor theorem to help solve each of the equations.

15. $x^3 - 3x^2 - 13x + 15 = 0$ **16.** $8x^3 + 26x^2 - 17x - 35 = 0$

17. $x^4 - 5x^3 + 34x^2 - 82x + 52 = 0$ **18.** $x^3 - 4x^2 - 10x + 4 = 0$

For Problems 19 and 20, use Descartes's rule of signs (Property 6.6) to list the possibilities for the nature of the solutions. *Do not solve* the equations.

19. $4x^4 - 3x^3 + 2x^2 + x + 4 = 0$ **20.** $x^5 + 3x^3 + x + 7 = 0$

For Problems 21–24, graph each of the polynomial functions.

21. $f(x) = -(x - 2)^3 + 3$ **22.** $f(x) = (x + 3)(x - 1)(3 - x)$

23. $f(x) = x^4 - 4x^2$ **24.** $f(x) = x^3 - 4x^2 + x + 6$

For Problems 25–28, graph each of the rational functions. Be sure to identify the asymptotes.

25. $f(x) = \dfrac{2x}{x - 3}$ **26.** $f(x) = \dfrac{-3}{x^2 + 1}$

27. $f(x) = \dfrac{-x^2}{x^2 - x - 6}$ **28.** $f(x) = \dfrac{x^2 + 3}{x + 1}$

For Problems 29 and 30, find the partial fraction decomposition.

29. $\dfrac{5x^2 - 4}{x^2(x + 2)}$ **30.** $\dfrac{x^2 - x - 21}{(x^2 + 4)(2x - 1)}$

Systems of Equations

7

7.1 Systems of Two Linear Equations in Two Variables

7.2 Systems of Three Linear Equations in Three Variables

7.3 A Matrix Approach to Solving Systems

7.4 Determinants

7.5 Cramer's Rule

In this chapter we will begin by reviewing some techniques for solving systems of linear equations involving two or three variables. Then, since many applications of mathematics require the use of large numbers of variables and equations, we will introduce some additional techniques for solving such extensive systems. These new techniques also form a basis for solving systems by using a computer.

7.1

Systems of Two Linear Equations in Two Variables

In Chapter 3 we stated that any equation of the form $Ax + By = C$, where A, B, and C are real numbers (A and B not both zero), is a **linear equation** in the two variables x and y, and its graph is a straight line. Two linear equations in two variables considered together form a **system of two linear equations in two variables**, as illustrated by the following examples:

$$\begin{pmatrix} x + y = 6 \\ x - y = 2 \end{pmatrix}, \qquad \begin{pmatrix} 3x + 2y = 1 \\ 5x - 2y = 23 \end{pmatrix}, \qquad \begin{pmatrix} 4x - 5y = 21 \\ -3x + y = -7 \end{pmatrix}.$$

To **solve** a system, such as any of these three examples, means to find all of the ordered pairs that simultaneously satisfy both equations in the system. For example, if we graph the two equations $x + y = 6$ and $x - y = 2$ on the same set of axes, as in Figure 7.1, then the ordered pair associated with the point of intersection of the two lines is the **solution of the system**. Thus, we say that $\{(4, 2)\}$ is the solution set of the system

$$\begin{pmatrix} x + y = 6 \\ x - y = 2 \end{pmatrix}.$$

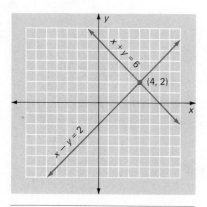

Figure 7.1

To check the solution, we substitute 4 for x and 2 for y in the two equations:

$x + y = 6$ becomes $4 + 2 = 6$, a true statement;

$x - y = 2$ becomes $4 - 2 = 2$, a true statement.

Because the graph of a linear equation in two variables is a straight line, there are three possible situations that can occur when solving a system of two linear equations in two variables. These situations are illustrated in Figure 7.2.

Case 1. The graphs of the two equations are two lines intersecting in *one* point. There is exactly one solution and the system is called a **consistent system**.

Case 2. The graphs of the two equations are parallel lines. There is *no solution* and the system is called an **inconsistent system**.

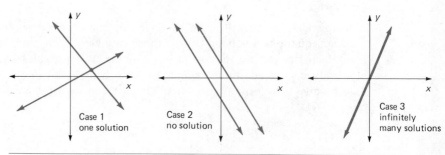

Figure 7.2

Case 3. The graphs of the two equations are the same line and there are *infinitely many solutions* of the system. Any pair of real numbers that satisfies one of the equations will also satisfy the other equation, and we say that the equations are **dependent**.

Thus, as we solve a system of two linear equations in two variables, we can expect one of three things: the system will have either *no* solutions, *one* ordered pair as a solution, or *infinitely many* ordered pairs as solutions.

The Substitution Method

Solving specific systems of equations by graphing requires accurate graphs. However, unless the solutions are integers, it is quite difficult to obtain exact solutions from a graph. Therefore we will consider some other techniques for solving systems of equations.

The **substitution method**, which works especially well with systems of two equations in two unknowns, can be described as follows:

Step 1. Solve one of the equations for one variable in terms of the other. (If possible, make a choice that will avoid fractions.)

Step 2. Substitute the expression obtained in Step 1 into the other equation, producing an equation in one variable.

Step 3. Solve the equation obtained in Step 2.

Step 4. Use the solution obtained in Step 3, along with the expression obtained in Step 1, to determine the solution of the system.

The next two examples will clarify the process.

EXAMPLE 1

Solve the system

$$\begin{pmatrix} x - 3y = -25 \\ 4x + 5y = 19 \end{pmatrix}.$$

Solution Solving the first equation for x in terms of y produces

$$x = 3y - 25.$$

Substituting $3y - 25$ for x in the second equation, we can now solve for y.

$$4x + 5y = 19$$
$$4(3y - 25) + 5y = 19$$
$$12y - 100 + 5y = 19$$
$$17y = 119$$
$$y = 7$$

Next, substituting 7 for y in the equation $x = 3y - 25$, we obtain

$$x = 3(7) - 25 = -4.$$

The solution set of the given system is $\{(-4, 7)\}$. (Perhaps you should check this solution in both of the original equations.)

EXAMPLE 2

Solve the system

$$\begin{pmatrix} 6x - 4y = 18 \\ y = \frac{3}{2}x - \frac{9}{2} \end{pmatrix}.$$

Solution The second equation is given in appropriate form to begin the substitution process. Substituting $\frac{3}{2}x - \frac{9}{2}$ for y in the first equation yields

$$6x - 4y = 18$$
$$6x - 4(\tfrac{3}{2}x - \tfrac{9}{2}) = 18$$
$$6x - 6x + 18 = 18$$
$$18 = 18.$$

Obtaining the true numerical statement $18 = 18$ indicates that the system has infinitely many solutions. Any ordered pair that satisfies one of the equations will also satisfy the other equation. Thus, in the second equation of the original system, if we let $x = k$, then $y = \frac{3}{2}k - \frac{9}{2}$. So the solution set can be expressed $\{(k, \frac{3}{2}k - \frac{9}{2}) \mid k$ is a real number$\}$. If some specific solutions are needed, they can be generated by the ordered pair $(k, \frac{3}{2}k - \frac{9}{2})$. For example, if we let $k = 1$, then we get $\frac{3}{2}(1) - \frac{9}{2} = -\frac{6}{2} = -3$. Thus, the ordered pair $(1, -3)$ is a member of the solution set of the given system.

The Elimination-by-Addition Method

Now let's consider the **elimination-by-addition method** for solving a system of equations. This is a very important method, since it is the basis for developing other techniques for solving systems containing many equations and variables. The method involves the replacement of systems of equations with *simpler equivalent systems* until we obtain a system for which the solutions are obvious. **Equivalent systems of equations are systems that have exactly the same solution set.** The following operations or transformations can be applied to a system of equations to produce an equivalent system:

1. Any two equations of the system can be interchanged.

2. Both sides of any equation of the system can be multiplied by any nonzero real number.

3. Any equation of the system can be replaced by the sum of that equation and a nonzero multiple of another equation.

EXAMPLE 3

Solve the system

$$\begin{pmatrix} 3x + 5y = -9 \\ 2x - 3y = 13 \end{pmatrix}. \qquad \begin{matrix} (1) \\ (2) \end{matrix}$$

Solution We can replace the given system with an equivalent system by multiplying equation (2) by -3.

$$\begin{pmatrix} 3x + 5y = -9 \\ -6x + 9y = -39 \end{pmatrix} \qquad \begin{matrix} (3) \\ (4) \end{matrix}$$

Now let's replace equation (4) with an equation formed by multiplying equation (3) by 2 and adding this result to equation (4).

$$\begin{pmatrix} 3x + 5y = -9 \\ 19y = -57 \end{pmatrix} \qquad \begin{matrix} (5) \\ (6) \end{matrix}$$

From equation (6) we can easily determine that $y = -3$. Then substituting -3 for y in equation (5) produces

$$3x + 5(-3) = -9$$
$$3x - 15 = -9$$
$$3x = 6$$
$$x = 2.$$

The solution set for the given system is $\{(2, -3)\}$.

Remark: We are using a format for the elimination-by-addition method that highlights the use of equivalent systems. In Section 7.3, this format will lead rather naturally into an approach using matrices. Thus, it is beneficial to stress the use of equivalent systems at this time.

EXAMPLE 4

Solve the system

$$\begin{pmatrix} \frac{1}{2}x + \frac{2}{3}y = -4 \\ \frac{1}{4}x - \frac{3}{2}y = 20 \end{pmatrix}. \qquad \begin{matrix} (7) \\ (8) \end{matrix}$$

Solution The given system can be replaced with an equivalent system by multiplying equation (7) by 6 and equation (8) by 4.

$$\begin{pmatrix} 3x + 4y = -24 \\ x - 6y = 80 \end{pmatrix} \qquad \begin{matrix} (9) \\ (10) \end{matrix}$$

Now let's exchange equations (9) and (10).

$$\begin{pmatrix} x - 6y = 80 \\ 3x + 4y = -24 \end{pmatrix} \qquad \begin{matrix} (11) \\ (12) \end{matrix}$$

We can replace equation (12) with an equation formed by multiplying equation (11) by -3 and adding this result to equation (12).

$$\begin{pmatrix} x - 6y = 80 \\ 22y = -264 \end{pmatrix} \qquad \begin{matrix} (13) \\ (14) \end{matrix}$$

From equation (14) we can determine that $y = -12$. Then substituting -12 for y in equation (13) produces

$$x - 6(-12) = 80$$
$$x + 72 = 80$$
$$x = 8.$$

The solution set of the given system is $\{(8, -12)\}$. (Check this claim!)

EXAMPLE 5

Solve the system

$$\left(\begin{array}{c} x - 4y = 9 \\ x - 4y = 3 \end{array}\right). \tag{15}$$
$$\tag{16}$$

Solution We can replace equation (16) with an equation formed by multiplying equation (15) by -1 and adding this result to equation (16).

$$\left(\begin{array}{c} x - 4y = 9 \\ 0 = -6 \end{array}\right) \tag{17}$$
$$\tag{18}$$

The statement $0 = -6$ is a contradiction and therefore the original system is *inconsistent*; it has no solution. The solution set is $\varnothing$.

Both the elimination-by-addition and substitution methods can be used to obtain exact solutions for any system of two linear equations in two unknowns. Sometimes the issue is one of deciding which method to use on a particular system. Some systems lend themselves to one or the other of the methods by the original format of the equations. We will illustrate this idea in a moment when we solve some word problems.

Using Systems to Solve Problems

Many word problems that we solved earlier in this text with one variable and one equation can also be solved by using a system of two linear equations in two variables. In fact, in many of these problems you may find it more natural to use two variables and two equations.

The two-variable expression $10t + u$ can be used to represent any two-digit whole number. The t represents the tens digit and the u represents the units digit. For example, if $t = 4$ and $u = 8$, then $10t + u$ becomes $10(4) + 8 = 48$. Now let's use this general representation for a two-digit number to help solve a problem.

PROBLEM 1

The units digit of a two-digit number is one more than twice the tens digit. The number with the digits reversed is 45 larger than the original number. Find the original number.

Solution Let u represent the units digit of the original number and let t represent the tens digit. Then $10t + u$ represents the original number and $10u + t$ represents the new number with the digits reversed. The problem translates into the following system.

$$\left(\begin{array}{c} u = 2t + 1 \\ 10u + t = 10t + u + 45 \end{array}\right)$$

The units digit is one more than twice the tens digit.

The number with the digits reversed is 45 larger than the original number.

Simplifying the second equation, the system becomes

$$\begin{pmatrix} u = 2t + 1 \\ 9u - 9t = 45 \end{pmatrix}.$$

Because of the form of the first equation, this system lends itself to solution by the substitution method. Substituting $2t + 1$ for u in the second equation produces

$$9(2t + 1) - 9t = 45$$
$$18t + 9 - 9t = 45$$
$$9t = 36$$
$$t = 4.$$

Now substituting 4 for t in the equation $u = 2t + 1$ produces

$$u = 2(4) + 1 = 9.$$

The tens digit is 4 and the units digit is 9, so the number is 49.

PROBLEM 2

Lucinda invested $950, part of it at 11% interest and the remainder at 12%. Her total yearly income from the two investments was $111.50. How much did she invest at each rate?

Solution Let x represent the amount invested at 11% and y the amount invested at 12%. The problem translates into the following system.

$$\begin{pmatrix} x + y = 950 \\ 0.11x + 0.12y = 111.50 \end{pmatrix} \begin{matrix} \text{The two investments total \$950.} \\ \text{The yearly interest from the two} \\ \text{investments totals \$111.50.} \end{matrix}$$

Multiplying the second equation by 100 produces an equivalent system.

$$\begin{pmatrix} x + y = 950 \\ 11x + 12y = 11150 \end{pmatrix}$$

Since neither equation is solved for one variable in terms of the other, let's use the elimination-by-addition method to solve the system. The second equation can be replaced by an equation formed by multiplying the first equation by -11 and adding this result to the second equation.

$$\begin{pmatrix} x + y = 950 \\ y = 700 \end{pmatrix}$$

Now we substitute 700 for y in the equation $x + y = 950$.

$$x + 700 = 950$$
$$x = 250$$

Therefore, Lucinda must have invested $250 at 11% and $700 at 12%.

Problem Set 7.1

Solve each of the following systems by using the substitution method.

1. $\begin{pmatrix} x + y = 16 \\ y = x + 2 \end{pmatrix}$

2. $\begin{pmatrix} 2x + 3y = -5 \\ y = 2x + 9 \end{pmatrix}$

3. $\begin{pmatrix} x = 3y - 25 \\ 4x + 5y = 19 \end{pmatrix}$

4. $\begin{pmatrix} 3x - 5y = 25 \\ x = y + 7 \end{pmatrix}$

5. $\begin{pmatrix} y = \frac{2}{3}x - 1 \\ 5x - 7y = 9 \end{pmatrix}$

6. $\begin{pmatrix} y = \frac{3}{4}x + 5 \\ 4x - 3y = -1 \end{pmatrix}$

7. $\begin{pmatrix} a = 4b + 13 \\ 3a + 6b = -33 \end{pmatrix}$

8. $\begin{pmatrix} 9a - 2b = 28 \\ b = -3a + 1 \end{pmatrix}$

9. $\begin{pmatrix} 2x - 3y = 4 \\ y = \frac{2}{3}x - \frac{4}{3} \end{pmatrix}$

10. $\begin{pmatrix} t + u = 11 \\ t = u + 7 \end{pmatrix}$

11. $\begin{pmatrix} u = t - 2 \\ t + u = 12 \end{pmatrix}$

12. $\begin{pmatrix} y = 5x - 9 \\ 5x - y = 9 \end{pmatrix}$

13. $\begin{pmatrix} 4x + 3y = -7 \\ 3x - 2y = 16 \end{pmatrix}$

14. $\begin{pmatrix} 5x - 3y = -34 \\ 2x + 7y = -30 \end{pmatrix}$

15. $\begin{pmatrix} 5x - y = 4 \\ y = 5x + 9 \end{pmatrix}$

16. $\begin{pmatrix} 2x + 3y = 3 \\ 4x - 9y = -4 \end{pmatrix}$

17. $\begin{pmatrix} 4x - 5y = 3 \\ 8x + 15y = -24 \end{pmatrix}$

18. $\begin{pmatrix} 4x + y = 9 \\ y = 15 - 4x \end{pmatrix}$

Solve each of the following systems by using the elimination-by-addition method.

19. $\begin{pmatrix} 3x + 2y = 1 \\ 5x - 2y = 23 \end{pmatrix}$

20. $\begin{pmatrix} 4x + 3y = -22 \\ 4x - 5y = 26 \end{pmatrix}$

21. $\begin{pmatrix} x - 3y = -22 \\ 2x + 7y = 60 \end{pmatrix}$

22. $\begin{pmatrix} 6x - y = 3 \\ 5x + 3y = -9 \end{pmatrix}$

23. $\begin{pmatrix} 4x - 5y = 21 \\ 3x + 7y = -38 \end{pmatrix}$

24. $\begin{pmatrix} 5x - 3y = -34 \\ 2x + 7y = -30 \end{pmatrix}$

25. $\begin{pmatrix} 5x - 2y = 19 \\ 5x - 2y = 7 \end{pmatrix}$

26. $\begin{pmatrix} 3a - 2b = 5 \\ 2a + 7b = 9 \end{pmatrix}$

27. $\begin{pmatrix} 6a - 3b = 4 \\ 5a + 2b = -1 \end{pmatrix}$

28. $\begin{pmatrix} 7x + 2y = 11 \\ 7x + 2y = -4 \end{pmatrix}$

29. $\begin{pmatrix} \frac{2}{3}s + \frac{1}{4}t = -1 \\ \frac{1}{2}s - \frac{1}{3}t = -7 \end{pmatrix}$

30. $\begin{pmatrix} \frac{1}{4}s - \frac{2}{3}t = -3 \\ \frac{1}{3}s + \frac{1}{3}t = 7 \end{pmatrix}$

31. $\begin{pmatrix} \dfrac{x}{2} - \dfrac{2y}{5} = \dfrac{-23}{60} \\ \dfrac{2x}{3} + \dfrac{y}{4} = \dfrac{-1}{4} \end{pmatrix}$

32. $\begin{pmatrix} \dfrac{2x}{3} - \dfrac{y}{2} = \dfrac{3}{5} \\ \dfrac{x}{4} + \dfrac{y}{2} = \dfrac{7}{80} \end{pmatrix}$

33. $\begin{pmatrix} \dfrac{4x}{5} - \dfrac{3y}{2} = \dfrac{1}{5} \\ -2x + y = -1 \end{pmatrix}$

34. $\begin{pmatrix} \dfrac{3x}{2} - \dfrac{2y}{7} = -1 \\ 4x + y = 2 \end{pmatrix}$

Solve each of the following systems by either the substitution method or the elimination-by-addition method, whichever seems more appropriate.

35. $\begin{pmatrix} 5x - y = -22 \\ 2x + 3y = -2 \end{pmatrix}$

36. $\begin{pmatrix} 4x + 5y = -41 \\ 3x - 2y = 21 \end{pmatrix}$

37. $\begin{pmatrix} x = 3y - 10 \\ x = -2y + 15 \end{pmatrix}$

38. $\begin{pmatrix} y = 4x - 24 \\ 7x + y = 42 \end{pmatrix}$

39. $\begin{pmatrix} 3x - 5y = 9 \\ 6x - 10y = -1 \end{pmatrix}$

40. $\begin{pmatrix} y = \frac{2}{5}x - 3 \\ 4x - 7y = 33 \end{pmatrix}$

41. $\begin{pmatrix} \frac{1}{2}x - \frac{2}{3}y = 22 \\ \frac{1}{2}x + \frac{1}{4}y = 0 \end{pmatrix}$

42. $\begin{pmatrix} \frac{2}{5}x - \frac{1}{3}y = -9 \\ \frac{3}{4}x + \frac{1}{3}y = -14 \end{pmatrix}$

43. $\begin{pmatrix} t = 2u + 2 \\ 9u - 9t = -45 \end{pmatrix}$

44. $\begin{pmatrix} 9u - 9t = 36 \\ u = 2t + 1 \end{pmatrix}$

45. $\begin{pmatrix} x + y = 1000 \\ 0.12x + 0.14y = 136 \end{pmatrix}$

46. $\begin{pmatrix} x + y = 10 \\ 0.3x + 0.7y = 4 \end{pmatrix}$

47. $\begin{pmatrix} y = 2x \\ 0.09x + 0.12y = 132 \end{pmatrix}$

48. $\begin{pmatrix} y = 3x \\ 0.1x + 0.11y = 64.5 \end{pmatrix}$

49. $\begin{pmatrix} x + y = 10.5 \\ 0.5x + 0.8y = 7.35 \end{pmatrix}$

50. $\begin{pmatrix} 2x + y = 7.75 \\ 3x + 2y = 12.5 \end{pmatrix}$

Solve each of the following problems by using a system of equations.

51. The sum of two numbers is 53 and their difference is 19. Find the numbers.

52. The sum of two numbers is -3 and their difference is 25. Find the numbers.

53. Find two numbers such that one of them is 3 times the other and their difference is 10.

54. One number is 2 less than 3 times the other number. Find the two numbers if their sum is 26.

55. The tens digit of a two-digit number is 1 more than 3 times the units digit. If the sum of the digits is 9, find the number.

56. The units digit of a two-digit number is 1 less than twice the tens digit. The sum of the digits is 8. Find the number.

57. The sum of the digits of a two-digit number is 7. If the digits are reversed, the newly formed number is 9 larger than the original number. Find the original number.

58. The units digit of a two-digit number is 1 less than twice the tens digit. If the digits are reversed, the newly formed number is 27 larger than the original number. Find the original number.

59. A motel rents double rooms at $32 per day and single rooms at $26 per day. If 23 rooms were rented one day for a total of $688, how many rooms of each kind were rented?

60. An apartment complex rents one-bedroom apartments for $325 per month and two-bedroom apartments for $375 per month. One month the number of one-bedroom apartments rented was twice the number of two-bedroom apartments. If the total income for that month was $12,300, how many apartments of each kind were rented?

61. The income from a student production was $10,000. The price of a student ticket was $3 and nonstudent tickets were sold at $5 each. Three thousand tickets were sold. How many tickets of each kind were sold?

62. Eric bought 50 stamps for $10.60. Some of them were 20-cent stamps and the rest were 22-cent stamps. How many of each kind did he buy?

63. Melinda invested three times as much money at 11% yearly interest as she did at 9%. Her total yearly interest from the two investments was $210. How much did she invest at each rate?

64. Sam invested $1950, part of it at 10% and the rest at 12% yearly interest. The yearly income on the 12% investment was $6 less than twice the income from the 10% investment. How much did he invest at each rate?

65. One solution contains 40% alcohol and another solution contains 60% alcohol. How many liters of each solution should be mixed to produce 20 liters of a 52% alcohol solution?

66. One solution contains 30% alcohol and a second solution contains 70% alcohol. How many liters of each solution should be mixed to make 10 liters containing 40% alcohol?

67. Bill bought 4 tennis balls and 3 golf balls for a total of $10.25. Bret went into the same store and bought 2 tennis balls and 5 golf balls for $11.25. What was the price for a tennis ball and the price for a golf ball?

68. Six cans of pop and 2 bags of potato chips cost $5.12. At the same prices, 8 cans of pop and 5 bags of potato chips cost $9.86. Find the price per can of pop and the price per bag of potato chips.

69. A cash drawer contains only five- and ten-dollar bills. There are 12 more five-dollar bills than there are ten-dollar bills. If the drawer contains $330, find the number of each kind of bill.

70. Brad has a collection of dimes and quarters totaling $47.50. The number of quarters is 10 more than twice the number of dimes. How many coins of each kind does he have?

Miscellaneous Problems

A system such as

$$\left(\begin{array}{l} \dfrac{2}{x} + \dfrac{3}{y} = \dfrac{19}{15} \\ -\dfrac{2}{x} + \dfrac{1}{y} = -\dfrac{7}{15} \end{array} \right)$$

is not a linear system, but it can be solved using the elimination-by-addition method as follows. Adding the first equation to the second produces the equivalent system

$$\left(\begin{array}{l} \dfrac{2}{x} + \dfrac{3}{y} = \dfrac{19}{15} \\ \dfrac{4}{y} = \dfrac{12}{15} \end{array} \right).$$

Now solving $4/y = \frac{12}{15}$ produces $y = 5$. Substituting 5 for y in the first equation and solving for x produces

$$\frac{2}{x} + \frac{3}{5} = \frac{19}{15}$$

$$\frac{2}{x} = \frac{10}{15}$$

$$10x = 30$$

$$x = 3.$$

The solution set of the original system is $\{(3, 5)\}$.

Solve each of the following systems in Problems 71–76.

71. $\left(\begin{array}{l} \dfrac{1}{x} + \dfrac{2}{y} = \dfrac{7}{12} \\ \dfrac{3}{x} - \dfrac{2}{y} = \dfrac{5}{12} \end{array} \right)$

72. $\left(\begin{array}{l} \dfrac{3}{x} + \dfrac{2}{y} = 2 \\ \dfrac{2}{x} - \dfrac{3}{y} = \dfrac{1}{4} \end{array} \right)$

73. $\left(\begin{array}{l} \dfrac{3}{x} - \dfrac{2}{y} = \dfrac{13}{6} \\ \dfrac{2}{x} + \dfrac{3}{y} = 0 \end{array} \right)$

74. $\left(\begin{array}{l} \dfrac{4}{x} + \dfrac{1}{y} = 11 \\ \dfrac{3}{x} - \dfrac{5}{y} = -9 \end{array}\right)$ **75.** $\left(\begin{array}{l} \dfrac{5}{x} - \dfrac{2}{y} = 23 \\ \dfrac{4}{x} + \dfrac{3}{y} = \dfrac{23}{2} \end{array}\right)$ **76.** $\left(\begin{array}{l} \dfrac{2}{x} - \dfrac{7}{y} = \dfrac{9}{10} \\ \dfrac{5}{x} + \dfrac{4}{y} = -\dfrac{41}{20} \end{array}\right)$

77. Consider the linear system

$$\left(\begin{array}{l} a_1x + b_1y = c_1 \\ a_2x + b_2y = c_2 \end{array}\right).$$

(a) Prove that this system has exactly one solution if and only if

$$\frac{a_1}{a_2} \neq \frac{b_1}{b_2}.$$

(b) Prove that this system has no solutions if and only if

$$\frac{a_1}{a_2} = \frac{b_1}{b_2} \neq \frac{c_1}{c_2}.$$

(c) Prove that this system has infinitely many solutions if and only if

$$\frac{a_1}{a_2} = \frac{b_1}{b_2} = \frac{c_1}{c_2}.$$

78. Use the results from Problem 77 to determine whether each of the following systems is consistent, inconsistent, or dependent.

(a) $\left(\begin{array}{l} 5x + y = 9 \\ x - 5y = 4 \end{array}\right)$ (b) $\left(\begin{array}{l} 3x - 2y = 14 \\ 2x + 3y = 9 \end{array}\right)$ (c) $\left(\begin{array}{l} x - 7y = 4 \\ x - 7y = 9 \end{array}\right)$

(d) $\left(\begin{array}{l} 3x - 5y = 10 \\ 6x - 10y = 1 \end{array}\right)$ (e) $\left(\begin{array}{l} 3x + 6y = 2 \\ \frac{3}{5}x + \frac{6}{5}y = \frac{2}{5} \end{array}\right)$ (f) $\left(\begin{array}{l} \frac{2}{3}x - \frac{3}{4}y = 2 \\ \frac{1}{2}x + \frac{2}{3}y = 9 \end{array}\right)$

(g) $\left(\begin{array}{l} 7x + 9y = 14 \\ 8x - 3y = 12 \end{array}\right)$ (h) $\left(\begin{array}{l} 4x - 5y = 3 \\ 12x - 15y = 9 \end{array}\right)$

7.2

Systems of Three Linear Equations in Three Variables

Consider a linear equation in three variables x, y, and z, such as $3x - 2y + z = 7$. Any **ordered triple** (x, y, z) that makes the equation a true numerical statement is said to be a **solution** of the equation. For example, the ordered triple $(2, 1, 3)$ is a solution because $3(2) - 2(1) + 3 = 7$. However, the ordered triple $(5, 2, 4)$ is not a solution because $3(5) - 2(2) + 4 \neq 7$. There are infinitely many solutions in the solution set.

> **Remark:** The idea of a linear equation is generalized to include equations of more than two variables. Thus, an equation such as $5x - 2y + 9z = 8$ is called a *linear equation in three variables*, the equation $5x - 7y + 2z - 11w = 1$ is called a *linear equation in four variables*, and so on.

To **solve** a system of three linear equations in three variables, such as

$$\begin{pmatrix} 3x - y + 2z = 13 \\ 4x + 2y + 5z = 30 \\ 5x - 3y - z = 3 \end{pmatrix},$$

means to find all of the ordered triples that satisfy all three equations. In other words, the solution set of the system is the intersection of the solution sets of all three equations in the system.

The graph of a linear equation in three variables is a *plane*, not a line. In fact, graphing equations in three variables requires the use of a three-dimensional coordinate system. Thus, using a graphing approach to solving systems of three linear equations in three variables is not at all practical. However, a simple graphic analysis does provide us with some direction as to what we can expect as we begin solving such systems.

In general, because each linear equation in three variables produces a plane, a system of three such equations produces three planes. There are various ways that three planes can be related. For example, they may be mutually parallel, or two of the planes may be parallel with the third one intersecting the other two. (You may want to analyze all of the other possibilities for the three planes!) However, for our purposes at this time we need to realize that from a solution-set viewpoint, a system of three linear equations in three variables produces one of the following possibilities.

1. There is *one ordered triple* that satisfies all three equations. The three planes have a common point of intersection as indicated in Figure 7.3.

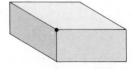

Figure 7.3

2. There are *infinitely many ordered triples* in the solution set, all of which are coordinates of *points on a line* common to the three planes. This can happen if the three planes have a common line of intersection (Figure 7.4(a)), or if two of the planes coincide and the third plane intersects them (Figure 7.4(b)).

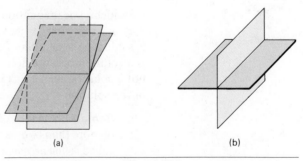

(a)

(b)

Figure 7.4

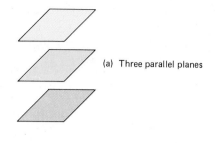

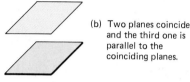

(a) Three parallel planes

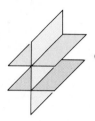

(b) Two planes coincide and the third one is parallel to the coinciding planes.

(c) Two planes are parallel and the third intersects them in parallel lines.

(d) No two planes are parallel, but two of them intersect in a line that is parallel to the third plane.

Figure 7.6

3. There are *infinitely many ordered triples* in the solution set, all of which are coordinates of *points on a plane*. This can happen if the three planes coincide, as illustrated in Figure 7.5.

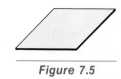

Figure 7.5

4. The solution set is *empty*; thus we write $\emptyset$. This can happen in various ways, as illustrated in Figure 7.6. Notice that in each situation there are no points common to all three planes.

Now that we know what possibilities exist, let's consider finding the solution sets for some systems. Our approach will be the elimination-by-addition method, whereby systems are replaced with equivalent systems until a system is obtained from which we can easily determine the solution set. The details of this approach will become apparent as we work a few examples.

EXAMPLE 1

Solve the system

$$\left(\begin{array}{rcl} 4x - 3y - 2z &=& 5 \\ 5y + z &=& -11 \\ 3z &=& 12 \end{array}\right).$$

$\qquad$ (1)
$\qquad$ (2)
$\qquad$ (3)

Solution The given form of this system makes it easy to solve. From equation (3) we obtain $z = 4$. Then substituting 4 for z in equation (2), we get

$$5y + 4 = -11$$
$$5y = -15$$
$$y = -3.$$

Finally, substituting 4 for z and -3 for y in equation (1) yields

$$4x - 3(-3) - 2(4) = 5$$
$$4x + 1 = 5$$
$$4x = 4$$
$$x = 1.$$

Thus, the solution set of the given system is $\{(1, -3, 4)\}$.

EXAMPLE 2

Solve the system

$$\left(\begin{array}{rcl} 3x - y + 2z &=& 13 \\ 5x - 3y - z &=& 3 \\ 4x + 2y + 5z &=& 30 \end{array}\right).$$

$\qquad$ (4)
$\qquad$ (5)
$\qquad$ (6)

Solution Equation (5) can be replaced with the equation formed by multiplying equation (4) by -3 and adding this result to equation (5). Equation (6) can be replaced with the equation formed by multiplying equation (4) by 2 and adding this result to equation (6). Thus, we produce the following equivalent system, in which equations (8) and (9) contain only the two variables x and z.

$$\begin{cases} 3x - y + 2z = 13 \\ -4x - 7z = -36 \\ 10x + 9z = 56 \end{cases}$$

$$\qquad (7)$$
$$\qquad (8)$$
$$\qquad (9)$$

Now, if we multiply equation (8) by 5 and equation (9) by 2, we get the equivalent system

$$\begin{cases} 3x - y + 2z = 13 \\ -20x - 35z = -180 \\ 20x + 18z = 112 \end{cases}.$$

$$\qquad (10)$$
$$\qquad (11)$$
$$\qquad (12)$$

Equation (12) can be replaced with the equation formed by adding equation (11) to equation (12).

$$\begin{cases} 3x - y + 2z = 13 \\ -20x - 35z = -180 \\ -17z = -68 \end{cases}$$

$$\qquad (13)$$
$$\qquad (14)$$
$$\qquad (15)$$

From equation (15) we obtain $z = 4$. Then, substituting 4 for z in equation (14), we obtain

$$-20x - 35(4) = -180$$
$$-20x = -40$$
$$x = 2.$$

Substituting 2 for x and 4 for z in equation (13) yields

$$3(2) - y + 2(4) = 13$$
$$-y + 14 = 13$$
$$-y = -1$$
$$y = 1.$$

The solution set of the original system is $\{(2, 1, 4)\}$. (Perhaps you should check this ordered triple in all three of the original equations.)

EXAMPLE 3

Solve the system

$$\begin{cases} x - 2y + 3z = 22 \\ 2x - 3y - z = 5 \\ 3x + y - 5z = -32 \end{cases}.$$

$$\qquad (16)$$
$$\qquad (17)$$
$$\qquad (18)$$

Solution Equation (17) can be replaced with the equation formed by multiplying equation (16) by -2 and adding this result to equation (17). Equation (18) can be replaced with the equation formed by multiplying equation (16) by -3 and adding this result to equation (18). The following equivalent system is produced, in which equations (20) and (21) contain only the two variables y and z.

$$\begin{pmatrix} x - 2y + 3z = 22 \\ y - 7z = -39 \\ 7y - 14z = -98 \end{pmatrix}$$

(19)
(20)
(21)

Equation (21) can be replaced with the equation formed by multiplying equation (20) by -7 and adding this result to equation (21). This produces the equivalent system

$$\begin{pmatrix} x - 2y + 3z = 22 \\ y - 7z = -39 \\ 35z = 175 \end{pmatrix}.$$

(22)
(23)
(24)

From equation (24) we obtain $z = 5$. Then, substituting 5 for z in equation (23), we obtain

$$y - 7(5) = -39$$
$$y = -4.$$

Finally, substituting -4 for y and 5 for z in equation (22) produces

$$x - 2(-4) + 3(5) = 22$$
$$x + 23 = 22$$
$$x = -1.$$

The solution set for the original system is $\{(-1, -4, 5)\}$.

EXAMPLE 4

Solve the system

$$\begin{pmatrix} 2x + 3y + z = 14 \\ 3x - 4y - 2z = -30 \\ 5x + 7y + 3z = 32 \end{pmatrix}.$$

(25)
(26)
(27)

Solution Equation (26) can be replaced with the equation formed by multiplying equation (25) by 2 and adding this result to equation (26). Equation (27) can be replaced with the equation formed by multiplying equation (25) by -3 and adding this result to equation (27). The following equivalent system is produced, in which equations (29) and (30) contain only the two variables x and y.

$$\begin{pmatrix} 2x + 3y + z = 14 \\ 7x + 2y = -2 \\ -x - 2y = -10 \end{pmatrix}$$

(28)
(29)
(30)

Now equation (30) can be replaced with the equation formed by adding equation (29) to equation (30).

$$\begin{pmatrix} 2x + 3y + z = 14 \\ 7x + 2y = -2 \\ 6x = -12 \end{pmatrix}$$

(31)
(32)
(33)

From equation (33) we obtain $x = -2$. Then, substituting -2 for x in equation (32), we obtain

$$7(-2) + 2y = -2$$
$$2y = 12$$
$$y = 6.$$

Finally, substituting 6 for y and -2 for x in equation (31) produces

$$2(-2) + 3(6) + z = 14$$
$$14 + z = 14$$
$$z = 0.$$

The solution set of the original system is $\{(-2, 6, 0)\}$.

Being able to solve systems of three linear equations in three unknowns enhances our problem-solving capabilities. Let's conclude this section with a problem that can be solved using such a system.

PROBLEM 1

A small company that manufactures sporting goods equipment produces three different styles of golf shirts. Each style of shirt requires the services of three departments, as indicated by the following table.

	STYLE A	STYLE B	STYLE C
CUTTING DEPARTMENT	0.1 hr	0.1 hr	0.3 hr
SEWING DEPARTMENT	0.3 hr	0.2 hr	0.4 hr
PACKAGING DEPARTMENT	0.1 hr	0.2 hr	0.1 hr

The cutting, sewing, and packaging departments have available a maximum of 340, 580, and 255 work-hours per week, respectively. How many of each style of golf shirt should be produced each week so that the company is operating at full capacity?

Solution Let a represent the number of shirts of style A produced per week, b the number of style B per week, and c the number of style C per week. Then the problem translates into the following system of equations.

$$\begin{cases} 0.1a + 0.1b + 0.3c = 340 & \leftarrow \text{ cutting department} \\ 0.3a + 0.2b + 0.4c = 580 & \leftarrow \text{ sewing department} \\ 0.1a + 0.2b + 0.1c = 255 & \leftarrow \text{ packaging department} \end{cases}$$

Solving this system (we will leave the details for you to carry out) produces $a = 500$, $b = 650$, and $c = 750$. Thus, the company should produce 500 golf shirts of style A, 650 of style B, and 750 of style C per week.

Problem Set 7.2

Solve each of the following systems.

1. $\begin{cases} x - 2y + 3z = 7 \\ 2x + y + 5z = 17 \\ 3x - 4y - 2z = 1 \end{cases}$

2. $\begin{cases} x - 2y + z = -4 \\ 2x + 4y - 3z = -1 \\ -3x - 6y + 7z = 4 \end{cases}$

3. $\begin{pmatrix} 2x - y + z = 0 \\ 3x - 2y + 4z = 11 \\ 5x + y - 6z = -32 \end{pmatrix}$

4. $\begin{pmatrix} 2x - y + 3z = -14 \\ 4x + 2y - z = 12 \\ 6x - 3y + 4z = -22 \end{pmatrix}$

5. $\begin{pmatrix} 3x + 2y - z = -11 \\ 2x - 3y + 4z = 11 \\ 5x + y - 2z = -17 \end{pmatrix}$

6. $\begin{pmatrix} 9x + 4y - z = 0 \\ 3x - 2y + 4z = 6 \\ 6x - 8y - 3z = 3 \end{pmatrix}$

7. $\begin{pmatrix} 2x + 3y - 4z = -10 \\ 4x - 5y + 3z = 2 \\ 2y + z = 8 \end{pmatrix}$

8. $\begin{pmatrix} x + 2y - 3z = 2 \\ 3x - z = -8 \\ 2x - 3y + 5z = -9 \end{pmatrix}$

9. $\begin{pmatrix} 3x + 2y - 2z = 14 \\ 2x - 5y + 3z = 7 \\ 4x - 3y + 7z = 5 \end{pmatrix}$

10. $\begin{pmatrix} 4x + 3y - 2z = -11 \\ 3x - 7y + 3z = 10 \\ 9x - 8y + 5z = 9 \end{pmatrix}$

11. $\begin{pmatrix} 2x - 3y + 4z = -12 \\ 4x + 2y - 3z = -13 \\ 6x - 5y + 7z = -31 \end{pmatrix}$

12. $\begin{pmatrix} 3x + 5y - 2z = -27 \\ 5x - 2y + 4z = 27 \\ 7x + 3y - 6z = -55 \end{pmatrix}$

13. $\begin{pmatrix} 5x - 3y - 6z = 22 \\ x - y + z = -3 \\ -3x + 7y - 5z = 23 \end{pmatrix}$

14. $\begin{pmatrix} 4x + 3y - 5z = -29 \\ 3x - 7y - z = -19 \\ 2x + 5y + 2z = -10 \end{pmatrix}$

Solve each of the following problems by setting up and solving a system of three linear equations in three variables.

15. The sum of three numbers is 43. The sum of the two smaller numbers is 3 larger than the largest number. Twice the smallest number plus 3 times the second number plus 4 times the largest number equals 141. Find the numbers.

16. The sum of three numbers is 20. The sum of the first and third numbers is 2 more than twice the second number. The third number minus the first yields 3 times the second number. Find the numbers.

17. A box contains $7.15 in nickels, dimes, and quarters. There are 42 coins in all and the sum of the numbers of nickels and dimes is 2 less than the number of quarters. How many coins of each kind are there?

18. A handful of 65 coins consists of pennies, nickels, and dimes. The number of nickels is 4 less than twice the number of pennies, and there are 13 more dimes than nickels. How many coins of each kind are there?

19. The measure of the largest angle of a triangle is twice the smallest angle. The sum of the smallest and the largest angle is twice the other angle. Find the measure of each angle.

20. The perimeter of a triangle is 45 centimeters. The longest side is 4 centimeters less than twice the shortest side. The sum of the lengths of the shortest and longest sides is 7 centimeters less than 3 times the length of the remaining side. Find the lengths of all three sides of the triangle.

21. Part of $3000 is invested at 12%, another part at 13%, and the remainder at 14% yearly interest. The total yearly income from the three investments is $400. The sum of the amounts invested at 12% and 13% equals the amount invested at 14%. How much is invested at each rate?

22. Different amounts of money are invested at 10%, 11%, and 12% yearly interest. The amount invested at 11% is $300 more than what is invested at 10%, and the total yearly income from all three investments is $324. If a total of $2900 is invested, find the amount invested at each rate.

23. A small company makes three different types of bird houses. Each type requires the services of three different departments according to the following table.

	TYPE A	TYPE B	TYPE C
CUTTING DEPARTMENT	0.1 hr	0.2 hr	0.1 hr
FINISHING DEPARTMENT	0.4 hr	0.4 hr	0.3 hr
ASSEMBLY DEPARTMENT	0.2 hr	0.1 hr	0.3 hr

The cutting, finishing, and assembly departments have available a maximum of 35, 95, and 62.5 work-hours per week, respectively. How many bird houses of each type should be made per week so that the company is operating at full capacity?

24. A certain diet consists of dishes A, B, and C. Each serving of A has 1 gram of fat, 2 grams of carbohydrate, and 4 grams of protein. Each serving of B has 2 grams of fat, 1 gram of carbohydrate, and 3 grams of protein. Each serving of C has 2 grams of fat, 4 grams of carbohydrate, and 3 grams of protein. The diet allows 15 grams of fat, 24 grams of carbohydrate, and 30 grams of protein. How many servings of each dish can be eaten?

25. Recall that one form of the equation of a circle is $x^2 + y^2 + Dx + Ey + F = 0$. Find the equation of the circle that passes through the points $(-3, 1)$, $(7, 1)$, and $(-7, 5)$.

7.3

A Matrix Approach to Solving Systems

In the first two sections of this chapter, we found that the techniques of substitution and elimination by addition worked effectively with two equations and two unknowns, but they started to get a bit cumbersome with three equations and three unknowns. Therefore, we shall now begin to analyze some techniques that lend themselves to use with larger systems of equations. Furthermore, some of these techniques form the basis for using a computer to solve systems. Even though these techniques are primarily designed for large systems of equations, we shall study them in the context of small systems so we won't become too bogged down with the computational aspects of the techniques.

Matrices

A **matrix** is an array of numbers arranged in horizontal rows and vertical columns and enclosed in brackets. For example, the matrix

$$2 \text{ rows} \longrightarrow \begin{bmatrix} 2 & 3 & -1 \\ -4 & 7 & 12 \end{bmatrix}$$

3 columns

has 2 rows and 3 columns and is called a 2 × 3 (read "two by three") matrix. Each number in a matrix is called an **element** of the matrix. Some additional examples of matrices ("matrices" is the plural of matrix) are as follows:

$$
\begin{array}{cccc}
\mathbf{3 \times 2} & \mathbf{2 \times 2} & \mathbf{1 \times 2} & \mathbf{4 \times 1} \\[6pt]
\begin{bmatrix} 2 & 1 \\ 1 & -4 \\ \frac{1}{2} & \frac{2}{3} \end{bmatrix} &
\begin{bmatrix} 17 & 18 \\ -14 & 16 \end{bmatrix} &
\begin{bmatrix} 7 & 14 \end{bmatrix} &
\begin{bmatrix} 3 \\ -2 \\ 1 \\ 19 \end{bmatrix}
\end{array}
$$

In general, a matrix of m rows and n columns is called a matrix of **dimension** $m \times n$ or **order** $m \times n$.

With every system of linear equations we can associate a matrix that consists of the coefficients and constant terms. For example, with the system

$$
\begin{pmatrix}
a_1 x + b_1 y + c_1 z = d_1 \\
a_2 x + b_2 y + c_2 z = d_2 \\
a_3 x + b_3 y + c_3 z = d_3
\end{pmatrix},
$$

we can associate the matrix

$$
\left[\begin{array}{ccc|c}
a_1 & b_1 & c_1 & d_1 \\
a_2 & b_2 & c_2 & d_2 \\
a_3 & b_3 & c_3 & d_3
\end{array}\right],
$$

which is commonly called the **augmented matrix** of the system of equations. The dashed line simply separates the coefficients from the constant terms and reminds us that we are working with an augmented matrix.

On page 308 we listed the operations or transformations that can be applied to a system of equations to produce an equivalent system. Since augmented matrices are essentially abbreviated forms of systems of linear equations, there are analogous transformations that can be applied to augmented matrices. These transformations are usually referred to as **elementary row operations** and can be stated as follows.

> For any augmented matrix of a system of linear equations, the following elementary row operations will produce a matrix of an equivalent system:
>
> **1.** Any two rows of the matrix can be interchanged.
>
> **2.** Any row of the matrix can be multiplied by a nonzero real number.
>
> **3.** Any row of the matrix can be replaced by the sum of a nonzero multiple of another row plus that row.

Let's illustrate the use of augmented matrices and elementary row operations to solve a system of two linear equations in two variables.

EXAMPLE 1

Solve the system

$$\begin{pmatrix} x - 3y = -17 \\ 2x + 7y = 31 \end{pmatrix}.$$

Solution The augmented matrix of the system is

$$\begin{bmatrix} 1 & -3 & | & -17 \\ 2 & 7 & | & 31 \end{bmatrix}.$$

We would like to change this matrix to one of the form

$$\begin{bmatrix} 1 & 0 & | & a \\ 0 & 1 & | & b \end{bmatrix},$$

from which we can easily determine that the solution is $x = a$ and $y = b$. Let's begin by adding -2 times row 1 to row 2 to produce a new row 2.

$$\begin{bmatrix} 1 & -3 & | & -17 \\ 0 & 13 & | & 65 \end{bmatrix}$$

Now we can multiply row 2 by $\frac{1}{13}$.

$$\begin{bmatrix} 1 & -3 & | & -17 \\ 0 & 1 & | & 5 \end{bmatrix}$$

Finally, we can add 3 times row 2 to row 1 to produce a new row 1.

$$\begin{bmatrix} 1 & 0 & | & -2 \\ 0 & 1 & | & 5 \end{bmatrix}.$$

From this last matrix we see that $x = -2$ and $y = 5$. In other words, the solution set of the original system is $\{(-2, 5)\}$.

It may seem as though the matrix approach does not provide us with much extra power for solving systems of two linear equations in two unknowns. However, as the systems become larger, the compactness of the matrix approach becomes more convenient. Let's now consider a system of three equations in three variables.

EXAMPLE 2

Solve the system

$$\begin{pmatrix} x + 2y - 3z = 15 \\ -2x - 3y + z = -15 \\ 4x + 9y - 4z = 49 \end{pmatrix}.$$

Solution The augmented matrix of this system is

$$\begin{bmatrix} 1 & 2 & -3 & | & 15 \\ -2 & -3 & 1 & | & -15 \\ 4 & 9 & -4 & | & 49 \end{bmatrix}.$$

If this system has a unique solution, then we will be able to change the augmented matrix to the form

$$\begin{bmatrix} 1 & 0 & 0 & | & a \\ 0 & 1 & 0 & | & b \\ 0 & 0 & 1 & | & c \end{bmatrix},$$

from which we will be able to read the solution $x = a$, $y = b$, and $z = c$.

Adding 2 times row 1 to row 2 produces a new row 2. Likewise, adding -4 times row 1 to row 3 produces a new row 3.

$$\begin{bmatrix} 1 & 2 & -3 & | & 15 \\ 0 & 1 & -5 & | & 15 \\ 0 & 1 & 8 & | & -11 \end{bmatrix}$$

Now adding -2 times row 2 to row 1 produces a new row 1. Also, adding -1 times row 2 to row 3 produces a new row 3.

$$\begin{bmatrix} 1 & 0 & 7 & | & -15 \\ 0 & 1 & -5 & | & 15 \\ 0 & 0 & 13 & | & -26 \end{bmatrix}$$

Now let's multiply row 3 by $\frac{1}{13}$.

$$\begin{bmatrix} 1 & 0 & 7 & | & -15 \\ 0 & 1 & -5 & | & 15 \\ 0 & 0 & 1 & | & -2 \end{bmatrix}$$

Finally, we can add -7 times row 3 to row 1 to produce a new row 1, and add 5 times row 3 to row 2 for a new row 2.

$$\begin{bmatrix} 1 & 0 & 0 & | & -1 \\ 0 & 1 & 0 & | & 5 \\ 0 & 0 & 1 & | & -2 \end{bmatrix}$$

From this last matrix we can see that the solution set of the original system is $\{(-1, 5, -2)\}$.

The final matrices of Examples 1 and 2,

$$\begin{bmatrix} 1 & 0 & | & -2 \\ 0 & 1 & | & 5 \end{bmatrix} \quad \text{and} \quad \begin{bmatrix} 1 & 0 & 0 & | & -1 \\ 0 & 1 & 0 & | & 5 \\ 0 & 0 & 1 & | & -2 \end{bmatrix},$$

are said to be in **reduced echelon form**. In general, a matrix is in reduced echelon form if the following conditions are satisfied.

1. Reading from left to right, the first nonzero entry of each row is 1.

2. In the *column* containing the leftmost 1 of a row, all the remaining entries are zeros.

3. The leftmost 1 of any row is to the right of the leftmost 1 of the preceding row.

4. Rows containing only zeros are below all the rows containing nonzero entries.

In addition to the final matrices of Examples 1 and 2, the following are also in reduced echelon form.

$$\begin{bmatrix} 1 & 2 & | & -3 \\ 0 & 0 & | & 0 \end{bmatrix}, \quad \begin{bmatrix} 1 & 0 & -2 & | & 5 \\ 0 & 1 & 4 & | & 7 \\ 0 & 0 & 0 & | & 0 \end{bmatrix}, \quad \begin{bmatrix} 1 & 0 & 0 & 0 & | & 8 \\ 0 & 1 & 0 & 0 & | & -9 \\ 0 & 0 & 1 & 0 & | & -2 \\ 0 & 0 & 0 & 1 & | & 12 \end{bmatrix}$$

In contrast, the following matrices are *not* in reduced echelon form for the reason indicated below each matrix.

$$\begin{bmatrix} 1 & 0 & 0 & | & 11 \\ 0 & 3 & 0 & | & -1 \\ 0 & 0 & 1 & | & -2 \end{bmatrix} \qquad \begin{bmatrix} 1 & 2 & -3 & | & 5 \\ 0 & 1 & 7 & | & 9 \\ 0 & 0 & 1 & | & -6 \end{bmatrix}$$

violates condition 1   violates condition 2

$$\begin{bmatrix} 1 & 0 & 0 & | & 7 \\ 0 & 0 & 1 & | & -8 \\ 0 & 1 & 0 & | & 14 \end{bmatrix} \qquad \begin{bmatrix} 1 & 0 & 0 & 0 & | & -1 \\ 0 & 0 & 0 & 0 & | & 0 \\ 0 & 0 & 1 & 0 & | & 7 \\ 0 & 0 & 0 & 0 & | & 0 \end{bmatrix}$$

violates condition 3   violates condition 4

Once we have an augmented matrix in reduced echelon form, it is easy to determine the solution set of the system. Furthermore, the procedure for changing a given augmented matrix to reduced echelon form can be described in a very systematic way. For example, if an augmented matrix of a system of three linear equations in three unknowns has a unique solution, it can be changed to reduced echelon form as follows.

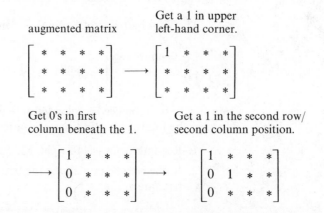

Get 0's above and below   Get a 1 in the third row/
the 1 in the second column.   third column position.

$$\longrightarrow \begin{bmatrix} 1 & 0 & * & * \\ 0 & 1 & * & * \\ 0 & 0 & * & * \end{bmatrix} \qquad \longrightarrow \begin{bmatrix} 1 & 0 & * & * \\ 0 & 1 & * & * \\ 0 & 0 & 1 & * \end{bmatrix}$$

Get 0's above the 1
in the third column.

$$\longrightarrow \begin{bmatrix} 1 & 0 & 0 & * \\ 0 & 1 & 0 & * \\ 0 & 0 & 1 & * \end{bmatrix}.$$

We can identify inconsistent and dependent systems while we are changing a matrix to reduced echelon form. We will show some examples of such cases in a moment, but first let's consider another example of a system of three linear equations in three unknowns where there's a unique solution.

EXAMPLE 3

Solve the system

$$\begin{pmatrix} 2x + 4y - 5z = 37 \\ x + 3y - 4z = 29 \\ 5x - y + 3z = -20 \end{pmatrix}.$$

The augmented matrix

$$\begin{bmatrix} 2 & 4 & -5 & | & 37 \\ 1 & 3 & -4 & | & 29 \\ 5 & -1 & 3 & | & -20 \end{bmatrix}$$

does not have a 1 in the upper left-hand corner, but this can be remedied by exchanging rows 1 and 2.

$$\begin{bmatrix} 1 & 3 & -4 & | & 29 \\ 2 & 4 & -5 & | & 37 \\ 5 & -1 & 3 & | & -20 \end{bmatrix}$$

Now we can get zeros in the first column beneath the 1 by adding -2 times row 1 to row 2 and by adding -5 times row 1 to row 3.

$$\begin{bmatrix} 1 & 3 & -4 & | & 29 \\ 0 & -2 & 3 & | & -21 \\ 0 & -16 & 23 & | & -165 \end{bmatrix}$$

Next, we can get a 1 for the first nonzero entry of the second row by multiplying the second row by $-\frac{1}{2}$.

$$\begin{bmatrix} 1 & 3 & -4 & | & 29 \\ 0 & 1 & -\frac{3}{2} & | & \frac{21}{2} \\ 0 & -16 & 23 & | & -165 \end{bmatrix}$$

Now we can get zeros above and below the 1 in the second column by adding -3 times row 2 to row 1 and by adding 16 times row 2 to row 3.

$$\begin{bmatrix} 1 & 0 & \frac{1}{2} & | & -\frac{5}{2} \\ 0 & 1 & -\frac{3}{2} & | & \frac{21}{2} \\ 0 & 0 & -1 & | & 3 \end{bmatrix}$$

Next, we can get a 1 as the first nonzero entry of the third row by multiplying the third row by -1.

$$\begin{bmatrix} 1 & 0 & \frac{1}{2} & | & -\frac{5}{2} \\ 0 & 1 & -\frac{3}{2} & | & \frac{21}{2} \\ 0 & 0 & 1 & | & -3 \end{bmatrix}$$

Finally, we can get zeros above the 1 in the third column by adding $-\frac{1}{2}$ times row 3 to row 1 and by adding $\frac{3}{2}$ times row 3 to row 2.

$$\begin{bmatrix} 1 & 0 & 0 & | & -1 \\ 0 & 1 & 0 & | & 6 \\ 0 & 0 & 1 & | & -3 \end{bmatrix}$$

From this last matrix, we see that the solution set of the original system is $\{(-1, 6, -3)\}$.

Example 3 illustrates that even though the process of changing to reduced echelon form can be systematically described, it can involve some rather messy calculations. However, with the aid of a computer, such calculations are not troublesome. For our purposes in this text, the examples and problems involve systems that minimize messy calculations. This will allow us to concentrate on the procedures being discussed.

We want to call your attention to another issue in the solution of Example 3. Consider the matrix

$$\begin{bmatrix} 1 & 3 & -4 & | & 29 \\ 0 & 1 & -\frac{3}{2} & | & \frac{21}{2} \\ 0 & -16 & 23 & | & -165 \end{bmatrix},$$

which is obtained about halfway through the solution. At this step it seems evident that the calculations are getting a little messy. Therefore, instead of continuing toward the reduced echelon form, let's add 16 times row 2 to row 3 to produce a new row 3.

$$\begin{bmatrix} 1 & 3 & -4 & | & 29 \\ 0 & 1 & -\frac{3}{2} & | & \frac{21}{2} \\ 0 & 0 & -1 & | & 3 \end{bmatrix}.$$

The system represented by this matrix is

$$\begin{pmatrix} x + 3y - 4z = 29 \\ y - \frac{3}{2}z = \frac{21}{2} \\ -z = 3 \end{pmatrix};$$

it is said to be in **triangular form**. The last equation determines the value for z; then we can use the process of back-substitution to determine the values for y and x.

Finally, let's consider two examples to illustrate what happens when we use the matrix approach on inconsistent and dependent systems.

EXAMPLE 4

Solve the system

$$\begin{pmatrix} x - 2y + 3z = 3 \\ 5x - 9y + 4z = 2 \\ 2x - 4y + 6z = -1 \end{pmatrix}.$$

Solution The augmented matrix of the system is

$$\begin{bmatrix} 1 & -2 & 3 & \vdots & 3 \\ 5 & -9 & 4 & \vdots & 2 \\ 2 & -4 & 6 & \vdots & -1 \end{bmatrix}.$$

We can get zeros below the 1 in the first column by adding -5 times row 1 to row 2 and by adding -2 times row 1 to row 3.

$$\begin{bmatrix} 1 & -2 & 3 & \vdots & 3 \\ 0 & 1 & -11 & \vdots & -13 \\ 0 & 0 & 0 & \vdots & -7 \end{bmatrix}$$

At this step we can stop, because the bottom row of the matrix represents the statement $0(x) + 0(y) + 0(z) = -7$, which is obviously false for all values of x, y, and z. Thus, the original system is inconsistent; its solution set is $\varnothing$.

EXAMPLE 5

Solve the system

$$\begin{pmatrix} x + 2y + 2z = 9 \\ x + 3y - 4z = 5 \\ 2x + 5y - 2z = 14 \end{pmatrix}.$$

Solution The augmented matrix of the system is

$$\begin{bmatrix} 1 & 2 & 2 & \vdots & 9 \\ 1 & 3 & -4 & \vdots & 5 \\ 2 & 5 & -2 & \vdots & 14 \end{bmatrix}.$$

We can get zeros in the first column below the 1 in the upper left-hand corner by adding -1 times row 1 to row 2 and by adding -2 times row 1 to row 3.

$$\begin{bmatrix} 1 & 2 & 2 & \vdots & 9 \\ 0 & 1 & -6 & \vdots & -4 \\ 0 & 1 & -6 & \vdots & -4 \end{bmatrix}$$

Now we can get zeros in the second column above and below the 1 in the second row by adding -2 times row 2 to row 1 and by adding -1 times row 2 to row 3.

$$\begin{bmatrix} 1 & 0 & 14 & | & 17 \\ 0 & 1 & -6 & | & -4 \\ 0 & 0 & 0 & | & 0 \end{bmatrix}$$

The bottom row of zeros represents the statement $0(x) + 0(y) + 0(z) = 0$, which is true for all values of x, y, and z. The second row represents the statement $y - 6z = -4$, which can be rewritten $y = 6z - 4$. The top row represents the statement $x + 14z = 17$, which can be rewritten $x = -14z + 17$. Therefore, if we let $z = k$, where k is any real number, the solution set of infinitely many ordered triples can be represented by $\{(-14k + 17, 6k - 4, k) \mid k \text{ is a real number}\}$. Specific solutions can be generated by letting k take on a value. For example, if $k = 2$, then $6k - 4$ becomes $6(2) - 4 = 8$ and $-14k + 17$ becomes $-14(2) + 17 = -11$. Thus the ordered triple $(-11, 8, 2)$ is a member of the solution set.

Problem Set 7.3

For Problems 1–10, indicate whether each matrix is in reduced echelon form.

1. $\begin{bmatrix} 1 & 0 & | & -4 \\ 0 & 1 & | & 14 \end{bmatrix}$

2. $\begin{bmatrix} 1 & 2 & | & 8 \\ 0 & 0 & | & 0 \end{bmatrix}$

3. $\begin{bmatrix} 1 & 0 & 2 & | & 5 \\ 0 & 1 & 3 & | & 7 \\ 0 & 0 & 0 & | & 0 \end{bmatrix}$

4. $\begin{bmatrix} 1 & 0 & 0 & | & 5 \\ 0 & 3 & 0 & | & 8 \\ 0 & 0 & 1 & | & -11 \end{bmatrix}$

5. $\begin{bmatrix} 1 & 0 & 0 & | & 17 \\ 0 & 0 & 0 & | & 0 \\ 0 & 1 & 0 & | & -14 \end{bmatrix}$

6. $\begin{bmatrix} 1 & 0 & 0 & | & -7 \\ 0 & 1 & 0 & | & 0 \\ 0 & 0 & 1 & | & 9 \end{bmatrix}$

7. $\begin{bmatrix} 1 & 1 & 0 & | & -3 \\ 0 & 1 & 2 & | & 5 \\ 0 & 0 & 1 & | & 7 \end{bmatrix}$

8. $\begin{bmatrix} 1 & 0 & 3 & | & 8 \\ 0 & 1 & 2 & | & -6 \\ 0 & 0 & 0 & | & 0 \end{bmatrix}$

9. $\begin{bmatrix} 1 & 0 & 0 & 3 & | & 4 \\ 0 & 1 & 0 & 5 & | & -3 \\ 0 & 0 & 1 & -1 & | & 7 \\ 0 & 0 & 0 & 0 & | & 0 \end{bmatrix}$

10. $\begin{bmatrix} 1 & 0 & 0 & 0 & | & 2 \\ 0 & 0 & 1 & 0 & | & 4 \\ 0 & 1 & 0 & 0 & | & -3 \\ 0 & 0 & 0 & 1 & | & 9 \end{bmatrix}$

Use a matrix approach to solve each of the following systems.

11. $\left(\begin{array}{l} x - 3y = 14 \\ 3x + 2y = -13 \end{array} \right)$

12. $\left(\begin{array}{l} x + 5y = -18 \\ -2x + 3y = -16 \end{array} \right)$

13. $\left(\begin{array}{l} 3x - 4y = 33 \\ x + 7y = -39 \end{array} \right)$

14. $\left(\begin{array}{l} 2x + 7y = -55 \\ x - 4y = 25 \end{array} \right)$

15. $\left(\begin{array}{l} x - 6y = -2 \\ 2x - 12y = 5 \end{array} \right)$

16. $\left(\begin{array}{l} 2x - 3y = -12 \\ 3x + 2y = 8 \end{array} \right)$

17. $\left(\begin{array}{l} 3x - 5y = 39 \\ 2x + 7y = -67 \end{array} \right)$

18. $\left(\begin{array}{l} 3x + 9y = -1 \\ x + 3y = 10 \end{array} \right)$

19. $\begin{pmatrix} x - 2y - 3z = -6 \\ 3x - 5y - z = 4 \\ 2x + y + 2z = 2 \end{pmatrix}$

20. $\begin{pmatrix} x + 3y - 4z = 13 \\ 2x + 7y - 3z = 11 \\ -2x - y + 2z = -8 \end{pmatrix}$

21. $\begin{pmatrix} -2x - 5y + 3z = 11 \\ x + 3y - 3z = -12 \\ 3x - 2y + 5z = 31 \end{pmatrix}$

22. $\begin{pmatrix} -3x + 2y + z = 17 \\ x - y + 5z = -2 \\ 4x - 5y - 3z = -36 \end{pmatrix}$

23. $\begin{pmatrix} x - 3y - z = 2 \\ 3x + y - 4z = -18 \\ -2x + 5y + 3z = 2 \end{pmatrix}$

24. $\begin{pmatrix} x - 4y + 3z = 16 \\ 2x + 3y - 4z = -22 \\ -3x + 11y - z = -36 \end{pmatrix}$

25. $\begin{pmatrix} x - y + 2z = 1 \\ -3x + 4y - z = 4 \\ -x + 2y + 3z = 6 \end{pmatrix}$

26. $\begin{pmatrix} x + 2y - 5z = -1 \\ 2x + 3y - 2z = 2 \\ 3x + 5y - 7z = 4 \end{pmatrix}$

27. $\begin{pmatrix} -2x + y + 5z = -5 \\ 3x + 8y - z = -34 \\ x + 2y + z = -12 \end{pmatrix}$

28. $\begin{pmatrix} 4x - 10y + 3z = -19 \\ 2x + 5y - z = -7 \\ x - 3y - 2z = -2 \end{pmatrix}$

29. $\begin{pmatrix} 2x + 3y - z = 7 \\ 3x + 4y + 5z = -2 \\ 5x + y + 3z = 13 \end{pmatrix}$

30. $\begin{pmatrix} 4x + 3y - z = 0 \\ 3x + 2y + 5z = 6 \\ 5x - y - 3z = 3 \end{pmatrix}$

Subscript notation is frequently used for working with large systems of equations. Use a matrix approach to solve each of the following systems. Express the solutions as 4-tuples of the form (x_1, x_2, x_3, x_4).

31. $\begin{pmatrix} x_1 - 3x_2 - 2x_3 + x_4 = -3 \\ -2x_1 + 7x_2 + x_3 - 2x_4 = -1 \\ 3x_1 - 7x_2 - 3x_3 + 3x_4 = -5 \\ 5x_1 + x_2 + 4x_3 - 2x_4 = 18 \end{pmatrix}$

32. $\begin{pmatrix} x_1 - 2x_2 + 2x_3 - x_4 = -2 \\ -3x_1 + 5x_2 - x_3 - 3x_4 = 2 \\ 2x_1 + 3x_2 + 3x_3 + 5x_4 = -9 \\ 4x_1 - x_2 - x_3 - 2x_4 = 8 \end{pmatrix}$

33. $\begin{pmatrix} x_1 + 3x_2 - x_3 + 2x_4 = -2 \\ 2x_1 + 7x_2 + 2x_3 - x_4 = 19 \\ -3x_1 - 8x_2 + 3x_3 + x_4 = -7 \\ 4x_1 + 11x_2 - 2x_3 - 3x_4 = 19 \end{pmatrix}$

34. $\begin{pmatrix} x_1 + 2x_2 - 3x_3 + x_4 = -2 \\ -2x_1 - 3x_2 + x_3 - x_4 = 5 \\ 4x_1 + 9x_2 - 2x_3 - 2x_4 = -28 \\ -5x_1 - 9x_2 + 2x_3 - 3x_4 = 14 \end{pmatrix}$

Each matrix in Problems 35–42 is the reduced echelon matrix for a system with variables x_1, x_2, x_3, and x_4. Find the solution set of each system.

35. $\begin{bmatrix} 1 & 0 & 0 & 0 & | & -2 \\ 0 & 1 & 0 & 0 & | & 4 \\ 0 & 0 & 1 & 0 & | & -3 \\ 0 & 0 & 0 & 1 & | & 0 \end{bmatrix}$

36. $\begin{bmatrix} 1 & 0 & 0 & 0 & | & 0 \\ 0 & 1 & 0 & 0 & | & -5 \\ 0 & 0 & 1 & 0 & | & 0 \\ 0 & 0 & 0 & 1 & | & 4 \end{bmatrix}$

37. $\begin{bmatrix} 1 & 0 & 0 & 0 & | & -8 \\ 0 & 1 & 0 & 0 & | & 5 \\ 0 & 0 & 1 & 0 & | & -2 \\ 0 & 0 & 0 & 0 & | & 1 \end{bmatrix}$

38. $\begin{bmatrix} 1 & 0 & 0 & 0 & | & 2 \\ 0 & 1 & 0 & 2 & | & -3 \\ 0 & 0 & 1 & 3 & | & 4 \\ 0 & 0 & 0 & 0 & | & 0 \end{bmatrix}$

39.
$$\begin{bmatrix} 1 & 0 & 0 & 3 & | & 5 \\ 0 & 1 & 0 & 0 & | & -1 \\ 0 & 0 & 1 & 4 & | & 2 \\ 0 & 0 & 0 & 0 & | & 0 \end{bmatrix}$$

40.
$$\begin{bmatrix} 1 & 3 & 0 & 0 & | & 0 \\ 0 & 0 & 1 & 0 & | & 0 \\ 0 & 0 & 0 & 0 & | & 1 \\ 0 & 0 & 0 & 0 & | & 0 \end{bmatrix}$$

41.
$$\begin{bmatrix} 1 & 3 & 0 & 0 & | & 9 \\ 0 & 0 & 1 & 0 & | & 2 \\ 0 & 0 & 0 & 1 & | & -3 \\ 0 & 0 & 0 & 0 & | & 0 \end{bmatrix}$$

42.
$$\begin{bmatrix} 1 & 0 & 0 & 0 & | & 7 \\ 0 & 1 & 0 & 0 & | & -3 \\ 0 & 0 & 1 & -2 & | & 5 \\ 0 & 0 & 0 & 0 & | & 0 \end{bmatrix}$$

Miscellaneous Problems

For Problems 43–48, change each augmented matrix of the system to reduced echelon form and then indicate the solutions of the system.

43. $\begin{pmatrix} x - 2y + 3z = 4 \\ 3x - 5y - z = 7 \end{pmatrix}$

44. $\begin{pmatrix} x + 3y - 2z = -1 \\ -2x - 5y + 7z = 4 \end{pmatrix}$

45. $\begin{pmatrix} 2x - 4y + 3z = 8 \\ 3x + 5y - z = 7 \end{pmatrix}$

46. $\begin{pmatrix} 3x + 6y - z = 9 \\ 2x - 3y + 4z = 1 \end{pmatrix}$

47. $\begin{pmatrix} x - 2y + 4z = 9 \\ 2x - 4y + 8z = 3 \end{pmatrix}$

48. $\begin{pmatrix} x + y - 2z = -1 \\ 3x + 3y - 6z = -3 \end{pmatrix}$

7.4

Determinants

Before introducing the concept of a determinant, let's agree upon some convenient new notation. A **general $m \times n$ (m-by-n) matrix** can be represented by

$$A = \begin{bmatrix} a_{11} & a_{12} & a_{13} & \cdots & a_{1n} \\ a_{21} & a_{22} & a_{23} & \cdots & a_{2n} \\ \vdots & \vdots & \vdots & & \vdots \\ a_{m1} & a_{m2} & a_{m3} & \cdots & a_{mn} \end{bmatrix},$$

where the double subscripts are used to identify the number of the row and the number of the column, in that order. For example, a_{23} is the entry at the intersection of the second row and the third column. In general, the entry at the intersection of row i and column j is denoted by a_{ij}.

A **square matrix** is one that has the same number of rows as columns. Each square matrix A with real-number entries can be associated with a real number called the **determinant** of the matrix, denoted by $|A|$. We will first define $|A|$ for a 2×2 matrix.

DEFINITION 7.1

If $A = \begin{bmatrix} a_{11} & a_{12} \\ a_{21} & a_{22} \end{bmatrix}$, then

$$|A| = \begin{vmatrix} a_{11} & a_{12} \\ a_{21} & a_{22} \end{vmatrix} = a_{11}a_{22} - a_{12}a_{21}.$$

EXAMPLE 1

If $A = \begin{bmatrix} 3 & -2 \\ 5 & 8 \end{bmatrix}$, find $|A|$.

Solution Using Definition 7.1, we obtain

$$|A| = \begin{vmatrix} 3 & -2 \\ 5 & 8 \end{vmatrix} = 3(8) - (-2)(5)$$

$$= 24 + 10$$
$$= 34.$$

Finding the determinant of a square matrix is commonly called **evaluating the determinant**, and the matrix notation is often omitted.

EXAMPLE 2

Evaluate $\begin{vmatrix} -3 & 6 \\ 2 & 8 \end{vmatrix}$.

Solution

$$\begin{vmatrix} -3 & 6 \\ 2 & 8 \end{vmatrix} = (-3)(8) - (6)(2)$$

$$= -24 - 12$$
$$= -36$$

To define determinants of 3×3 and larger square matrices, it is convenient to introduce some additional terminology.

DEFINITION 7.2

If A is a 3×3 matrix, then the **minor** (denoted by M_{ij}) of the a_{ij} element is the determinant of the 2×2 matrix obtained by deleting row i and column j of A.

EXAMPLE 3

If $A = \begin{bmatrix} 2 & 1 & 4 \\ -6 & 3 & -2 \\ 4 & 2 & 5 \end{bmatrix}$, find (**a**) M_{11} and (**b**) M_{23}.

Solutions

(a) To find M_{11} we first delete row 1 and column 1 of A.

$$\begin{bmatrix} 2 & 1 & 4 \\ -6 & 3 & -2 \\ 4 & 2 & 5 \end{bmatrix}$$

Thus,

$$M_{11} = \begin{vmatrix} 3 & -2 \\ 2 & 5 \end{vmatrix} = 3(5) - (-2)(2) = 19.$$

(b) To find M_{23} we first delete row 2 and column 3 of A.

$$\begin{bmatrix} 2 & 1 & 4 \\ -6 & 3 & -2 \\ 4 & 2 & 5 \end{bmatrix}$$

Thus,

$$M_{23} = \begin{vmatrix} 2 & 1 \\ 4 & 2 \end{vmatrix} = 2(2) - (1)(4) = 0.$$

The following definition will also be used.

DEFINITION 7.3

If A is a 3×3 matrix, then the **cofactor** (denoted by C_{ij}) of the element a_{ij} is defined by

$$C_{ij} = (-1)^{i+j} M_{ij}.$$

According to Definition 7.3, to find the cofactor of any element a_{ij} of a square matrix A, we find the minor of a_{ij} and multiply it by 1 if $i + j$ is even, or multiply it by -1 if $i + j$ is odd.

EXAMPLE 4

If $A = \begin{bmatrix} 3 & 2 & -4 \\ 1 & 5 & 4 \\ 2 & -3 & 1 \end{bmatrix}$, find C_{32}.

Solution First, let's find M_{32} by deleting row 3 and column 2 of A.

$$\begin{bmatrix} 3 & 2 & -4 \\ 1 & 5 & 4 \\ 2 & -3 & 1 \end{bmatrix}$$

Thus,

$$M_{32} = \begin{vmatrix} 3 & -4 \\ 1 & 4 \end{vmatrix} = 3(4) - (-4)(1) = 16.$$

Therefore,

$$C_{32} = (-1)^{3+2}M_{32} = (-1)^5(16) = -16.$$

The concept of a cofactor can be used to define the determinant of a 3×3 matrix as follows.

DEFINITION 7.4

If $A = \begin{bmatrix} a_{11} & a_{12} & a_{13} \\ a_{21} & a_{22} & a_{23} \\ a_{31} & a_{32} & a_{33} \end{bmatrix}$, then

$$|A| = a_{11}C_{11} + a_{21}C_{21} + a_{31}C_{31}.$$

Definition 7.4 simply states that the determinant of a 3×3 matrix can be found by multiplying each element of the first column by its corresponding cofactor and then adding the three results. Let's illustrate this procedure.

EXAMPLE 5

Find $|A|$ if $A = \begin{bmatrix} -2 & 1 & 4 \\ 3 & 0 & 5 \\ 1 & -4 & -6 \end{bmatrix}$.

Solution

$$|A| = a_{11}C_{11} + a_{21}C_{21} + a_{31}C_{31}$$

$$= (-2)(-1)^{1+1}\begin{vmatrix} 0 & 5 \\ -4 & -6 \end{vmatrix} + (3)(-1)^{2+1}\begin{vmatrix} 1 & 4 \\ -4 & -6 \end{vmatrix}$$

$$+ (1)(-1)^{3+1}\begin{vmatrix} 1 & 4 \\ 0 & 5 \end{vmatrix}$$

$$= (-2)(1)(20) + (3)(-1)(10) + (1)(1)(5)$$

$$= -40 - 30 + 5$$

$$= -65$$

When we use Definition 7.4, we often say that *the determinant is being expanded about the first column*. It can be shown that **any row or column can be used to expand a determinant**. For example, for matrix A in Example 5, the expansion of the determinant about the *second row* is as follows:

$$\begin{vmatrix} -2 & 1 & 4 \\ 3 & 0 & 5 \\ 1 & -4 & -6 \end{vmatrix} = (3)(-1)^{2+1}\begin{vmatrix} 1 & 4 \\ -4 & -6 \end{vmatrix} + (0)(-1)^{2+2}\begin{vmatrix} -2 & 4 \\ 1 & -6 \end{vmatrix}$$

$$+ (5)(-1)^{2+3}\begin{vmatrix} -2 & 1 \\ 1 & -4 \end{vmatrix}$$

$$= (3)(-1)(10) + (0)(1)(8) + (5)(-1)(7)$$

$$= -30 + 0 - 35$$

$$= -65.$$

Notice that when we expanded about the second row, the computation was simplified by the presence of a zero. In general, it is usually helpful to expand about the row or column that contains the most zeros.

The concepts of minor and cofactor have been defined in terms of 3×3 matrices. Analogous definitions can be given for any square matrix (that is, any $n \times n$ matrix with $n \geq 2$), and the determinant can then be expanded about any row or column. Certainly as the matrices become larger than 3×3, the computations get more and more tedious. We will concentrate most of our efforts in this text on 2×2 and 3×3 matrices.

Properties of Determinants

Determinants have several interesting properties, some of which are primarily important from a theoretical standpoint. But some of the properties are also very useful when evaluating determinants. We will state these properties for square matrices in general, but we will use 2×2 or 3×3 matrices as examples. We can demonstrate some of the proofs of these properties by evaluating the determinants involved, and some of the proofs for 3×3 matrices will be left for you to verify in the next problem set.

PROPERTY 7.1

> If any row (or column) of a square matrix A contains only zeros, then $|A| = 0$.

If every element of a row (or column) of a square matrix A is 0, then it should be evident that expanding the determinant about that row (or column) of zeros will produce 0.

PROPERTY 7.2

> If square matrix B is obtained from square matrix A by interchanging two rows (or two columns), then $|B| = -|A|$.

Property 7.2 states that **interchanging two rows (or columns) changes the sign of the determinant**. As an example of this property suppose that

$$A = \begin{bmatrix} 2 & 5 \\ -1 & 6 \end{bmatrix}$$

and that rows 1 and 2 are interchanged to form

$$B = \begin{bmatrix} -1 & 6 \\ 2 & 5 \end{bmatrix}.$$

Calculating $|A|$ and $|B|$ we obtain

$$|A| = \begin{vmatrix} 2 & 5 \\ -1 & 6 \end{vmatrix} = 2(6) - (5)(-1) = 17$$

and

$$|B| = \begin{vmatrix} -1 & 6 \\ 2 & 5 \end{vmatrix} = (-1)(5) - (6)(2) = -17.$$

PROPERTY 7.3

> If square matrix B is obtained from square matrix A by multiplying each element of any row (or column) of A by some real number k, then $|B| = k|A|$.

Property 7.3 states that **multiplying any row (or column) by a factor of k affects the value of the determinant by a factor of k**. As an example of this property, suppose that

$$A = \begin{bmatrix} 1 & -2 & 8 \\ 2 & 1 & 12 \\ 3 & 2 & -16 \end{bmatrix}$$

and that B is formed by multiplying each element of the third column by $\frac{1}{4}$:

$$B = \begin{bmatrix} 1 & -2 & 2 \\ 2 & 1 & 3 \\ 3 & 2 & -4 \end{bmatrix}.$$

Now let's calculate $|A|$ and $|B|$ by expanding about the third column in each case.

$$|A| = \begin{vmatrix} 1 & -2 & 8 \\ 2 & 1 & 12 \\ 3 & 2 & -16 \end{vmatrix} = (8)(-1)^{1+3}\begin{vmatrix} 2 & 1 \\ 3 & 2 \end{vmatrix} + (12)(-1)^{2+3}\begin{vmatrix} 1 & -2 \\ 3 & 2 \end{vmatrix}$$

$$+ (-16)(-1)^{3+3}\begin{vmatrix} 1 & -2 \\ 2 & 1 \end{vmatrix}$$

$$= (8)(1)(1) + (12)(-1)(8) + (-16)(1)(5)$$
$$= -168$$

$$|B| = \begin{vmatrix} 1 & -2 & 2 \\ 2 & 1 & 3 \\ 3 & 2 & -4 \end{vmatrix} = (2)(-1)^{1+3}\begin{vmatrix} 2 & 1 \\ 3 & 2 \end{vmatrix} + (3)(-1)^{2+3}\begin{vmatrix} 1 & -2 \\ 3 & 2 \end{vmatrix}$$

$$+ (-4)(-1)^{3+3}\begin{vmatrix} 1 & -2 \\ 2 & 1 \end{vmatrix}$$

$$= (2)(1)(1) + (3)(-1)(8) + (-4)(1)(5)$$
$$= -42$$

We see that $|B| = \frac{1}{4}|A|$. This example also illustrates the usual computational use of Property 7.3: we can factor out a common factor from a row or column and then adjust the value of the determinant by that factor. For example,

$$\begin{vmatrix} 2 & 6 & 8 \\ -1 & 2 & 7 \\ 5 & 2 & 1 \end{vmatrix} = 2\begin{vmatrix} 1 & 3 & 4 \\ -1 & 2 & 7 \\ 5 & 2 & 1 \end{vmatrix}$$

factor a 2 from
the top row

PROPERTY 7.4

> If square matrix B is obtained from square matrix A by adding k times a row (or column) of A to another row (or column) of A, then $|B| = |A|$.

Property 7.4 states that **adding the product of k times a row (or column) to another row (or column) does not affect the value of the determinant**. As an example of this property, suppose that

$$A = \begin{bmatrix} 1 & 2 & 4 \\ 2 & 4 & 7 \\ -1 & 3 & 5 \end{bmatrix}.$$

Now let's form B by replacing row 2 with the result of adding -2 times row 1 to row 2.

$$B = \begin{bmatrix} 1 & 2 & 4 \\ 0 & 0 & -1 \\ -1 & 3 & 5 \end{bmatrix}$$

Next, let's evaluate $|A|$ and $|B|$ by expanding about the second row in each case.

$$|A| = \begin{vmatrix} 1 & 2 & 4 \\ 2 & 4 & 7 \\ -1 & 3 & 5 \end{vmatrix} = (2)(-1)^{2+1}\begin{vmatrix} 2 & 4 \\ 3 & 5 \end{vmatrix} + (4)(-1)^{2+2}\begin{vmatrix} 1 & 4 \\ -1 & 5 \end{vmatrix}$$

$$+ (7)(-1)^{2+3}\begin{vmatrix} 1 & 2 \\ -1 & 3 \end{vmatrix}$$

$$= (2)(-1)(-2) + (4)(1)(9) + (7)(-1)(5)$$

$$= 5$$

$$|B| = \begin{vmatrix} 1 & 2 & 4 \\ 0 & 0 & -1 \\ -1 & 3 & 5 \end{vmatrix} = (0)(-1)^{2+1}\begin{vmatrix} 2 & 4 \\ 3 & 5 \end{vmatrix} + (0)(-1)^{2+2}\begin{vmatrix} 1 & 4 \\ -1 & 5 \end{vmatrix}$$

$$+ (-1)(-1)^{2+3}\begin{vmatrix} 1 & 2 \\ -1 & 3 \end{vmatrix}$$

$$= 0 + 0 + (-1)(-1)(5)$$

$$= 5$$

Notice that $|B| = |A|$. Furthermore, notice that because of the zeros in the second row, evaluating $|B|$ is much easier than evaluating $|A|$. Property 7.4 can often be used to obtain some zeros before evaluating a determinant.

A word of caution is in order at this time. Be careful not to confuse Properties 7.2, 7.3, and 7.4 with the three elementary row transformations of augmented matrices that were used in Section 7.3. The statements of the two sets of properties do resemble each other, but the properties pertain to *two different concepts*, so be sure to keep them separate.

One final property of determinants should be mentioned.

PROPERTY 7.5

> If two rows (or columns) of a square matrix A are identical, then $|A| = 0$.

Property 7.5 is a direct consequence of Property 7.2. Suppose that A is a square matrix (any size) with two identical rows. Square matrix B can be formed from A by interchanging the two identical rows. Since identical rows were interchanged, $|B| = |A|$. *But* by Property 7.2, $|B| = -|A|$. For both of these statements to hold, $|A| = 0$.

Let's conclude this section by evaluating a 4×4 determinant using Properties 7.3 and 7.4 to facilitate the computation.

EXAMPLE 6

Evaluate $\begin{vmatrix} 6 & 2 & 1 & -2 \\ 9 & -1 & 4 & 1 \\ 12 & -2 & 3 & -1 \\ 0 & 0 & 9 & 3 \end{vmatrix}.$

Solution   First, let's add -3 times the fourth column to the third column.

$$\begin{vmatrix} 6 & 2 & 7 & -2 \\ 9 & -1 & 1 & 1 \\ 12 & -2 & 6 & -1 \\ 0 & 0 & 0 & 3 \end{vmatrix}$$

Now if we expand about the fourth row, we get only one nonzero product.

$$(3)(-1)^{4+4} \begin{vmatrix} 6 & 2 & 7 \\ 9 & -1 & 1 \\ 12 & -2 & 6 \end{vmatrix}$$

Factoring a 3 out of the first column of the 3×3 determinant, we obtain

$$(3)(-1)^{8}(3) \begin{vmatrix} 2 & 2 & 7 \\ 3 & -1 & 1 \\ 4 & -2 & 6 \end{vmatrix}.$$

Now working with this 3×3 determinant, we can first add column 3 to column 2 and then add -3 times column 3 to column 1.

$$(3)(-1)^{8}(3) \begin{vmatrix} -19 & 9 & 7 \\ 0 & 0 & 1 \\ -14 & 4 & 6 \end{vmatrix}$$

Finally, by expanding this 3×3 determinant about the second row, we obtain

$$(3)(-1)^{8}(3)(1)(-1)^{2+3} \begin{vmatrix} -19 & 9 \\ -14 & 4 \end{vmatrix}.$$

Our final result is

$$(3)(-1)^{8}(3)(1)(-1)^{5}(50) = -450.$$

Problem Set 7.4

Evaluate each of the following 2×2 determinants by using Definition 7.1.

1. $\begin{vmatrix} 4 & 3 \\ 2 & 7 \end{vmatrix}$
2. $\begin{vmatrix} 3 & 5 \\ 6 & 4 \end{vmatrix}$
3. $\begin{vmatrix} -3 & 2 \\ 7 & 5 \end{vmatrix}$
4. $\begin{vmatrix} 5 & 3 \\ 6 & -1 \end{vmatrix}$

5. $\begin{vmatrix} 2 & -3 \\ 8 & -2 \end{vmatrix}$
6. $\begin{vmatrix} -5 & 5 \\ -6 & 2 \end{vmatrix}$
7. $\begin{vmatrix} -2 & -3 \\ -1 & -4 \end{vmatrix}$
8. $\begin{vmatrix} -4 & -3 \\ -5 & -7 \end{vmatrix}$

9. $\begin{vmatrix} \frac{1}{2} & \frac{1}{3} \\ -3 & -6 \end{vmatrix}$
10. $\begin{vmatrix} \frac{2}{3} & \frac{3}{4} \\ 8 & 6 \end{vmatrix}$
11. $\begin{vmatrix} \frac{1}{2} & \frac{2}{3} \\ \frac{3}{4} & -\frac{1}{3} \end{vmatrix}$
12. $\begin{vmatrix} \frac{2}{3} & \frac{1}{5} \\ -\frac{1}{4} & \frac{3}{2} \end{vmatrix}$

Evaluate each of the following 3×3 determinants. Use the properties of determinants to your advantage.

13. $\begin{vmatrix} 1 & 2 & -1 \\ 3 & 1 & 2 \\ 2 & 4 & 3 \end{vmatrix}$
14. $\begin{vmatrix} 1 & -2 & 1 \\ 2 & 1 & -1 \\ 3 & 2 & 4 \end{vmatrix}$
15. $\begin{vmatrix} 1 & -4 & 1 \\ 2 & 5 & -1 \\ 3 & 3 & 4 \end{vmatrix}$

16. $\begin{vmatrix} 3 & -2 & 1 \\ 2 & 1 & 4 \\ -1 & 3 & 5 \end{vmatrix}$
17. $\begin{vmatrix} 6 & 12 & 3 \\ -1 & 5 & 1 \\ -3 & 6 & 2 \end{vmatrix}$
18. $\begin{vmatrix} 2 & 35 & 5 \\ 1 & -5 & 1 \\ -4 & 15 & 2 \end{vmatrix}$

19. $\begin{vmatrix} 2 & -1 & 3 \\ 0 & 3 & 1 \\ 1 & -2 & -1 \end{vmatrix}$
20. $\begin{vmatrix} 2 & -17 & 3 \\ 0 & 5 & 1 \\ 1 & -3 & -1 \end{vmatrix}$
21. $\begin{vmatrix} -3 & -2 & 1 \\ 5 & 0 & 6 \\ 2 & 1 & -4 \end{vmatrix}$

22. $\begin{vmatrix} -5 & 1 & -1 \\ 3 & 4 & 2 \\ 0 & 2 & -3 \end{vmatrix}$
23. $\begin{vmatrix} 3 & -4 & -2 \\ 5 & -2 & 1 \\ 1 & 0 & 0 \end{vmatrix}$
24. $\begin{vmatrix} -6 & 5 & 3 \\ 2 & 0 & -1 \\ 4 & 0 & 7 \end{vmatrix}$

25. $\begin{vmatrix} 24 & -1 & 4 \\ 40 & 2 & 0 \\ -16 & 6 & 0 \end{vmatrix}$
26. $\begin{vmatrix} 2 & -1 & 3 \\ 0 & 3 & 1 \\ 4 & -8 & -4 \end{vmatrix}$
27. $\begin{vmatrix} 2 & 3 & -4 \\ 4 & 6 & -1 \\ -6 & 1 & -2 \end{vmatrix}$

28. $\begin{vmatrix} 1 & 2 & -3 \\ -3 & -1 & 1 \\ 4 & 5 & 4 \end{vmatrix}$

Evaluate each of the following 4×4 determinants. Use the properties of determinants to your advantage.

29. $\begin{vmatrix} 1 & -2 & 3 & 2 \\ 2 & -1 & 0 & 4 \\ -3 & 4 & 0 & -2 \\ -1 & 1 & 1 & 5 \end{vmatrix}$
30. $\begin{vmatrix} 1 & 2 & 5 & 7 \\ -6 & 3 & 0 & 9 \\ -3 & 5 & 2 & 7 \\ 2 & 1 & 4 & 3 \end{vmatrix}$

31. $\begin{vmatrix} 3 & -1 & 2 & 3 \\ 1 & 0 & 2 & 1 \\ 2 & 3 & 0 & 1 \\ 5 & 2 & 4 & -5 \end{vmatrix}$
32. $\begin{vmatrix} 1 & 2 & 0 & 0 \\ 3 & -1 & 4 & 5 \\ -2 & 4 & 1 & 6 \\ 2 & -1 & -2 & -3 \end{vmatrix}$

Use the appropriate property of determinants from this section to justify each of the following true statements. *Do not* evaluate the determinants.

33. $(-4)\begin{vmatrix} 2 & 1 & -1 \\ 3 & 2 & 1 \\ 2 & 1 & 3 \end{vmatrix} = \begin{vmatrix} 2 & -4 & -1 \\ 3 & -8 & 1 \\ 2 & -4 & 3 \end{vmatrix}$

34. $\begin{vmatrix} 1 & -2 & 3 \\ 4 & -6 & -8 \\ 0 & 2 & 7 \end{vmatrix} = (-2)\begin{vmatrix} 1 & -2 & 3 \\ -2 & 3 & 4 \\ 0 & 2 & 7 \end{vmatrix}$

35. $\begin{vmatrix} 4 & 7 & 9 \\ 6 & -8 & 2 \\ 4 & 3 & -1 \end{vmatrix} = -\begin{vmatrix} 4 & 9 & 7 \\ 6 & 2 & -8 \\ 4 & -1 & 3 \end{vmatrix}$

36. $\begin{vmatrix} 3 & -1 & 4 \\ 5 & 2 & 7 \\ 3 & -1 & 4 \end{vmatrix} = 0$

37. $\begin{vmatrix} 1 & 3 & 4 \\ -2 & 5 & 7 \\ -3 & -1 & 2 \end{vmatrix} = \begin{vmatrix} 1 & 3 & 4 \\ -2 & 5 & 7 \\ 0 & 8 & 14 \end{vmatrix}$

38. $\begin{vmatrix} 3 & 2 & 0 \\ 1 & 4 & 1 \\ -4 & 9 & 2 \end{vmatrix} = \begin{vmatrix} 3 & 2 & -3 \\ 1 & 4 & 0 \\ -4 & 9 & 6 \end{vmatrix}$

39. $\begin{vmatrix} 6 & 2 & 2 \\ 3 & -1 & 4 \\ 9 & -3 & 6 \end{vmatrix} = 6\begin{vmatrix} 2 & 2 & 1 \\ 1 & -1 & 2 \\ 3 & -3 & 3 \end{vmatrix} = 18\begin{vmatrix} 2 & 2 & 1 \\ 1 & -1 & 2 \\ 1 & -1 & 1 \end{vmatrix}$

40. $\begin{vmatrix} 2 & 1 & -3 \\ 0 & 2 & -4 \\ -5 & 1 & 3 \end{vmatrix} = -\begin{vmatrix} 2 & 1 & -3 \\ -5 & 1 & 3 \\ 0 & 2 & -4 \end{vmatrix}$

41. $\begin{vmatrix} 2 & -3 & 2 \\ 1 & -4 & 1 \\ 7 & 8 & 7 \end{vmatrix} = 0$

42. $\begin{vmatrix} 3 & 1 & 2 \\ -4 & 5 & -1 \\ 2 & -2 & -4 \end{vmatrix} = \begin{vmatrix} 3 & 1 & 0 \\ -4 & 5 & -11 \\ 2 & -2 & 0 \end{vmatrix}$

Miscellaneous Problems

For Problems 43–45, use

$$A = \begin{bmatrix} a_{11} & a_{12} & a_{13} \\ a_{21} & a_{22} & a_{23} \\ a_{31} & a_{32} & a_{33} \end{bmatrix}$$

as a general representation for any 3×3 matrix.

43. Verify Property 7.2 for 3×3 matrices.

44. Verify Property 7.3 for 3×3 matrices.

45. Verify Property 7.4 for 3×3 matrices.

46. If

$$A = \begin{bmatrix} a_{11} & a_{12} & a_{13} & a_{14} \\ 0 & a_{22} & a_{23} & a_{24} \\ 0 & 0 & a_{33} & a_{34} \\ 0 & 0 & 0 & a_{44} \end{bmatrix},$$

then show that $|A| = a_{11}a_{22}a_{33}a_{44}$.

7.5

Cramer's Rule

Determinants provide the basis for another method of solving linear systems. Consider the following linear system of two equations and two unknowns:

$$\begin{pmatrix} a_1 x + b_1 y = c_1 \\ a_2 x + b_2 y = c_2 \end{pmatrix}.$$

The augmented matrix of this system is

$$\begin{bmatrix} a_1 & b_1 & \vdots & c_1 \\ a_2 & b_2 & \vdots & c_2 \end{bmatrix}.$$

Using the elementary row transformations of augmented matrices, we can change this matrix to the following reduced echelon form. (The details of this are left for you to do as an exercise.)

$$\begin{bmatrix} 1 & 0 & \vdots & \dfrac{c_1 b_2 - c_2 b_1}{a_1 b_2 - a_2 b_1} \\ 0 & 1 & \vdots & \dfrac{a_1 c_2 - a_2 c_1}{a_1 b_2 - a_2 b_1} \end{bmatrix}, \quad a_1 b_2 - a_2 b_1 \neq 0$$

The solutions for x and y can be expressed in determinant form as follows:

$$x = \frac{c_1 b_2 - c_2 b_1}{a_1 b_2 - a_2 b_1} = \frac{\begin{vmatrix} c_1 & b_1 \\ c_2 & b_2 \end{vmatrix}}{\begin{vmatrix} a_1 & b_1 \\ a_2 & b_2 \end{vmatrix}};$$

$$y = \frac{a_1 c_2 - a_2 c_1}{a_1 b_2 - a_2 b_1} = \frac{\begin{vmatrix} a_1 & c_1 \\ a_2 & c_2 \end{vmatrix}}{\begin{vmatrix} a_1 & b_1 \\ a_2 & b_2 \end{vmatrix}}.$$

This method of using determinants to solve a system of two linear equations in two variables is called **Cramer's rule** and can be stated as follows.

Cramer's Rule
(2 × 2 case)

Given the system

$$\begin{pmatrix} a_1x + b_1y = c_1 \\ a_2x + b_2y = c_2 \end{pmatrix},$$

with

$$D = \begin{vmatrix} a_1 & b_1 \\ a_2 & b_2 \end{vmatrix} \neq 0,$$

$$D_x = \begin{vmatrix} c_1 & b_1 \\ c_2 & b_2 \end{vmatrix} \quad \text{and} \quad D_y = \begin{vmatrix} a_1 & c_1 \\ a_2 & c_2 \end{vmatrix},$$

then the solution for this system is given by

$$x = \frac{D_x}{D} \quad \text{and} \quad y = \frac{D_y}{D}.$$

Notice that the elements of D are the coefficients of the variables in the given system. In D_x the coefficients of x are replaced by the corresponding constants, and in D_y the coefficients of y are replaced by the corresponding constants. Let's illustrate the use of Cramer's rule to solve some systems.

EXAMPLE 1

Solve the system

$$\begin{pmatrix} y = -2x - 2 \\ 4x - 5y = 17 \end{pmatrix}.$$

Solution To begin, we must change the form of the first equation so that the system fits the form given in Cramer's rule. The equation $y = -2x - 2$ can be rewritten $2x + y = -2$. The system now becomes

$$\begin{pmatrix} 2x + y = -2 \\ 4x - 5y = 17 \end{pmatrix},$$

and we can proceed to determine D, D_x, and D_y.

$$D = \begin{vmatrix} 2 & 1 \\ 4 & -5 \end{vmatrix} = -10 - 4 = -14$$

$$D_x = \begin{vmatrix} -2 & 1 \\ 17 & -5 \end{vmatrix} = 10 - 17 = -7$$

$$D_y = \begin{vmatrix} 2 & -2 \\ 4 & 17 \end{vmatrix} = 34 - (-8) = 42$$

Thus,

$$x = \frac{D_x}{D} = \frac{-7}{-14} = \frac{1}{2}$$

and

$$y = \frac{D_y}{D} = \frac{42}{-14} = -3.$$

The solution set is $\{(\frac{1}{2}, -3)\}$, which can be verified, as always, by substituting back into the original equations.

EXAMPLE 2

Solve the system

$$\left(\begin{array}{l}\frac{1}{2}x + \frac{2}{3}y = -4 \\ \frac{1}{4}x - \frac{3}{2}y = 20\end{array}\right).$$

Solution With such a system either we can first produce an equivalent system with integral coefficients and then apply Cramer's rule, or we can apply the rule immediately. Let's avoid some work with fractions by multiplying the first equation by 6 and the second equation by 4, producing the following equivalent system.

$$\left(\begin{array}{l}3x + 4y = -24 \\ x - 6y = 80\end{array}\right)$$

Now we can proceed as before.

$$D = \begin{vmatrix} 3 & 4 \\ 1 & -6 \end{vmatrix} = -18 - 4 = -22$$

$$D_x = \begin{vmatrix} -24 & 4 \\ 80 & -6 \end{vmatrix} = 144 - 320 = -176$$

$$D_y = \begin{vmatrix} 3 & -24 \\ 1 & 80 \end{vmatrix} = 240 - (-24) = 264$$

Therefore,

$$x = \frac{D_x}{D} = \frac{-176}{-22} = 8$$

and

$$y = \frac{D_y}{D} = \frac{264}{-22} = -12.$$

The solution set is $\{(8, -12)\}$.

In the statement of Cramer's rule, the condition was imposed that $D \neq 0$. If $D = 0$ and either D_x or D_y (or both) is nonzero, then the system is inconsistent and has no solution. If $D = 0$, $D_x = 0$, and $D_y = 0$, then the equations are dependent and there are infinitely many solutions.

Cramer's Rule Extended

Without showing the details, we will simply state that Cramer's rule also applies to solving systems of three linear equations in three variables. It can be stated as follows.

Cramer's Rule
(3 × 3 case)

Given the system

$$\begin{cases} a_1x + b_1y + c_1z = d_1 \\ a_2x + b_2y + c_2z = d_2 \\ a_3x + b_3y + c_3z = d_3 \end{cases},$$

with

$$D = \begin{vmatrix} a_1 & b_1 & c_1 \\ a_2 & b_2 & c_2 \\ a_3 & b_3 & c_3 \end{vmatrix} \neq 0, \qquad D_x = \begin{vmatrix} d_1 & b_1 & c_1 \\ d_2 & b_2 & c_2 \\ d_3 & b_3 & c_3 \end{vmatrix},$$

$$D_y = \begin{vmatrix} a_1 & d_1 & c_1 \\ a_2 & d_2 & c_2 \\ a_3 & d_3 & c_3 \end{vmatrix}, \qquad D_z = \begin{vmatrix} a_1 & b_1 & d_1 \\ a_2 & b_2 & d_2 \\ a_3 & b_3 & d_3 \end{vmatrix},$$

then

$$x = \frac{D_x}{D}, \qquad y = \frac{D_y}{D}, \qquad \text{and} \qquad z = \frac{D_z}{D}.$$

Again notice the restriction that $D \neq 0$. If $D = 0$ and at least one of D_x, D_y, and D_z is not zero, then the system is inconsistent. If D, D_x, D_y, and D_z are all zero, then the equations are dependent and there are infinitely many solutions.

EXAMPLE 3

Solve the system

$$\begin{cases} x - 2y + z = -4 \\ 2x + y - z = 5 \\ 3x + 2y + 4z = 3 \end{cases}.$$

Solution We will simply indicate the values of D, D_x, D_y, and D_z, leaving the computations for you to check.

$$D = \begin{vmatrix} 1 & -2 & 1 \\ 2 & 1 & -1 \\ 3 & 2 & 4 \end{vmatrix} = 29 \qquad D_x = \begin{vmatrix} -4 & -2 & 1 \\ 5 & 1 & -1 \\ 3 & 2 & 4 \end{vmatrix} = 29$$

$$D_y = \begin{vmatrix} 1 & -4 & 1 \\ 2 & 5 & -1 \\ 3 & 3 & 4 \end{vmatrix} = 58 \qquad D_z = \begin{vmatrix} 1 & -2 & -4 \\ 2 & 1 & 5 \\ 3 & 2 & 3 \end{vmatrix} = -29$$

Therefore,

$$x = \frac{D_x}{D} = \frac{29}{29} = 1,$$

$$y = \frac{D_y}{D} = \frac{58}{29} = 2,$$

and

$$z = \frac{D_z}{D} = \frac{-29}{29} = -1.$$

The solution set is $\{(1, 2, -1)\}$. (Be sure to check it!)

EXAMPLE 4

Solve the system

$$\begin{pmatrix} x + 3y - z = 4 \\ 3x - 2y + z = 7 \\ 2x + 6y - 2z = 1 \end{pmatrix}.$$

Solution

$$D = \begin{vmatrix} 1 & 3 & -1 \\ 3 & -2 & 1 \\ 2 & 6 & -2 \end{vmatrix}$$

Since the third row is twice the first row, we know that $D = 0$. We can also establish that

$$D_x = \begin{vmatrix} 4 & 3 & -1 \\ 7 & -2 & 1 \\ 1 & 6 & -2 \end{vmatrix} = -7.$$

Therefore, since $D = 0$ and at least one of D_x, D_y, and D_z is not zero, the system is inconsistent. The solution set is $\emptyset$.

Example 4 illustrates the reason that D should be determined first. Once we found that $D = 0$ and $D_x \neq 0$, we knew that the system was inconsistent and there was no need to find D_y and D_z.

Finally, it should be noted that Cramer's rule can be extended to systems of n linear equations in n variables; however, that method is not considered to be a very efficient way of solving a large system of linear equations.

Problem Set 7.5

Use Cramer's rule to find the solution set for each of the following systems. If the equations are dependent, simply indicate that there are infinitely many solutions.

1. $\begin{pmatrix} 2x - y = -2 \\ 3x + 2y = 11 \end{pmatrix}$

2. $\begin{pmatrix} 3x + y = -9 \\ 4x - 3y = 1 \end{pmatrix}$

3. $\begin{pmatrix} 5x + 2y = 5 \\ 3x - 4y = 29 \end{pmatrix}$

4. $\begin{pmatrix} 4x - 7y = -23 \\ 2x + 5y = -3 \end{pmatrix}$ **5.** $\begin{pmatrix} 5x - 4y = 14 \\ -x + 2y = -4 \end{pmatrix}$ **6.** $\begin{pmatrix} -x + 2y = 10 \\ 3x - y = -10 \end{pmatrix}$

7. $\begin{pmatrix} y = 2x - 4 \\ 6x - 3y = 1 \end{pmatrix}$ **8.** $\begin{pmatrix} -3x - 4y = 14 \\ -2x + 3y = -19 \end{pmatrix}$ **9.** $\begin{pmatrix} -4x + 3y = 3 \\ 4x - 6y = -5 \end{pmatrix}$

10. $\begin{pmatrix} x = 4y - 1 \\ 2x - 8y = -2 \end{pmatrix}$ **11.** $\begin{pmatrix} 9x - y = -2 \\ 8x + y = 4 \end{pmatrix}$ **12.** $\begin{pmatrix} 6x - 5y = 1 \\ 4x + 7y = 2 \end{pmatrix}$

13. $\begin{pmatrix} -\frac{2}{3}x + \frac{1}{2}y = -7 \\ \frac{1}{3}x - \frac{3}{2}y = 6 \end{pmatrix}$ **14.** $\begin{pmatrix} \frac{1}{2}x + \frac{2}{3}y = -6 \\ \frac{1}{4}x - \frac{1}{3}y = -1 \end{pmatrix}$ **15.** $\begin{pmatrix} 2x + 7y = -1 \\ x = 2 \end{pmatrix}$

16. $\begin{pmatrix} 5x - 3y = 2 \\ y = 4 \end{pmatrix}$ **17.** $\begin{pmatrix} x - y + 2z = -8 \\ 2x + 3y - 4z = 18 \\ -x + 2y - z = 7 \end{pmatrix}$

18. $\begin{pmatrix} x - 2y + z = 3 \\ 3x + 2y + z = -3 \\ 2x - 3y - 3z = -5 \end{pmatrix}$ **19.** $\begin{pmatrix} 2x - 3y + z = -7 \\ -3x + y - z = -7 \\ x - 2y - 5z = -45 \end{pmatrix}$

20. $\begin{pmatrix} 3x - y - z = 18 \\ 4x + 3y - 2z = 10 \\ -5x - 2y + 3z = -22 \end{pmatrix}$ **21.** $\begin{pmatrix} 4x + 5y - 2z = -14 \\ 7x - y + 2z = 42 \\ 3x + y + 4z = 28 \end{pmatrix}$

22. $\begin{pmatrix} -5x + 6y + 4z = -4 \\ -7x - 8y + 2z = -2 \\ 2x + 9y - z = 1 \end{pmatrix}$ **23.** $\begin{pmatrix} 2x - y + 3z = -17 \\ 3y + z = 5 \\ x - 2y - z = -3 \end{pmatrix}$

24. $\begin{pmatrix} 2x - y + 3z = -5 \\ 3x + 4y - 2z = -25 \\ -x + z = 6 \end{pmatrix}$ **25.** $\begin{pmatrix} x + 3y - 4z = -1 \\ 2x - y + z = 2 \\ 4x + 5y - 7z = 0 \end{pmatrix}$

26. $\begin{pmatrix} x - 2y + z = 1 \\ 3x + y - z = 2 \\ 2x - 4y + 2z = -1 \end{pmatrix}$ **27.** $\begin{pmatrix} 3x - 2y - 3z = -5 \\ x + 2y + 3z = -3 \\ -x + 4y - 6z = 8 \end{pmatrix}$

28. $\begin{pmatrix} 3x - 2y + z = 11 \\ 5x + 3y = 17 \\ x + y - 2z = 6 \end{pmatrix}$ **29.** $\begin{pmatrix} x - 2y + 3z = 1 \\ -2x + 4y - 3z = -3 \\ 5x - 6y + 6z = 10 \end{pmatrix}$

30. $\begin{pmatrix} 2x - y + 2z = -1 \\ 4x + 3y - 4z = 2 \\ x + 5y - z = 9 \end{pmatrix}$ **31.** $\begin{pmatrix} -x - y + 3z = -2 \\ -2x + y + 7z = 14 \\ 3x + 4y - 5z = 12 \end{pmatrix}$

32. $\begin{pmatrix} -2x + y - 3z = -4 \\ x + 5y - 4z = 13 \\ 7x - 2y - z = 37 \end{pmatrix}$

Miscellaneous Problems

33. A linear system in which the constant terms are all zero is called a **homogeneous system**.
   (a) Verify that for a 3×3 homogeneous system, if $D \neq 0$, then $(0,0,0)$ is the only solution for the system.

(b) Verify that for a 3×3 homogeneous system, if $D = 0$, then the equations are dependent.

For Problems 34–37, solve each of the homogeneous systems (see Problem 33). If the equations are dependent, indicate that the system has infinitely many solutions.

34. $\begin{pmatrix} x - 2y + 5z = 0 \\ 3x + y - 2z = 0 \\ 4x - y + 3z = 0 \end{pmatrix}$
$\qquad$
35. $\begin{pmatrix} 2x - y + z = 0 \\ 3x + 2y + 5z = 0 \\ 4x - 7y + z = 0 \end{pmatrix}$

36. $\begin{pmatrix} 3x + y - z = 0 \\ x - y + 2z = 0 \\ 4x - 5y - 2z = 0 \end{pmatrix}$
$\qquad$
37. $\begin{pmatrix} 2x - y + 2z = 0 \\ x + 2y + z = 0 \\ x - 3y + z = 0 \end{pmatrix}$

Chapter 7 Summary

The primary focus of this entire chapter is the development of different techniques for solving systems of linear equations.

The Substitution Method

With the aid of an example, we can describe the substitution method as follows. Suppose we want to solve the system

$$\begin{pmatrix} x - 2y = 22 \\ 3x + 4y = -24 \end{pmatrix}.$$

Step 1. Solve the first equation for x in terms of y.

$$x - 2y = 22$$
$$x = 2y + 22$$

Step 2. Substitute $2y + 22$ for x in the second equation.

$$3(2y + 22) + 4y = -24$$

Step 3. Solve the equation obtained in Step 2.

$$6y + 66 + 4y = -24$$
$$10y + 66 = -24$$
$$10y = -90$$
$$y = -9$$

Step 4. Substitute -9 for y in the equation of Step 1.

$$x = 2(-9) + 22 = 4$$

The solution set is $\{(4, -9)\}$.

The Elimination-by-Addition Method

This method allows us to replace systems of equations with *simpler equivalent systems* until we obtain a system for which we can easily determine the solution. The following operations produce equivalent systems.

1. Any two equations of a system can be interchanged.

2. Both sides of any equation of the system can be multiplied by any nonzero real number.

3. Any equation of the system can be replaced by the sum of a nonzero multiple of another equation plus that equation.

For example, through a sequence of operations, we can transform the system

$$\begin{pmatrix} 5x + 3y = -28 \\ \frac{1}{2}x - y = -8 \end{pmatrix}$$

to the equivalent system

$$\begin{pmatrix} x - 2y = -16 \\ 13y = 52 \end{pmatrix},$$

from which we can easily determine the solution set $\{(-8, 4)\}$.

The Matrix Approach

We can change the augmented matrix of a system to reduced echelon form by applying the following elementary row operations.

1. Any two rows of the matrix can be interchanged.

2. Any row of the matrix can be multiplied by a nonzero real number.

3. Any row of the matrix can be replaced by the sum of a nonzero multiple of another row plus that row.

For example, the augmented matrix of the system

$$\begin{pmatrix} x - 2y + 3z = 4 \\ 2x + y - 4z = 3 \\ -3x + 4y - z = -2 \end{pmatrix}$$

is

$$\begin{bmatrix} 1 & -2 & 3 & | & 4 \\ 2 & 1 & -4 & | & 3 \\ -3 & 4 & -1 & | & -2 \end{bmatrix}.$$

We can change this matrix to the reduced echelon form

$$\begin{bmatrix} 1 & 0 & 0 & | & 4 \\ 0 & 1 & 0 & | & 3 \\ 0 & 0 & 1 & | & 2 \end{bmatrix},$$

from which the solution set of $\{(4, 3, 2)\}$ is obvious.

Cramer's Rule

Cramer's rule for solving systems of linear equations involves the use of determinants. It is stated for the 2×2 case on page 343 and for the 3×3 case on page 345. For example, the solution set of the system

$$\begin{pmatrix} 3x - y - z = 2 \\ 2x + y + 3z = 9 \\ -x + 5y - 6z = -29 \end{pmatrix}$$

is determined by

$$x = \frac{\begin{vmatrix} 2 & -1 & -1 \\ 9 & 1 & 3 \\ -29 & 5 & -6 \end{vmatrix}}{\begin{vmatrix} 3 & -1 & -1 \\ 2 & 1 & 3 \\ -1 & 5 & -6 \end{vmatrix}} = \frac{-83}{-83} = 1,$$

$$y = \frac{\begin{vmatrix} 3 & 2 & -1 \\ 2 & 9 & 3 \\ -1 & -29 & -6 \end{vmatrix}}{\begin{vmatrix} 3 & -1 & -1 \\ 2 & 1 & 3 \\ -1 & 5 & -6 \end{vmatrix}} = \frac{166}{-83} = -2,$$

and

$$z = \frac{\begin{vmatrix} 3 & -1 & 2 \\ 2 & 1 & 9 \\ -1 & 5 & -29 \end{vmatrix}}{\begin{vmatrix} 3 & -1 & -1 \\ 2 & 1 & 3 \\ -1 & 5 & -6 \end{vmatrix}} = \frac{-249}{-83} = 3.$$

Chapter 7 Review Problem Set

Solve each of the following systems by using the *substitution* method.

1. $\begin{pmatrix} 3x - y = 16 \\ 5x + 7y = -34 \end{pmatrix}$ 2. $\begin{pmatrix} 6x + 5y = -21 \\ x - 4y = 11 \end{pmatrix}$

3. $\begin{pmatrix} 2x - 3y = 12 \\ 3x + 5y = -20 \end{pmatrix}$ 4. $\begin{pmatrix} 5x + 8y = 1 \\ 4x + 7y = -2 \end{pmatrix}$

Solve each of the following systems by using the *elimination-by-addition* method.

5. $\begin{pmatrix} 4x - 3y = 34 \\ 3x + 2y = 0 \end{pmatrix}$ 6. $\begin{pmatrix} \frac{1}{2}x - \frac{2}{3}y = 1 \\ \frac{3}{4}x + \frac{1}{6}y = -1 \end{pmatrix}$

7. $\begin{pmatrix} 2x - y + 3z = -19 \\ 3x + 2y - 4z = 21 \\ 5x - 4y - z = -8 \end{pmatrix}$ 8. $\begin{pmatrix} 3x + 2y - 4z = 4 \\ 5x + 3y - z = 2 \\ 4x - 2y + 3z = 11 \end{pmatrix}$

Solve each of the following systems by *changing the augmented matrix to reduced echelon form.*

9. $\begin{pmatrix} x - 3y = 17 \\ -3x + 2y = -23 \end{pmatrix}$

10. $\begin{pmatrix} 2x + 3y = 25 \\ 3x - 5y = -29 \end{pmatrix}$

11. $\begin{pmatrix} x - 2y + z = -7 \\ 2x - 3y + 4z = -14 \\ -3x + y - 2z = 10 \end{pmatrix}$

12. $\begin{pmatrix} -2x - 7y + z = 9 \\ x + 3y - 4z = -11 \\ 4x + 5y - 3z = -11 \end{pmatrix}$

Solve each of the following systems by using *Cramer's rule.*

13. $\begin{pmatrix} 5x + 3y = -18 \\ 4x - 9y = -3 \end{pmatrix}$

14. $\begin{pmatrix} 0.2x + 0.3y = 2.6 \\ 0.5x - 0.1y = 1.4 \end{pmatrix}$

15. $\begin{pmatrix} 2x - 3y - 3z = 25 \\ 3x + y + 2z = -5 \\ 5x - 2y - 4z = 32 \end{pmatrix}$

16. $\begin{pmatrix} 3x - y + z = -10 \\ 6x - 2y + 5z = -35 \\ 7x + 3y - 4z = 19 \end{pmatrix}$

Solve each of the following systems by using whichever method you think is most appropriate.

17. $\begin{pmatrix} 4x + 7y = -15 \\ 3x - 2y = 25 \end{pmatrix}$

18. $\begin{pmatrix} \frac{3}{4}x - \frac{1}{2}y = -15 \\ \frac{2}{3}x + \frac{1}{4}y = -5 \end{pmatrix}$

19. $\begin{pmatrix} x + 4y = 3 \\ 3x - 2y = 1 \end{pmatrix}$

20. $\begin{pmatrix} 7x - 3y = -49 \\ y = \frac{3}{5}x - 1 \end{pmatrix}$

21. $\begin{pmatrix} x - y - z = 4 \\ -3x + 2y + 5z = -21 \\ 5x - 3y - 7z = 30 \end{pmatrix}$

22. $\begin{pmatrix} 2x - y + z = -7 \\ -5x + 2y - 3z = 17 \\ 3x + y + 7z = -5 \end{pmatrix}$

23. $\begin{pmatrix} 3x - 2y - 5z = 2 \\ -4x + 3y + 11z = 3 \\ 2x - y + z = -1 \end{pmatrix}$

24. $\begin{pmatrix} 7x - y + z = -4 \\ -2x + 9y - 3z = -50 \\ x - 5y + 4z = 42 \end{pmatrix}$

Evaluate each of the following determinants.

25. $\begin{vmatrix} -2 & 6 \\ 3 & 8 \end{vmatrix}$

26. $\begin{vmatrix} 5 & -4 \\ 7 & -3 \end{vmatrix}$

27. $\begin{vmatrix} 2 & 3 & -1 \\ 3 & 4 & -5 \\ 6 & 4 & 2 \end{vmatrix}$

28. $\begin{vmatrix} 3 & -2 & 4 \\ 1 & 0 & 6 \\ 3 & -3 & 5 \end{vmatrix}$

29. $\begin{vmatrix} 5 & 4 & 3 \\ 2 & -7 & 0 \\ 3 & -2 & 0 \end{vmatrix}$

30. $\begin{vmatrix} 5 & -4 & 2 & 1 \\ 3 & 7 & 6 & -2 \\ 2 & 1 & -5 & 0 \\ 3 & -2 & 4 & 0 \end{vmatrix}$

Solve each of the following problems by setting up and solving a system of linear equations.

31. The sum of the digits of a two-digit number is 9. If the digits are reversed, the newly formed number is 45 less than the original number. Find the original number.

32. Sara invested $2500, part of it at 10% and the rest at 12% yearly interest. The yearly income on the 12% investment was $102 more than the income on the 10% investment. How much money did she invest at each rate?

33. A box contains $17.70 in nickels, dimes, and quarters. The number of dimes is 8 less than twice the number of nickels. The number of quarters is 2 more than the sum of the number of nickels and dimes. How many coins of each kind are there in the box?

34. The measure of the largest angle of a triangle is 10° more than 4 times the smallest angle. The sum of the smallest and largest angles is 3 times the measure of the other angle. Find the measure of each angle of the triangle.

The Algebra of Matrices

8.1 The Algebra of 2 × 2 Matrices

8.2 Multiplicative Inverses

8.3 $m \times n$ Matrices

8.4 Systems Involving Linear Inequalities:
Linear Programming

In Section 7.3, matrices were used strictly as a device to help solve systems of linear equations. Our primary objective was the development of techniques for solving systems of equations, not the study of matrices. However, matrices can be studied from an algebraic viewpoint, much as we study the set of real numbers. That is, we can define certain operations on matrices and verify properties of those operations. This algebraic approach to matrices is the focal point of this chapter. In order to get a simplified view of the algebra of matrices, we will begin by studying 2 × 2 matrices, and then later we will extend to $m \times n$ matrices. As a bonus, another technique for solving systems of equations will emerge from our study.

8.1

The Algebra of 2 × 2 Matrices

Throughout these next two sections, we will be working primarily with 2×2 matrices. Therefore, any reference to matrices means 2×2 matrices unless stated otherwise. The following general 2×2 matrix notation will be used frequently.

$$A = \begin{bmatrix} a_{11} & a_{12} \\ a_{21} & a_{22} \end{bmatrix} \quad B = \begin{bmatrix} b_{11} & b_{12} \\ b_{21} & b_{22} \end{bmatrix} \quad C = \begin{bmatrix} c_{11} & c_{12} \\ c_{21} & c_{22} \end{bmatrix}$$

Two matrices are **equal** if and only if all elements in corresponding positions are equal. Thus, $A = B$ if and only if $a_{11} = b_{11}$, $a_{12} = b_{12}$, $a_{21} = b_{21}$, and $a_{22} = b_{22}$.

Addition of Matrices

To **add** two matrices, we add the elements that appear in corresponding positions. Therefore, the sum of A and B is defined as follows.

DEFINITION 8.1

$$A + B = \begin{bmatrix} a_{11} & a_{12} \\ a_{21} & a_{22} \end{bmatrix} + \begin{bmatrix} b_{11} & b_{12} \\ b_{21} & b_{22} \end{bmatrix}$$

$$= \begin{bmatrix} a_{11} + b_{11} & a_{12} + b_{12} \\ a_{21} + b_{21} & a_{22} + b_{22} \end{bmatrix}$$

For example,

$$\begin{bmatrix} 2 & -1 \\ -3 & 4 \end{bmatrix} + \begin{bmatrix} -5 & 4 \\ -1 & 7 \end{bmatrix} = \begin{bmatrix} -3 & 3 \\ -4 & 11 \end{bmatrix}.$$

It is not difficult to show that **the commutative and associative properties are valid for the addition of matrices**. Thus, we can state that

$$A + B = B + A \quad \text{and} \quad (A + B) + C = A + (B + C).$$

Since

$$\begin{bmatrix} a_{11} & a_{12} \\ a_{21} & a_{22} \end{bmatrix} + \begin{bmatrix} 0 & 0 \\ 0 & 0 \end{bmatrix} = \begin{bmatrix} a_{11} & a_{12} \\ a_{21} & a_{22} \end{bmatrix},$$

we see that $\begin{bmatrix} 0 & 0 \\ 0 & 0 \end{bmatrix}$, which is called the **zero matrix** and represented by O, is the **additive identity element**. Thus, we can state that

$$A + O = O + A = A.$$

Since every real number has an additive inverse, it follows that any matrix A has an **additive inverse**, $-A$, that is formed by taking the additive inverse of each element of A. For example, if

$$A = \begin{bmatrix} 4 & -2 \\ -1 & 0 \end{bmatrix} \quad \text{then} \quad -A = \begin{bmatrix} -4 & 2 \\ 1 & 0 \end{bmatrix}$$

and

$$A + (-A) = \begin{bmatrix} 4 & -2 \\ -1 & 0 \end{bmatrix} + \begin{bmatrix} -4 & 2 \\ 1 & 0 \end{bmatrix} = \begin{bmatrix} 0 & 0 \\ 0 & 0 \end{bmatrix}.$$

In general, we can state that **every matrix A has an additive inverse $-A$ such that**

$$A + (-A) = (-A) + A = 0.$$

Subtraction of Matrices

Again paralleling the algebra of real numbers, *subtraction* of matrices can be defined in terms of *adding the additive inverse*. Therefore, we can define subtraction as follows.

DEFINITION 8.2

$$A - B = A + (-B)$$

For example,

$$\begin{bmatrix} 2 & -7 \\ -6 & 5 \end{bmatrix} - \begin{bmatrix} 3 & 4 \\ -2 & -1 \end{bmatrix} = \begin{bmatrix} 2 & -7 \\ -6 & 5 \end{bmatrix} + \begin{bmatrix} -3 & -4 \\ 2 & 1 \end{bmatrix}$$
$$= \begin{bmatrix} -1 & -11 \\ -4 & 6 \end{bmatrix}.$$

Scalar Multiplication

When we work with matrices, we commonly refer to a single real number as a **scalar** to distinguish it from a matrix. Then the **product** of a scalar and a matrix (often referred to as **scalar multiplication**) can be formed by multiplying each element of the matrix by the scalar. For example,

$$3 \begin{bmatrix} -4 & -6 \\ 1 & -2 \end{bmatrix} = \begin{bmatrix} 3(-4) & 3(-6) \\ 3(1) & 3(-2) \end{bmatrix} = \begin{bmatrix} -12 & -18 \\ 3 & -6 \end{bmatrix}.$$

In general, scalar multiplication can be defined as follows.

DEFINITION 8.3

$$kA = k \begin{bmatrix} a_{11} & a_{12} \\ a_{21} & a_{22} \end{bmatrix} = \begin{bmatrix} ka_{11} & ka_{12} \\ ka_{21} & ka_{22} \end{bmatrix},$$

where k is any real number.

EXAMPLE 1

If $A = \begin{bmatrix} -4 & 3 \\ 2 & -5 \end{bmatrix}$ and $B = \begin{bmatrix} 2 & -3 \\ 7 & -6 \end{bmatrix}$, find

(a) $-2A$,   **(b)** $3A + 2B$, and   **(c)** $A - 4B$.

Solutions

(a) $-2A = -2 \begin{bmatrix} -4 & 3 \\ 2 & -5 \end{bmatrix} = \begin{bmatrix} 8 & -6 \\ -4 & 10 \end{bmatrix}$

(b) $3A + 2B = 3 \begin{bmatrix} -4 & 3 \\ 2 & -5 \end{bmatrix} + 2 \begin{bmatrix} 2 & -3 \\ 7 & -6 \end{bmatrix}$

$\qquad = \begin{bmatrix} -12 & 9 \\ 6 & -15 \end{bmatrix} + \begin{bmatrix} 4 & -6 \\ 14 & -12 \end{bmatrix}$

$\qquad = \begin{bmatrix} -8 & 3 \\ 20 & -27 \end{bmatrix}$

(c) $A - 4B = \begin{bmatrix} -4 & 3 \\ 2 & -5 \end{bmatrix} - 4 \begin{bmatrix} 2 & -3 \\ 7 & -6 \end{bmatrix}$

$\qquad = \begin{bmatrix} -4 & 3 \\ 2 & -5 \end{bmatrix} - \begin{bmatrix} 8 & -12 \\ 28 & -24 \end{bmatrix}$

$\qquad = \begin{bmatrix} -4 & 3 \\ 2 & -5 \end{bmatrix} + \begin{bmatrix} -8 & 12 \\ -28 & 24 \end{bmatrix}$

$\qquad = \begin{bmatrix} -12 & 15 \\ -26 & 19 \end{bmatrix}$

The following properties, which are easy to check, pertain to scalar multiplication and matrix addition (where k and l represent any real numbers).

$$k(A + B) = kA + kB$$
$$(k + l)A = kA + lA$$
$$(kl)A = k(lA)$$

Multiplication of Matrices

At this time, it probably would seem quite natural to define matrix multiplication by multiplying corresponding elements of two matrices. However, it turns out that such a definition does not have many worthwhile applications. Therefore a special type of **matrix multiplication** is used, sometimes referred to as "row-by-column" multiplication. We will state the definition, paraphrase what it says, and then use some examples to be sure of its meaning.

DEFINITION 8.4

$$AB = \begin{bmatrix} a_{11} & a_{12} \\ a_{21} & a_{22} \end{bmatrix}\begin{bmatrix} b_{11} & b_{12} \\ b_{21} & b_{22} \end{bmatrix}$$

$$= \begin{bmatrix} a_{11}b_{11} + a_{12}b_{21} & a_{11}b_{12} + a_{12}b_{22} \\ a_{21}b_{11} + a_{22}b_{21} & a_{21}b_{12} + a_{22}b_{22} \end{bmatrix}$$

Notice the row-by-column pattern of Definition 8.4. We multiply the rows of A times the columns of B in a pairwise entry fashion, adding the results. For example, the element in the first row and second column of the product is obtained by multiplying the elements of the first row of A times the elements of the second column of B and adding the results.

$$\begin{bmatrix} \boxed{a_{11} \quad a_{12}} \\ a_{21} \quad a_{22} \end{bmatrix}\begin{bmatrix} b_{11} & \boxed{b_{12}} \\ b_{21} & \boxed{b_{22}} \end{bmatrix} = \begin{bmatrix} & a_{11}b_{12} + a_{12}b_{22} \end{bmatrix}$$

Now let's look at some specific examples.

EXAMPLE 2

If $A = \begin{bmatrix} -2 & 1 \\ 4 & 5 \end{bmatrix}$ and $B = \begin{bmatrix} 3 & -2 \\ -1 & 7 \end{bmatrix}$, find **(a)** AB and **(b)** BA.

Solutions

(a) $AB = \begin{bmatrix} -2 & 1 \\ 4 & 5 \end{bmatrix}\begin{bmatrix} 3 & -2 \\ -1 & 7 \end{bmatrix}$

$= \begin{bmatrix} (-2)(3) + (1)(-1) & (-2)(-2) + (1)(7) \\ (4)(3) + (5)(-1) & (4)(-2) + (5)(7) \end{bmatrix}$

$= \begin{bmatrix} -7 & 11 \\ 7 & 27 \end{bmatrix}$

(b) $BA = \begin{bmatrix} 3 & -2 \\ -1 & 7 \end{bmatrix}\begin{bmatrix} -2 & 1 \\ 4 & 5 \end{bmatrix}$

$= \begin{bmatrix} (3)(-2) + (-2)(4) & (3)(1) + (-2)(5) \\ (-1)(-2) + (7)(4) & (-1)(1) + (7)(5) \end{bmatrix} = \begin{bmatrix} -14 & -7 \\ 30 & 34 \end{bmatrix}$

Already from Example 2 we see that **matrix multiplication is not a commutative operation**.

EXAMPLE 3

If $A = \begin{bmatrix} 2 & -6 \\ -3 & 9 \end{bmatrix}$ and $B = \begin{bmatrix} -3 & 6 \\ -1 & 2 \end{bmatrix}$, find AB.

Solution Once you feel comfortable with Definition 8.4, the additions can be done mentally.

$$AB = \begin{bmatrix} 2 & -6 \\ -3 & 9 \end{bmatrix}\begin{bmatrix} -3 & 6 \\ -1 & 2 \end{bmatrix} = \begin{bmatrix} 0 & 0 \\ 0 & 0 \end{bmatrix}$$

Example 3 illustrates that the product of two matrices can be the zero matrix even though neither of the two matrices is the zero matrix. This is different from the property of real numbers that states "$ab = 0$ if and only if $a = 0$ or $b = 0$."

As illustrated and stated earlier, matrix multiplication is *not* a commutative operation. However, it is an **associative operation** and it does abide by two **distributive properties**. These properties can be stated as follows.

$$(AB)C = A(BC)$$
$$A(B + C) = AB + AC$$
$$(B + C)A = BA + CA$$

We will have you verify these properties in the next set of problems.

Problem Set 8.1

For Problems 1–12, compute the indicated matrix using the following matrices:

$$A = \begin{bmatrix} 1 & -2 \\ 3 & 4 \end{bmatrix}, \qquad B = \begin{bmatrix} 2 & -3 \\ 5 & -1 \end{bmatrix}, \qquad C = \begin{bmatrix} 0 & 6 \\ -4 & 2 \end{bmatrix},$$

$$D = \begin{bmatrix} -2 & 3 \\ 5 & -4 \end{bmatrix}, \qquad \text{and} \qquad E = \begin{bmatrix} 2 & 5 \\ 7 & 3 \end{bmatrix}.$$

1. $A + B$ **2.** $B - C$ **3.** $3C + D$ **4.** $2D - E$

5. $4A - 3B$ **6.** $2B + 3D$ **7.** $(A - B) - C$ **8.** $B - (D - E)$

9. $2D - 4E$ **10.** $3A - 4E$ **11.** $B - (D + E)$ **12.** $A - (B + C)$

For Problems 13–26, compute AB and BA.

13. $A = \begin{bmatrix} 1 & -1 \\ 2 & -2 \end{bmatrix}, \quad B = \begin{bmatrix} 3 & -4 \\ -1 & 2 \end{bmatrix}$ **14.** $A = \begin{bmatrix} -3 & 4 \\ 2 & 1 \end{bmatrix}, \quad B = \begin{bmatrix} -2 & 5 \\ 6 & -1 \end{bmatrix}$

15. $A = \begin{bmatrix} 1 & -3 \\ -4 & 6 \end{bmatrix}, \quad B = \begin{bmatrix} 7 & -3 \\ 4 & 5 \end{bmatrix}$ **16.** $A = \begin{bmatrix} 5 & 0 \\ -2 & 3 \end{bmatrix}, \quad B = \begin{bmatrix} -3 & 6 \\ 4 & 1 \end{bmatrix}$

17. $A = \begin{bmatrix} 2 & -4 \\ 1 & -2 \end{bmatrix}, \quad B = \begin{bmatrix} 1 & -2 \\ -3 & 6 \end{bmatrix}$ **18.** $A = \begin{bmatrix} 1 & 2 \\ 1 & 2 \end{bmatrix}, \quad B = \begin{bmatrix} 2 & 2 \\ -1 & -1 \end{bmatrix}$

19. $A = \begin{bmatrix} -3 & -2 \\ -4 & -1 \end{bmatrix}, \quad B = \begin{bmatrix} 2 & -1 \\ 4 & 5 \end{bmatrix}$ **20.** $A = \begin{bmatrix} -2 & 3 \\ -1 & 7 \end{bmatrix}, \quad B = \begin{bmatrix} -1 & -3 \\ -5 & -7 \end{bmatrix}$

21. $A = \begin{bmatrix} 2 & -1 \\ -5 & 3 \end{bmatrix}, \quad B = \begin{bmatrix} 3 & 1 \\ 5 & 2 \end{bmatrix}$ **22.** $A = \begin{bmatrix} -8 & -5 \\ 3 & 2 \end{bmatrix}, \quad B = \begin{bmatrix} -2 & -5 \\ 3 & 8 \end{bmatrix}$

23. $A = \begin{bmatrix} \frac{1}{2} & -\frac{1}{3} \\ \frac{1}{3} & \frac{1}{4} \end{bmatrix}$, $B = \begin{bmatrix} 4 & -6 \\ 6 & -4 \end{bmatrix}$ **24.** $A = \begin{bmatrix} \frac{1}{3} & -\frac{1}{2} \\ \frac{3}{2} & -\frac{2}{3} \end{bmatrix}$, $B = \begin{bmatrix} -6 & -18 \\ 12 & -12 \end{bmatrix}$

25. $A = \begin{bmatrix} 5 & 6 \\ 2 & 3 \end{bmatrix}$, $B = \begin{bmatrix} 1 & -2 \\ -\frac{2}{3} & \frac{5}{3} \end{bmatrix}$ **26.** $A = \begin{bmatrix} -3 & -5 \\ 2 & 4 \end{bmatrix}$, $B = \begin{bmatrix} -2 & -\frac{5}{2} \\ 1 & \frac{3}{2} \end{bmatrix}$

For Problems 27–30, use the following matrices.

$$A = \begin{bmatrix} -2 & 3 \\ 5 & 4 \end{bmatrix}, \quad B = \begin{bmatrix} 0 & 1 \\ 1 & 0 \end{bmatrix}, \quad C = \begin{bmatrix} 1 & 0 \\ 1 & 0 \end{bmatrix},$$

$$D = \begin{bmatrix} 1 & 1 \\ 1 & 1 \end{bmatrix}, \quad I = \begin{bmatrix} 1 & 0 \\ 0 & 1 \end{bmatrix}$$

27. Compute AB and BA.

28. Compute AC and CA.

29. Compute AD and DA.

30. Compute AI and IA.

For Problems 31–34, use the following matrices.

$$A = \begin{bmatrix} 2 & 4 \\ 5 & -3 \end{bmatrix}, \quad B = \begin{bmatrix} -2 & 3 \\ -1 & 2 \end{bmatrix}, \quad C = \begin{bmatrix} 2 & 1 \\ 3 & 7 \end{bmatrix}$$

31. Show that $(AB)C = A(BC)$.

32. Show that $A(B + C) = AB + AC$.

33. Show that $(A + B)C = AC + BC$.

34. Show that $(3 + 2)A = 3A + 2A$.

For Problems 35–43, use the following matrices.

$$A = \begin{bmatrix} a_{11} & a_{12} \\ a_{21} & a_{22} \end{bmatrix}, \quad B = \begin{bmatrix} b_{11} & b_{12} \\ b_{21} & b_{22} \end{bmatrix}, \quad C = \begin{bmatrix} c_{11} & c_{12} \\ c_{21} & c_{22} \end{bmatrix}, \quad O = \begin{bmatrix} 0 & 0 \\ 0 & 0 \end{bmatrix}$$

35. Show that $A + B = B + A$.

36. Show that $(A + B) + C = A + (B + C)$.

37. Show that $A + (-A) = O$.

38. Show that $k(A + B) = kA + kB$ for any real number k.

39. Show that $(k + l)A = kA + lA$ for any real numbers k and l.

40. Show that $(kl)A = k(lA)$ for any real numbers k and l.

41. Show that $(AB)C = A(BC)$.

42. Show that $A(B + C) = AB + AC$.

43. Show that $(A + B)C = AC + BC$.

Miscellaneous Problems

44. If $A = \begin{bmatrix} 2 & 0 \\ 0 & 3 \end{bmatrix}$, calculate A^2 and A^3, where A^2 means AA and A^3 means AAA.

45. If $A = \begin{bmatrix} 1 & -1 \\ 2 & 3 \end{bmatrix}$, calculate A^2 and A^3.

46. Does $(A + B)(A - B) = A^2 - B^2$ for all 2 × 2 matrices? Defend your answer.

8.2

Multiplicative Inverses

We know that 1 is a multiplicative identity element for the set of real numbers. That is, $a(1) = 1(a) = a$ for any real number a. Is there a multiplicative identity element for 2×2 matrices? Yes, the matrix

$$I = \begin{bmatrix} 1 & 0 \\ 0 & 1 \end{bmatrix}$$

is the **multiplicative identity element**, because

$$\begin{bmatrix} 1 & 0 \\ 0 & 1 \end{bmatrix} \begin{bmatrix} a_{11} & a_{12} \\ a_{21} & a_{22} \end{bmatrix} = \begin{bmatrix} a_{11} & a_{12} \\ a_{21} & a_{22} \end{bmatrix}$$

and

$$\begin{bmatrix} a_{11} & a_{12} \\ a_{21} & a_{22} \end{bmatrix} \begin{bmatrix} 1 & 0 \\ 0 & 1 \end{bmatrix} = \begin{bmatrix} a_{11} & a_{12} \\ a_{21} & a_{22} \end{bmatrix}.$$

Therefore, we can state that

$$AI = IA = A$$

for all 2×2 matrices.

Again referring to the real numbers, we know that every nonzero real number a has a multiplicative inverse $1/a$ such that $a(1/a) = (1/a)a = 1$. Does every 2×2 matrix have a multiplicative inverse? To help answer this question, let's think about finding the multiplicative inverse (if one exists) for a specific matrix. This should give us some clues about a general approach.

EXAMPLE 1

Find the multiplicative inverse of $A = \begin{bmatrix} 3 & 5 \\ 2 & 4 \end{bmatrix}$.

Solution We are looking for a matrix A^{-1} such that $AA^{-1} = A^{-1}A = I$. In other words, we want to solve the following matrix equation:

$$\begin{bmatrix} 3 & 5 \\ 2 & 4 \end{bmatrix} \begin{bmatrix} x & y \\ z & w \end{bmatrix} = \begin{bmatrix} 1 & 0 \\ 0 & 1 \end{bmatrix}.$$

We need to multiply the two matrices on the left side of this equation and then set the elements of the product matrix equal to the corresponding elements of the identity matrix. We obtain the following system of equations.

$$\begin{cases} 3x + 5z = 1 \\ 3y + 5w = 0 \\ 2x + 4z = 0 \\ 2y + 4w = 1 \end{cases}$$

(1)
(2)
(3)
(4)

Solving equations (1) and (3) simultaneously produces $x = 2$ and $z = -1$. Likewise, solving equations (2) and (4) simultaneously produces $y = -\frac{5}{2}$ and $w = \frac{3}{2}$. Therefore,

$$A^{-1} = \begin{bmatrix} x & y \\ z & w \end{bmatrix} = \begin{bmatrix} 2 & -\frac{5}{2} \\ -1 & \frac{3}{2} \end{bmatrix}.$$

To check this, we perform the following multiplication:

$$\begin{bmatrix} 3 & 5 \\ 2 & 4 \end{bmatrix}\begin{bmatrix} 2 & -\frac{5}{2} \\ -1 & \frac{3}{2} \end{bmatrix} = \begin{bmatrix} 2 & -\frac{5}{2} \\ -1 & \frac{3}{2} \end{bmatrix}\begin{bmatrix} 3 & 5 \\ 2 & 4 \end{bmatrix} = \begin{bmatrix} 1 & 0 \\ 0 & 1 \end{bmatrix}.$$

Now let's use the approach in Example 1 on the general matrix

$$A = \begin{bmatrix} a_{11} & a_{12} \\ a_{21} & a_{22} \end{bmatrix}.$$

We want to find

$$A^{-1} = \begin{bmatrix} x & y \\ z & w \end{bmatrix}$$

such that $AA^{-1} = I$. Therefore, we need to solve the matrix equation

$$\begin{bmatrix} a_{11} & a_{12} \\ a_{21} & a_{22} \end{bmatrix}\begin{bmatrix} x & y \\ z & w \end{bmatrix} = \begin{bmatrix} 1 & 0 \\ 0 & 1 \end{bmatrix}$$

for x, y, z, and w. Once again, we multiply the two matrices on the left side of the equation and set the elements of this product matrix equal to the corresponding elements of the identity matrix. We then obtain the following system of equations.

$$\begin{cases} a_{11}x + a_{12}z = 1 \\ a_{11}y + a_{12}w = 0 \\ a_{21}x + a_{22}z = 0 \\ a_{21}y + a_{22}w = 1 \end{cases}$$

Solving this system produces

$$x = \frac{a_{22}}{a_{11}a_{22} - a_{12}a_{21}}, \qquad y = \frac{-a_{12}}{a_{11}a_{22} - a_{12}a_{21}},$$

$$z = \frac{-a_{21}}{a_{11}a_{22} - a_{12}a_{21}}, \qquad w = \frac{a_{11}}{a_{11}a_{22} - a_{12}a_{21}}.$$

Notice that the number in each denominator, $a_{11}a_{22} - a_{12}a_{21}$, is the determinant of the matrix A. Thus, if $|A| \neq 0$,

$$A^{-1} = \frac{1}{|A|}\begin{bmatrix} a_{22} & -a_{12} \\ -a_{21} & a_{11} \end{bmatrix}.$$

Matrix multiplication will show that $AA^{-1} = A^{-1}A = I$. **If $|A| = 0$, then the matrix A has no multiplicative inverse.**

EXAMPLE 2

Find A^{-1} if $A = \begin{bmatrix} 3 & 5 \\ -2 & -4 \end{bmatrix}$.

Solution First, let's find $|A|$.

$$|A| = (3)(-4) - (5)(-2) = -2$$

Therefore,

$$A^{-1} = \frac{1}{-2} \begin{bmatrix} -4 & -5 \\ 2 & 3 \end{bmatrix} = \begin{bmatrix} 2 & \frac{5}{2} \\ -1 & -\frac{3}{2} \end{bmatrix}.$$

It is easily checked that $AA^{-1} = A^{-1}A = I$.

EXAMPLE 3

Find A^{-1} if $A = \begin{bmatrix} 8 & -2 \\ -12 & 3 \end{bmatrix}$.

Solution

$$|A| = (8)(3) - (-2)(-12) = 0$$

Therefore, A has no multiplicative inverse.

More about Multiplication of Matrices

Thus far we have found the product only of 2×2 matrices. The "row-by-column" multiplication pattern can be applied to many different kinds of matrices. We shall do this in the next section, but for now let's consider only the additional situation of finding the product of a 2×2 matrix and a 2×1 matrix, with the 2×2 matrix on the left. This is done as follows:

$$\begin{bmatrix} a_{11} & a_{12} \\ a_{21} & a_{22} \end{bmatrix} \begin{bmatrix} b_{11} \\ b_{21} \end{bmatrix} = \begin{bmatrix} a_{11}b_{11} + a_{12}b_{21} \\ a_{21}b_{11} + a_{22}b_{21} \end{bmatrix}.$$

Notice that the product matrix is a 2×1 matrix. The following example illustrates this pattern.

$$\begin{bmatrix} -2 & 3 \\ 1 & -4 \end{bmatrix} \begin{bmatrix} 5 \\ 7 \end{bmatrix} = \begin{bmatrix} (-2)(5) + (3)(7) \\ (1)(5) + (-4)(7) \end{bmatrix} = \begin{bmatrix} 11 \\ -23 \end{bmatrix}$$

Back to Solving Systems of Equations

The linear system of equations

$$\begin{pmatrix} a_{11}x + a_{12}y = d_1 \\ a_{21}x + a_{22}y = d_2 \end{pmatrix}$$

can be represented by the matrix equation

$$\begin{bmatrix} a_{11} & a_{12} \\ a_{21} & a_{22} \end{bmatrix} \begin{bmatrix} x \\ y \end{bmatrix} = \begin{bmatrix} d_1 \\ d_2 \end{bmatrix}.$$

If we let

$$A = \begin{bmatrix} a_{11} & a_{12} \\ a_{21} & a_{22} \end{bmatrix}, \qquad X = \begin{bmatrix} x \\ y \end{bmatrix}, \qquad \text{and} \qquad B = \begin{bmatrix} d_1 \\ d_2 \end{bmatrix},$$

then the previous matrix equation can be written $AX = B$.

If A^{-1} exists, then we can multiply both sides of $AX = B$ by A^{-1} (on the left) and simplify as follows.

$$AX = B$$
$$A^{-1}(AX) = A^{-1}(B)$$
$$(A^{-1}A)X = A^{-1}B$$
$$IX = A^{-1}B$$
$$X = A^{-1}B$$

Therefore, the product $A^{-1}B$ is the solution of the system.

EXAMPLE 4

Solve the system

$$\begin{pmatrix} 5x + 4y = 10 \\ 6x + 5y = 13 \end{pmatrix}.$$

Solution By letting

$$A = \begin{bmatrix} 5 & 4 \\ 6 & 5 \end{bmatrix}, \qquad X = \begin{bmatrix} x \\ y \end{bmatrix}, \qquad \text{and} \qquad B = \begin{bmatrix} 10 \\ 13 \end{bmatrix},$$

the given system can be represented by the matrix equation $AX = B$. From our previous discussion, we know that the solution of this equation is $X = A^{-1}B$, so we need to find A^{-1} and the product $A^{-1}B$.

$$A^{-1} = \frac{1}{|A|} \begin{bmatrix} 5 & -4 \\ -6 & 5 \end{bmatrix} = \frac{1}{1} \begin{bmatrix} 5 & -4 \\ -6 & 5 \end{bmatrix} = \begin{bmatrix} 5 & -4 \\ -6 & 5 \end{bmatrix}$$

Therefore,

$$A^{-1}B = \begin{bmatrix} 5 & -4 \\ -6 & 5 \end{bmatrix} \begin{bmatrix} 10 \\ 13 \end{bmatrix} = \begin{bmatrix} -2 \\ 5 \end{bmatrix}.$$

The solution set of the given system is $\{(-2, 5)\}$.

EXAMPLE 5

Solve the system

$$\begin{pmatrix} 3x - 2y = 9 \\ 4x + 7y = -17 \end{pmatrix}.$$

Solution If we let

$$A = \begin{bmatrix} 3 & -2 \\ 4 & 7 \end{bmatrix}, \qquad X = \begin{bmatrix} x \\ y \end{bmatrix}, \qquad \text{and} \qquad B = \begin{bmatrix} 9 \\ -17 \end{bmatrix},$$

then the system is represented by $AX = B$, where $X = A^{-1}B$ and

$$A^{-1} = \frac{1}{|A|} \begin{bmatrix} 7 & 2 \\ -4 & 3 \end{bmatrix} = \frac{1}{29} \begin{bmatrix} 7 & 2 \\ -4 & 3 \end{bmatrix} = \begin{bmatrix} \frac{7}{29} & \frac{2}{29} \\ -\frac{4}{29} & \frac{3}{29} \end{bmatrix}.$$

Therefore,

$$A^{-1}B = \begin{bmatrix} \frac{7}{29} & \frac{2}{29} \\ -\frac{4}{29} & \frac{3}{29} \end{bmatrix} \begin{bmatrix} 9 \\ -17 \end{bmatrix} = \begin{bmatrix} 1 \\ -3 \end{bmatrix}.$$

The solution set of the given system is $\{(1, -3)\}$.

This technique of using matrix inverses to solve systems of linear equations is especially useful when there are many systems to be solved that have the same coefficients but different constant terms.

Problem Set 8.2

Find the multiplicative inverse (if it exists) of each of the following matrices.

1. $\begin{bmatrix} 5 & 7 \\ 2 & 3 \end{bmatrix}$
2. $\begin{bmatrix} 3 & 4 \\ 2 & 3 \end{bmatrix}$
3. $\begin{bmatrix} 3 & 8 \\ 2 & 5 \end{bmatrix}$
4. $\begin{bmatrix} 2 & 9 \\ 3 & 13 \end{bmatrix}$

5. $\begin{bmatrix} -1 & 2 \\ 3 & 4 \end{bmatrix}$
6. $\begin{bmatrix} 1 & -2 \\ 4 & -3 \end{bmatrix}$
7. $\begin{bmatrix} -2 & -3 \\ 4 & 6 \end{bmatrix}$
8. $\begin{bmatrix} 5 & -1 \\ 3 & 4 \end{bmatrix}$

9. $\begin{bmatrix} -3 & 2 \\ -4 & 5 \end{bmatrix}$
10. $\begin{bmatrix} 3 & -4 \\ 6 & -8 \end{bmatrix}$
11. $\begin{bmatrix} 0 & 1 \\ 5 & 3 \end{bmatrix}$
12. $\begin{bmatrix} -2 & 0 \\ -3 & 5 \end{bmatrix}$

13. $\begin{bmatrix} -2 & -3 \\ -1 & -4 \end{bmatrix}$
14. $\begin{bmatrix} -2 & -5 \\ -3 & -6 \end{bmatrix}$
15. $\begin{bmatrix} -2 & 5 \\ -3 & 6 \end{bmatrix}$
16. $\begin{bmatrix} -3 & 4 \\ 1 & -2 \end{bmatrix}$

17. $\begin{bmatrix} 1 & 1 \\ 1 & -1 \end{bmatrix}$
18. $\begin{bmatrix} 1 & -1 \\ 1 & 1 \end{bmatrix}$

For Problems 19–26, compute AB.

19. $A = \begin{bmatrix} 4 & 3 \\ 2 & 5 \end{bmatrix}$, $B = \begin{bmatrix} 3 \\ 6 \end{bmatrix}$
20. $A = \begin{bmatrix} 5 & -2 \\ 3 & 1 \end{bmatrix}$, $B = \begin{bmatrix} 5 \\ 8 \end{bmatrix}$

21. $A = \begin{bmatrix} -3 & -4 \\ 2 & 1 \end{bmatrix}$, $B = \begin{bmatrix} 4 \\ -3 \end{bmatrix}$
22. $A = \begin{bmatrix} 5 & 2 \\ -1 & -3 \end{bmatrix}$, $B = \begin{bmatrix} 3 \\ -5 \end{bmatrix}$

23. $A = \begin{bmatrix} -4 & 2 \\ 7 & -5 \end{bmatrix}$, $B = \begin{bmatrix} -1 \\ -4 \end{bmatrix}$
24. $A = \begin{bmatrix} 0 & -3 \\ 2 & 9 \end{bmatrix}$, $B = \begin{bmatrix} -3 \\ -6 \end{bmatrix}$

25. $A = \begin{bmatrix} -2 & -3 \\ -5 & -6 \end{bmatrix}$, $B = \begin{bmatrix} 5 \\ -2 \end{bmatrix}$
26. $A = \begin{bmatrix} -3 & -5 \\ 4 & -7 \end{bmatrix}$, $B = \begin{bmatrix} -3 \\ -10 \end{bmatrix}$

Use the method of matrix inverses to solve each of the following systems.

27. $\begin{pmatrix} 2x + 3y = 13 \\ x + 2y = 8 \end{pmatrix}$
28. $\begin{pmatrix} 3x + 2y = 10 \\ 7x + 5y = 23 \end{pmatrix}$
29. $\begin{pmatrix} 4x - 3y = -23 \\ -3x + 2y = 16 \end{pmatrix}$

30. $\begin{pmatrix} 6x - y = -14 \\ 3x + 2y = -17 \end{pmatrix}$
31. $\begin{pmatrix} x - 7y = 7 \\ 6x + 5y = -5 \end{pmatrix}$
32. $\begin{pmatrix} x + 9y = -5 \\ 4x - 7y = -20 \end{pmatrix}$

33. $\begin{pmatrix} 3x - 5y = 2 \\ 4x - 3y = -1 \end{pmatrix}$ **34.** $\begin{pmatrix} 5x - 2y = 6 \\ 7x - 3y = 8 \end{pmatrix}$ **35.** $\begin{pmatrix} y = 19 - 3x \\ 9x - 5y = 1 \end{pmatrix}$

36. $\begin{pmatrix} 4x + 3y = 31 \\ x = 5y + 2 \end{pmatrix}$ **37.** $\begin{pmatrix} 3x + 2y = 0 \\ 30x - 18y = -19 \end{pmatrix}$ **38.** $\begin{pmatrix} 12x + 30y = 23 \\ 12x - 24y = -13 \end{pmatrix}$

39. $\begin{pmatrix} \frac{1}{3}x - \frac{3}{4}y = -18 \\ \frac{2}{3}x + \frac{1}{5}y = -2 \end{pmatrix}$ **40.** $\begin{pmatrix} \frac{3}{2}x + \frac{1}{6}y = 11 \\ \frac{2}{3}x - \frac{1}{4}y = 1 \end{pmatrix}$

8.3

$m \times n$ Matrices

Now let's see how much of the algebra of 2×2 matrices extends to $m \times n$ matrices, that is, to matrices of any dimension. In Section 7.4 we represented a general $m \times n$ matrix by

$$A = \begin{bmatrix} a_{11} & a_{12} & a_{13} & \cdots & a_{1n} \\ a_{21} & a_{22} & a_{23} & \cdots & a_{2n} \\ \vdots & \vdots & \vdots & & \vdots \\ a_{m1} & a_{m2} & a_{m3} & \cdots & a_{mn} \end{bmatrix}.$$

We denote the element at the intersection of row i and column j by a_{ij}. It is also customary to denote a matrix A with the abbreviated notation (a_{ij}).

Addition of matrices can be extended to matrices of any dimension by the following definition.

DEFINITION 8.5

Let $A = (a_{ij})$ and $B = (b_{ij})$ be two matrices *of the same dimension*. Then

$$A + B = (a_{ij}) + (b_{ij}) = (a_{ij} + b_{ij}).$$

Definition 8.5 simply states that to add two matrices, we add the elements that appear in corresponding positions in the matrices. The matrices must be of the same dimension for this to work.

An example of the sum of two 3×2 matrices is

$$\begin{bmatrix} 3 & 2 \\ 4 & -1 \\ -3 & 8 \end{bmatrix} + \begin{bmatrix} -2 & 1 \\ -3 & -7 \\ 5 & 9 \end{bmatrix} = \begin{bmatrix} 1 & 3 \\ 1 & -8 \\ 2 & 17 \end{bmatrix}.$$

The **commutative** and **associative properties** hold for any matrices that can be added. The $m \times n$ **zero matrix**, denoted by O, is the matrix containing all zeros. It is the **identity element for addition**. For example,

$$\begin{bmatrix} 2 & 3 & -1 & -5 \\ -7 & 6 & 2 & 8 \end{bmatrix} + \begin{bmatrix} 0 & 0 & 0 & 0 \\ 0 & 0 & 0 & 0 \end{bmatrix} = \begin{bmatrix} 2 & 3 & -1 & -5 \\ -7 & 6 & 2 & 8 \end{bmatrix}.$$

Every matrix A has an **additive inverse**, $-A$, that can be found by changing the sign of each element of A. For example, if

$$A = [2 \quad -3 \quad 0 \quad 4 \quad -7],$$

then

$$-A = [-2 \quad 3 \quad 0 \quad -4 \quad 7].$$

Furthermore, $A + (-A) = O$ for all matrices.

The definition we gave earlier for subtraction, $A - B = A + (-B)$, can be extended to any two matrices of the same dimension. For example,

$$[-4 \quad 3 \quad -5] - [7 \quad -4 \quad -1] = [-4 \quad 3 \quad -5] + [-7 \quad 4 \quad 1]$$
$$= [-11 \quad 7 \quad -4].$$

The **scalar product** of any real number k and any $m \times n$ matrix $A = (a_{ij})$ is defined by

$$kA = (ka_{ij}).$$

In other words, to find kA, we simply multiply each element of A by k. For example,

$$(-4) \begin{bmatrix} 1 & -1 \\ -2 & 3 \\ 4 & 5 \\ 0 & -8 \end{bmatrix} = \begin{bmatrix} -4 & 4 \\ 8 & -12 \\ -16 & -20 \\ 0 & 32 \end{bmatrix}.$$

The properties $k(A + B) = kA + kB$, $(k + l)A = kA + lA$, and $(kl)A = k(lA)$ hold for all matrices. The matrices A and B must be of the same dimension to be added.

The "row-by-column" definition for multiplying two matrices can be extended, but a little care must be taken. In order to define the product AB of two matrices A and B, **the number of columns of A must equal the number of rows of B**. Therefore, suppose $A = (a_{ij})$ is $m \times n$ and $B = (b_{ij})$ is $n \times p$. Then

$$AB = \begin{bmatrix} a_{11} & a_{12} & \cdots & a_{1n} \\ \vdots & \vdots & & \vdots \\ a_{i1} & a_{i2} & \cdots & a_{in} \\ \vdots & \vdots & & \vdots \\ a_{m1} & a_{m2} & \cdots & a_{mn} \end{bmatrix} \begin{bmatrix} b_{11} & \cdots & b_{1j} & \cdots & b_{1p} \\ b_{21} & \cdots & b_{2j} & \cdots & b_{2p} \\ \vdots & & \vdots & & \vdots \\ b_{n1} & \cdots & b_{nj} & \cdots & b_{np} \end{bmatrix} = C.$$

The product matrix C is of dimension $m \times p$ and the general element, c_{ij}, is determined as follows:

$$c_{ij} = a_{i1}b_{1j} + a_{i2}b_{2j} + \cdots + a_{in}b_{nj}.$$

A specific element of the product matrix, such as c_{23}, is the result of multiplying the elements in row 2 of matrix A times the elements in column 3 of matrix B and adding the results. Therefore,

$$c_{23} = a_{21}b_{13} + a_{22}b_{23} + \cdots + a_{2n}b_{n3}.$$

The following example illustrates the product of a 2×3 matrix and a 3×2 matrix.

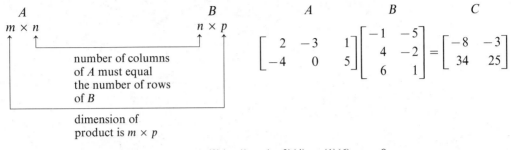

$$
\begin{array}{ccc}
A & B & C \\
\end{array}
$$

$$
\begin{bmatrix} 2 & -3 & 1 \\ -4 & 0 & 5 \end{bmatrix}
\begin{bmatrix} -1 & -5 \\ 4 & -2 \\ 6 & 1 \end{bmatrix}
=
\begin{bmatrix} -8 & -3 \\ 34 & 25 \end{bmatrix}
$$

$$
\begin{aligned}
c_{11} &= (2)(-1) + (-3)(4) + (1)(6) = -8 \\
c_{12} &= (2)(-5) + (-3)(-2) + (1)(1) = -3 \\
c_{21} &= (-4)(-1) + (0)(4) + (5)(6) = 34 \\
c_{22} &= (-4)(-5) + (0)(-2) + (5)(1) = 25
\end{aligned}
$$

Recall that **matrix multiplication is not commutative**. In fact, it may be that AB is defined and BA is not defined. For example, if A is a 2×3 matrix and B is a 3×4 matrix, then the product AB is a 2×4 matrix but the product BA is not defined, because the number of columns of B does not equal the number of rows of A.

The **associative property for multiplication** and the two **distributive properties** hold if the matrices have the proper number of rows and columns for the operations to be defined. In that case, we have $(AB)C = A(BC)$, $A(B + C) = AB + AC$, and $(A + B)C = AC + BC$.

Square Matrices

Now let's extend some of the algebra of 2×2 matrices to all square matrices (where the number of rows equals the number of columns). For example, the general **multiplicative identity element** for square matrices contains 1's in the main diagonal from the upper left-hand corner to the lower right-hand corner and 0's elsewhere. Therefore, for 3×3 and 4×4 matrices, the multiplicative identity elements are as follows:

$$
I_3 = \begin{bmatrix} 1 & 0 & 0 \\ 0 & 1 & 0 \\ 0 & 0 & 1 \end{bmatrix},
\qquad
I_4 = \begin{bmatrix} 1 & 0 & 0 & 0 \\ 0 & 1 & 0 & 0 \\ 0 & 0 & 1 & 0 \\ 0 & 0 & 0 & 1 \end{bmatrix}.
$$

We saw in Section 8.2 that some, but not all, 2×2 matrices have multiplicative inverses. In general, some, but not all, square matrices of a particular dimension have multiplicative inverses. If an $n \times n$ square matrix A does have a multiplicative inverse A^{-1}, then

$$AA^{-1} = A^{-1}A = I_n.$$

The technique used in Section 8.2 for finding multiplicative inverses of 2×2 matrices does generalize, but it becomes quite complicated. Therefore, we shall now describe another technique that works for all square matrices. Given an $n \times n$ matrix A, we begin by forming the $n \times 2n$ matrix

$$\begin{bmatrix} a_{11} & a_{12} & \cdots & a_{1n} & 1 & 0 & 0 & \cdots & 0 \\ a_{21} & a_{22} & \cdots & a_{2n} & 0 & 1 & 0 & \cdots & 0 \\ \vdots & \vdots & & \vdots & \vdots & \vdots & & \ddots & \vdots \\ a_{n1} & a_{n2} & \cdots & a_{nn} & 0 & 0 & 0 & \cdots & 1 \end{bmatrix},$$

where the identity matrix I_n appears to the right of A. Now we apply a succession of elementary row transformations to this double matrix until we obtain a matrix of the form

$$\begin{bmatrix} 1 & 0 & 0 & \cdots & 0 & b_{11} & b_{12} & \cdots & b_{1n} \\ 0 & 1 & 0 & \cdots & 0 & b_{21} & b_{22} & \cdots & b_{2n} \\ \vdots & \vdots & \ddots & & \vdots & \vdots & \vdots & & \vdots \\ 0 & 0 & 0 & \cdots & 1 & b_{n1} & b_{n2} & \cdots & b_{nn} \end{bmatrix}.$$

The B matrix in this matrix is the desired inverse A^{-1}. If A does not have an inverse, then it is impossible to change the original matrix to this final form.

EXAMPLE 1

Find A^{-1} if $A = \begin{bmatrix} 2 & 4 \\ 3 & 5 \end{bmatrix}$.

Solution First, we form the matrix

$$\begin{bmatrix} 2 & 4 & 1 & 0 \\ 3 & 5 & 0 & 1 \end{bmatrix}.$$

Now multiply row 1 by $\frac{1}{2}$.

$$\begin{bmatrix} 1 & 2 & \frac{1}{2} & 0 \\ 3 & 5 & 0 & 1 \end{bmatrix}$$

Next, add -3 times row 1 to row 2 to form a new row 2.

$$\begin{bmatrix} 1 & 2 & \frac{1}{2} & 0 \\ 0 & -1 & -\frac{3}{2} & 1 \end{bmatrix}$$

Next, multiply row 2 by -1.

$$\begin{bmatrix} 1 & 2 & \frac{1}{2} & 0 \\ 0 & 1 & \frac{3}{2} & -1 \end{bmatrix}$$

Finally, add -2 times row 2 to row 1 to form a new row 1.

$$\begin{bmatrix} 1 & 0 & \boxed{\begin{matrix} -\frac{5}{2} & 2 \\ \frac{3}{2} & -1 \end{matrix}} \\ 0 & 1 \end{bmatrix}$$

The matrix inside the box is A^{-1}, that is,

$$A^{-1} = \begin{bmatrix} -\frac{5}{2} & 2 \\ \frac{3}{2} & -1 \end{bmatrix}.$$

This can be checked, as always, by showing that $AA^{-1} = A^{-1}A = I_2$.

EXAMPLE 2

Find A^{-1} if $A = \begin{bmatrix} 1 & 1 & 2 \\ 2 & 3 & -1 \\ -3 & 1 & -2 \end{bmatrix}$.

Solution Form the matrix

$$\left[\begin{array}{ccc|ccc} 1 & 1 & 2 & 1 & 0 & 0 \\ 2 & 3 & -1 & 0 & 1 & 0 \\ -3 & 1 & -2 & 0 & 0 & 1 \end{array}\right].$$

Add -2 times row 1 to row 2, and add 3 times row 1 to row 3.

$$\left[\begin{array}{ccc|ccc} 1 & 1 & 2 & 1 & 0 & 0 \\ 0 & 1 & -5 & -2 & 1 & 0 \\ 0 & 4 & 4 & 3 & 0 & 1 \end{array}\right]$$

Add -1 times row 2 to row 1, and add -4 times row 2 to row 3.

$$\left[\begin{array}{ccc|ccc} 1 & 0 & 7 & 3 & -1 & 0 \\ 0 & 1 & -5 & -2 & 1 & 0 \\ 0 & 0 & 24 & 11 & -4 & 1 \end{array}\right]$$

Multiply row 3 by $\frac{1}{24}$.

$$\left[\begin{array}{ccc|ccc} 1 & 0 & 7 & 3 & -1 & 0 \\ 0 & 1 & -5 & -2 & 1 & 0 \\ 0 & 0 & 1 & \frac{11}{24} & -\frac{1}{6} & \frac{1}{24} \end{array}\right]$$

Add -7 times row 3 to row 1, and add 5 times row 3 to row 2.

$$\left[\begin{array}{ccc|ccc} 1 & 0 & 0 & -\frac{5}{24} & \frac{1}{6} & -\frac{7}{24} \\ 0 & 1 & 0 & \frac{7}{24} & \frac{1}{6} & \frac{5}{24} \\ 0 & 0 & 1 & \frac{11}{24} & -\frac{1}{6} & \frac{1}{24} \end{array}\right]$$

Therefore

$$A^{-1} = \begin{bmatrix} -\frac{5}{24} & \frac{1}{6} & -\frac{7}{24} \\ \frac{7}{24} & \frac{1}{6} & \frac{5}{24} \\ \frac{11}{24} & -\frac{1}{6} & \frac{1}{24} \end{bmatrix}.$$

(Be sure to check this!)

Systems of Equations

In Section 8.2 we used the concept of the multiplicative inverse to solve systems of two linear equations in two variables. This same technique can be applied to general systems of n linear equations in n variables. Let's consider one such example involving three equations in three variables.

EXAMPLE 3

Solve the system

$$\left(\begin{array}{rl} x + y + 2z &= -8 \\ 2x + 3y - z &= 3 \\ -3x + y - 2z &= 4 \end{array}\right).$$

Solution By letting

$$A = \begin{bmatrix} 1 & 1 & 2 \\ 2 & 3 & -1 \\ -3 & 1 & -2 \end{bmatrix}, \quad X = \begin{bmatrix} x \\ y \\ z \end{bmatrix}, \quad \text{and} \quad B = \begin{bmatrix} -8 \\ 3 \\ 4 \end{bmatrix},$$

the given system can be represented by the matrix equation $AX = B$. Therefore we know that $X = A^{-1}B$, so we need to find A^{-1} and the product $A^{-1}B$. The matrix A^{-1} was found in Example 2, so let's use that result and find $A^{-1}B$.

$$A^{-1}B = \begin{bmatrix} -\frac{5}{24} & \frac{1}{6} & -\frac{7}{24} \\ \frac{7}{24} & \frac{1}{6} & \frac{5}{24} \\ \frac{11}{24} & -\frac{1}{6} & \frac{1}{24} \end{bmatrix} \begin{bmatrix} -8 \\ 3 \\ 4 \end{bmatrix} = \begin{bmatrix} 1 \\ -1 \\ -4 \end{bmatrix}$$

The solution set of the given system is $\{(1, -1, -4)\}$.

Problem Set 8.3

For Problems 1–8, find $A + B$, $A - B$, $2A + 3B$, and $4A - 2B$.

1. $A = \begin{bmatrix} 2 & -1 & 4 \\ -2 & 0 & 5 \end{bmatrix}$, $B = \begin{bmatrix} -1 & 4 & -7 \\ 5 & -6 & 2 \end{bmatrix}$

2. $A = \begin{bmatrix} 3 & -6 \\ 2 & -1 \\ -4 & 5 \end{bmatrix}$, $B = \begin{bmatrix} 1 & 0 \\ 5 & -7 \\ -6 & 9 \end{bmatrix}$

3. $A = \begin{bmatrix} 2 & -1 & 4 & 12 \end{bmatrix}$, $B = \begin{bmatrix} -3 & -6 & 9 & -5 \end{bmatrix}$

4. $A = \begin{bmatrix} 3 \\ -9 \\ 7 \end{bmatrix}$, $B = \begin{bmatrix} -6 \\ 12 \\ 9 \end{bmatrix}$

5. $A = \begin{bmatrix} 3 & -2 & 1 \\ -1 & 4 & -7 \\ 0 & 5 & 9 \end{bmatrix}$, $B = \begin{bmatrix} 5 & -1 & -3 \\ 10 & -2 & 4 \\ 7 & 0 & 12 \end{bmatrix}$

6. $A = \begin{bmatrix} 7 & -4 \\ -5 & 9 \\ -1 & 2 \end{bmatrix}$, $B = \begin{bmatrix} 12 & 3 \\ -2 & -4 \\ -6 & 7 \end{bmatrix}$

7. $A = \begin{bmatrix} -1 & 0 \\ 2 & 3 \\ -5 & -4 \\ -7 & 11 \end{bmatrix}$, $B = \begin{bmatrix} 1 & 2 \\ -3 & 7 \\ 6 & -5 \\ 9 & -2 \end{bmatrix}$

8. $A = \begin{bmatrix} 0 & -1 & -2 \\ 3 & -4 & 6 \\ 5 & 4 & -9 \end{bmatrix}$, $B = \begin{bmatrix} 2 & 1 & -7 \\ -6 & 4 & 5 \\ 3 & -2 & -1 \end{bmatrix}$

For Problems 9–20, find AB and BA, whenever they exist.

9. $A = \begin{bmatrix} 2 & -1 \\ 0 & -4 \\ -5 & 3 \end{bmatrix}$, $B = \begin{bmatrix} 5 & -2 & 6 \\ -1 & 4 & -2 \end{bmatrix}$

10. $A = \begin{bmatrix} -2 & 3 & -1 \\ 7 & -4 & 5 \end{bmatrix}$, $B = \begin{bmatrix} 1 & -1 \\ -2 & 3 \\ -5 & -6 \end{bmatrix}$

11. $A = \begin{bmatrix} 2 & -1 & -3 \\ 0 & -4 & 7 \end{bmatrix}$, $B = \begin{bmatrix} 2 & 1 & -1 & 4 \\ 0 & -2 & 3 & 5 \\ -6 & 4 & -2 & 0 \end{bmatrix}$

12. $A = \begin{bmatrix} 3 & -1 & -4 \\ -5 & 2 & 2 \end{bmatrix}$, $B = \begin{bmatrix} 3 & -2 \\ -4 & 1 \end{bmatrix}$

13. $A = \begin{bmatrix} 1 & -1 & 2 \\ 0 & 1 & -2 \\ 3 & 1 & 4 \end{bmatrix}$, $B = \begin{bmatrix} 2 & 3 & -1 \\ 4 & 0 & 2 \\ -5 & 1 & -1 \end{bmatrix}$

14. $A = \begin{bmatrix} 1 & 0 & 1 \\ 0 & 1 & 1 \\ -1 & 2 & 3 \end{bmatrix}$, $B = \begin{bmatrix} -1 & -1 & 1 \\ 0 & 1 & 0 \\ 2 & -3 & 1 \end{bmatrix}$

15. $A = \begin{bmatrix} 2 & -1 & 3 & 4 \end{bmatrix}$, $B = \begin{bmatrix} -1 \\ -3 \\ 2 \\ -4 \end{bmatrix}$

16. $A = \begin{bmatrix} -2 \\ 3 \\ -5 \end{bmatrix}$, $B = \begin{bmatrix} 3 & -4 & -5 \end{bmatrix}$

17. $A = \begin{bmatrix} 2 \\ -7 \end{bmatrix}$, $B = \begin{bmatrix} 3 & -2 \\ 1 & 0 \\ -1 & 4 \end{bmatrix}$

18. $A = \begin{bmatrix} 3 & -2 & 2 & -4 \\ 1 & 0 & -1 & 2 \end{bmatrix}$, $B = \begin{bmatrix} 3 & -2 & 1 \\ -3 & 1 & 4 \\ 5 & 2 & 0 \\ -4 & -1 & -2 \end{bmatrix}$

19. $A = \begin{bmatrix} 3 \\ -4 \\ 2 \end{bmatrix}$, $B = \begin{bmatrix} 3 & -4 \end{bmatrix}$

20. $A = \begin{bmatrix} 3 & -7 \end{bmatrix}$, $B = \begin{bmatrix} 8 \\ -9 \end{bmatrix}$

For Problems 21–36, use the technique discussed in this section to find the multiplicative inverse (if it exists) of each of the following matrices.

21. $\begin{bmatrix} 1 & 3 \\ 4 & 2 \end{bmatrix}$ **22.** $\begin{bmatrix} 1 & 2 \\ 2 & -3 \end{bmatrix}$ **23.** $\begin{bmatrix} 2 & 1 \\ 7 & 4 \end{bmatrix}$

24. $\begin{bmatrix} 3 & 7 \\ 2 & 5 \end{bmatrix}$ **25.** $\begin{bmatrix} -2 & 1 \\ 3 & -4 \end{bmatrix}$ **26.** $\begin{bmatrix} -3 & 1 \\ 3 & -2 \end{bmatrix}$

27. $\begin{bmatrix} 1 & 2 & 3 \\ 1 & 3 & 4 \\ 1 & 4 & 3 \end{bmatrix}$ **28.** $\begin{bmatrix} 1 & 3 & -2 \\ 1 & 4 & -1 \\ -2 & -7 & 5 \end{bmatrix}$ **29.** $\begin{bmatrix} 1 & -2 & 1 \\ -2 & 5 & 3 \\ 3 & -5 & 7 \end{bmatrix}$

30. $\begin{bmatrix} 1 & 4 & -2 \\ -3 & -11 & 1 \\ 2 & 7 & 3 \end{bmatrix}$ **31.** $\begin{bmatrix} 2 & 3 & -4 \\ 3 & -1 & -2 \\ 1 & -4 & 2 \end{bmatrix}$ **32.** $\begin{bmatrix} -2 & 2 & 3 \\ 1 & -1 & 0 \\ 0 & 1 & 4 \end{bmatrix}$

33. $\begin{bmatrix} 1 & 2 & 3 \\ -3 & -4 & 3 \\ 2 & 4 & -1 \end{bmatrix}$ **34.** $\begin{bmatrix} 1 & -2 & 3 \\ -1 & 3 & -2 \\ -2 & 6 & 1 \end{bmatrix}$ **35.** $\begin{bmatrix} 2 & 0 & 0 \\ 0 & 4 & 0 \\ 0 & 0 & 10 \end{bmatrix}$

36. $\begin{bmatrix} 1 & -3 & 5 \\ 0 & 1 & 2 \\ 0 & 0 & 1 \end{bmatrix}$

Use the method of matrix inverses to solve each of the following systems. The required multiplicative inverses have been found in Problems 21–36.

37. $\begin{pmatrix} 2x + y = -4 \\ 7x + 4y = -13 \end{pmatrix}$ **38.** $\begin{pmatrix} 3x + 7y = -38 \\ 2x + 5y = -27 \end{pmatrix}$

39. $\begin{pmatrix} -2x + y = 1 \\ 3x - 4y = -14 \end{pmatrix}$ **40.** $\begin{pmatrix} -3x + y = -18 \\ 3x - 2y = 15 \end{pmatrix}$

41. $\begin{pmatrix} x + 2y + 3z = -2 \\ x + 3y + 4z = -3 \\ x + 4y + 3z = -6 \end{pmatrix}$ **42.** $\begin{pmatrix} x + 3y - 2z = 5 \\ x + 4y - z = 3 \\ -2x - 7y + 5z = -12 \end{pmatrix}$

43. $\begin{pmatrix} x - 2y + z = -3 \\ -2x + 5y + 3z = 34 \\ 3x - 5y + 7z = 14 \end{pmatrix}$ **44.** $\begin{pmatrix} x + 4y - 2z = 2 \\ -3x - 11y + z = -2 \\ 2x + 7y + 3z = -2 \end{pmatrix}$

45. $\begin{pmatrix} x + 2y + 3z = 2 \\ -3x - 4y + 3z = 0 \\ 2x + 4y - z = 4 \end{pmatrix}$

46. $\begin{pmatrix} x - 2y + 3z = -39 \\ -x + 3y - 2z = 40 \\ -2x + 6y + z = 45 \end{pmatrix}$

47. Five systems of linear equations can be generated from the system

$$\begin{pmatrix} x + y + 2z = a \\ 2x + 3y - z = b \\ -3x + y - 2z = c \end{pmatrix}$$

by letting *a*, *b*, and *c* assume five different sets of values. Solve the system for each set of values. The inverse of the coefficient matrix of these systems is given in Example 2 of this section.

(a) $a = 7$, $b = 1$, and $c = -1$
(b) $a = -7$, $b = 5$, and $c = 1$
(c) $a = -9$, $b = -8$, and $c = 19$
(d) $a = -1$, $b = -13$, and $c = -17$
(e) $a = -2$, $b = 0$, and $c = -2$.

Miscellaneous Problems

48. Matrices can be used to code and decode messages. For example, suppose that we set up a one-to-one correspondence between the letters of the alphabet and the first 26 counting numbers, as follows.

$$\begin{matrix} A & B & C & & Z \\ \updownarrow & \updownarrow & \updownarrow & \cdots & \updownarrow \\ 1 & 2 & 3 & & 26 \end{matrix}$$

Now suppose that we want to code the message "PLAY IT BY EAR." We can partition the letters of the message into groups of two. Since the last group will contain only one letter, let's arbitrarily stick in a Z to form a group of two. Let's also assign a number to each letter based on the letter/number association we exhibited.

$$\begin{matrix} P & L & A & Y & I & T & B & Y & E & A & R & Z \\ \updownarrow & \updownarrow & \updownarrow & \updownarrow & \updownarrow & \updownarrow & \updownarrow & \updownarrow & \updownarrow & \updownarrow & \updownarrow & \updownarrow \\ 16 & 12 & 1 & 25 & 9 & 20 & 2 & 25 & 5 & 1 & 18 & 26 \end{matrix}$$

Each pair of numbers can be recorded as columns in a 2 × 6 matrix *B*.

$$B = \begin{bmatrix} 16 & 1 & 9 & 2 & 5 & 18 \\ 12 & 25 & 20 & 25 & 1 & 26 \end{bmatrix}$$

Now let's choose a 2 × 2 matrix that contains only integers and such that its inverse also contains only integers. For example, we can use

$$A = \begin{bmatrix} 3 & 1 \\ 5 & 2 \end{bmatrix}; \quad \text{then} \quad A^{-1} = \begin{bmatrix} 2 & -1 \\ -5 & 3 \end{bmatrix}.$$

Next, let's find the product *AB*.

$$\begin{aligned} AB &= \begin{bmatrix} 3 & 1 \\ 5 & 2 \end{bmatrix} \begin{bmatrix} 16 & 1 & 9 & 2 & 5 & 18 \\ 12 & 25 & 20 & 25 & 1 & 26 \end{bmatrix} \\ &= \begin{bmatrix} 60 & 28 & 47 & 31 & 16 & 80 \\ 104 & 55 & 85 & 60 & 27 & 142 \end{bmatrix} \end{aligned}$$

Now we have our **coded message**:

60   104   28   55   47   85   31   60   16   27   80   142.

A person decoding the message would put the numbers back into a 2 × 6 matrix, multiply it on the left by A^{-1}, and convert the numbers back to letters.

Each of the following coded messages has been formed by using the matrix $A = \begin{bmatrix} 2 & 3 \\ 1 & 2 \end{bmatrix}$. Decode each of the messages.

(a) 68   40   77   51   78   49   23   15   29   19   85   52   41   27
(b) 62   40   78   47   64   36   19   11   93   57   93   56   88   57
(c) 64   36   58   37   63   36   21   13   75   47   63   36   38   23   118   72
(d) 61   38   115   69   93   57   36   20   78   49   68   40   77   51   60   37   47
    26   84   51   21   11

49. Suppose that the ordered pair (x, y) of a rectangular coordinate system is recorded as a 2 × 1 matrix and then multiplied on the left by the matrix $\begin{bmatrix} 1 & 0 \\ 0 & -1 \end{bmatrix}$. We would obtain

$$\begin{bmatrix} 1 & 0 \\ 0 & -1 \end{bmatrix} \begin{bmatrix} x \\ y \end{bmatrix} = \begin{bmatrix} x \\ -y \end{bmatrix}.$$

The point $(x, -y)$ is an x-axis reflection of the point (x, y). Therefore, the matrix $\begin{bmatrix} 1 & 0 \\ 0 & -1 \end{bmatrix}$ performs an x-axis reflection. What type of geometric transformation is performed by each of the following matrices?

(a) $\begin{bmatrix} -1 & 0 \\ 0 & 1 \end{bmatrix}$

(b) $\begin{bmatrix} -1 & 0 \\ 0 & -1 \end{bmatrix}$

(c) $\begin{bmatrix} 0 & -1 \\ 1 & 0 \end{bmatrix}$ (Hint: Check the slopes of lines through the origin.)

(d) $\begin{bmatrix} 0 & 1 \\ -1 & 0 \end{bmatrix}$

8.4

Systems Involving Linear Inequalities: Linear Programming

Finding solution sets for **systems of linear inequalities** relies heavily on the graphing approach. (Recall that we discussed graphing of linear inequalities in Section 3.3.) The solution set of the system

$$\begin{pmatrix} x + y > 2 \\ x - y < 2 \end{pmatrix}$$

is the intersection of the solution sets of the individual inequalities. In Figure 8.1(a) we have indicated the solution set for $x + y > 2$, and in Figure 8.1(b) we indicate the solution set for $x - y < 2$. Then, in Figure 8.1(c), the shaded region represents the intersection of the two solution sets; therefore it is the graph of the system. Remember that dashed lines are used to indicate that the points on

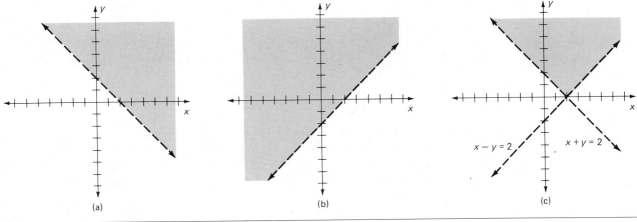

Figure 8.1

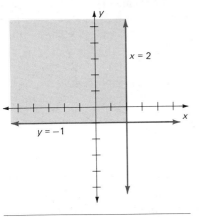

Figure 8.2

the lines are not included in the solution set. In the following examples, we have indicated only the final solution set for the system.

EXAMPLE 1

Solve the following system by graphing.

$$\begin{pmatrix} 2x - y \geq 4 \\ x + 2y < 2 \end{pmatrix}$$

Solution The graph of $2x - y \geq 4$ consists of all points *on or below* the line $2x - y = 4$. The graph of $x + 2y < 2$ consists of all points *below* the line $x + 2y = 2$. The graph of the system is indicated by the shaded region in Figure 8.2. Notice that all points in the shaded region are on or below the line $2x - y = 4$ and below the line $x + 2y = 2$.

EXAMPLE 2

Solve the following system by graphing.

$$\begin{pmatrix} x \leq 2 \\ y \geq -1 \end{pmatrix}$$

Solution Remember that even though each inequality contains only one variable, we are working in a rectangular coordinate system involving ordered pairs. That is to say, the system could also be written

$$\begin{pmatrix} x + 0(y) \leq 2 \\ 0(x) + y \geq -1 \end{pmatrix}.$$

The graph of this system is the shaded region in Figure 8.3. Notice that all points in the shaded region are *on or to the left* of the line $x = 2$ and *on or above* the line $y = -1$.

A system may contain more than two inequalities, as the next example illustrates.

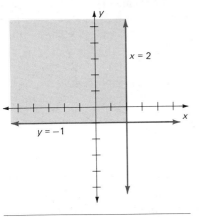

Figure 8.3

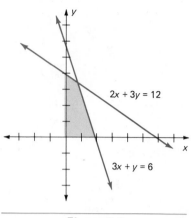

Figure 8.4

EXAMPLE 3

Solve the following system by graphing.

$$\begin{pmatrix} x \geq 0 \\ y \geq 0 \\ 2x + 3y \leq 12 \\ 3x + y \leq 6 \end{pmatrix}$$

Solution The solution set for the system is the intersection of the solution sets of the four inequalities. The shaded region in Figure 8.4 indicates the solution set for the system. Notice that all points in the shaded region are *on or to the right* of the y-axis, *on or above* the x-axis, *on or below* the line $2x + 3y = 12$, and *on or below* the line $3x + y = 6$.

Linear Programming: Another Look at Problem Solving

Throughout this text, problem solving is a unifying theme. Therefore, it seems appropriate at this time to give you a brief glimpse of an area of mathematics that was developed in the 1940's specifically as a problem-solving tool. Many applied problems involve the idea of *maximizing* or *minimizing* a certain function that is subject to various constraints; these can be expressed as linear inequalities. **Linear programming** was developed as one method for solving such problems.

> **Remark:** The term *programming* refers to the distribution of limited resources in order to maximize or minimize a certain function, such as cost, profit, distance, and so on. Thus, it is not synonymous with its meaning in computer programming. The constraints under which the distribution of resources is to be made determine the linear inequalities and equations; thus, the term *linear programming* is used.

Before introducing a linear programming type of problem, we need to extend one mathematical concept a bit. A **linear function in two variables** x and y is a function of the form $f(x, y) = ax + by + c$, where a, b, and c are real numbers. In other words, with each ordered pair (x, y) we associate a third number by the rule $ax + by + c$. For example, suppose the function f is described by $f(x, y) = 4x + 3y + 5$. Then $f(2, 1) = 4(2) + 3(1) + 5 = 16$.

First, let's take a look at some mathematical ideas that form the basis for solving a linear programming problem. Consider the shaded region in Figure 8.5 and the following linear functions in two variables.

$$f(x, y) = 4x + 3y + 5$$

$$f(x, y) = 2x + 7y - 1$$

$$f(x, y) = x - 2y$$

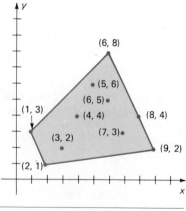

Figure 8.5

Suppose that we need to find the maximum and minimum value achieved by each of the functions in the indicated region. The following chart summarizes the values for the ordered pairs indicated in Figure 8.5.

	ORDERED PAIRS	VALUE OF $f(x, y) = 4x + 3y + 5$	VALUE OF $f(x, y) = 2x + 7y - 1$	VALUE OF $f(x, y) = x - 2y$
vertex	$(2, 1)$	16 (*minimum*)	10 (*minimum*)	0
	$(3, 2)$	23	19	-1
vertex	$(9, 2)$	47	31	5 (*maximum*)
vertex	$(1, 3)$	18	22	-5
	$(7, 3)$	42	34	1
	$(4, 4)$	33	35	-4
	$(8, 4)$	49	43	0
	$(6, 5)$	44	46	-4
	$(5, 6)$	43	51	-7
vertex	$(6, 8)$	53 (*maximum*)	67 (*maximum*)	-10 (*minimum*)

Notice that for each function, the maximum and minimum values are obtained at vertices of the region.

We claim that, for linear functions, maximum and minimum functional values are *always* obtained at vertices of the region. To substantiate this, let's consider the family of lines $x - 2y = k$, where k is an arbitrary constant. (We are now working only with the function $f(x, y) = x - 2y$.) In slope–intercept form, $x - 2y = k$ becomes $y = \frac{1}{2}x - \frac{1}{2}k$; so we have a family of parallel lines each having a slope of $\frac{1}{2}$. In Figure 8.6 we have sketched some of these lines, so that each line has at least one point in common with the given region. Note that $x - 2y$ reaches a minimum value of -10 at the vertex $(6, 8)$ and a maximum value of 5 at the vertex $(9, 2)$.

In general, **suppose that f is a linear function in two variables x and y and that S is a region of the xy-plane. If f attains a maximum (minimum) value in S, then that maximum (minimum) value is obtained at a vertex of S.**

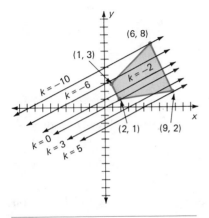

Figure 8.6

Remark: A subset of the xy-plane is said to be **bounded** if there is a circle that contains all of its points; otherwise the subset is said to be **unbounded**. A bounded set will contain maximum and minimum values for a function, but an unbounded set may not contain such values.

Now we will consider two examples that illustrate a general graphing approach to solving a linear programming problem in two variables. In the first example we will illustrate the general makeup of such a problem without the "story" from which the function and inequalities evolve. The second example will illustrate the type of setting from which such a problem arises.

EXAMPLE 4

Find the maximum value and the minimum value of the function $f(x, y) = 9x + 13y$ in the region determined by the following system of inequalities.

$$\begin{pmatrix} x \geq 0 \\ y \geq 0 \\ 2x + 3y \leq 18 \\ 2x + y \leq 10 \end{pmatrix}$$

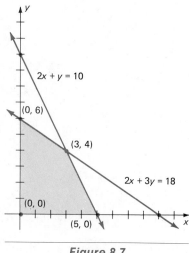

Figure 8.7

Solution First, let's graph the inequalities to determine the region, as indicated in Figure 8.7. (Such a region is called the **set of feasible solutions** and the inequalities are referred to as **constraints.**) The point $(3, 4)$ is determined by solving the system

$$\begin{pmatrix} 2x + 3y = 18 \\ 2x + y = 10 \end{pmatrix}.$$

Next we can determine the values of the given function at the vertices of the region. (Such a function to be maximized or minimized is called the **objective function.**) A minimum value of 0 is obtained at $(0, 0)$ and a maximum value of 79 is obtained at $(3, 4)$.

VERTICES	VALUE OF $f(x, y) = 9x + 13y$
$(0, 0)$	0 (*minimum*)
$(5, 0)$	45
$(3, 4)$	79 (*maximum*)
$(0, 6)$	78

EXAMPLE 5

A company that manufactures gidgets and gadgets has the following production information available:

1. To produce a gidget requires 3 hours of working time on machine A and 1 hour on machine B.

2. To produce a gadget requires 2 hours on machine A and 1 hour on machine B.

3. Machine A is available for no more than 120 hours per week and machine B is available for no more than 50 hours per week.

4. Gidgets can be sold at a profit of $3.75 each, while a profit of $3 each can be realized on a gadget.

How many gidgets and how many gadgets should be produced each week to maximize profit? What would the maximum profit be?

Solution Let x be the number of gidgets and y the number of gadgets. Thus, the profit function is $P(x, y) = 3.75x + 3y$. The constraints for the problem can be represented by the following inequalities.

$3x + 2y \leq 120$ Machine A is available for no more than 120 hours.

$x + y \leq 50$ Machine B is available for no more than 50 hours.

$x \geq 0$
$y \geq 0$ The number of gidgets and gadgets must be represented by a nonnegative number.

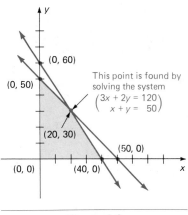

Figure 8.8

When we graph these inequalities, we obtain the set of feasible solutions indicated by the shaded region in Figure 8.8. Next we find the value of the profit function at the vertices; this produces the following chart.

VERTICES	VALUE OF $P(x, y) = 3.75x + 3y$
$(0, 0)$	0
$(40, 0)$	150
$(20, 30)$	165 (*maximum*)
$(0, 50)$	150

Thus, a maximum profit of \$165 is realized by producing 20 gidgets and 30 gadgets.

Problem Set 8.4

Indicate the solution set for each of the following systems of inequalities by graphing the system and shading the appropriate region.

1. $\begin{pmatrix} x + y > 3 \\ x - y > 1 \end{pmatrix}$
2. $\begin{pmatrix} x - y < 2 \\ x + y < 1 \end{pmatrix}$
3. $\begin{pmatrix} x - 2y \le 4 \\ x + 2y > 4 \end{pmatrix}$

4. $\begin{pmatrix} 3x - y > 6 \\ 2x + y \le 4 \end{pmatrix}$
5. $\begin{pmatrix} 2x + 3y \le 6 \\ 3x - 2y \le 6 \end{pmatrix}$
6. $\begin{pmatrix} 4x + 3y \ge 12 \\ 3x - 4y \ge 12 \end{pmatrix}$

7. $\begin{pmatrix} 2x - y \ge 4 \\ x + 3y < 3 \end{pmatrix}$
8. $\begin{pmatrix} 3x - y < 3 \\ x + y \ge 1 \end{pmatrix}$
9. $\begin{pmatrix} x + 2y > -2 \\ x - y < -3 \end{pmatrix}$

10. $\begin{pmatrix} x - 3y < -3 \\ 2x - 3y > -6 \end{pmatrix}$
11. $\begin{pmatrix} y > x - 4 \\ y < x \end{pmatrix}$
12. $\begin{pmatrix} y \le x + 2 \\ y \ge x \end{pmatrix}$

13. $\begin{pmatrix} x - y > 2 \\ x - y > -1 \end{pmatrix}$
14. $\begin{pmatrix} x + y > 1 \\ x + y > 3 \end{pmatrix}$
15. $\begin{pmatrix} y \ge x \\ x > -1 \end{pmatrix}$

16. $\begin{pmatrix} y < x \\ y \leq 2 \end{pmatrix}$

17. $\begin{pmatrix} y < x \\ y > x + 3 \end{pmatrix}$

18. $\begin{pmatrix} x \leq 3 \\ y \leq -1 \end{pmatrix}$

19. $\begin{pmatrix} y > -2 \\ x > 1 \end{pmatrix}$

20. $\begin{pmatrix} x + 2y > 4 \\ x + 2y < 2 \end{pmatrix}$

21. $\begin{pmatrix} x \geq 0 \\ y \geq 0 \\ x + y \leq 4 \\ 2x + y \leq 6 \end{pmatrix}$

22. $\begin{pmatrix} x \geq 0 \\ y \geq 0 \\ x - y \leq 5 \\ 4x + 7y \leq 28 \end{pmatrix}$

23. $\begin{pmatrix} x \geq 0 \\ y \geq 0 \\ 2x + y \leq 4 \\ 2x - 3y \leq 6 \end{pmatrix}$

24. $\begin{pmatrix} x \geq 0 \\ y \geq 0 \\ 3x + 5y \geq 15 \\ 5x + 3y \geq 15 \end{pmatrix}$

For Problems 25–28, find the maximum and minimum values of the given function in the indicated region.

25. $f(x, y) = 3x + 5y$

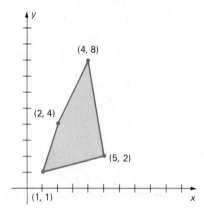

(4, 8)

(2, 4)

(5, 2)

(1, 1)

26. $f(x, y) = 8x + 3y$

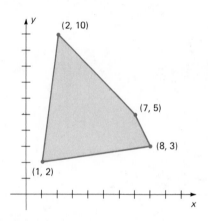

(2, 10)

(7, 5)

(8, 3)

(1, 2)

27. $f(x, y) = x + 4y$

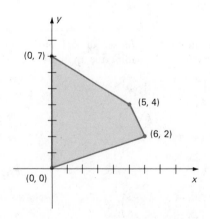

(0, 7)

(5, 4)

(6, 2)

(0, 0)

28. $f(x, y) = 2.5x + 3.5y$

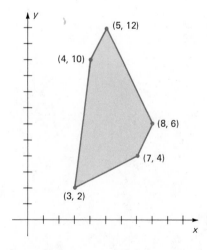

(5, 12)

(4, 10)

(8, 6)

(7, 4)

(3, 2)

29. Maximize the function $f(x, y) = 3x + 7y$ in the region determined by the following constraints.

$$3x + 2y \leq 18$$
$$3x + 4y \geq 12$$
$$x \geq 0$$
$$y \geq 0$$

30. Maximize the function $f(x, y) = 1.5x + 2y$ in the region determined by the following constraints.

$$3x + 2y \leq 36$$
$$3x + 10y \leq 60$$
$$x \geq 0$$
$$y \geq 0$$

31. Maximize the function $f(x, y) = 40x + 55y$ in the region determined by the following constraints.

$$2x + y \leq 10$$
$$x + y \leq 7$$
$$2x + 3y \leq 18$$
$$x \geq 0$$
$$y \geq 0$$

32. Maximize the function $f(x, y) = 0.08x + 0.09y$ in the region determined by the following constraints.

$$x + y \leq 8000$$
$$y \leq \tfrac{1}{3}x$$
$$y \geq 500$$
$$x \leq 7000$$
$$x \geq 0$$

33. Minimize the function $f(x, y) = 0.2x + 0.5y$ in the region determined by the following constraints.

$$2x + y \geq 12$$
$$2x + 5y \geq 20$$
$$x \geq 0$$
$$y \geq 0$$

34. Minimize the function $f(x, y) = 3x + 7y$ in the region determined by the following constraints.

$$x + y \geq 9$$
$$6x + 11y \geq 84$$
$$x \geq 0$$
$$y \geq 0$$

35. Maximize the function $f(x, y) = 9x + 2y$ in the region determined by the following constraints.

$$5y - 4x \leq 20$$
$$4x + 5y \leq 60$$
$$x \geq 0$$
$$x \leq 10$$
$$y \geq 0$$

36. Maximize the function $f(x, y) = 3x + 4y$ in the region determined by the following constraints.

$$2y - x \leq 6$$
$$x + y \leq 12$$
$$x \geq 2$$
$$x \leq 8$$
$$y \geq 0$$

Solve each of the following linear programming problems by using the graphing method illustrated in Example 5.

37. Suppose that an investor wants to invest up to $10,000. She plans to buy one speculative type of stock and one conservative type. The speculative stock is paying a 12% return and the conservative stock is paying a 9% return. She has decided to invest at least $2000 in the conservative stock and no more than $6000 in the speculative stock. Furthermore, she does not want the speculative investment to exceed the conservative one. How much should she invest at each rate to maximize her return?

38. A manufacturer of golf clubs makes a profit of $50 per set on a model A set and $45 per set on a model B set. Daily production of the model A clubs is between 30 and 50 sets, inclusive, and that of the model B clubs is between 10 and 20 sets, inclusive. The total daily production is not to exceed 50 sets. How many sets of each model should be manufactured per day to maximize the profit?

39. A company makes two types of calculators. Type A sells for $12 and type B sells for $10. It costs the company $9 to produce one type A calculator and $8 to produce one type B calculator. In one month, the company is equipped to produce between 200 and 300, inclusive, of the type A calculator and between 100 and 250, inclusive, of the type B calculator, but not more than 500 altogether. How many calculators of each type should be produced per month to maximize the difference between the total selling price and the total cost of production?

40. A manufacturer of small copiers makes a profit of $200 on a deluxe model and $250 on a standard model. The company wants to produce at least 50 deluxe models per week and at least 75 standard models per week. However, the weekly production is not to exceed 150 copiers. How many copiers of each kind should be produced in order to maximize the profit?

41. Products A and B are produced by a company according to the following production information.
   (a) To produce one unit of product A requires 1 hour of working time on machine I, 2 hours on machine II, and 1 hour on machine III.
   (b) To produce one unit of product B requires 1 hour of working time on machine I, one hour on machine II, and 3 hours on machine III.

(c) Machine I is available for no more than 40 hours per week, machine II is available for no more than 40 hours per week, and machine III for no more than 60 hours per week.

(d) Product A can be sold at a profit of $2.75 per unit and product B at a profit of $3.50 per unit.

How many units each of product A and product B should be produced per week to maximize profit?

42. Suppose that the company referred to in Example 5 also manufactures widgets and wadgets and has the following production information available.

(a) To produce a widget requires 4 hours of working time on machine A and 2 hours on machine B.

(b) To produce a wadget requires 5 hours of working time on machine A and 5 hours on machine B.

(c) Machine A is available for no more than 200 hours in a month and machine B is available for no more than 150 hours per month.

(d) Widgets can be sold at a profit of $7 each and wadgets at a profit of $8 each. How many widgets and wadgets should be produced per month in order to maximize profit?

Chapter 8 Summary

Be sure that you understand the following ideas pertaining to the algebra of matrices.

1. Matrices of the same dimension are added by adding elements in corresponding positions.

2. Matrix addition is a commutative and an associative operation.

3. Matrices of any specific dimension have an additive identity element, namely, the matrix of that same dimension containing all zeros.

4. Every matrix A has an additive inverse, A^{-1}, that can be found by changing the sign of each element of A.

5. Matrices of the same dimension can be subtracted by the definition $A - B = A + (-B)$.

6. The scalar product of a real number k and a matrix A can be found by multiplying each element of A by k.

7. The following properties hold for scalar multiplication and matrix addition:

$$k(A + B) = kA + kB$$
$$(k + l)A = kA + lA$$
$$(kl)A = k(lA)$$

8. If A is an $m \times n$ matrix and B is an $n \times p$ matrix, then the product AB is an $n \times p$ matrix. The general term, c_{ij}, of the product matrix $C = AB$ is determined by the equation

$$c_{ij} = a_{i1}b_{1j} + a_{i2}b_{2j} + \cdots + a_{in}b_{nj}.$$

9. Matrix multiplication is not a commutative operation, but it is an associative operation.

10. Matrix multiplication has two distributive properties:

$$A(B + C) = AB + AC \quad \text{and} \quad (A + B)C = AC + BC.$$

11. The general multiplicative identity element, I_n, for square $n \times n$ matrices contains only 1's in the main diagonal and 0's elsewhere. For example,

$$I_2 = \begin{bmatrix} 1 & 0 \\ 0 & 1 \end{bmatrix} \quad \text{and} \quad I_3 = \begin{bmatrix} 1 & 0 & 0 \\ 0 & 1 & 0 \\ 0 & 0 & 1 \end{bmatrix}.$$

12. If a square matrix A has a multiplicative inverse A^{-1}, then $AA^{-1} = A^{-1}A = I_n$.

13. The multiplicative inverse of the 2×2 matrix

$$A = \begin{bmatrix} a_{11} & a_{12} \\ a_{21} & a_{22} \end{bmatrix}$$

is

$$A^{-1} = \frac{1}{|A|} \begin{bmatrix} a_{22} & -a_{12} \\ -a_{21} & a_{11} \end{bmatrix}$$

for $|A| \neq 0$. If $|A| = 0$, then the matrix A has no inverse.

14. When the inverse of a square matrix exists, a general technique for finding it is described on page 368.

15. The solution set of a system of n linear equations in n variables can be found by multiplying the inverse of the coefficient matrix times the column matrix consisting of the constant terms. For example, the solution set of the system

$$\begin{pmatrix} 2x + 3y - z = 4 \\ 3x - y + 2z = 5 \\ 5x - 7y - 4z = -1 \end{pmatrix}$$

can be found by the product

$$\begin{bmatrix} 2 & 3 & -1 \\ 3 & -1 & 2 \\ 5 & -7 & -4 \end{bmatrix}^{-1} \begin{bmatrix} 4 \\ 5 \\ -1 \end{bmatrix}.$$

The solution set of a system of linear inequalities is the intersection of the solution sets of the individual inequalities. Such solution sets are easily determined by the graphing approach.

Linear programming problems deal with the idea of maximizing or minimizing a certain linear function that is subject to various constraints. The constraints are expressed as linear inequalities. Look over Examples 4 and 5 of Section 8.4 to help summarize the general approach to linear programming problems in this chapter.

Chapter 8 Review Problem Set

For Problems 1–10, compute the indicated matrix, if it exists, using the following matrices.

$$A = \begin{bmatrix} 2 & -4 \\ -3 & 8 \end{bmatrix}, \quad B = \begin{bmatrix} 5 & -1 \\ 0 & 2 \end{bmatrix}, \quad C = \begin{bmatrix} 3 & -1 \\ -2 & 4 \\ 5 & -6 \end{bmatrix}$$

$$D = \begin{bmatrix} -2 & -1 & 4 \\ 5 & 0 & -3 \end{bmatrix}, \quad E = \begin{bmatrix} 1 \\ -3 \\ -7 \end{bmatrix}, \quad F = \begin{bmatrix} 1 & -2 \\ 4 & -4 \\ 7 & -8 \end{bmatrix}$$

1. $A + B$

2. $B - A$

3. $C - F$

4. $2A + 3B$

5. $3C - 2F$

6. CD

7. DC

8. $DC + AB$

9. DE

10. EF

11. Use A and B from the preceding problems and show that $AB \neq BA$.

12. Use C, D, and F from the preceding problems and show that $D(C + F) = DC + DF$.

13. Use C, D, and F from the preceding problems and show that $(C + F)D = CD + FD$.

For each of the matrices in Problems 14–23, find the multiplicative inverse, if it exists.

14. $\begin{bmatrix} 9 & 5 \\ 7 & 4 \end{bmatrix}$

15. $\begin{bmatrix} 9 & 4 \\ 7 & 3 \end{bmatrix}$

16. $\begin{bmatrix} -2 & 1 \\ 2 & 3 \end{bmatrix}$

17. $\begin{bmatrix} 4 & -6 \\ 2 & -3 \end{bmatrix}$

18. $\begin{bmatrix} -1 & -3 \\ -4 & -5 \end{bmatrix}$

19. $\begin{bmatrix} 0 & -3 \\ 7 & 6 \end{bmatrix}$

20. $\begin{bmatrix} 1 & -2 & 1 \\ 2 & -5 & 2 \\ -3 & 7 & 5 \end{bmatrix}$

21. $\begin{bmatrix} 1 & 3 & -2 \\ 4 & 13 & -7 \\ 5 & 16 & -8 \end{bmatrix}$

22. $\begin{bmatrix} -2 & 4 & 7 \\ 1 & -3 & 5 \\ 1 & -5 & 22 \end{bmatrix}$

23. $\begin{bmatrix} -1 & 2 & 3 \\ 2 & -5 & -7 \\ -3 & 5 & 11 \end{bmatrix}$

For Problems 24–28, use the multiplicative inverse matrix approach to solve each of the systems. The required inverses have been found in Problems 14–23.

24. $\begin{pmatrix} 9x + 5y = 12 \\ 7x + 4y = 10 \end{pmatrix}$

25. $\begin{pmatrix} -2x + y = -9 \\ 2x + 3y = 5 \end{pmatrix}$

26. $\begin{pmatrix} x - 2y + z = 7 \\ 2x - 5y + 2z = 17 \\ -3x + 7y + 5z = -32 \end{pmatrix}$

27. $\begin{pmatrix} x + 3y - 2z = -7 \\ 4x + 13y - 7z = -21 \\ 5x + 16y - 8z = -23 \end{pmatrix}$

28. $\begin{pmatrix} -x + 2y + 3z = 22 \\ 2x - 5y - 7z = -51 \\ -3x + 5y + 11z = 71 \end{pmatrix}$

For Problems 29–32, indicate the solution set for each of the systems of linear inequalities by graphing the system and shading the appropriate region.

29. $\begin{pmatrix} 3x - 4y \geq 0 \\ 2x + 3y \leq 0 \end{pmatrix}$

30. $\begin{pmatrix} 3x - 2y < 6 \\ 2x - 3y < 6 \end{pmatrix}$

31. $\begin{pmatrix} x - 4y < 4 \\ 2x + y \geq 2 \end{pmatrix}$

32. $\begin{pmatrix} x \geq 0 \\ y \geq 0 \\ x + 2y \leq 4 \\ 2x - y \leq 4 \end{pmatrix}$

33. Maximize the function $f(x, y) = 8x + 5y$ in the region determined by the following constraints.

$$y \le 4x$$
$$x + y \le 5$$
$$x \ge 0$$
$$y \ge 0$$
$$x \le 4$$

34. Maximize the function $f(x, y) = 2x + 7y$ in the region determined by the following constraints.

$$x \ge 0$$
$$y \ge 0$$
$$x + 2y \le 16$$
$$x + y \le 9$$
$$3x + 2y \le 24$$

35. Maximize the function $f(x, y) = 7x + 5y$ in the region determined by the constraints of Problem 34.

36. Maximize the function $f(x, y) = 150x + 200y$ in the region determined by the constraints of Problem 34.

37. A manufacturer of electric ice cream freezers makes a profit of $4.50 on a one-gallon freezer and a profit of $5.25 on a two-gallon freezer. The company wants to produce at least 75 one-gallon and at least 100 two-gallon freezers per week. However, the weekly production is not to exceed a total of 250 freezers. How many freezers of each type should be produced per week in order to maximize the profit?

Conic Sections

9

9.1 Parabolas

9.2 Ellipses

9.3 Hyperbolas

9.4 Systems Involving Nonlinear Equations

Parabolas, circles, ellipses, and hyperbolas can be formed by intersecting a right circular conical surface with a plane, as shown in Figure 9.1. Hence, these figures are often referred to as **conic sections**.

The conic sections are not new to you. You did some graphing of circles, parabolas, ellipses, and hyperbolas in Chapters 3 and 4. However, at that time, except for the circle, we did not present any formal definitions or standard forms of equations. In Chapter 3 we developed the standard form for the equation of a circle, $(x - h)^2 + (y - k)^2 = r^2$. We used this equation to solve a variety of problems pertaining to circles. It is now time to study the other conic sections in the same manner. We will define each conic section and derive the standard form of an equation. We will then use the standard forms to study specific conic sections.

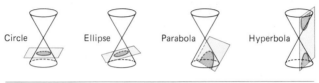

Figure 9.1

9.1

Parabolas

We discussed parabolas in Sections 4.2 and 4.3 as the graphs of quadratic functions. All parabolas in those sections had vertical lines as axes of symmetry. Furthermore, we did not state the definition for a parabola at that time. We shall now define a parabola and derive standard forms of equations for those that have either vertical or horizontal axes of symmetry.

DEFINITION 9.1

A **parabola** is the set of all points in a plane such that the distance of each point from a fixed point F (the **focus**) is equal to its distance from a fixed line d (the **directrix**) in the plane.

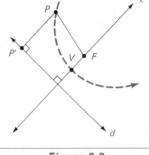

Figure 9.2

Using Definition 9.1, we can sketch a parabola by starting with a fixed line d and a fixed point F, not on d. Then a point P is on the parabola if and only if $PF = PP'$, where PP' is perpendicular to d (Figure 9.2). The dashed curved line in Figure 9.2 indicates the possible positions of P; it is the parabola. The line l, through F and perpendicular to the directrix d, is called the **axis of symmetry**. The point V, on the axis of symmetry halfway from F to the directrix d, is the **vertex** of the parabola.

We can derive a standard form for the equation of a parabola by coordinatizing the plane so that the origin is at the vertex of the parabola and the y-axis is the axis of symmetry (Figure 9.3). If the focus is at $(0, p)$, where $p \neq 0$, then the equation of the directrix is $y = -p$. Therefore, for any point P on the parabola, $PF = PP'$, and using the distance formula we obtain

$$\sqrt{(x - 0)^2 + (y - p)^2} = \sqrt{(x - x)^2 + (y + p)^2}.$$

Squaring both sides and simplifying produces

$$(x - 0)^2 + (y - p)^2 = (x - x)^2 + (y + p)^2$$
$$x^2 + y^2 - 2py + p^2 = y^2 + 2py + p^2$$
$$x^2 = 4py.$$

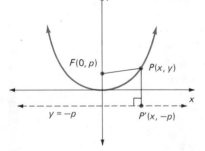

Figure 9.3

Thus, the **standard form for the equation of a parabola**, with its vertex at the origin and having the y-axis as its axis of symmetry, is

$$x^2 = 4py.$$

If $p > 0$ the parabola opens upward, and if $p < 0$ the parabola opens downward.

In Figure 9.4 the line segment $\overline{QP}$ is called the **latus rectum**. It contains the focus and is parallel to the directrix. Since $FP = PP' = |2p|$, the entire length of the latus rectum is $|4p|$ units. You will see in a moment how this fact can be used when graphing parabolas.

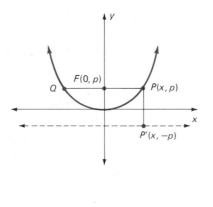

Figure 9.4

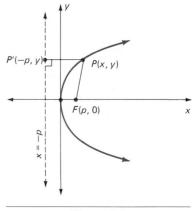

Figure 9.5

In a similar fashion, we can develop the standard form for the equation of a parabola with its vertex at the origin and having the x-axis as its axis of symmetry. By choosing a focus at $F(p, 0)$ and a directrix with an equation of $x = -p$ (see Figure 9.5), and by applying the definition of a parabola, we obtain the standard form for the equation:

$$y^2 = 4px.$$

If $p > 0$ the parabola opens to the right, as in Figure 9.5, and if $p < 0$ it opens to the left.

The concept of symmetry can be used to decide which of the two equations, $x^2 = 4py$ or $y^2 = 4px$, is to be used. The graph of $x^2 = 4py$ is symmetric with respect to the y-axis, since replacing x with $-x$ does not change the equation. Likewise, the graph of $y^2 = 4px$ is symmetric with respect to the x-axis, since replacing y with $-y$ leaves the equation unchanged.

The following property summarizes our previous discussion.

PROPERTY 9.1

The graph of each of the following equations is a parabola that has its vertex at the origin and has the indicated focus, directrix, and symmetry:

1. $x^2 = 4py$:   focus $(0, p)$, directrix $y = -p$, y-axis symmetry;

2. $y^2 = 4px$:   focus $(p, 0)$, directrix $x = -p$, x-axis symmetry.

Now let's illustrate some uses of the equations $x^2 = 4py$ and $y^2 = 4px$.

EXAMPLE 1

Find the focus and directrix of the parabola $x^2 = -8y$ and sketch its graph.

Solution Comparing $x^2 = -8y$ to the standard form $x^2 = 4py$, we have $4p = -8$. Therefore $p = -2$ and the parabola opens downward. The focus is

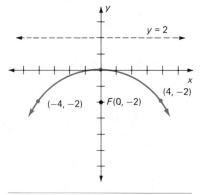

Figure 9.6

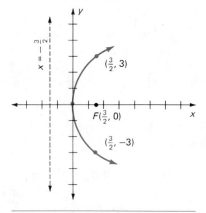

Figure 9.7

at $(0, -2)$ and the equation of the directrix is $y = -(-2) = 2$. The latus rectus is $|4p| = |-8| = 8$ units long. Therefore, the endpoints of the latus rectum are at $(4, -2)$ and $(-4, -2)$. The graph is sketched in Figure 9.6.

EXAMPLE 2

Write the equation of the parabola that is symmetric with respect to the y-axis, has its vertex at the origin, and contains the point $P(6, 3)$.

Solution   The standard form of the parabola is $x^2 = 4py$. Since P is on the parabola, the ordered pair $(6, 3)$ must satisfy the equation. Therefore,

$$6^2 = 4p(3)$$
$$36 = 12p$$
$$3 = p.$$

If $p = 3$, the equation becomes

$$x^2 = 4(3)y$$
$$x^2 = 12y.$$

EXAMPLE 3

Find the focus and directrix of the parabola $y^2 = 6x$ and sketch its graph.

Solution   Comparing $y^2 = 6x$ to the standard form $y^2 = 4px$, we see that $4p = 6$, and therefore $p = \frac{3}{2}$. So the focus is at $(\frac{3}{2}, 0)$ and the equation of the directrix is $x = -\frac{3}{2}$. The parabola opens to the right. The latus rectum is $|4p| = |6| = 6$ units long. Thus, the endpoints of the latus rectum are at $(\frac{3}{2}, 3)$ and $(\frac{3}{2}, -3)$. The graph is sketched in Figure 9.7.

Other Parabolas

In much the same way we can develop the standard form for an equation of a parabola that is symmetric with respect to a line parallel to a coordinate axis. In Figure 9.8 we have taken the vertex V at (h, k) and the focus F at $(h, k + p)$; the equation of the directrix is $y = k - p$. By the definition of a parabola, we know that $FP = PP'$. Therefore, applying the distance formula, we obtain

$$\sqrt{(x - h)^2 + (y - (k + p))^2} = \sqrt{(x - x)^2 + (y - (k - p))^2}.$$

We leave it to the reader to show that this equation simplifies to

$$(x - h)^2 = 4p(y - k),$$

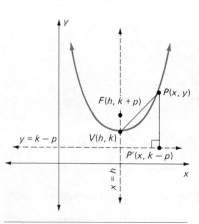

Figure 9.8

which is called the standard form for the equation of a parabola with its vertex at (h, k) and symmetric with respect to the line $x = h$. If $p > 0$ the parabola opens upward, and if $p < 0$ the parabola opens downward.

In a similar fashion, we can show that the standard form for the equation of a parabola, with its vertex at (h, k) and symmetric with respect to the line

$y = k$, is

$$(y - k)^2 = 4p(x - h).$$

If $p > 0$ the parabola opens to the right, and if $p < 0$ it opens to the left.

Let's summarize our discussion of parabolas having lines of symmetry parallel to the x-axis or y-axis by stating the following property.

PROPERTY 9.2

The graph of each of the following equations is a parabola that has its vertex at (h, k) and has the indicated focus, directrix, and symmetry:

1. $(x - h)^2 = 4p(y - k)$:   focus $(h, k + p)$, directrix $y = k - p$, line of symmetry $x = h$;

2. $(y - k)^2 = 4p(x - h)$:   focus $(h + p, k)$, directrix $x = h - p$, line of symmetry $y = k$.

EXAMPLE 4

Find the vertex, focus, and directrix of the parabola $y^2 + 4y - 4x + 16 = 0$, and sketch its graph.

Solution Writing the given equation as $y^2 + 4y = 4x - 16$, we can complete the square on the left side by adding 4 to both sides.

$$y^2 + 4y + 4 = 4x - 16 + 4$$
$$(y + 2)^2 = 4x - 12$$
$$(y + 2)^2 = 4(x - 3)$$

Now let's compare this final equation to the form $(y - k)^2 = 4p(x - h)$.

$$(y - (-2))^2 = 4(x - 3)$$

$$k = -2 \qquad \begin{array}{c} 4p = 4 \\ p = 1 \end{array} \qquad h = 3$$

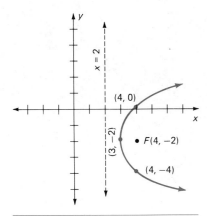

Figure 9.9

The vertex is at $(3, -2)$ and because $p > 0$, the parabola opens to the right and the focus is at $(4, -2)$. The equation of the directrix is $x = 2$. The latus rectum is $|4p| = |4| = 4$ units long and its endpoints are at $(4, 0)$ and $(4, -4)$. The graph is sketched in Figure 9.9.

EXAMPLE 5

Write the equation of the parabola if its focus is at $(-4, 1)$ and the equation of its directrix is $y = 5$.

Solution Since the directrix is a horizontal line, we know that the equation of the parabola is of the form $(x - h)^2 = 4p(y - k)$. The vertex is halfway between the focus and directrix, so the vertex is at $(-4, 3)$. This means that $h = -4$ and

$k = 3$. The parabola opens downward because the focus is below the directrix and the distance between the focus and the vertex is 2 units; so $p = -2$. Substituting -4 for h, 3 for k, and -2 for p in the equation $(x - h)^2 = 4p(y - k)$, we obtain

$$(x - (-4))^2 = 4(-2)(y - 3).$$

This simplifies to

$$(x + 4)^2 = -8(y - 3)$$
$$x^2 + 8x + 16 = -8y + 24$$
$$x^2 + 8x + 8y - 8 = 0.$$

Remark: For a problem such as Example 5, you may find it helpful to put the given information on a set of axes and draw a rough sketch of the parabola to help you with the analysis of the problem.

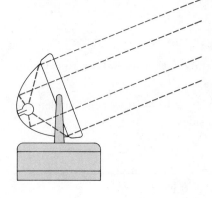

Parabolas possess various properties that make them very useful. For example, if a parabola is rotated about its axis, a parabolic surface is formed. The rays from a source of light placed at the focus of the surface will reflect from the surface parallel to the axis. It is for this reason that parabolic reflectors are used on searchlights. Likewise, rays of light coming into a parabolic surface parallel to the axis will be reflected through the focus. This property of parabolas is useful in the design of mirrors for telescopes and in the construction of radar antennae (see figures at left and below).

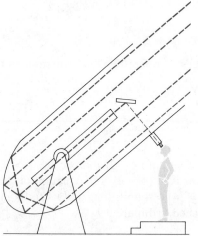

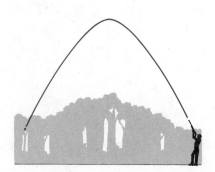

A bullet fired into the air will follow the curvature of a parabola if air resistance and any other outside factors are ignored or, in other words, if only the force of gravity is considered (see figure above).

Different types of bridges have constructions based on the concept of a parabola. One example is shown in the figure for Problem 43 of the next set of problems.

Problem Set 9.1

For Problems 1–22, find the vertex, focus, and directrix of the given parabola and sketch its graph.

1. $y^2 = 8x$ **2.** $y^2 = -4x$ **3.** $x^2 = -12y$

4. $x^2 = 8y$ **5.** $y^2 = -2x$ **6.** $y^2 = 6x$

7. $x^2 = 6y$ **8.** $x^2 = -7y$ **9.** $x^2 - 4y + 8 = 0$

10. $x^2 - 8y - 24 = 0$ **11.** $x^2 + 8y + 16 = 0$ **12.** $x^2 + 4y - 4 = 0$

13. $y^2 - 12x + 24 = 0$ **14.** $y^2 + 8x - 24 = 0$

15. $x^2 - 2x - 4y + 9 = 0$ **16.** $x^2 + 4x - 8y - 4 = 0$

17. $x^2 + 6x + 8y + 1 = 0$ **18.** $x^2 - 4x + 4y - 4 = 0$

19. $y^2 - 2y + 12x - 35 = 0$ **20.** $y^2 + 4y + 8x - 4 = 0$

21. $y^2 + 6y - 4x + 1 = 0$ **22.** $y^2 - 6y - 12x + 21 = 0$

For Problems 23–42, find an equation of the parabola that satisfies the given conditions.

23. Focus $(0, 3)$, directrix $y = -3$ **24.** Focus $(0, -\frac{1}{2})$, directrix $y = \frac{1}{2}$

25. Focus $(-1, 0)$, directrix $x = 1$ **26.** Focus $(5, 0)$, directrix $x = 1$

27. Focus $(0, 1)$, directrix $y = 7$ **28.** Focus $(0, -2)$, directrix $y = -10$

29. Focus $(3, 4)$, directrix $y = -2$ **30.** Focus $(-3, -1)$, directrix $y = 7$

31. Focus $(-4, 5)$, directrix $x = 0$ **32.** Focus $(5, -2)$, directrix $x = -1$

33. Vertex $(0, 0)$, symmetric with respect to the x-axis, and contains the point $(-3, 5)$

34. Vertex $(0, 0)$, symmetric with respect to the y-axis, and contains the point $(-2, -4)$

35. Vertex $(0, 0)$, focus $(\frac{5}{2}, 0)$

36. Vertex $(0, 0)$, focus $(0, -\frac{7}{2})$

37. Vertex $(7, 3)$, focus $(7, 5)$, and symmetric with respect to the line $x = 7$

38. Vertex $(-4, -6)$, focus $(-7, -6)$, and symmetric with respect to the line $y = -6$

39. Vertex $(8, -3)$, focus $(11, -3)$, and symmetric with respect to the line $y = -3$

40. Vertex $(-2, 9)$, focus $(-2, 5)$, and symmetric with respect to the line $x = -2$

41. Vertex $(-9, 1)$, symmetric with respect to the line $x = -9$, and contains the point $(-8, 0)$

42. Vertex $(6, -4)$, symmetric with respect to the line $y = -4$, and contains the point $(8, -3)$

Miscellaneous Problems

43. One section of a suspension bridge is suspended between two towers that are 40 feet high above the surface and 300 feet apart (see the accompanying figure). A cable strung between the tops of the two towers is in the shape of a parabola with its vertex 10 feet above the surface. With axes drawn as indicated in the figure, find the equation of the parabola.

44. In the figure for Problem 43, suppose that 5 equally spaced vertical cables are used to support the bridge. Find the total length of these supports.

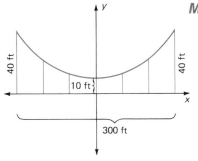

Figure for Problem 43

9.2

Ellipses

Let's begin by defining the concept of an ellipse.

DEFINITION 9.2

> An **ellipse** is the set of all points in a plane such that the sum of the distances of each point from two fixed points F and F' (the **foci**) in the plane is constant.

Using two thumbtacks, a piece of string, and a pencil, it is easy to draw an ellipse by satisfying the conditions of Definition 9.2. First, insert two thumbtacks in a piece of cardboard at points F and F' and fasten the ends of the piece of string to the thumbtacks, as in Figure 9.10. Then loop the string around the point of a pencil and hold the pencil so that the string is taut. Finally, move the pencil around the tacks, always keeping the string taut: you will draw an ellipse. The two points F and F' are the foci referred to in Definition 9.2, and the sum of the distances FP and $F'P$ is constant, since it represents the length of the piece of string. With the same piece of string, you can vary the shape of the ellipse by changing the positions of the foci. By moving F and F' further apart, the ellipse will become flatter. Likewise, by moving F and F' closer together, the ellipse will more resemble a circle. In fact, if $F = F'$, you will obtain a circle.

We can derive a standard form for the equation of an ellipse by coordinatizing the plane so that the foci are on the x-axis and equidistant from the origin (Figure 9.11). If F has coordinates $(c, 0)$, where $c > 0$, then F' has coordinates $(-c, 0)$, and the distance between F and F' is $2c$ units. We will let $2a$ represent the constant sum of $FP + F'P$. Note that $2a > 2c$ and therefore $a > c$. For any point P on the ellipse,

$$FP + F'P = 2a.$$

Using the distance formula, we can write this as

$$\sqrt{(x - c)^2 + (y - 0)^2} + \sqrt{(x + c)^2 + (y - 0)^2} = 2a.$$

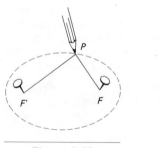

Figure 9.10

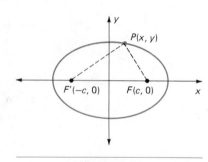

Figure 9.11

Let's change the form of this equation to

$$\sqrt{(x - c)^2 + y^2} = 2a - \sqrt{(x + c)^2 + y^2}$$

and square both sides:

$$(x - c)^2 + y^2 = 4a^2 - 4a\sqrt{(x + c)^2 + y^2} + (x + c)^2 + y^2.$$

This can be simplified to

$$a^2 + cx = a\sqrt{(x + c)^2 + y^2}.$$

Again squaring both sides produces

$$a^4 + 2a^2cx + c^2x^2 = a^2[(x + c)^2 + y^2],$$

which can be rewritten in the form

$$x^2(a^2 - c^2) + a^2y^2 = a^2(a^2 - c^2).$$

Dividing both sides by $a^2(a^2 - c^2)$ leads to the form

$$\frac{x^2}{a^2} + \frac{y^2}{a^2 - c^2} = 1.$$

Letting $b^2 = a^2 - c^2$, where $b > 0$, produces the equation

$$\frac{x^2}{a^2} + \frac{y^2}{b^2} = 1. \tag{1}$$

Since $c > 0$, $a > c$, and $b^2 = a^2 - c^2$, it follows that $a^2 > b^2$ and hence $a > b$. This equation that we have derived is called the **standard form for the equation of an ellipse** with its foci on the x-axis and its center at the origin.

The x-intercepts of equation (1) can be found by letting $y = 0$. Doing this produces $x^2/a^2 = 1$ or $x^2 = a^2$; consequently, the x-intercepts are a and $-a$. The corresponding points on the graph (See Figure 9.12) are $A(a, 0)$ and $A'(-a, 0)$, and the line segment $\overline{A'A}$, which is of length $2a$, is called the **major axis** of the ellipse. The endpoints of the major axis are also referred to as the **vertices** of the ellipse. Similarly, letting $x = 0$ produces $y^2/b^2 = 1$ or $y^2 = b^2$; consequently the y-intercepts are b and $-b$. The corresponding points on the graph are $B(0, b)$ and $B'(0, -b)$, and the line segment $\overline{BB'}$, which is of length $2b$, is called the **minor axis**. Since $a > b$, **the major axis is always longer than the minor axis**. The point of intersection of the major and minor axis is called the **center** of the ellipse.

Let's summarize this discussion by stating the following property.

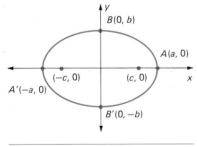

Figure 9.12

PROPERTY 9.3

> The graph of the equation
>
> $$\frac{x^2}{a^2} + \frac{y^2}{b^2} = 1,$$
>
> for $a^2 > b^2$, is an ellipse with the endpoints of its major axis (the vertices) at $(a, 0)$ and $(-a, 0)$, and the endpoints of its minor axis at $(0, b)$ and $(0, -b)$. The foci are at $(c, 0)$ and $(-c, 0)$, where $c^2 = a^2 - b^2$.

Notice that replacing y with $-y$, or x with $-x$, or both x and y with $-x$ and $-y$, leaves the equation unchanged. Thus the graph of

$$\frac{x^2}{a^2} + \frac{y^2}{b^2} = 1$$

is symmetric with respect to the x-axis, the y-axis, and the origin.

EXAMPLE 1

Find the vertices, the endpoints of the minor axis, and the foci, and sketch the ellipse $4x^2 + 9y^2 = 36$.

Solution The given equation can be changed to standard form by dividing both sides by 36.

$$\frac{4x^2}{36} + \frac{9y^2}{36} = \frac{36}{36}$$

$$\frac{x^2}{9} + \frac{y^2}{4} = 1$$

Therefore, $a^2 = 9$ and $b^2 = 4$; hence, the vertices are at $(3, 0)$ and $(-3, 0)$, and the ends of the minor axis are at $(0, 2)$ and $(0, -2)$. Since $c^2 = a^2 - b^2$, we have

$$c^2 = 9 - 4 = 5.$$

So the foci are at $(\sqrt{5}, 0)$ and $(-\sqrt{5}, 0)$. The ellipse is sketched in Figure 9.13.

> **Remark:** For a problem such as Example 1, it is not necessary to change to standard form to find the values for a and b. After all, $\pm a$ are the x-intercepts and $\pm b$ are the y-intercepts. These values can be found quite easily from the given form of the equation.

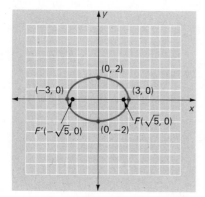

Figure 9.13

EXAMPLE 2

Find the equation of the ellipse with vertices at $(\pm 6, 0)$ and foci at $(\pm 4, 0)$.

Solution From the given information we know that $a = 6$ and $c = 4$. Therefore,

$$b^2 = a^2 - c^2 = 36 - 16 = 20.$$

Substituting 36 for a^2 and 20 for b^2 in the standard form produces

$$\frac{x^2}{36} + \frac{y^2}{20} = 1.$$

Multiplying both sides by 180 leads to

$$5x^2 + 9y^2 = 180.$$

Ellipses with Foci on y-Axis

We can develop a standard form for the equation of an ellipse with foci on the y-axis in a similar fashion. The following property summarizes the results of such a development with the foci at $(0, c)$ and $(0, -c)$, where $c > 0$.

PROPERTY 9.4

The graph of the equation

$$\frac{x^2}{b^2} + \frac{y^2}{a^2} = 1,$$

for $a^2 > b^2$, is an ellipse with the endpoints of its major axis (vertices) at $(0, a)$ and $(0, -a)$, and the endpoints of its minor axis at $(b, 0)$ and $(-b, 0)$. The foci are at $(0, c)$ and $(0, -c)$, where $c^2 = a^2 - b^2$.

From Properties 9.3 and 9.4 it is evident that an equation of an ellipse with its center at the origin and foci on a coordinate axis can be written in the form

$$\frac{x^2}{p} + \frac{y^2}{q} = 1 \qquad \text{or} \qquad qx^2 + py^2 = pq,$$

where p and q are positive. If $p > q$ the major axis lies on the x-axis, but if $q > p$ the major axis is on the y-axis. It is not necessary to memorize these facts, since for any specific problem the endpoints of the major and minor axes are determined by the x- and y-intercepts. However, it is necessary to remember the relationship $c^2 = a^2 - b^2$.

EXAMPLE 3

Find the vertices, the endpoints of the minor axis, and the foci, and sketch the ellipse $18x^2 + 4y^2 = 36$.

Solution To find the x-intercepts we let $y = 0$, obtaining

$$18x^2 = 36$$
$$x^2 = 2$$
$$x = \pm\sqrt{2}.$$

Similarly, to find the y-intercepts we let $x = 0$, obtaining

$$4y^2 = 36$$
$$y^2 = 9$$
$$y = \pm 3.$$

Since $3 > \sqrt{2}$, we know that $a = 3$ and $b = \sqrt{2}$. Therefore, the vertices are at $(0, 3)$ and $(0, -3)$, and the endpoints of the minor axis are at $(\sqrt{2}, 0)$ and $(-\sqrt{2}, 0)$. From the relationship $c^2 = a^2 - b^2$ we get

$$c^2 = 9 - 2 = 7.$$

So the foci are at $(0, \sqrt{7})$ and $(0, -\sqrt{7})$. The ellipse is sketched in Figure 9.14.

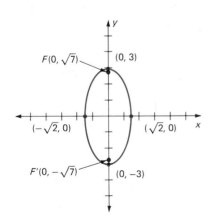

Figure 9.14

Other Ellipses

In the same way, we can develop the standard form for an equation of an ellipse that is symmetric with respect to a line parallel to a coordinate axis. We will not show such developments in this text, but Figures 9.15 and 9.16 indicate the basic facts needed to develop and use the resulting equations. Notice that in each case, the center of the ellipse is at a point (h, k). Furthermore, the physical significance of a, b, and c is the same as before. However, these values are used relative to the center (h, k) to find the endpoints of the major and minor axis, and the foci. Let's see how this works in a specific example.

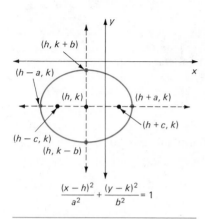

$$\frac{(x - h)^2}{a^2} + \frac{(y - k)^2}{b^2} = 1$$

Figure 9.15

EXAMPLE 4

Find the vertices, the endpoints of the minor axis, and the foci, and sketch the ellipse $9x^2 + 54x + 4y^2 - 8y + 49 = 0$.

Solution First, we need to change to a standard form by completing the square on both x and y.

$$9(x^2 + 6x + \underline{\quad}) + 4(y^2 - 2y + \underline{\quad}) = -49$$
$$9(x^2 + 6x + 9) + 4(y^2 - 2y + 1) = -49 + 81 + 4$$
$$9(x + 3)^2 + 4(y - 1)^2 = 36$$
$$\frac{(x + 3)^2}{4} + \frac{(y - 1)^2}{9} = 1$$

Since $a > b$, this last equation is of the form

$$\frac{(x - h)^2}{b^2} + \frac{(y - k)^2}{a^2} = 1,$$

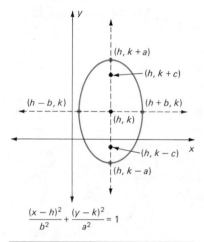

$$\frac{(x - h)^2}{b^2} + \frac{(y - k)^2}{a^2} = 1$$

Figure 9.16

where $h = -3$, $k = 1$, $a = 3$, and $b = 2$. Thus, the endpoints of the major axis (vertices) are up 3 units and down 3 units from the center, $(-3, 1)$, so they are at $(-3, 4)$ and $(-3, -2)$. Likewise, the endpoints of the minor axis are 2 units to the right and 2 units to the left of the center. Thus, they are at $(-1, 1)$ and $(-5, 1)$. From the relationship $c^2 = a^2 - b^2$, we get

$$c^2 = 9 - 4 = 5.$$

Thus, the foci are at $(-3, 1 + \sqrt{5})$ and $(-3, 1 - \sqrt{5})$. The ellipse is sketched in Figure 9.17.

EXAMPLE 5

Write the equation of the ellipse that has vertices at $(-3, -5)$ and $(7, -5)$, and foci at $(-1, -5)$ and $(5, -5)$.

Solution Since the vertices and foci are on the same horizontal line $(y = -5)$, this ellipse has an equation of the form

$$\frac{(x - h)^2}{a^2} + \frac{(y - k)^2}{b^2} = 1.$$

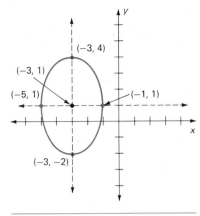

Figure 9.17

The center of the ellipse is at the midpoint of the major axis. Therefore,

$$h = \frac{-3 + 7}{2} = 2 \quad \text{and} \quad k = \frac{-5 + (-5)}{2} = -5.$$

The distance between the center $(2, -5)$ and a vertex $(7, -5)$ is 5 units; thus, $a = 5$. The distance between the center $(2, -5)$ and a focus $(5, -5)$ is 3 units; thus, $c = 3$. Using the relationship $c^2 = a^2 - b^2$, we obtain

$$b^2 = a^2 - c^2 = 25 - 9 = 16.$$

Now let's substitute 2 for h, -5 for k, 25 for a^2, and 16 for b^2 in the general form, and then we can simplify.

$$\frac{(x - 2)^2}{25} + \frac{(y + 5)^2}{16} = 1$$
$$16(x - 2)^2 + 25(y + 5)^2 = 400$$
$$16(x^2 - 4x + 4) + 25(y^2 + 10y + 25) = 400$$
$$16x^2 - 64x + 64 + 25y^2 + 250y + 625 = 400$$
$$16x^2 - 64x + 25y^2 + 250y + 289 = 0$$

As with parabolas, ellipses also possess properties that make them very useful. For example, the elliptical surface formed by rotating an ellipse about its major axis has the following property: light or sound waves emitted at one focus will reflect off of the surface and converge at the other focus. This is the principle behind "whispering galleries," such as the Rotunda of the Capitol Building in Washington, D.C. In such buildings, two people standing at two specific spots that are the foci of the elliptical ceiling can whisper and yet hear each other clearly, even though they may be quite far apart.

Ellipses also play an important role in astronomy. Johannes Kepler (1571–1630) showed that the orbit of a planet is an ellipse with the sun at one focus. For example, the orbit of the earth is elliptical but nearly circular; at the same time, the moon moves about the earth in an elliptical path (see figure at left).

The arches for concrete bridges are sometimes elliptical. One example is shown in the figure for Problem 39 of the next set of problems. Also, elliptical gears are used in certain kinds of machinery that require a slow but powerful force at impact, such as a heavy-duty punch (see accompanying figure).

Problem Set 9.2

For Problems 1–22, find the vertices, the endpoints of minor axis, and the foci of the given ellipse, and sketch its graph.

1. $\dfrac{x^2}{4} + \dfrac{y^2}{1} = 1$

2. $\dfrac{x^2}{16} + \dfrac{y^2}{1} = 1$

3. $\dfrac{x^2}{4} + \dfrac{y^2}{9} = 1$

4. $\dfrac{x^2}{4} + \dfrac{y^2}{16} = 1$

5. $9x^2 + 3y^2 = 27$

6. $4x^2 + 3y^2 = 36$

7. $2x^2 + 5y^2 = 50$

8. $5x^2 + 36y^2 = 180$

9. $12x^2 + y^2 = 36$

10. $8x^2 + y^2 = 16$

11. $7x^2 + 11y^2 = 77$

12. $4x^2 + y^2 = 12$

13. $4x^2 - 8x + 9y^2 - 36y + 4 = 0$

14. $x^2 + 6x + 9y^2 - 36y + 36 = 0$

15. $4x^2 + 16x + y^2 + 2y + 1 = 0$

16. $9x^2 - 36x + 4y^2 + 16y + 16 = 0$

17. $x^2 - 6x + 4y^2 + 5 = 0$

18. $16x^2 + 9y^2 + 36y - 108 = 0$

19. $9x^2 - 72x + 2y^2 + 4y + 128 = 0$

20. $5x^2 + 10x + 16y^2 + 160y + 325 = 0$

21. $2x^2 + 12x + 11y^2 - 88y + 172 = 0$

22. $9x^2 + 72x + y^2 + 6y + 135 = 0$

For Problems 23–36, find an equation of the ellipse that satisfies the given conditions.

23. Vertices $(\pm 5, 0)$, foci $(\pm 3, 0)$

24. Vertices $(\pm 4, 0)$, foci $(\pm 2, 0)$

25. Vertices $(0, \pm 6)$, foci $(0, \pm 5)$

26. Vertices $(0, \pm 3)$, foci $(0, \pm 2)$

27. Vertices $(\pm 3, 0)$, length of minor axis is 2

28. Vertices $(0, \pm 5)$, length of minor axis is 4

29. Foci $(0, \pm 2)$, length of minor axis is 3

30. Foci $(\pm 1, 0)$, length of minor axis is 2

31. Vertices $(0, \pm 5)$, contains the point $(3, 2)$

32. Vertices $(\pm 6, 0)$, contains the point $(5, 1)$

33. Vertices $(5, 1)$ and $(-3, 1)$, foci $(3, 1)$ and $(-1, 1)$

34. Vertices $(2, 4)$ and $(2, -6)$, foci $(2, 3)$ and $(2, -5)$

35. Center $(0, 1)$, one focus at $(-4, 1)$, length of minor axis is 6

36. Center $(3, 0)$, one focus at $(3, 2)$, length of minor axis is 4

37. Find an equation of the set of points in a plane such that the sum of the distances between each point of the set and the points $(2, 0)$ and $(-2, 0)$ is 8 units.

38. Find an equation of the set of points in a plane such that the sum of the distances between each point of the set and the points $(0, 3)$ and $(0, -3)$ is 10 units.

Miscellaneous Problems

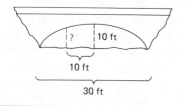

39. An arch of a bridge is semielliptical and the major axis is horizontal (see accompanying figure). The arch is 30 feet wide and 10 feet high. Find the height of the arch 10 feet from the center of the base.

40. In the figure for Problem 39, how much clearance is there 10 feet from the bank?

Figure for Problem 39

9.3

Hyperbolas

The definition of a hyperbola resembles that of an ellipse except the *difference* of distances from two fixed points is involved instead of the *sum* of distances.

DEFINITION 9.3

A **hyperbola** is the set of all points in a plane such that the difference of the distances of each point from two fixed points F and F' (the **foci**) in the plane is a positive constant.

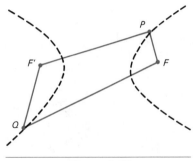

Figure 9.18

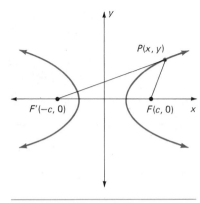

Figure 9.19

Using Definition 9.3, we can sketch a hyperbola by starting with two fixed points F and F' (see Figure 9.18). Then we locate all points P such that $PF' - PF$ is a positive constant. Likewise, all points Q are located so that $QF - QF'$ is the same positive constant. The two dashed curved lines in Figure 9.18 make up the hyperbola. The two curves are called the **branches** of the hyperbola.

To develop a standard form for the equation of a hyperbola, let's coordinatize the plane so that the foci are located at $F(c, 0)$ and $F'(-c, 0)$, as indicated in Figure 9.19. Using the distance formula and setting $2a$ equal to the difference of the distances from any point P on the hyperbola to the foci, we have the following equation.

$$\left| \sqrt{(x - c)^2 + (y - 0)^2} - \sqrt{(x + c)^2 + (y - 0)^2} \right| = 2a$$

(The absolute value sign is used to allow the point P to be on either branch of the hyperbola.) Using the same type of simplification procedure that we used for deriving the standard form for the equation of an ellipse, this equation simplifies to

$$\frac{x^2}{a^2} - \frac{y^2}{c^2 - a^2} = 1.$$

Letting $b^2 = c^2 - a^2$, where $b > 0$, we obtain the standard form

$$\frac{x^2}{a^2} - \frac{y^2}{b^2} = 1. \qquad (1)$$

Equation (1) indicates that this hyperbola is symmetric with respect to both axes and the origin. Furthermore, by letting $y = 0$, we obtain $x^2/a^2 = 1$ or $x^2 = a^2$, and therefore the x-intercepts are a and $-a$. The corresponding points $A(a, 0)$ and $A'(-a, 0)$ are the **vertices** of the hyperbola and the line segment $\overline{AA'}$ is called the **transverse axis**; it is of length $2a$ (see Figure 9.20). The midpoint of the transverse axis is called the **center** of the hyperbola; it is located at the origin. By letting $x = 0$ in equation (1), we obtain $-y^2/b^2 = 1$ or $y^2 = -b^2$. This implies that there are no y-intercepts, as indicated in Figure 9.20.

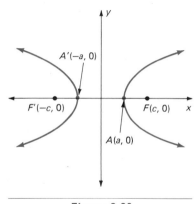

Figure 9.20

The following property summarizes the previous discussion.

PROPERTY 9.5

The graph of the equation

$$\frac{x^2}{a^2} - \frac{y^2}{b^2} = 1$$

is a hyperbola with vertices at $(a, 0)$ and $(-a, 0)$. The foci are at $(c, 0)$ and $(-c, 0)$, where $c^2 = a^2 + b^2$.

In conjunction with every hyperbola there are two intersecting lines that pass through the center of the hyperbola. These lines, referred to as *asymptotes*, are very helpful when sketching a hyperbola. Their equations are easily determined by using the following type of reasoning: Solving the equation

$$\frac{x^2}{a^2} - \frac{y^2}{b^2} = 1$$

for y produces $y = \pm\dfrac{b}{a}\sqrt{x^2 - a^2}$. From this form it is evident that there are no points on the graph for $x^2 - a^2 < 0$, that is, if $-a < x < a$. However, there are points on the graph if $x \geq a$ or $x \leq -a$. If $x \geq a$, then $y = \pm\dfrac{b}{a}\sqrt{x^2 - a^2}$ can be written

$$y = \pm\frac{b}{a}\sqrt{x^2\left(1 - \frac{a^2}{x^2}\right)}$$

$$= \pm\frac{b}{a}\sqrt{x^2}\sqrt{1 - \frac{a^2}{x^2}}$$

$$= \pm\frac{b}{a}x\sqrt{1 - \frac{a^2}{x^2}}.$$

Now suppose that we are going to determine some y-values for very large values of x. (Remember that a and b are arbitrary constants; they have specific values for a particular hyperbola.) When x is very large, $\dfrac{a^2}{x^2}$ will be close to zero, so the radicand will be close to 1. Therefore, the y-value will be close to either $\dfrac{b}{a}x$ or $-\dfrac{b}{a}x$. In other words, as x becomes larger and larger, the point $P(x, y)$ gets closer and closer to either the line $y = \dfrac{b}{a}x$ or the line $y = -\dfrac{b}{a}x$. A corresponding situation occurs when $x \leq a$. The lines with equations

$$y = \pm\frac{b}{a}x$$

are called the **asymptotes** of the hyperbola.

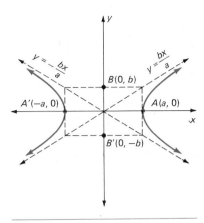

Figure 9.21

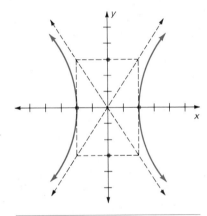

Figure 9.22

As we mentioned earlier, the asymptotes are very helpful for sketching hyperbolas. An easy way to sketch the asymptotes is first to plot the vertices $A(a, 0)$ and $A'(-a, 0)$, and the points $B(0, b)$ and $B'(0, -b)$, as in Figure 9.21. The line segment $\overline{BB'}$ is of length $2b$ and is called the **conjugate axis** of the hyperbola. The horizontal line segments drawn through B and B', together with the vertical line segments drawn through A and A', form a rectangle. The diagonals of this rectangle have slopes $\dfrac{b}{a}$ and $-\dfrac{b}{a}$. Therefore, by extending the diagonals, we obtain the asymptotes $y = \dfrac{b}{a}x$ and $y = -\dfrac{b}{a}x$. The two branches of the hyperbola can be sketched using the asymptotes as guidelines, as shown in Figure 9.21.

EXAMPLE 1

Find the vertices, the foci, and the equations of the asymptotes, and sketch the hyperbola $9x^2 - 4y^2 = 36$.

Solution Dividing both sides of the given equation by 36 and simplifying changes the equation to the standard form:

$$\frac{x^2}{4} - \frac{y^2}{9} = 1,$$

where $a^2 = 4$ and $b^2 = 9$. Hence, $a = 2$ and $b = 3$. The vertices are $(\pm 2, 0)$ and the endpoints of the conjugate axis are $(0, \pm 3)$; these points determine the rectangle whose diagonals extend to become the asymptotes. Using $a = 2$ and $b = 3$, the equations of the asymptotes are $y = \frac{3}{2}x$ and $y = -\frac{3}{2}x$. Then, using the relationship $c^2 = a^2 + b^2$, we obtain $c^2 = 4 + 9 = 13$. So the foci are at $(\sqrt{13}, 0)$ and $(-\sqrt{13}, 0)$. Using the vertices and the asymptotes, the hyperbola has been sketched in Figure 9.22.

EXAMPLE 2

Find the equation of the hyperbola with vertices at $(\pm 4, 0)$ and foci at $(\pm 2\sqrt{5}, 0)$.

Solution From the given information we know that $a = 4$ and $c = 2\sqrt{5}$. Then, using the relationship $b^2 = c^2 - a^2$, we obtain

$$b^2 = (2\sqrt{5})^2 - 4^2 = 20 - 16 = 4.$$

Substituting 16 for a^2 and 4 for b^2 in the standard form produces

$$\frac{x^2}{16} - \frac{y^2}{4} = 1.$$

Multiplying both sides of this equation by 16 produces

$$x^2 - 4y^2 = 16.$$

Hyperbolas with Foci on the y-axis

In a similar fashion, we can develop a standard form for the equation of a hyperbola with foci on the y-axis. The following property summarizes the results of such a development, where the foci are at $(0, c)$ and $(0, -c)$.

PROPERTY 9.6

The graph of the equation

$$\frac{y^2}{a^2} - \frac{x^2}{b^2} = 1$$

is a hyperbola with vertices at $(0, a)$ and $(0, -a)$. The foci are at $(0, c)$ and $(0, -c)$, where $c^2 = a^2 + b^2$.

For this type of hyperbola, the endpoints of the conjugate axis are at $(b, 0)$ and $(-b, 0)$. In this case we can find the asymptotes by extending the diagonals of the rectangle determined by the horizontal lines through the vertices and the vertical lines through the endpoints of the conjugate axis. The slopes of these diagonals are a/b and $-a/b$; thus, the equations of these asymptotes are

$$y = \frac{a}{b}x \qquad \text{and} \qquad y = -\frac{a}{b}x.$$

EXAMPLE 3

Find the vertices, the foci, and the equations of the asymptotes, and sketch the hyperbola $4y^2 - x^2 = 12$.

Solution Dividing both sides of the given equation by 12 changes the equation to the standard form:

$$\frac{y^2}{3} - \frac{x^2}{12} = 1,$$

where $a^2 = 3$ and $b^2 = 12$. Hence, $a = \sqrt{3}$ and $b = 2\sqrt{3}$. The vertices, $(0, \pm\sqrt{3})$, and the endpoints of the conjugate axis, $(\pm 2\sqrt{3}, 0)$, determine the rectangle whose diagonals extend to become the asymptotes. Using $a = \sqrt{3}$ and $b = 2\sqrt{3}$, the equations of the asymptotes are $y = (\sqrt{3}/2\sqrt{3})x = \frac{1}{2}x$ and $y = -\frac{1}{2}x$. Then, using the relationship $c^2 = a^2 + b^2$, we obtain $c^2 = 3 + 12 = 15$, so the foci are at $(0, \sqrt{15})$ and $(0, -\sqrt{15})$. The hyperbola is sketched in Figure 9.23.

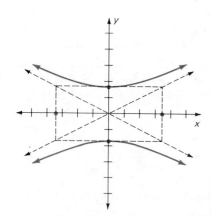

Figure 9.23

Other Hyperbolas

In the same way, we can develop the standard form for an equation of a hyperbola that is symmetric with respect to a line parallel to a coordinate axis. We will not show such developments in this text but will simply state and use the results.

$$\frac{(x-h)^2}{a^2} - \frac{(y-k)^2}{b^2} = 1$$

A hyperbola with center at (h, k) and transverse axis on the horizontal line $y = k$

$$\frac{(y-k)^2}{a^2} - \frac{(x-h)^2}{b^2} = 1$$

A hyperbola with center at (h, k) and transverse axis on the vertical line $x = h$

The relationship $c^2 = a^2 + b^2$ still holds and the physical significance of a, b, and c, remains the same. However, these values are used relative to the center (h, k) to find the endpoints of the transverse and conjugate axes, and the foci. Furthermore, the slopes of the asymptotes are as before, but these lines now contain the new center, (h, k). Let's see how all of this works in a specific example.

EXAMPLE 4

Find the vertices, the foci, the equations of the asymptotes, and sketch the hyperbola $9x^2 - 36x - 16y^2 + 96y - 252 = 0$.

Solution First, we need to change to a standard form by completing the square on both x and y.

$$9(x^2 - 4x + \_\_) - 16(y^2 - 6y + \_\_) = 252$$
$$9(x^2 - 4x + 4) - 16(y^2 - 6y + 9) = 252 + 36 - 144$$
$$9(x - 2)^2 - 16(y - 3)^2 = 144$$
$$\frac{(x-2)^2}{16} - \frac{(y-3)^2}{9} = 1$$

The center is at $(2, 3)$ and the transverse axis is on the line $y = 3$. Since $a^2 = 16$, we know that $a = 4$. Therefore, the vertices are 4 units to the right and 4 units to the left of the center, $(2, 3)$, so they are at $(6, 3)$ and $(-2, 3)$. Likewise, since $b^2 = 9$ or $b = 3$, the endpoints of the conjugate axis are 3 units up and 3 units down from the center, so they are at $(2, 6)$ and $(2, 0)$. Using $a = 4$ and $b = 3$, the slopes of the asymptotes are $\frac{3}{4}$ and $-\frac{3}{4}$. Then, using these slopes, the center $(2, 3)$, and the point–slope form for writing the equation of a line, the equations of the asymptotes can be determined to be $3x - 4y = -6$ and $3x + 4y = 18$. From the relationship $c^2 = a^2 + b^2$ we obtain $c^2 = 16 + 9 = 25$. Thus, the foci are at $(7, 3)$ and $(-3, 3)$. The hyperbola is sketched in Figure 9.24.

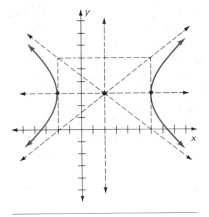

Figure 9.24

EXAMPLE 5

Find the equation of the hyperbola with vertices at $(-4, 2)$ and $(-4, -4)$, and with foci at $(-4, 3)$ and $(-4, -5)$.

Solution Since the vertices and foci are on the same vertical line ($x = -4$), this hyperbola has an equation of the form

$$\frac{(y-k)^2}{a^2} - \frac{(x-h)^2}{b^2} = 1.$$

The center of the hyperbola is at the midpoint of the transverse axis. Therefore,

$$h = \frac{-4 + (-4)}{2} = -4 \qquad \text{and} \qquad k = \frac{2 + (-4)}{2} = -1.$$

The distance between the center, $(-4, -1)$, and a vertex, $(-4, 2)$, is 3 units, so $a = 3$. The distance between the center, $(-4, -1)$, and a focus, $(-4, 3)$, is 4 units, so $c = 4$. Then, using the relationship $c^2 = a^2 + b^2$, we obtain

$$b^2 = c^2 - a^2 = 16 - 9 = 7.$$

Now we can substitute -4 for h, -1 for k, 9 for a^2, and 7 for b^2 in the general form and simplify.

$$\frac{(y + 1)^2}{9} - \frac{(x + 4)^2}{7} = 1$$

$$7(y + 1)^2 - 9(x + 4)^2 = 63$$
$$7(y^2 + 2y + 1) - 9(x^2 + 8x + 16) = 63$$
$$7y^2 + 14y + 7 - 9x^2 - 72x - 144 = 63$$
$$7y^2 + 14y - 9x^2 - 72x - 200 = 0$$

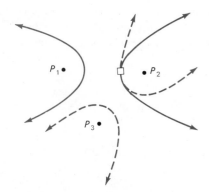

Figure 9.25

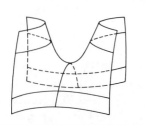

Figure 9.26

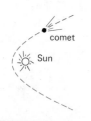

Figure 9.27

The hyperbola also has numerous applications, many of which you may not be aware. For example, one method of artillery range-finding is based on the concept of a hyperbola. If each of two listening posts, P_1 and P_2 in Figure 9.25, records the time that an artillery blast is heard, then the difference between the times multiplied by the speed of sound gives the difference of the distances of the gun from the two fixed points. Thus, the gun is located somewhere on the hyperbola whose foci are the two listening posts. Now by bringing in a third listening post, P_3, another hyperbola can be formed with foci at P_2 and P_3. Then the location of the gun must be at one of the intersections of the two hyperbolas.

This same principle of intersecting hyperbolas is used in a long-range navigation system known as LORAN. Radar stations serve as the foci of the hyperbolas and of course computers are used for the many calculations that are necessary to fix the location of a plane or a ship. At the present time, LORAN is probably used mostly for coastal navigation in connection with small pleasure boats.

Some rather unique architectural creations have used the concept of a hyperbolic paraboloid, pictured in Figure 9.26. For example, the TWA building at Kennedy Airport is so-designed. Some comets, upon entering the Sun's gravitational field, will follow a hyperbolic path, with the Sun as one of the foci (see Figure 9.27).

Problem Set 9.3

For Problems 1–22, find the vertices, the foci, and the equations of the asymptotes, and sketch each of the hyperbolas.

1. $\dfrac{x^2}{9} - \dfrac{y^2}{4} = 1$ **2.** $\dfrac{x}{4} - \dfrac{y^2}{16} = 1$ **3.** $\dfrac{y^2}{4} - \dfrac{x^2}{9} = 1$

4. $\dfrac{y^2}{16} - \dfrac{x^2}{4} = 1$ **5.** $9y^2 - 16x^2 = 144$ **6.** $4y^2 - x^2 = 4$

7. $x^2 - y^2 = 9$ **8.** $x^2 - y^2 = 1$ **9.** $5y^2 - x^2 = 25$

10. $y^2 - 2x^2 = 8$ **11.** $y^2 - 9x^2 = -9$ **12.** $16y^2 - x^2 = -16$

13. $4x^2 - 24x - 9y^2 - 18y - 9 = 0$ **14.** $9x^2 + 72x - 4y^2 - 16y + 92 = 0$

15. $y^2 - 4y - 4x^2 - 24x - 36 = 0$ **16.** $9y^2 + 54y - x^2 + 6x + 63 = 0$

17. $2x^2 - 8x - y^2 + 4 = 0$ **18.** $x^2 + 6x - 3y^2 = 0$

19. $y^2 + 10y - 9x^2 + 16 = 0$ **20.** $4y^2 - 16y - x^2 + 12 = 0$

21. $x^2 + 4x - y^2 - 4y - 1 = 0$ **22.** $y^2 + 8y - x^2 + 2x + 14 = 0$

For Problems 23–38, find an equation of the hyperbola that satisifies the given equations.

23. Vertices $(\pm 2, 0)$, foci $(\pm 3, 0)$ **24.** Vertices $(\pm 1, 0)$, foci $(\pm 4, 0)$

25. Vertices $(0, \pm 3)$, foci $(0, \pm 5)$ **26.** Vertices $(0, \pm 2)$, foci $(0, \pm 6)$

27. Vertices $(\pm 1, 0)$, contains the point $(2, 3)$

28. Vertices $(0, \pm 1)$, contains the point $(-3, 5)$

29. Vertices $(0, \pm\sqrt{3})$, length of conjugate axis is 4

30. Vertices $(\pm\sqrt{5}, 0)$, length of conjugate axis is 6

31. Foci $(\pm 3, 0)$, length of transverse axis is 8

32. Foci $(0, \pm 3\sqrt{2})$, length of conjugate axis is 4

33. Vertices $(6, -3)$ and $(2, -3)$, foci $(7, -3)$ and $(1, -3)$

34. Vertices $(-7, -4)$ and $(-5, -4)$, foci $(-8, -4)$ and $(-4, -4)$

35. Vertices $(-3, 7)$ and $(-3, 3)$, foci $(-3, 9)$ and $(-3, 1)$

36. Vertices $(7, 5)$ and $(7, -1)$, foci $(7, 7)$ and $(7, -3)$

37. Vertices $(0, 0)$ and $(4, 0)$, foci $(5, 0)$ and $(-1, 0)$

38. Vertices $(0, 0)$ and $(0, -6)$, foci $(0, 2)$ and $(0, -8)$

Miscellaneous Problems

For Problems 39–48, identify the graph of each of the equations as a straight line, a circle, a parabola, an ellipse, or a hyperbola. Do not sketch the graphs.

39. $x^2 - 7x + y^2 + 8y - 2 = 0$ **40.** $x^2 - 7x - y^2 + 8y - 2 = 0$

41. $5x - 7y = 9$ **42.** $4x^2 - x + y^2 + 2y - 3 = 0$

43. $10x^2 + y^2 = 8$ **44.** $-3x - 2y = 9$

45. $5x^2 + 3x - 2y^2 - 3y - 1 = 0$ **46.** $x^2 + y^2 - 3y - 6 = 0$

47. $x^2 - 3x + y - 4 = 0$ **48.** $5x + y^2 - 2y - 1 = 0$

9.4

Systems Involving Nonlinear Equations

In Chapters 7 and 8, we used several techniques to solve systems of linear equations. We will use two of those techniques in this section to solve some systems that contain at least one nonlinear equation. Furthermore, we will use our knowledge of graphing lines, circles, parabolas, ellipses, and hyperbolas to get a pictorial view of the systems. That will give us a basis for predicting approximate real-number solutions if there are any. In other words, we have once again arrived at a topic that vividly illustrates the merging of mathematical ideas. Let's begin by considering a system that contains one linear and one nonlinear equation.

EXAMPLE 1

Solve the system

$$\begin{pmatrix} x^2 + y^2 = 13 \\ 3x + 2y = 0 \end{pmatrix}.$$

Solution From our previous graphing experiences, we should recognize that $x^2 + y^2 = 13$ is a circle and $3x + 2y = 0$ is a straight line. Thus, the system can be pictured as in Figure 9.28. The graph indicates that the solution set of this system should consist of two ordered pairs of real numbers, representing the points of intersection in the second and fourth quadrants.

Now let's solve the system analytically by using the *substitution method.* Changing the form of $3x + 2y = 0$ to $y = -3x/2$ and then substituting $-3x/2$ for y in the other equation produces

$$x^2 + \left(-\frac{3x}{2}\right)^2 = 13.$$

This equation can now be solved for x.

$$x^2 + \frac{9x^2}{4} = 13$$

$$4x^2 + 9x^2 = 52$$

$$13x^2 = 52$$

$$x^2 = 4$$

$$x = \pm 2$$

Substituting 2 for x and then -2 for x in the second equation of the system produces two values for y:

$3x + 2y = 0$	$3x + 2y = 0$
$3(2) + 2y = 0$	$3(-2) + 2y = 0$
$2y = -6$	$2y = 6$
$y = -3$	$y = 3$

Therefore, the solution set of the system is $\{(2, -3),(-2, 3)\}$.

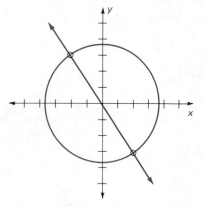

Figure 9.28

Remark: Don't forget that, as always, the solutions can be checked by substituting them back into the original equations. Graphing the system only permits you to approximate any possible real-number solutions before solving the system. Then, after solving the system, the graph can be used again to check the reasonableness of the answers.

EXAMPLE 2

Solve the system

$$\begin{pmatrix} x^2 + y^2 = 16 \\ y^2 - x^2 = 4 \end{pmatrix}.$$

Solution Graphing the system produces Figure 9.29. This figure indicates that there should be four ordered pairs of real numbers in the solution set of the system. Solving the system by using the *elimination method* works nicely. We can simply add the two equations, which eliminates the x's.

$$
\begin{aligned}
x^2 + y^2 &= 16 \\
-x^2 + y^2 &= 4 \\
\hline
2y^2 &= 20 \\
y^2 &= 10 \\
y &= \pm\sqrt{10}
\end{aligned}
$$

Substituting $\sqrt{10}$ for y in the first equation yields

$$
\begin{aligned}
x^2 + y^2 &= 16 \\
x^2 + (\sqrt{10})^2 &= 16 \\
x^2 + 10 &= 16 \\
x^2 &= 6 \\
x &= \pm\sqrt{6}.
\end{aligned}
$$

Thus, $(\sqrt{6}, \sqrt{10})$ and $(-\sqrt{6}, \sqrt{10})$ are solutions. Substituting $-\sqrt{10}$ for y in the first equation yields

$$
\begin{aligned}
x^2 + y^2 &= 16 \\
x^2 + (-\sqrt{10})^2 &= 16 \\
x^2 + 10 &= 16 \\
x^2 &= 6 \\
x &= \pm\sqrt{6}.
\end{aligned}
$$

Thus, $(\sqrt{6}, -\sqrt{10})$ and $(-\sqrt{6}, -\sqrt{10})$ are also solutions. The solution set is thus $\{(-\sqrt{6}, \sqrt{10}), (-\sqrt{6}, -\sqrt{10}), (\sqrt{6}, \sqrt{10}), (\sqrt{6}, -\sqrt{10})\}$.

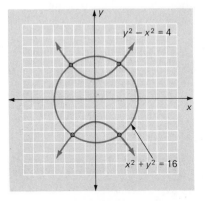

Figure 9.29

Sometimes a sketch of the graph of a system may not clearly indicate whether or not the system contains any real-number solutions. The next example illustrates such a situation.

EXAMPLE 3

Solve the system

$$\begin{pmatrix} y = x^2 + 2 \\ 3x - 2y = -2 \end{pmatrix}.$$

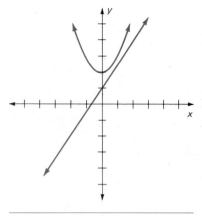

Figure 9.30

Solution From previous graphing experiences, we should recognize that $y = x^2 + 2$ represents a parabola and $3x - 2y = -2$ is a straight line (see Figure 9.30). Because of the close proximity of the curves, it is difficult to tell whether or not they intersect. In other words, the graph does not definitely indicate any real-number solutions to the system.

Let's solve the system using the substitution method. Substituting $x^2 + 2$ for y in the second equation produces two values for x.

$$3x - 2(x^2 + 2) = -2$$
$$3x - 2x^2 - 4 = -2$$
$$-2x^2 + 3x - 2 = 0$$
$$2x^2 - 3x + 2 = 0$$

$$x = \frac{3 \pm \sqrt{9 - 16}}{4}$$

$$x = \frac{3 \pm i\sqrt{7}}{4}$$

It is now obvious that the system has no real-number solutions. The line and the parabola do not intersect in the real-number plane. However, there will be two pairs of complex numbers in the solution set. Substituting $(3 + i\sqrt{7})/4$ for x in the first equation yields

$$y = \left(\frac{3 + i\sqrt{7}}{4}\right)^2 + 2$$

$$= \frac{2 + 6i\sqrt{7}}{16} + 2$$

$$= \frac{17 + 3i\sqrt{7}}{8}.$$

Likewise, substituting $(3 - i\sqrt{7})/8$ for x in the first equation yields

$$y = \left(\frac{3 - i\sqrt{7}}{4}\right)^2 + 2$$

$$= \frac{2 - 6i\sqrt{7}}{16} + 2$$

$$= \frac{17 - 3i\sqrt{7}}{8}.$$

The solution set of the given system is

$$\left\{\left(\frac{3 + i\sqrt{7}}{4}, \frac{17 + 3i\sqrt{7}}{8}\right), \left(\frac{3 - i\sqrt{7}}{4}, \frac{17 - 3i\sqrt{7}}{8}\right)\right\}.$$

Problem Set 9.4

For each of the following systems, **(a)** graph the system so that approximate real-number solutions (if there are any) can be predicted, and **(b)** solve the system by the substitution or elimination method.

1. $\begin{pmatrix} x^2 + y^2 = 5 \\ x + 2y = 5 \end{pmatrix}$

2. $\begin{pmatrix} x^2 + y^2 = 13 \\ 2x + 3y = 13 \end{pmatrix}$

3. $\begin{pmatrix} x^2 + y^2 = 26 \\ x + y = -4 \end{pmatrix}$

4. $\begin{pmatrix} x^2 + y^2 = 10 \\ x + y = -2 \end{pmatrix}$

5. $\begin{pmatrix} x^2 + y^2 = 2 \\ x - y = 4 \end{pmatrix}$

6. $\begin{pmatrix} x^2 + y^2 = 3 \\ x - y = -5 \end{pmatrix}$

7. $\begin{pmatrix} y = x^2 + 6x + 7 \\ 2x + y = -5 \end{pmatrix}$

8. $\begin{pmatrix} y = x^2 - 4x + 5 \\ y - x = 1 \end{pmatrix}$

9. $\begin{pmatrix} 2x + y = -2 \\ y = x^2 + 4x + 7 \end{pmatrix}$

10. $\begin{pmatrix} 2x + y = 0 \\ y = -x^2 + 2x - 4 \end{pmatrix}$

11. $\begin{pmatrix} y = x^2 - 3 \\ x + y = -4 \end{pmatrix}$

12. $\begin{pmatrix} y = -x^2 + 1 \\ x + y = 2 \end{pmatrix}$

13. $\begin{pmatrix} x^2 + 2y^2 = 9 \\ x - 4y = -9 \end{pmatrix}$

14. $\begin{pmatrix} 2x - y = 7 \\ 3x^2 + y^2 = 21 \end{pmatrix}$

15. $\begin{pmatrix} x + y = -3 \\ x^2 + 2y^2 - 12y - 18 = 0 \end{pmatrix}$

16. $\begin{pmatrix} 4x^2 + 9y^2 = 25 \\ 2x + 3y = 7 \end{pmatrix}$

17. $\begin{pmatrix} x - y = 2 \\ x^2 - y^2 = 16 \end{pmatrix}$

18. $\begin{pmatrix} x^2 - 4y^2 = 16 \\ 2y - x = 2 \end{pmatrix}$

19. $\begin{pmatrix} y = -x^2 + 3 \\ y = x^2 + 1 \end{pmatrix}$

20. $\begin{pmatrix} y = x^2 \\ y = x^2 - 4x + 4 \end{pmatrix}$

21. $\begin{pmatrix} y = x^2 + 2x - 1 \\ y = x^2 + 4x + 5 \end{pmatrix}$

22. $\begin{pmatrix} y = -x^2 + 1 \\ y = x^2 - 2 \end{pmatrix}$

23. $\begin{pmatrix} x^2 - y^2 = 4 \\ x^2 + y^2 = 4 \end{pmatrix}$

24. $\begin{pmatrix} 2x^2 + y^2 = 8 \\ x^2 + y^2 = 4 \end{pmatrix}$

25. $\begin{pmatrix} 8y^2 - 9x^2 = 6 \\ 8x^2 - 3y^2 = 7 \end{pmatrix}$

26. $\begin{pmatrix} 2x^2 + y^2 = 11 \\ x^2 - y^2 = 4 \end{pmatrix}$

27. $\begin{pmatrix} 2x^2 - 3y^2 = -1 \\ 2x^2 + 3y^2 = 5 \end{pmatrix}$

28. $\begin{pmatrix} 4x^2 + 3y^2 = 9 \\ y^2 - 4x^2 = 7 \end{pmatrix}$

Miscellaneous Problems

29. **(a)** Solve the system $\begin{pmatrix} 7x^2 + 8y^2 = 36 \\ 11x^2 + 5y^2 = -4 \end{pmatrix}$.

   (b) What happens if you try to graph the system?

30. For what value(s) of k will the line $x + y = k$ touch the ellipse $x^2 + 2y^2 = 6$ in one and only one point?

31. The system

$$\begin{pmatrix} x^2 - 6x + y^2 - 4y + 4 = 0 \\ x^2 - 4x + y^2 + 8y - 5 = 0 \end{pmatrix}$$

represents two circles that intersect in two points. An equivalent system can be formed by replacing the second equation with the result of adding -1 times the first equation to the second equation. Thus, we obtain the system

$$\begin{pmatrix} x^2 - 6x + y^2 - 4y + 4 = 0 \\ 2x + 12y - 9 = 0 \end{pmatrix}.$$

Explain why the linear equation in this system is the equation of the common chord of the original two intersecting circles.

Chapter 9 Summary

The following standard forms for the equations of conic sections were developed in this chapter.

Parabolas

$x^2 = 4py$:   focus $(0, p)$, directrix $y = -p$, y-axis symmetry

$y^2 = 4px$:   focus $(p, 0)$, directrix $x = -p$, x-axis symmetry

$(x - h)^2 = 4p(y - k)$:   focus $(h, k + p)$, directrix $y = k - p$, symmetric with respect to the line $x = h$

$(y - k)^2 = 4p(x - h)$:   focus $(h + p, k)$, directrix $x = h - p$, symmetric with respect to the line $y = k$

Ellipses

$\dfrac{x^2}{a^2} + \dfrac{y^2}{b^2} = 1$:   center $(0, 0)$, vertices $(\pm a, 0)$, endpoints of minor axis $(0, \pm b)$, foci $(\pm c, 0)$, $c^2 = a^2 - b^2$, $a^2 > b^2$

$\dfrac{x^2}{b^2} + \dfrac{y^2}{a^2} = 1$:   center $(0, 0)$, vertices $(0, \pm a)$, endpoints of minor axis $(\pm b, 0)$, foci $(0, \pm c)$, $c^2 = a^2 - b^2$, $a^2 > b^2$

$\dfrac{(x - h)^2}{a^2} + \dfrac{(y - k)^2}{b^2} = 1$:   center (h, k), vertices $(h \pm a, k)$, endpoints of minor axis $(h, k \pm b)$, foci $(h \pm c, k)$, $c^2 = a^2 - b^2$, $a^2 > b^2$

$\dfrac{(x - h)^2}{b^2} + \dfrac{(y - k)^2}{a^2} = 1$:   center (h, k), vertices $(h, k \pm a)$, endpoints of minor axis $(h \pm b, k)$, foci $(h, k \pm c)$, $c^2 = a^2 - b^2$, $a^2 > b^2$

Hyperbolas

$\dfrac{x^2}{a^2} - \dfrac{y^2}{b^2} = 1$:   center $(0, 0)$, vertices $(\pm a, 0)$, endpoints of conjugate axis $(0, \pm b)$, foci $(\pm c, 0)$, $c^2 = a^2 + b^2$, asymptotes $y = \pm \dfrac{b}{a} x$

$\dfrac{y^2}{a^2} - \dfrac{x^2}{b^2} = 1$:   center $(0, 0)$ vertices $(0, \pm a)$, endpoints of conjugate axis $(\pm b, 0)$, foci $(0, \pm c)$, $c^2 = a^2 + b^2$, asymptotes $y = \pm \dfrac{a}{b} x$

$\dfrac{(x - h)^2}{a^2} - \dfrac{(y - k)^2}{b^2} = 1$:   center (h, k), vertices $(h \pm a, k)$, endpoints of conjugate axis $(h, k \pm b)$, foci $(h \pm c, k)$, $c^2 = a^2 + b^2$, asymptotes $y - k = \pm \dfrac{b}{a}(x - h)$

$\dfrac{(y - k)^2}{a^2} - \dfrac{(x - h)^2}{b^2} = 1$:   center (h, k), vertices $(h, k \pm a)$, endpoints of conjugate axis $(h \pm b, k)$, foci $(h, k \pm c)$, $c^2 = a^2 + b^2$, asymptotes $y - k = \pm \dfrac{a}{b}(x - h)$

Systems containing at least one nonlinear equation can often be solved by substitution or by the elimination method. Graphing the system will often provide a basis for predicting approximate real-number solutions, if there are any.

Chapter 9 Review Problem Set

For Problems 1–12, **(a)** identify the conic section as a parabola, an ellipse, or a hyperbola, **(b)** if it is a parabola, find its vertex, focus, and directrix; if it is an ellipse, find its vertices, endpoints of minor axis, and foci; if it is a hyperbola, find its vertices, endpoints of conjugate axis, foci, and asymptotes, and **(c)** sketch each of the curves.

1. $x^2 + 2y^2 = 32$ **2.** $y^2 = -12x$ **3.** $3y^2 - x^2 = 9$

4. $2x^2 - 3y^2 = 18$ **5.** $5x^2 + 2y^2 = 20$ **6.** $x^2 = 2y$

7. $x^2 - 8x - 2y^2 + 4y + 10 = 0$ **8.** $9x^2 - 54x + 2y^2 + 8y + 71 = 0$

9. $y^2 - 2y + 4x + 9 = 0$ **10.** $x^2 + 2x + 8y + 25 = 0$

11. $x^2 + 10x + 4y^2 - 16y + 25 = 0$ **12.** $3y^2 + 12y - 2x^2 - 8x - 8 = 0$

For Problems 13–24, find the equation of the indicated conic section satisfying the given conditions.

13. Parabola with vertex $(0,0)$, focus $(-5,0)$, directrix $x = 5$

14. Ellipse with vertices $(0, \pm 4)$, foci $(0, \pm\sqrt{15})$

15. Hyperbola with vertices $(\pm\sqrt{2}, 0)$, length of conjugate axis 10

16. Ellipse with vertices $(\pm 2, 0)$, contains the point $(1, -2)$

17. Parabola with vertex $(0,0)$, symmetric with respect to the y-axis, contains the point $(2, 6)$

18. Hyperbola with vertices $(0, \pm 1)$, foci $(0, \pm\sqrt{10})$

19. Ellipse with vertices $(6, 1)$ and $(6, 7)$, length of minor axis 2 units

20. Parabola with vertex $(4, -2)$, focus $(6, -2)$

21. Hyperbola with vertices $(-5, -3)$ and $(-5, -5)$, foci $(-5, -2)$ and $(-5, -6)$

22. Parabola with vertex $(-6, -3)$, symmetric with respect to the line $x = -6$, contains the point $(-5, -2)$

23. Ellipse with endpoints of minor axis $(-5, 2)$ and $(-5, -2)$, length of major axis 10 units

24. Hyperbola with vertices $(2, 0)$ and $(6, 0)$, length of conjugate axis 8 units

For Problems 25–30, **(a)** graph the system, and **(b)** solve the system by using the substitution or elimination method.

25. $\begin{pmatrix} x^2 + y^2 = 17 \\ x - 4y = -17 \end{pmatrix}$ **26.** $\begin{pmatrix} x^2 - y^2 = 8 \\ 3x - y = 8 \end{pmatrix}$ **27.** $\begin{pmatrix} x - y = 1 \\ y = x^2 + 4x + 1 \end{pmatrix}$

28. $\begin{pmatrix} 4x^2 - y^2 = 16 \\ 9x^2 + 9y^2 = 16 \end{pmatrix}$ **29.** $\begin{pmatrix} x^2 + 2y^2 = 8 \\ 2x^2 + 3y^2 = 12 \end{pmatrix}$ **30.** $\begin{pmatrix} y^2 - x^2 = 1 \\ 4x^2 + y^2 = 4 \end{pmatrix}$

10

Sequences and Mathematical Induction

10.1   Arithmetic Sequences

10.2   Geometric Sequences

10.3   Another Look at Problem Solving

10.4   Mathematical Induction

Suppose that an auditorium has 35 seats in the first row, 40 seats in the second row, 45 seats in the third row, and so on for 10 rows. The numbers 35, 40, 45, 50, ..., 80 represent the number of seats per row from row 1 through row 10. This list of numbers has a constant difference of 5 between any two successive numbers in the list; such a list is called an **arithmetic sequence**.

Suppose that a fungus culture growing under controlled conditions doubles in size each day. If today the size of the culture is 6 units, then the numbers 12, 24, 48, 96, 192 represent the size of the culture for the next five days. In this list of numbers, each number after the first is two times the previous number; such a list is called a **geometric sequence**. Arithmetic sequences and geometric sequences will be at the center of our attention in this chapter.

10.1

Arithmetic Sequences

An **infinite sequence** is a function whose domain is the set of positive integers. For example, consider the function defined by the equation

$$f(n) = 5n + 1,$$

where the domain is the set of positive integers. If we substitute the numbers of the domain in order, starting with 1, we can list the resulting ordered pairs

$$(1, 6), \quad (2, 11), \quad (3, 16), \quad (4, 21), \quad (5, 26),$$

and so on. However, since we know we are using the domain of positive integers in order, starting with 1, there is no need to use ordered pairs. We can simply express the infinite sequence as

$$6, 11, 16, 21, 26, \ldots.$$

Frequently the letter a is used to represent sequential functions and the functional value of a at n is written a_n (read "a sub n") instead of $a(n)$. The sequence is then expressed

$$a_1, a_2, a_3, a_4, \ldots,$$

where a_1 is the **first term**, a_2 is the **second term**, a_3 is the **third term**, and so on. The expression a_n, which defines the sequence, is called the **general term** of the sequence. Knowing the general term of a sequence allows us to find as many terms of the sequence as needed and also to find any specific terms. Consider the following example.

EXAMPLE 1

Find the first five terms of the sequence for which $a_n = 2n^2 - 3$; find the twentieth term.

Solution The first five terms are generated by replacing n with 1, 2, 3, 4, and 5.

$$a_1 = 2(1)^2 - 3 = -1, \qquad a_2 = 2(2)^2 - 3 = 5$$
$$a_3 = 2(3)^2 - 3 = 15, \qquad a_4 = 2(4)^2 - 3 = 29$$
$$a_5 = 2(5)^2 - 3 = 47$$

The first five terms are thus $-1, 5, 15, 29,$ and 47. The twentieth term is

$$a_{20} = 2(20)^2 - 3 = 797.$$

Arithmetic Sequences

An **arithmetic sequence** (also called an arithmetic progression) is a sequence that has a common difference between successive terms. The following are examples of arithmetic sequences.

(a) 1, 8, 15, 22, 29, . . .

(b) 4, 7, 10, 13, 16, . . .

(c) 4, 1, -5, -5, -8, . . .

(d) -1, -6, -11, -16, -21, . . .

The common difference in sequence (a) is 7. That is to say, $8 - 1 = 7$, $15 - 8 = 7$, $22 - 15 = 7$, $29 - 22 = 7$, and so on. The common differences for sequences (b), (c), and (d) are 3, -3, and -5, respectively.

In a more general setting we say that the sequence

$$a_1, a_2, a_3, a_4, \ldots, a_n, \ldots$$

is an arithmetic sequence if and only if there is a real number d such that

$$a_{k+1} - a_k = d$$

for every positive integer k. The number d is called the **common difference**.

From the definition we see that $a_{k+1} = a_k + d$. In other words, we can generate an arithmetic sequence having a common difference of d by starting with a first term a_1 and then simply adding d to each successive term.

first term: a_1

second term: $a_1 + d$

third term: $a_1 + 2d$ $(a_1 + d) + d = a_1 + 2d$

fourth term: $a_1 + 3d$

$\vdots$

nth term: $a_1 + (n - 1)d$

Thus, the **general term** of an arithmetic sequence is given by

$$a_n = a_1 + (n - 1)d,$$

where a_1 is the first term and d is the common difference. This formula for the general term can be used to solve a variety of problems involving arithmetic sequences.

EXAMPLE 2

Find the general-term expression for the arithmetic sequence 6, 2, -2, -6, ...

Solution The common difference, d, is $2 - 6 = -4$ and the first term, a_1, is 6. Substituting these values into $a_n = a_1 + (n - 1)d$ and simplifying, we obtain

$$a_n = a_1 + (n - 1)d$$
$$= 6 + (n - 1)(-4)$$
$$= 6 - 4n + 4$$
$$= -4n + 10.$$

EXAMPLE 3

Find the fortieth term of the arithmetic sequence $1, 5, 9, 13, \ldots$.

Solution Using $a_n = a_1 + (n - 1)d$, we obtain

$$a_{40} = 1 + (40 - 1)4$$
$$= 1 + (39)(4)$$
$$= 157.$$

EXAMPLE 4

Find the first term of the arithmetic sequence for which the fourth term is 26 and the ninth term is 61.

Solution Using $a_n = a_1 + (n - 1)d$ with $a_4 = 26$ (the fourth term is 26) and $a_9 = 61$ (the ninth term is 61), we have

$$26 = a_1 + (4 - 1)d = a_1 + 3d,$$
$$61 = a_1 + (9 - 1)d = a_1 + 8d.$$

Solving the system of equations

$$\left(\begin{matrix} a_1 + 3d = 26 \\ a_1 + 8d = 61 \end{matrix} \right)$$

yields $a_1 = 5$ and $d = 7$. Thus, the first term is 5.

Sums of Arithmetic Sequences

We often use sequences to solve problems, so we need to be able to find the sum of a certain number of terms of the sequence. Before developing a general sum formula for arithmetic sequences, let's consider an approach to a specific problem that we can then use in a general setting.

EXAMPLE 5

Find the sum of the first one hundred positive integers.

Solution We are being asked to find the sum of $1 + 2 + 3 + 4 + \cdots + 100$. Rather than adding in the usual way, let's find the sum in the following manner.

$$\begin{array}{r} 1 + \quad 2 + \quad 3 + \quad 4 + \cdots + 100 \\ 100 + \quad 99 + \quad 98 + \quad 97 + \cdots + \quad 1 \\ \hline 101 + 101 + 101 + 101 + \cdots + 101 \end{array}$$

$$\frac{\overset{50}{\cancel{100}}(101)}{\underset{1}{\cancel{2}}} = 5050$$

Note that we simply wrote the indicated sum forward and backward, and then we added the results. In so doing, we produced 100 sums of 101, but half of them are "repeats." For example, $100 + 1$ and $1 + 100$ are both counted in this process. Thus, we divide the product $(100)(101)$ by 2, yielding the final result of 5050.

The "forward-backward" approach used in Example 5 can be used to develop a formula for finding the sum of the first n terms of any arithmetic sequence. Consider an arithmetic sequence $a_1, a_2, a_3, a_4, \ldots, a_n$ with a common difference of d. Using S_n to represent the sum of the first n terms, we can proceed as follows:

$$S_n = a_1 + (a_1 + d) + (a_1 + 2d) + \cdots + (a_n - 2d) + (a_n - d) + a_n.$$

Now write this sum in reverse:

$$S_n = a_n + (a_n - d) + (a_n - 2d) + \cdots + (a_1 + 2d) + (a_1 + d) + a_1.$$

Adding the two equations produces

$$2S_n = (a_1 + a_n) + (a_1 + a_n) + (a_1 + a_n) + \cdots + (a_1 + a_n) \\ + (a_1 + a_n) + (a_1 + a_n).$$

That is, we have n sums $a_1 + a_n$, so

$$2S_n = n(a_1 + a_n),$$

from which we obtain a **sum formula**:

$$S_n = \frac{n(a_1 + a_n)}{2}.$$

Using the nth-term formula and/or the sum formula, we can solve a variety of problems involving arithmetic sequences.

EXAMPLE 6

Find the sum of the first thirty terms of the arithmetic sequence $3, 7, 11, 15, \ldots$.

Solution Using $a_n = a_1 + (n - 1)d$, we can find the 30th term.

$$a_{30} = 3 + (30 - 1)4 = 3 + 29(4) = 119$$

Now we can use the sum formula.

$$S_{30} = \frac{30(3 + 119)}{2} = 1830.$$

EXAMPLE 7

Find the sum $7 + 10 + 13 + \cdots + 157$.

Solution To use the sum formula, we need to know the number of terms. The nth-term formula will do that for us.

$$a_n = a_1 + (n - 1)d$$
$$157 = 7 + (n - 1)3$$
$$157 = 7 + 3n - 3$$
$$157 = 3n + 4$$
$$153 = 3n$$
$$51 = n$$

Now we can use the sum formula.

$$S_{51} = \frac{51(7 + 157)}{2} = 4182$$

Keep in mind that we developed the sum formula for an arithmetic sequence by using the forward-backward technique, which we had previously used on a specific problem. Now that we have the sum formula, we have two choices when solving problems to which the formula applies. We can either memorize the formula and use it, or we can simply use the forward-backward technique. However, we should emphasize that even if you choose to use the formula and some day you forget it, don't panic: just use the forward-backward technique. In other words, understanding the development of a formula often allows you to do problems even when you forget the formula itself.

Summation Notation

Sometimes a special notation is used to indicate the sum of a certain number of terms of a sequence. The capital Greek letter *sigma*, $\sum$, is used as a **summation symbol**. For example,

$$\sum_{i=1}^{5} a_i$$

represents the sum $a_1 + a_2 + a_3 + a_4 + a_5$. The letter i is frequently used as the **index of summation**: it takes on all integer values from the lower limit to the upper limit, inclusively. Thus we can write the following examples.

$$\sum_{i=1}^{4} b_i = b_1 + b_2 + b_3 + b_4$$

$$\sum_{i=3}^{7} a_i = a_3 + a_4 + a_5 + a_6 + a_7$$

$$\sum_{i=1}^{15} i^2 = 1^2 + 2^2 + 3^2 + \cdots + 15^2$$

$$\sum_{i=1}^{n} a_i = a_1 + a_2 + a_3 + \cdots + a_n$$

If $a_1, a_2, a_3, \ldots$ represents an arithmetic sequence, we can now write the sum formula

$$\sum_{i=1}^{n} a_i = \frac{n}{2}(a_1 + a_n).$$

EXAMPLE 8

Find the sum $\sum_{i=1}^{50} (3i + 4)$.

Solution This indicated sum means

$$\sum_{i=1}^{50} (3i + 4) = [3(1) + 4] + [3(2) + 4] + [3(3) + 4] + \cdots + [3(50) + 4]$$

$$= 7 + 10 + 13 + \cdots + 154.$$

Since this is an indicated sum of an arithmetic sequence, we can use our sum formula.

$$S_{50} = \frac{50}{2}(7 + 154) = 4025$$

EXAMPLE 9

Find the sum $\sum_{i=2}^{7} 2i^2$.

Solution This indicated sum means

$$\sum_{i=2}^{7} 2i^2 = 2(2)^2 + 2(3)^2 + 2(4)^2 + 2(5)^2 + 2(6)^2 + 2(7)^2$$

$$= 8 + 18 + 32 + 50 + 72 + 98.$$

This is not the indicated sum of an *arithmetic* sequence; therefore, let's simply add the numbers in the usual way. The sum is 278.

Example 9 should serve as a word of caution. Be sure that you analyze the sequence of numbers that is represented by the summation symbol. You may or may not be able to use a formula for adding the numbers.

Problem Set 10.1

For Problems 1–10, write the first five terms of the sequence that has the indicated general term.

1. $a_n = 3n - 7$ **2.** $a_n = 5n - 2$ **3.** $a_n = -2n + 4$ **4.** $a_n = -4n + 7$

5. $a_n = 3n^2 - 1$ **6.** $a_n = 2n^2 - 6$ **7.** $a_n = n(n - 1)$

8. $a_n = (n + 1)(n + 2)$ **9.** $a_n = 2^{n+1}$ **10.** $a_n = 3^{n-1}$

11. Find the 15th and 30th terms of the sequence for which $a_n = -5n - 4$.

12. Find the 20th and 50th terms of the sequence for which $a_n = -n - 3$.

13. Find the 25th and 50th terms of the sequence for which $a_n = (-1)^{n+1}$.

14. Find the 10th and 15th terms of the sequence for which $a_n = -n^2 - 10$.

For Problems 15–24, find the general term (the nth term) for each of the arithmetic sequences.

15. 11, 13, 15, 17, 19, ...   **16.** 7, 10, 13, 16, 19, ...

17. 2, −1, −4, −7, −10, ...   **18.** 4, 2, 0, −2, −4, ...

19. $\frac{3}{2}, 2, \frac{5}{2}, 3, \frac{7}{2}, \ldots$

20. $0, \frac{1}{2}, 1, \frac{3}{2}, 2, \ldots$

21. 2, 6, 10, 14, 18, ...

22. 2, 7, 12, 17, 22, ...

23. $-3, -6, -9, -12, -15, \ldots$

24. $-4, -8, -12, -16, -20, \ldots$

For Problems 25–30, find the required term for each of the arithmetic sequences.

25. The 15th term of 3, 8, 13, 18, ...

26. The 20th term of 4, 11, 18, 25, ...

27. The 30th term of 15, 26, 37, 48, ...

28. The 35th term of 9, 17, 25, 33, ...

29. The 52nd term of $1, \frac{5}{3}, \frac{7}{3}, 3, \ldots$

30. The 47th term of $\frac{1}{2}, \frac{5}{4}, 2, \frac{11}{4}, \ldots$

31. If the 6th term of an arithmetic sequence is 12 and the 10th term is 16, find the first term.

32. If the 5th term of an arithmetic sequence is 14 and the 12th term is 42, find the first term.

33. If the 3rd term of an arithmetic sequence is 20 and the 7th term is 32, find the 25th term.

34. If the 5th term of an arithmetic sequence is -5 and the 15th term is -25, find the 50th term.

35. Find the sum of the first 50 terms of the arithmetic sequence 5, 7, 9, 11, 13, ...

36. Find the sum of the first 30 terms of the arithmetic sequence 0, 2, 4, 6, 8, ...

37. Find the sum of the first 40 terms of the arithmetic sequence 2, 6, 10, 14, 18, ...

38. Find the sum of the first 60 terms of the arithmetic sequence $-2, 3, 8, 13, 18, \ldots$.

39. Find the sum of the first 75 terms of the arithmetic sequence $5, 2, -1, -4, -7, \ldots$.

40. Find the sum of the first 80 terms of the arithmetic sequence $7, 3, -1, -5, -9, \ldots$.

41. Find the sum of the first 50 terms of the arithmetic sequence $\frac{1}{2}, 1, \frac{3}{2}, 2, \frac{5}{2}, \ldots$.

42. Find the sum of the first 100 terms of the arithmetic sequence $-\frac{1}{3}, \frac{1}{3}, 1, \frac{5}{3}, \frac{7}{3}, \ldots$.

For Problems 43–50, find each of the indicated sums.

43. $1 + 5 + 9 + 13 + \cdots + 197$

44. $3 + 8 + 13 + 18 + \cdots + 398$

45. $2 + 8 + 14 + 20 + \cdots + 146$

46. $6 + 9 + 12 + 15 + \cdots + 93$

47. $(-7) + (-10) + (-13) + (-16) + \cdots + (-109)$

48. $(-5) + (-9) + (-13) + (-17) + \cdots + (-169)$

49. $(-5) + (-3) + (-1) + 1 + \cdots + 119$

50. $(-7) + (-4) + (-1) + 2 + \cdots + 131$

51. Find the sum of the first 200 odd whole numbers.

52. Find the sum of the first 175 positive even whole numbers.

53. Find the sum of all even numbers between 18 and 482, inclusive.

54. Find the sum of all odd numbers between 17 and 379, inclusive.

55. Find the sum of the first 30 terms of the arithmetic sequence with the general term $a_n = 5n - 4$.

56. Find the sum of the first 40 terms of the arithmetic sequence with the general term $a_n = 4n - 7$.

57. Find the sum of the first 25 terms of the arithmetic sequence with the general term $a_n = -4n - 1$.

58. Find the sum of the first 35 terms of the arithmetic sequence with the general term $a_n = -5n - 3$.

For Problems 59–70, find each of the following sums.

59. $\displaystyle\sum_{i=1}^{45} (5i + 2)$ **60.** $\displaystyle\sum_{i=1}^{38} (3i + 6)$ **61.** $\displaystyle\sum_{i=1}^{30} (-2i + 4)$ **62.** $\displaystyle\sum_{i=1}^{40} (-3i + 3)$

63. $\displaystyle\sum_{i=4}^{32} (3i - 10)$ **64.** $\displaystyle\sum_{i=6}^{47} (4i - 9)$ **65.** $\displaystyle\sum_{i=10}^{20} 4i$ **66.** $\displaystyle\sum_{i=15}^{30} (-5i)$

67. $\displaystyle\sum_{i=1}^{5} i^2$ **68.** $\displaystyle\sum_{i=1}^{6} (i^2 + 1)$ **69.** $\displaystyle\sum_{i=3}^{8} (2i^2 + i)$ **70.** $\displaystyle\sum_{i=4}^{7} (3i^2 - 2)$

Miscellaneous Problems

The general term of a sequence can consist of one expression for certain values of n and another expression (or expressions) for other values of n. That is to say, a **multiple description** of the sequence can be given. For example,

$$a_n = \begin{cases} 2n + 3 & \text{for } n \text{ odd} \\ 3n - 2 & \text{for } n \text{ even} \end{cases}$$

means that we use $a_n = 2n + 3$ for $n = 1, 3, 5, 7, \ldots$, and we use $a_n = 3n - 2$ for $n = 2, 4, 6, 8, \ldots$. The first six terms of this sequence are 5, 4, 9, 10, 13, and 16.

For Problems 71–74, write the first six terms of each sequence.

71. $a_n = \begin{cases} 2n + 1 & \text{for } n \text{ odd} \\ 2n - 1 & \text{for } n \text{ even} \end{cases}$ **72.** $a_n = \begin{cases} \dfrac{1}{n} & \text{for } n \text{ odd} \\ n^2 & \text{for } n \text{ even} \end{cases}$

73. $a_n = \begin{cases} 3n + 1 & \text{for } n \leq 3 \\ 4n - 3 & \text{for } n > 3 \end{cases}$ **74.** $a_n = \begin{cases} 5n - 1 & \text{for } n \text{ a multiple of } 3 \\ 2n & \text{otherwise} \end{cases}$

The multiple-description approach can also be used to give a **recursive description** for a sequence. A sequence is said to be **described recursively** if the first n terms are stated and then each succeeding term is defined as a function of one or more of the preceding terms. For example,

$$\begin{cases} a_1 = 2 \\ a_n = 2a_{n-1} & \text{for } n \geq 2 \end{cases}$$

means that the first term, a_1, is 2 and each succeeding term is 2 times the previous term. Thus, the first six terms are 2, 4, 8, 16, 32, and 64.

For Problems 75–80, write the first six terms of each sequence.

75. $\begin{cases} a_1 = 4 \\ a_n = 3a_{n-1} & \text{for } n \geq 2 \end{cases}$ **76.** $\begin{cases} a_1 = 3 \\ a_n = a_{n-1} + 2 & \text{for } n \geq 2 \end{cases}$

77. $\begin{cases} a_1 = 1 \\ a_2 = 1 \\ a_n = a_{n-2} + a_{n-1} & \text{for } n \geq 3 \end{cases}$ **78.** $\begin{cases} a_1 = 2 \\ a_2 = 3 \\ a_n = 2a_{n-2} + 3a_{n-1} & \text{for } n \geq 3 \end{cases}$

79. $\begin{cases} a_1 = 3 \\ a_2 = 1 \\ a_n = (a_{n-1} - a_{n-2})^2 & \text{for } n \geq 3 \end{cases}$

80. $\begin{cases} a_1 = 1 \\ a_2 = 2 \\ a_3 = 3 \\ a_n = a_{n-1} + a_{n-2} + a_{n-3} & \text{for } n \geq 4 \end{cases}$

10.2

Geometric Sequences

A **geometric sequence** or **geometric progression** is a sequence in which we obtain each term after the first by multiplying the preceding term by a common multiplier, called the **common ratio** of the sequence. We can find the common ratio of a geometric sequence by dividing any term (other than the first) by the preceding term. The following geometric sequences have common ratios of $3, 2, \frac{1}{2},$ and -4, respectively.

$1, 3, 9, 27, 81, \ldots$

$3, 6, 12, 24, 48, \ldots$

$16, 8, 4, 2, 1, \ldots$

$-1, 4, -16, 64, -256, \ldots$

In a more general setting we say that the sequence $a_1, a_2, a_3, \ldots, a_n, \ldots$ is a geometric sequence if and only if there is a nonzero real number r such that

$$a_{k+1} = ra_k$$

for every positive integer k. The nonzero real number r is called the common ratio of the sequence.

The previous equation can be used to generate a general geometric sequence that has a_1 as a first term and r as a common ratio. We can proceed as follows:

first term: a_1

second term: $a_1 r$

third term: $a_1 r^2$ $(a_1 r)(r) = a_1 r^2$

fourth term: $a_1 r^3$

$\vdots$

nth term: $a_1 r^{n-1}$

Thus, the **general term** of a geometric sequence is given by

$$a_n = a_1 r^{n-1},$$

where a_1 is the first term and r is the common ratio.

EXAMPLE 1

Find the general term for the geometric sequence $8, 16, 32, 64, \ldots$.

Solution Using $a_n = a_1 r^{n-1}$, we obtain

$$a_n = 8(2)^{n-1} = (2^3)(2)^{n-1} = 2^{n+2}.$$

EXAMPLE 2

Find the ninth term of the geometric sequence $27, 9, 3, 1, \ldots$.

Solution Using $a_n = a_1 r^{n-1}$, we can find the ninth term as follows.

$$a_9 = 27\left(\frac{1}{3}\right)^{9-1} = 27\left(\frac{1}{3}\right)^8 = \frac{3^3}{3^8} = \frac{1}{3^5} = \frac{1}{243}$$

Sums of Geometric Sequences

As with arithmetic sequences, we often need to find the sum of a certain number of terms of a geometric sequence. Before developing a general sum formula for geometric sequences, let's consider an approach to a specific problem that we can then use in a general setting.

EXAMPLE 3

Find the sum $1 + 3 + 9 + 27 + \cdots + 6561$.

Solution Letting S represent the sum, we can proceed as follows.

$$S = 1 + 3 + 9 + 27 + \cdots + 6561 \tag{1}$$
$$3S = 3 + 9 + 27 + \cdots + 6561 + 19683 \tag{2}$$

Equation (2) is the result of multiplying equation (1) by the common ratio, 3. Subtracting equation (1) from equation (2) produces

$$2S = 19683 - 1 = 19682$$
$$S = 9841.$$

Now let's consider a general geometric sequence $a_1, a_1 r, a_1 r^2, \ldots, a_1 r^{n-1}$. By applying a procedure similar to the one we used in Example 3, we can develop a formula for finding the sum of the first n terms of any geometric sequence.

We let S_n represent the sum of the first n terms:

$$S_n = a_1 + a_1 r + a_1 r^2 + \cdots + a_1 r^{n-1}. \tag{3}$$

Next we multiply both sides of equation (3) by the common ratio r:

$$rS_n = a_1 r + a_1 r^2 + a_1 r^3 + \cdots + a_1 r^n. \tag{4}$$

We then subtract equation (3) from equation (4):

$$rS_n - S_n = a_1 r^n - a_1.$$

When we apply the distributive property to the left side and then solve for S_n, we obtain

$$S_n(r - 1) = a_1 r^n - a_1$$

$$S_n = \frac{a_1 r^n - a_1}{r - 1}, \qquad r \neq 1.$$

Therefore, the sum of the first n terms of a geometric sequence with a first term a_1 and a common ratio r is given by

$$S_n = \frac{a_1 r^n - a_1}{r - 1}, \qquad r \neq 1.$$

EXAMPLE 4

Find the sum of the first eight terms of the geometric sequence $1, 2, 4, 8, \ldots$.

Solution Using the sum formula, we obtain

$$S_8 = \frac{1(2)^8 - 1}{2 - 1} = \frac{2^8 - 1}{1} = 255.$$

If the common ratio of a geometric sequence is less than 1, it may be more convenient to change the form of the sum formula. That is, the fraction

$$\frac{a_1 r^n - a_1}{r - 1}$$

can be changed to

$$\frac{a_1 - a_1 r^n}{1 - r}$$

by multiplying both the numerator and denominator by -1. Thus, using

$$S_n = \frac{a_1 - a_1 r^n}{1 - r}$$

can sometimes avoid unnecessary work with negative numbers when $r < 1$, as the next example illustrates.

EXAMPLE 5

Find the sum $1 + \dfrac{1}{2} + \dfrac{1}{4} + \cdots + \dfrac{1}{256}$.

Solution A To use the sum formula, we need to know the number of terms, which can be found by counting them or by applying the nth-term formula, as follows.

$$a_n = a_1 r^{n-1}$$

$$\frac{1}{256} = 1 \left(\frac{1}{2}\right)^{n-1}$$

$$\left(\frac{1}{2}\right)^8 = \left(\frac{1}{2}\right)^{n-1}$$

$$8 = n - 1 \qquad \text{If } b^n = b^m, \text{ then } n = m.$$

$$9 = n$$

Now we use $n = 9$, $a_1 = 1$, and $r = \frac{1}{2}$ in the sum formula of the form

$$S_n = \frac{a_1 - a_1 r^n}{1 - r}.$$

$$S_9 = \frac{1 - 1(\frac{1}{2})^9}{1 - \frac{1}{2}} = \frac{1 - \frac{1}{512}}{\frac{1}{2}} = \frac{\frac{511}{512}}{\frac{1}{2}} = 1\frac{255}{256}.$$

We can also do a problem like Example 5 without finding the number of terms; we use the general approach illustrated in Example 3. Solution B demonstrates this idea.

Solution B Let S represent the desired sum.

$$S = 1 + \tfrac{1}{2} + \tfrac{1}{4} + \cdots + \tfrac{1}{256}$$

Multiply both sides by the common ratio, $\frac{1}{2}$.

$$\tfrac{1}{2}S = \tfrac{1}{2} + \tfrac{1}{4} + \tfrac{1}{8} + \cdots + \tfrac{1}{256} + \tfrac{1}{512}$$

Subtract the second equation from the first and solve for S.

$$\tfrac{1}{2}S = 1 - \tfrac{1}{512} = \tfrac{511}{512}$$
$$S = \tfrac{511}{256} = 1\tfrac{255}{256}$$

Summation notation can also be used to indicate the sum of a certain number of terms of a geometric sequence.

EXAMPLE 6

Find the sum $\displaystyle\sum_{i=1}^{10} 2^i$.

Solution This indicated sum means

$$\sum_{i=1}^{10} 2^i = 2^1 + 2^2 + 2^3 + \cdots + 2^{10}$$

$$= 2 + 4 + 8 + \cdots + 1024.$$

Since this is the indicated sum of a geometric sequence, we can use the sum formula, with $a_1 = 2$, $r = 2$, and $n = 10$.

$$S_{10} = \frac{2(2)^{10} - 2}{2 - 1} = \frac{2(2^{10} - 1)}{1} = 2046$$

The Sum of an Infinite Geometric Sequence

Let's take the formula

$$S_n = \frac{a_1 - a_1 r^n}{1 - r}$$

and rewrite the right side by applying the property

$$\frac{a - b}{c} = \frac{a}{c} - \frac{b}{c}.$$

Thus we obtain

$$S_n = \frac{a_1}{1 - r} - \frac{a_1 r^n}{1 - r}. \tag{1}$$

Now let's examine the behavior of r^n for $|r| < 1$, that is, for $-1 < r < 1$. For example, suppose that $r = \frac{1}{2}$; then

$$r^2 = (\tfrac{1}{2})^2 = \tfrac{1}{4}, \qquad r^3 = (\tfrac{1}{2})^3 = \tfrac{1}{8},$$
$$r^4 = (\tfrac{1}{2})^4 = \tfrac{1}{16}, \qquad r^5 = (\tfrac{1}{2})^5 = \tfrac{1}{32},$$

and so on. We can make $(\tfrac{1}{2})^n$ as close to zero as we please by choosing sufficiently large values for n. In general, for values of r such that $|r| < 1$, the expression r^n will approach zero as n gets larger and larger. Therefore, the fraction $a_1 r^n/(1 - r)$ in equation (1) will approach zero as n increases. We say that **the sum of the infinite geometric sequence** is given by

$$S_\infty = \frac{a_1}{1 - r}, \qquad |r| < 1.$$

EXAMPLE 7

Find the sum of the infinite geometric sequence

$$1, \tfrac{1}{2}, \tfrac{1}{4}, \tfrac{1}{8}, \ldots .$$

Solution Since $a_1 = 1$ and $r = \tfrac{1}{2}$, we obtain

$$S_\infty = \frac{1}{1 - \tfrac{1}{2}} = \frac{1}{\tfrac{1}{2}} = 2.$$

When we state that $S_\infty = 2$ in Example 7, we mean that as we add more and more terms, the sum approaches 2. Observe what happens when we calculate the sums up to five terms.

first term: 1

sum of first two terms: $1 + \tfrac{1}{2} = 1\tfrac{1}{2}$

sum of first three terms: $1 + \tfrac{1}{2} + \tfrac{1}{4} = 1\tfrac{3}{4}$

sum of first four terms: $1 + \tfrac{1}{2} + \tfrac{1}{4} + \tfrac{1}{8} = 1\tfrac{7}{8}$

sum of first five terms: $1 + \tfrac{1}{2} + \tfrac{1}{4} + \tfrac{1}{8} + \tfrac{1}{16} = 1\tfrac{15}{16}$

If $|r| \geq 1$, the absolute value of r^n increases without bound as n increases. Consider the following two examples and notice the unbounded growth of the absolute value of r^n.

Let $r = 3$.

$r^2 = 3^2 = 9$

$r^3 = 3^3 = 27$

$r^4 = 3^4 = 81$

$r^5 = 3^5 = 243$

Let $r = -2$.

$r^2 = (-2)^2 = 4$

$r^3 = (-2)^3 = -8$ $\qquad |-8| = 8$

$r^4 = (-2)^4 = 16$

$r^5 = (-2)^5 = -32$ $\qquad |-32| = 32$

We say that **the sum of any infinite geometric sequence for which $|r| \geq 1$ does not exist.**

Repeating Decimals as Sums of Infinite Geometric Sequences

In Section 1.1, rational numbers were defined to be numbers that have either a terminating or repeating decimal representation. For example,

$$2.23, \quad 0.147, \quad 0.\overline{3}, \quad 0.\overline{14}, \quad \text{and} \quad 0.5\overline{6}$$

are examples of rational numbers. (Remember that $0.\overline{3}$ means $0.3333\dots$.) Place value provides the basis for changing terminating decimals such as 2.23 and 0.147 to a/b form, where a and b are integers and $b \neq 0$:

$$2.23 = \frac{223}{100} \quad \text{and} \quad 0.147 = \frac{147}{1000}.$$

However, changing repeating decimals to a/b form requires a different technique, and our work with sums of infinite geometric sequences provides the basis for one such approach. Consider the following examples.

EXAMPLE 8

Change $0.\overline{14}$ to a/b form, where a and b are integers and $b \neq 0$.

Solution The repeating decimal $0.\overline{14}$ can be written as the indicated sum of an infinite geometric sequence with first term 0.14 and common ratio 0.01.

$$0.14 + 0.0014 + 0.000014 + \cdots$$

Using $S_\infty = a_1/(1 - r)$, we obtain

$$S_\infty = \frac{0.14}{1 - 0.01} = \frac{0.14}{0.99} = \frac{14}{99}.$$

Thus $0.\overline{14} = \frac{14}{99}$.

If the repeating block of digits does not begin immediately after the decimal point, as in $0.5\overline{6}$, we can make a slight adjustment in the technique used in Example 8.

EXAMPLE 9

Change $0.5\overline{6}$ to a/b form, where a and b are integers and $b \neq 0$.

Solution The repeating decimal $0.5\overline{6}$ can be written

$$(0.5) + (0.06 + 0.006 + 0.0006 + \cdots),$$

where

$$0.06 + 0.006 + 0.0006 + \cdots$$

is the indicated sum of the infinite geometric sequence with $a_1 = 0.06$ and $r = 0.1$. Therefore,

$$S_\infty = \frac{0.06}{1 - 0.1} = \frac{0.06}{0.9} = \frac{6}{90} = \frac{1}{15}.$$

Now we can add 0.5 and $\frac{1}{15}$.

$$0.5\overline{6} = 0.5 + \frac{1}{15} = \frac{1}{2} + \frac{1}{15} = \frac{15}{30} + \frac{2}{30} = \frac{17}{30}$$

Problem Set 10.2

For Problems 1–12, find the general term (the nth term) for each of the geometric sequences.

1. $3, 6, 12, 24, \ldots$

2. $2, 6, 18, 54, \ldots$

3. $3, 9, 27, 81, \ldots$

4. $2, 4, 8, 16, \ldots$

5. $\frac{1}{4}, \frac{1}{8}, \frac{1}{16}, \frac{1}{32}, \ldots$

6. $8, 4, 2, 1, \ldots$

7. $4, 16, 64, 256, \ldots$

8. $6, 2, \frac{2}{3}, \frac{2}{9}, \ldots$

9. $1, 0.3, 0.09, 0.027, \ldots$

10. $0.2, 0.04, 0.008, 0.0016, \ldots$

11. $1, -2, 4, -8, \ldots$

12. $-3, 9, -27, 81, \ldots$

For Problems 13–20, find the required term for each of the given geometric sequences.

13. The 8th term of $\frac{1}{2}, 1, 2, 4, \ldots$

14. The 7th term of $2, 6, 18, 54, \ldots$

15. The 9th term of $729, 243, 81, 27, \ldots$

16. The 11th term of $768, 384, 192, 96, \ldots$

17. The 10th term of $1, -2, 4, -8, \ldots$

18. The 8th term of $-1, -\frac{3}{2}, -\frac{9}{4}, -\frac{27}{8}, \ldots$

19. The 8th term of $\frac{1}{2}, \frac{1}{6}, \frac{1}{18}, \frac{1}{54}, \ldots$

20. The 9th term of $\frac{16}{81}, \frac{8}{27}, \frac{4}{9}, \frac{2}{3}, \ldots$

21. Find the first term of the geometric sequence with 5th term $\frac{32}{3}$ and common ratio 2.

22. Find the first term of the geometric sequence with 4th term $\frac{27}{128}$ and common ratio $\frac{3}{4}$.

23. Find the common ratio of the geometric sequence with 3rd term 12 and 6th term 96.

24. Find the common ratio of the geometric sequence with 2nd term $\frac{8}{3}$ and 5th term $\frac{64}{81}$.

25. Find the sum of the first 10 terms of the geometric sequence $1, 2, 4, 8, \ldots$.

26. Find the sum of the first 7 terms of the geometric sequence $3, 9, 27, 81, \ldots$.

27. Find the sum of the first 9 terms of the geometric sequence $2, 6, 18, 54, \ldots$.

28. Find the sum of the first 10 terms of the geometric sequence $5, 10, 20, 40, \ldots$.

29. Find the sum of the first 8 terms of the geometric sequence $8, 12, 18, 27, \ldots$.

30. Find the sum of the first 8 terms of the geometric sequence $9, 12, 16, \frac{64}{3}, \ldots$.

31. Find the sum of the first 10 terms of the geometric sequence $-4, 8, -16, 32, \ldots$.

32. Find the sum of the first 9 terms of the geometric sequence $-2, 6, -18, 54, \ldots$.

For Problems 33–38, find each of the indicated sums.

33. $9 + 27 + 81 + \cdots + 729$

34. $2 + 8 + 32 + \cdots + 8192$

35. $4 + 2 + 1 + \cdots + \frac{1}{512}$

36. $1 + (-2) + 4 + \cdots + 256$

37. $(-1) + 3 + (-9) + \cdots + (-729)$

38. $16 + 8 + 4 + \cdots + \frac{1}{32}$

For Problems 39–44, find each of the indicated sums.

39. $\displaystyle\sum_{i=1}^{9} 2^{i-3}$

40. $\displaystyle\sum_{i=1}^{6} 3^{i}$

41. $\displaystyle\sum_{i=2}^{5} (-3)^{i+1}$

42. $\displaystyle\sum_{i=3}^{8} (-2)^{i-1}$

43. $\displaystyle\sum_{i=1}^{6} 3\left(\frac{1}{2}\right)^{i}$

44. $\displaystyle\sum_{i=1}^{5} 2\left(\frac{1}{3}\right)^{i}$

For Problems 45–56, find the sum of each infinite geometric sequence. If the sequence has no sum, so state.

45. $2, 1, \frac{1}{2}, \frac{1}{4}, \ldots$

46. $9, 3, 1, \frac{1}{3}, \ldots$

47. $1, \frac{2}{3}, \frac{4}{9}, \frac{8}{27}, \ldots$

48. $5, 3, \frac{9}{5}, \frac{27}{25}, \ldots$

49. $4, 8, 16, 32, \ldots$

50. $32, 16, 8, 4, \ldots$

51. $9, -3, 1, -\frac{1}{3}, \ldots$

52. $2, -6, 18, -54, \ldots$

53. $\frac{1}{2}, \frac{3}{8}, \frac{9}{32}, \frac{27}{128}, \ldots$

54. $4, -\frac{4}{3}, \frac{4}{9}, -\frac{4}{27}, \ldots$

55. $8, -4, 2, -1, \ldots$

56. $7, \frac{14}{5}, \frac{28}{25}, \frac{56}{125}, \ldots$

For Problems 57–68, change each repeating decimal to a/b form, where a and b are integers and $b \neq 0$. Express a/b in reduced form.

57. $0.\overline{3}$

58. $0.\overline{4}$

59. $0.\overline{26}$

60. $0.\overline{18}$

61. $0.\overline{123}$

62. $0.\overline{273}$

63. $0.2\overline{6}$

64. $0.4\overline{3}$

65. $0.2\overline{14}$

66. $0.3\overline{71}$

67. $2.\overline{3}$

68. $3.\overline{7}$

10.3

Another Look at Problem Solving

In the previous two sections, many of the exercises fell into one of the following four categories.

1. Finding the nth term of an arithmetic sequence

$$a_n = a_1 + (n - 1)d$$

2. Finding the sum of the first n terms of an arithmetic sequence

$$S_n = \frac{n(a_1 + a_n)}{2}$$

3. Finding the nth term of a geometric sequence

$$a_n = a_1 r^{n-1}$$

4. Finding the sum of the first n terms of a geometric sequence

$$S_n = \frac{a_1 r^n - a_1}{r - 1}$$

In this section we want to use this knowledge of arithmetic sequences and geometric sequences to expand our problem-solving capabilities. Let's begin by restating some "old" problem-solving suggestions that continue to apply here; we will also consider some other suggestions that are directly related to problems involving sequences of numbers. (We will indicate the "new" suggestions with an asterisk.)

Suggestions for Solving Word Problems

1. Read the problem carefully, making certain that you understand the meanings of all the words. Be especially alert for any technical terms used in the statement of the problem.

2. Read the problem a second time (perhaps even a third time) to get an overview of the situation being described and to determine the known facts, as well as what you are to find.

3. Sketch a figure, diagram, or chart that might be helpful in analyzing the problem.

*4. Write down the first few terms of the sequence to describe what is taking place in the problem. Be sure that you understand, term by term, what the sequence represents in the problem.

*5. Determine whether the sequence is arithmetic or geometric.

*6. Determine whether the problem is asking for a specific term of the sequence or for the sum of a certain number of terms.

*7. Carry out the necessary calculations and check your answer for reasonableness.

As we solve some problems, these suggestions will become more meaningful.

PROBLEM 1

Domenica started to work in 1975 at an annual salary of $14,500. She received a $1050 raise each year. What was her annual salary in 1984?

Solution The following sequence represents her annual salary beginning in 1975.

14500, 15550, 16600, 17650, . . .

This is an arithmetic sequence, with $a_1 = 14500$ and $d = 1050$. Since each term of the sequence represents her annual salary, we are looking for the tenth term.

$$a_{10} = 14500 + (10 - 1)1050 = 14500 + 9(1050) = 23950$$

Her annual salary in 1984 was $23,950.

PROBLEM 2

An auditorium has 20 seats in the front row, 24 seats in the second row, 28 seats in the third row, and so on, for 15 rows. How many seats are there in the auditorium?

Solution The following sequence represents the number of seats per row starting with the first row.

$$20, 24, 28, 32, \ldots$$

This is an arithmetic sequence, with $a_1 = 20$ and $d = 4$. Therefore the fifteenth term, which represents the number of seats in the fifteenth row, is given by

$$a_{15} = 20 + (15 - 1)4 = 20 + 14(4) = 76.$$

The total number of seats in the auditorium is represented by

$$20 + 24 + 28 + \cdots + 76.$$

Using the sum formula for an arithmetic sequence, we obtain

$$S_{15} = \tfrac{15}{2}(20 + 76) = 720.$$

There are 720 seats in the auditorium.

PROBLEM 3

Suppose that you save 25 cents the first day of a week, 50 cents the second day, one dollar the third day, and continue to double your savings each day. How much will you save on the seventh day? What will be your total savings for the week?

Solution The following sequence represents your savings per day, expressed in cents.

$$25, 50, 100, \ldots$$

This is a geometric sequence, with $a_1 = 25$ and $r = 2$. Your savings on the seventh day is the seventh term of this sequence. Therefore, using $a_n = a_1 r^{n-1}$, we obtain

$$a_7 = 25(2)^6 = 1600.$$

So you will save $16 on the seventh day. Your total savings for the 7 days is given by

$$25 + 50 + 100 + \cdots + 1600.$$

Using the sum formula for a geometric sequence, we obtain

$$S_7 = \frac{25(2)^7 - 25}{2 - 1} = \frac{25(2^7 - 1)}{1} = 3175.$$

So your savings for the entire week are $31.75.

PROBLEM 4

A pump is attached to a container for the purpose of creating a vacuum. For each stroke of the pump, $\tfrac{1}{4}$ of the air remaining in the container is removed. To the nearest tenth of a percent, how much of the air remains in the container after 6 strokes?

Solution Let's draw a diagram to help with the analysis of this problem.

first stroke:	$\frac{1}{4}$ of the air is removed	$1 - \frac{1}{4} = \frac{3}{4}$ of the air remains
second stroke:	$\frac{1}{4}(\frac{3}{4}) = \frac{3}{16}$ of the air is removed	$\frac{3}{4} - \frac{3}{16} = \frac{9}{16}$ of the air remains
third stroke:	$\frac{1}{4}(\frac{9}{16}) = \frac{9}{64}$ of the air is removed	$\frac{9}{16} - \frac{9}{64} = \frac{27}{64}$ of the air remains

The diagram suggests two approaches to the problem.

Approach A. The sequence

$$\frac{1}{4}, \frac{3}{16}, \frac{9}{64}, \cdots$$

represents, term by term, the fractional amount of air that is removed with each successive stroke. Therefore, we can find the total amount removed and subtract it from 100%. The sequence is geometric, with $a_1 = \frac{1}{4}$ and $r = \frac{3}{4}$. First, let's find the sixth term.

$$a_6 = \frac{1}{4}(\frac{3}{4})^{6-1} = \frac{1}{4}(\frac{3}{4})^5 = \frac{243}{4096}$$

Now, using the sum formula, we obtain

$$S_6 = \frac{\frac{1}{4} - \frac{1}{4}(\frac{3}{4})^6}{1 - \frac{3}{4}} = \frac{\frac{1}{4}[1 - (\frac{3}{4})^6]}{\frac{1}{4}}$$

$$= 1 - \frac{729}{4096} = \frac{3367}{4096} = 82.2\%$$

Therefore, $100\% - 82.2\% = 17.8\%$ of the air remains after 6 strokes.

Approach B. The sequence

$$\frac{3}{4}, \frac{9}{16}, \frac{27}{64}, \cdots$$

represents, term by term, the amount of air that remains in the container after each stroke. Therefore, when we find the sixth term of this geometric sequence, we will have the answer to the problem. Since $a_1 = \frac{3}{4}$ and $r = \frac{3}{4}$, we obtain

$$a_6 = \frac{3}{4}(\frac{3}{4})^{6-1} = (\frac{3}{4})^6 = \frac{729}{4096} = 17.8\%.$$

Therefore, 17.8% of the air remains after 6 strokes.

It will be helpful for you to take another look at the two approaches used to solve Problem 4. Notice in Approach B that finding the sixth term of the sequence produced the answer to the problem without any further calculations. In Approach A, however, we had to find the sum of 6 terms of the sequence (which also required that we find the sixth term) and then subtract that amount from 100%. Obviously Approach B requires less computation, but both approaches are meaningful. As we solve problems involving sequences, it is necessary that we understand what a particular sequence physically represents on a term-by-term basis.

Problem Set 10.3

Use your knowledge of arithmetic sequences and geometric sequences to help solve each of the following problems.

1. A man started to work in 1960 at an annual salary of $9500. He received a $700 raise each year. How much was his annual salary in 1981?

2. A woman started to work in 1970 at an annual salary of $13,400. She received a $900 raise per year. How much was her annual salary in 1985?

3. State University had an enrollment of 9600 students in 1960. Each year the enrollment increased by 150 students. What was the enrollment in 1973?

4. Math University had an enrollment of 12,800 students in 1977. Each year the enrollment decreased by 75 students. What was the enrollment in 1984?

5. The enrollment at University X is predicted to increase at the rate of 10% per year. If the enrollment for 1982 was 5000 students, find the predicted enrollment for 1986. Express your answer to the nearest whole number.

6. If you pay $12,000 for a car and its value depreciates 20% per year, how much will it be worth in 5 years? Express your answer to the nearest dollar.

7. A tank contains 16,000 liters of water. Each day one-half of the water in the tank is removed and not replaced. How much water remains in the tank at the end of 7 days?

8. If the price of a pound of coffee is $3.20 and the projected rate of inflation is 5% per year, how much per pound will coffee cost in 5 years? Express your answer to the nearest cent.

9. A tank contains 5832 gallons of water. Each day one-third of the water in the tank is removed and not replaced. How much water remains in the tank at the end of 6 days?

10. A fungus culture growing under controlled conditions doubles in size each day. How many units will the culture contain after 7 days if it originally contains 4 units?

11. Sue is saving quarters. She saves 1 quarter the first day, 2 quarters the second day, 3 quarters the third day, and so on for 30 days. How much money will she have saved in 30 days?

12. Suppose you save a penny the first day of a month, 2 cents the second day, 3 cents the third day, and so on for 31 days. What will be your total savings for the 31 days?

13. Suppose you save a penny the first day of a month, 2 cents the second day, 4 cents the third day, and continue to double your savings each day. How much will you save on the 15th day of the month? How much will your total savings be for the 15 days?

14. Eric saved a nickel the first day of a month, a dime the second day, 20 cents the third day, and continued to double his daily savings each day for 14 days. What were his daily savings on the 14th day? What were his total savings for the 14 days?

15. Ms. Bryan invested $1500 at 12% simple interest at the beginning of each year for a period of 10 years. Find the total accumulated value of all the investments at the end of the 10-year period.

16. Mr. Woodley invested $1200 at 11% simple interest at the beginning of each year for a period of 8 years. Find the total accumulated value of all the investments at the end of the 8-year period.

17. An object falling from rest in a vacuum falls approximately 16 feet the first second, 48 feet the second second, 80 feet the third second, 112 feet the fourth second, and so on. How far will it fall in 11 seconds?

18. A raffle is organized so that the amount paid for each ticket is determined by the number on the ticket. The tickets are numbered with the consecutive odd whole numbers 1, 3, 5, 7, ... Each contestant pays as many cents as the number on the ticket drawn. How much money will the raffle take in if 1000 tickets are sold?

19. Suppose an element has a half-life of 4 hours. This means that if n grams of it exist at a specific time, then only $\frac{1}{2}n$ grams remain 4 hours later. If at a particular moment we have 60 grams of the element, how much of it will remain 24 hours later?

20. Suppose an element has a half-life of 3 hours. (See Problem 19 for a definition of half-life.) If at a particular moment we have 768 grams of the element, how much of it will remain 24 hours later?

21. A rubber ball is dropped from a height of 1458 feet, and at each bounce it rebounds one-third of the height from which it last fell. How far has the ball traveled by the time it strikes the ground for the 6th time?

22. A rubber ball is dropped from a height of 100 feet, and at each bounce it rebounds one-half of the height from which it last fell. What distance has the ball traveled up to the instant it hits the ground for the 8th time?

23. A pile of logs has 25 logs in the bottom layer, 24 logs in the next layer, 23 logs in the next layer, and so on, until the top layer has 1 log. How many logs are in the pile?

24. A well-driller charges $9.00 per foot for the first 10 feet, $9.10 per foot for the next 10 feet, $9.20 per foot for the next 10 feet, and so on, continuing to increase the price by $.10 per foot for succeeding intervals of 10 feet. How much does it cost to drill a well to a depth of 150 feet?

25. A pump is attached to a container for the purpose of creating a vacuum. For each stroke of the pump, $\frac{1}{3}$ of the air remaining in the container is removed. To the nearest tenth of a percent, how much of the air remains in the container after 7 strokes?

26. Suppose that in Problem 25, each stroke of the pump removes $\frac{1}{2}$ of the air remaining in the container. What fractional part of the air has been removed after 6 strokes?

27. A tank contains 20 gallons of water. One-half of the water is removed and replaced with antifreeze. Then one-half of this mixture is removed and replaced with antifreeze. This process is continued 8 times. How much water remains in the tank after the 8th replacement process?

28. The radiator of a truck contains 10 gallons of water. Suppose we remove 1 gallon of water and replace it with antifreeze. Then we remove 1 gallon of this mixture and replace it with antifreeze. This process is continued 7 times. To the nearest tenth of a gallon, how much antifreeze is in the final mixture?

10.4

Mathematical Induction

Is $2^n > n$ for all positive integer values of n? In an attempt to answer this question we might proceed as follows:

If $n = 1$, then $2^n > n$ becomes $2^1 > 1$, a true statement.

If $n = 2$, then $2^n > n$ becomes $2^2 > 2$, a true statement.

If $n = 3$, then $2^n > n$ becomes $2^3 > 3$, a true statement.

We can continue in this way as long as we want, but obviously we can never show that $2^n > n$ for *every* positive integer n in this manner. However, we do have a form of proof, called **proof by mathematical induction**, that can be used to verify the truth of many mathematical statements involving positive integers. This form of proof is based on the following principle.

Principle of Mathematical Induction

Let P_n be a statement in terms of n, where n is a positive integer. If

1. P_1 is true, and
2. the truth of P_k implies the truth of P_{k+1} for every positive integer k,

then P_n is true for every positive integer n.

Using the principle of mathematical induction, a proof that some statement is true for all positive integers consists of two parts. First, we must show that the statement is true for the positive integer 1. Then we must show that if the statement is true for some positive integer, then it follows that it is also true for the next positive integer. Let's illustrate what this means.

EXAMPLE 1

Prove that $2^n > n$ for all positive integer values of n.

Proof

Part 1. If $n = 1$, then $2^n > n$ becomes $2^1 > 1$, which is a true statement.

Part 2. We must prove the statement, "If $2^k > k$, then $2^{k+1} > k + 1$ for all positive integer values of k." In other words, we should be able to start with $2^k > k$ and from that deduce $2^{k+1} > k + 1$. This can be done as follows:

$$2^k > k$$
$$2(2^k) > 2(k) \qquad \text{Multiply both sides by 2.}$$
$$2^{k+1} > 2k$$

We know that $k \geq 1$, since we are working with positive integers.

Therefore,

$$k + k \geq k + 1 \qquad \text{Add } k \text{ to both sides.}$$
$$2k \geq k + 1$$

Since $2^{k+1} > 2k$ and $2k \geq k + 1$, by the transitive property we conclude that

$$2^{k+1} > k + 1.$$

It will be helpful for you to look back over the proof in Example 1. Notice that in Part 1 we established that $2^n > n$ is true for $n = 1$. Then in Part 2 we established that if $2^n > n$ is true for any positive integer, then it must be true for the next consecutive positive integer. Therefore, since $2^n > n$ is true for $n = 1$, then it must be true for $n = 2$. Likewise, if $2^n > n$ is true for $n = 2$, then it must be true for $n = 3$, and so on, for *all* positive integers.

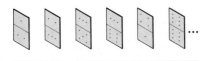

Figure 10.1

Proof by mathematical induction is sometimes physically illustrated with dominoes. Suppose that in Figure 10.1 we have infinitely many dominoes lined up. If we can push the first domino over (Part 1 of a mathematical induction proof) and if the dominoes are spaced so that each time one falls over, it causes the next one to fall over (Part 2 of a mathematical induction proof), then by pushing the first one over we will cause a chain reaction that will topple all of the dominoes (Figure 10.2).

Figure 10.2

Now let's consider another proof by mathematical induction.

EXAMPLE 2

Prove that for all positive integers n, the number $3^{2n} - 1$ is divisible by 8.

Proof

Part 1. If $n = 1$, then $3^{2n} - 1$ becomes $3^{2(1)} - 1 = 3^2 - 1 = 8$, and of course 8 is divisible by 8.

Part 2. We need to prove the statement, "If $3^{2k} - 1$ is divisible by 8, then $3^{2k+2} - 1$ is divisible by 8 for all integer values of k." This can be verified as follows. If $3^{2k} - 1$ is divisible by 8, then for some integer x, we have $3^{2k} - 1 = 8x$. Therefore,

$$3^{2k} - 1 = 8x$$
$$3^{2k} = 1 + 8x$$
$$3^2(3^{2k}) = 3^2(1 + 8x) \qquad \text{Multiply both sides by } 3^2.$$
$$3^{2k+2} = 9(1 + 8x)$$
$$3^{2k+2} = 9 + 9(8x)$$
$$3^{2k+2} = 1 + 8 + 9(8x) \qquad 9 = 1 + 8$$
$$3^{2k+2} = 1 + 8(1 + 9x) \qquad \text{Apply distributive property to } 8 + 9(8x).$$
$$3^{2k+2} - 1 = 8(1 + 9x)$$

Therefore, $3^{2k+2} - 1$ is divisible by 8.

Recall that in the first three sections of this chapter, we used a_n to represent the nth term of a sequence and S_n to represent the sum of the first n terms of a sequence. For example, if $a_n = 2n$, then the first three terms of the sequence are $a_1 = 2(1) = 2$, $a_2 = 2(2) = 4$, and $a_3 = 2(3) = 6$. Furthermore, the kth term is $a_k = 2(k) = 2k$ and the $(k + 1)$st term is $a_{k+1} = 2(k + 1) = 2k + 2$. Relative to this same sequence, we can state that $S_1 = 2$, $S_2 = 2 + 4 = 6$, and $S_3 = 2 + 4 + 6 = 12$.

There are numerous sum formulas for sequences that can be verified by mathematical induction. For such proofs, the following property of sequences is used:

$$S_{k+1} = S_k + a_{k+1}$$

This property states that **the sum of the first $k + 1$ terms is equal to the sum of the first k terms plus the $(k + 1)$st term.** Let's see how this can be used in a specific example.

EXAMPLE 3

Prove that $S_n = n(n + 1)$ for the sequence $a_n = 2n$, where n is any positive integer.

Proof

Part 1. If $n = 1$, then $1(1 + 1) = 2$ and 2 is the first term of the sequence $a_n = 2n$, so $S_1 = a_1 = 2$.

Part 2. We need to prove that if $S_k = k(k + 1)$, then $S_{k+1} = (k + 1)(k + 2)$. Using the property $S_{k+1} = S_k + a_{k+1}$, we can proceed as follows:

$$S_{k+1} = S_k + a_{k+1}$$
$$= k(k + 1) + 2(k + 1)$$
$$= (k + 1)(k + 2).$$

EXAMPLE 4

Prove that

$$S_n = \frac{5n(n + 1)}{2}$$

for the sequence $a_n = 5n$, where n is any positive integer.

Proof

Part 1. Since $5(1)(1 + 1)/2 = 5$ and 5 is the first term of the sequence $a_n = 5n$, we have $S_1 = a_1 = 5$.

Part 2. We need to prove that if $S_k = 5k(k + 1)/2$, then

$$S_{k+1} = \frac{5(k + 1)(k + 2)}{2}.$$

$$S_{k+1} = S_k + a_{k+1}$$

$$= \frac{5k(k+1)}{2} + 5(k+1)$$

$$= \frac{5k(k+1)}{2} + 5k + 5$$

$$= \frac{5k(k+1) + 2(5k+5)}{2}$$

$$= \frac{5k^2 + 5k + 10k + 10}{2}$$

$$= \frac{5k^2 + 15k + 10}{2}$$

$$= \frac{5(k^2 + 3k + 2)}{2}$$

$$= \frac{5(k+1)(k+2)}{2}$$

We conclude this section with a few final comments about proof by mathematical induction. Every mathematical induction proof is a two-part proof, and both parts are absolutely necessary. There can be mathematical statements that hold for one or the other of the two parts but not for both. For example, $(a + b)^n = a^n + b^n$ is true for $n = 1$, but it is false for every positive integer greater than 1. Therefore, if we were to attempt a mathematical induction proof for $(a + b)^n = a^n + b^n$, we could establish Part 1 but Part 2 would break down. Another example of this type is the statement, "$n^2 - n + 41$ produces a prime number for all positive integer values of n." This statement is true for $n = 1, 2, 3, 4, \ldots, 40$, but it is false when $n = 41$ (since $41^2 - 41 + 41 = 41^2$, which is not a prime number).

It is also possible that Part 2 of a mathematical induction proof can be established but Part 1 breaks down. For example, consider the sequence $a_n = n$ and the sum formula $S_n = (n + 3)(n - 2)/2$. If $n = 1$, then $a_1 = 1$ but $S_1 = (4)(-1)/2 = -2$, so Part 1 does not hold. However, it is possible to show that $S_k = (k + 3)(k - 2)/2$ implies $S_{k+1} = (k + 4)(k - 1)/2$. We will leave the details of this for you to do.

Finally, it is important to realize that some mathematical statements are true for all positive integers greater than some fixed positive integer other than 1. (Back in Figure 10.1, perhaps we cannot knock down the first four dominoes, whereas we can knock down the fifth domino and every one thereafter.) For example, we can prove by mathematical induction that $2^n > n^2$ for all positive integers $n > 4$. It requires a slight variation in the statement of the principle of mathematical induction. We will not concern ourselves with such problems in this text, but we do want you to be aware of their existence.

Problem Set 10.4

In Problems 1–10, use mathematical induction to prove that each statement is true for all positive integers n.

1. $3^n \geq 2n + 1$ **2.** $4^n \geq 4n$ **3.** $n^2 \geq n$ **4.** $2^n \geq n + 1$

5. $4^n - 1$ is divisible by 3   **6.** $5^n - 1$ is divisible by 4

7. $6^n - 1$ is divisible by 5   **8.** $9^n - 1$ is divisible by 4

9. $n^2 + n$ is divisible by 2   **10.** $n^2 - n$ is divisible by 2

In Problems 11–20, use mathematical induction to prove each of the sum formulas for the indicated sequences. They are to hold for all positive integers n.

11. $S_n = \dfrac{n(n + 1)}{2}$ for $a_n = n$

12. $S_n = n^2$ for $a_n = 2n - 1$

13. $S_n = \dfrac{n(3n + 1)}{2}$ for $a_n = 3n - 1$

14. $S_n = \dfrac{n(5n + 9)}{2}$ for $a_n = 5n + 2$

15. $S_n = 2(2^n - 1)$ for $a_n = 2^n$

16. $S_n = \dfrac{3(3^n - 1)}{2}$ for $a_n = 3^n$

17. $S_n = \dfrac{n(n + 1)(2n + 1)}{6}$ for $a_n = n^2$

18. $S_n = \dfrac{n^2(n + 1)^2}{4}$ for $a_n = n^3$

19. $S_n = \dfrac{n}{n + 1}$ for $a_n = \dfrac{1}{n(n + 1)}$

20. $S_n = \dfrac{n(n + 1)(n + 2)}{3}$ for $a_n = n(n + 1)$

Chapter 10 Summary

There are four main topics in this chapter: arithmetic sequences, geometric sequences, problem solving, and mathematical induction.

Arithmetic Sequences

The sequence $a_1, a_2, a_3, a_4, \ldots$ is called **arithmetic** if and only if

$$a_{k+1} - a_k = d$$

for every positive integer k. In other words, there is a **common difference**, d, between successive terms.

The **general term** of an arithmetic sequence is given by the formula

$$a_n = a_1 + (n - 1)d,$$

where a_1 is the first term, n is the number of terms, and d is the common difference.

The **sum** of the first n terms of an arithmetic sequence is given by the formula

$$S_n = \frac{n(a_1 + a_n)}{2}.$$

Summation notation can be used to indicate the sum of a certain number of terms of a sequence. For example,

$$\sum_{i=1}^{5} 4^i = 4^1 + 4^2 + 4^3 + 4^4 + 4^5.$$

Geometric Sequences

The sequence $a_1, a_2, a_3, a_4, \ldots$ is called **geometric** if and only if

$$a_{k+1} = ra_k$$

for every positive integer k. There is a **common ratio**, r, between successive terms.

The **general term** of a geometric sequence is given by the formula

$$a_n = a_1 r^{n-1},$$

where a_1 is the first term, n is the number of terms, and r is the common ratio.

The **sum** of the first n terms of a geometric sequence is given by the formula

$$S_n = \frac{a_1 r^n - a_1}{r - 1}, \qquad r \neq 1.$$

The **sum of an infinite geometric sequence** is given by the formula

$$S_\infty = \frac{a_1}{1 - r}, \qquad \text{for } |r| < 1.$$

If $|r| \geq 1$, the sequence has no sum.

Repeating decimals (such as $0.\overline{4}$) can be changed to a/b form, where a and b are integers and $b \neq 0$, by treating them as the sum of an infinite geometric sequence. For example, the repeating decimal $0.\overline{4}$ can be written $0.4 + 0.04 + 0.004 + 0.0004 + \cdots$.

Problem Solving

Many of the problem-solving suggestions offered earlier in this text are still appropriate when solving problems dealing with sequences. However, there are also some special suggestions pertaining to sequence problems.

1. Write down the first few terms of the sequence to describe what is taking place in the problem. A picture or diagram may help with this step.

2. Be sure that you understand, term by term, what the sequence represents in the problem.

3. Determine whether the sequence is arithmetic or geometric. (Those are the only kinds of sequences we are working with in this text.)

4. Determine whether the problem is asking for a specific term or whether it is asking for the sum of a certain number of terms.

Mathematical Induction

Proof by mathematical induction relies on the following **principle of induction**:

Let P_n be a statement in terms of n, where n is a positive integer. If

1. P_1 is true, and

2. the truth of P_k implies the truth of P_{k+1}, for every positive integer k,

then P_n is true for every positive integer n.

Chapter 10 Review Problem Set

For Problems 1–10, find the general term (the nth term) for each of the following sequences. These problems contain a mixture of arithmetic sequences and geometric sequences.

1. $3, 9, 15, 21, \ldots$ **2.** $\frac{1}{3}, 1, 3, 9, \ldots$ **3.** $10, 20, 40, 80, \ldots$

4. $5, 2, -1, -4, \ldots$ **5.** $-5, -3, -1, 1, \ldots$ **6.** $9, 3, 1, \frac{1}{3}, \ldots$

7. $-1, 2, -4, 8, \ldots$ **8.** $12, 15, 18, 21, \ldots$ **9.** $\frac{2}{3}, 1, \frac{4}{3}, \frac{5}{3}, \ldots$

10. $1, 4, 16, 64, \ldots$

For Problems 11–16, find the required term of each of the sequences.

11. The 19th term of $1, 5, 9, 13, \ldots$ **12.** The 28th term of $-2, 2, 6, 10, \ldots$

13. The 9th term of $8, 4, 2, 1, \ldots$ **14.** The 8th term of $\frac{243}{32}, \frac{81}{16}, \frac{27}{8}, \frac{9}{4}, \ldots$

15. The 34th term of $7, 4, 1, -2, \ldots$ **16.** The 10th term of $-32, 16, -8, 4, \ldots$

17. If the 5th term of an arithmetic sequence is -19 and the 8th term is -34, find the common difference of the sequence.

18. If the 8th term of an arithmetic sequence is 37 and the 13th term is 57, find the 20th term.

19. Find the first term of a geometric sequence if the 3rd term is 5 and the 6th term is 135.

20. Find the common ratio of a geometric sequence if the 2nd term is $\frac{1}{2}$ and the 6th term is 8.

21. Find the sum of the first 9 terms of the sequence $81, 27, 9, 3, \ldots$.

22. Find the sum of the first 70 terms of the sequence $-3, 0, 3, 6, \ldots$.

23. Find the sum of the first 75 terms of the sequence $5, 1, -3, -7, \ldots$.

24. Find the sum of the first 10 terms of the sequence for which $a_n = 2^{5-n}$.

25. Find the sum of the first 95 terms of the sequence for which $a_n = 7n + 1$.

26. Find the sum $5 + 7 + 9 + \cdots + 137$.

27. Find the sum $64 + 16 + 4 + \cdots + \frac{1}{64}$.

28. Find the sum of all even numbers between 8 and 384, inclusive.

29. Find the sum of all multiples of 3 between 27 and 276, inclusive.

For Problems 30–33, find each of the indicated sums.

30. $\sum_{i=1}^{45} (-2i + 5)$ **31.** $\sum_{i=1}^{5} i^3$ **32.** $\sum_{i=1}^{8} 2^{8-i}$ **33.** $\sum_{i=4}^{75} (3i - 4)$

34. Find the sum of the infinite geometric sequence $64, 16, 4, 1, \ldots$.

35. Change $0.\overline{36}$ to reduced a/b form, where a and b are integers and $b \neq 0$.

36. Change $0.4\overline{5}$ to reduced a/b form, where a and b are integers and $b \neq 0$.

Solve each of Problems 37–40 by using your knowledge of arithmetic sequences and geometric sequences.

37. Suppose that at the beginning of a year, your savings account contains $3750. If you withdraw $250 per month from the account, how much will it contain at the end of the year?

38. Sonya decides to start saving dimes. She plans to save 1 dime the first day of April, 2 dimes the second day, 3 dimes the third day, 4 dimes the fourth day, and so on for the 30 days of April. How much money will she save in April?

39. Nancy decides to start saving dimes. She plans to save 1 dime the first day of April, 2 dimes the second day, 4 dimes the third day, 8 dimes the fourth day, and so on for the first 15 days of April. How much will she save in 15 days?

40. A tank contains 61,440 gallons of water. Each day one-fourth of the water is drained out. How much remains in the tank at the end of 6 days?

For Problems 41–43, show a mathematical induction proof.

41. Prove that $5^n > 5n - 1$ for all positive integer values of n.

42. Prove that $n^3 - n + 3$ is divisible by 3 for all positive integer values of n.

43. Prove that

$$S_n = \frac{n(n + 3)}{4(n + 1)(n + 2)}$$

is the sum formula for the sequence $a_n = \dfrac{1}{n(n + 1)(n + 2)}$, where n is any positive integer.

Counting Techniques, Probability, and the Binomial Theorem

11.1 The Fundamental Principle of Counting

11.2 Permutations and Combinations

11.3 Probability

11.4 Some Properties of Probability; Expected Values

11.5 Conditional Probability; Dependent and Independent Events

11.6 The Binomial Theorem

Using an ordinary deck of 52 playing cards, there is 1 chance out of 54,145 that you will be dealt four aces in a five-card hand. The weatherman is predicting a 40% chance of locally severe thunderstorms by late afternoon. The odds in favor of the Cubs winning the pennant are 2 to 3. Suppose that in a box containing 50 light bulbs, 45 are good ones and 5 are burned out. If two bulbs are chosen at random, the probability of getting at least one good bulb is $\frac{243}{245}$. Historically, many basic probability concepts have been developed as a result of studying various games of chance. However, in recent years, probability applications have been surfacing at a phenomenal rate in a large variety of fields, such as physics, biology, psychology, economics, insurance, military science, manufacturing, and politics. It is our purpose in this chapter first to introduce some counting techniques and then to use those techniques to motivate some basic concepts of probability. The last section of the chapter will be devoted to the binomial theorem.

11.1

The Fundamental Principle of Counting

One very useful counting principle is referred to as the **fundamental principle of counting**. We will motivate this property with some examples, state the property, and then use it to solve a variety of counting-type problems. Let's consider two examples to lead up to the statement of the property.

EXAMPLE 1

A woman has 4 skirts and 5 blouses. Assuming that each blouse can be worn with each skirt, how many different skirt–blouse outfits does she have?

Solution For *each* of the 4 skirts she has a choice of 5 blouses. Therefore, she has $4(5) = 20$ different skirt–blouse outfits from which to choose.

EXAMPLE 2

Eric is shopping for a new bicycle and has 2 different models (5-speed or 10-speed) and 4 different colors (red, white, blue, or silver) from which to choose. How many different choices does he have?

Solution His different choices can be counted with the help of a **tree diagram** as follows.

Models	Colors	Choices
	red	5-speed red
	white	5-speed white
5-speed	blue	5-speed blue
	silver	5-speed silver
	red	10-speed red
	white	10-speed white
10-speed	blue	10-speed blue
	silver	10-speed silver

For *each* of the 2 model choices, there are 4 choices of color. So altogether Eric has $2 \cdot 4 = 8$ choices.

The two previous examples motivate the following general principle.

Fundamental Principle of Counting

> If one task can be accomplished in x different ways and, following this task, a second task can be accomplished in y different ways, then the first task followed by the second task can be accomplished in $x \cdot y$ different ways. (This counting principle can be extended to any finite number of tasks.)

As you apply the fundamental principle of counting, it is often helpful to analyze a problem systematically in terms of the tasks to be accomplished. Let's consider some examples to illustrate this idea.

EXAMPLE 3

How many numbers of 3 different digits each can be formed by choosing from the digits 1, 2, 3, 4, 5, and 6?

Solution Let's analyze this problem in terms of 3 tasks.

Task 1. Choose the hundred's digit, for which there are 6 choices.

Task 2. Now choose the ten's digit, for which there are only 5 choices, since one digit was used in the hundred's place.

Task 3. Now choose the unit's digit, for which there are only 4 choices, since two digits have been used for the other places.

Therefore, Task 1 followed by Task 2 followed by Task 3 can be accomplished in $(6)(5)(4) = 120$ ways. In other words, there are 120 numbers of 3 different digits that can be formed by choosing from the 6 given digits.

Now look back over the solution for Example 3 and think about each of the following questions.

1. Can we solve the problem by choosing the unit's digit first, then the ten's digit, and finally the hundred's digit?

2. How many 3-digit numbers can be formed from 1, 2, 3, 4, 5, and 6 if we do not require each number to have 3 *different* digits? (Your answer should be 216.)

3. Suppose that the digits from which to choose are 0, 1, 2, 3, 4, and 5. Now how many numbers of 3 different digits each can be formed, assuming that we do not want zero in the hundred's place? (Your answer should be 100.)

4. Suppose that we want to know the number of *even* numbers with 3 different digits each that can be formed by choosing from 1, 2, 3, 4, 5, and 6. How many are there? (Your answer should be 60.)

EXAMPLE 4

Employee ID numbers at a certain factory consist of one capital letter followed by a 3-digit number containing no repeated digits. For example, A-014 is an ID number. How many such ID numbers can be formed? How many can be formed if repeated digits are allowed?

Solution Again, let's analyze in terms of tasks to be completed.

Task 1. Choose the letter part of the ID number: there are 26 choices.

Task 2. Choose the first digit of the 3-digit number: there are 10 choices.

Task 3. Choose the second digit: there are 9 choices.

Task 4. Choose the third digit: there are 8 choices.

Therefore, applying the fundamental principle, we obtain $(26)(10)(9)(8) = 18,720$ possible ID numbers.

If repeat digits were allowed, then there would be $(26)(10)(10)(10) = 26,000$ possible ID numbers.

EXAMPLE 5

How many ways can Al, Barb, Chad, Dan, and Edna be seated in a row of 5 seats so that Al and Barb are seated side by side?

Solution This problem can be analyzed in terms of three tasks.

> **Task 1.** Choose the 2 adjacent seats to be occupied by Al and Barb. An illustration such as

> helps us to see that there are 4 choices for the two adjacent seats.

> **Task 2.** Determine the number of ways that Al and Barb can be seated. Since Al can be seated on the left and Barb on the right, or vice versa, there are 2 ways to seat Al and Barb for each pair of adjacent seats.

> **Task 3.** The remaining 3 people must be seated in the remaining 3 seats. This can be done in $(3)(2)(1) = 6$ different ways.

Therefore, by the fundamental principle, Task 1 followed by Task 2 followed by Task 3 can be done in $(4)(2)(6) = 48$ ways.

Suppose instead in Example 5 that we wanted the number of ways the five people can sit so that Al and Barb are *not* side by side. We can determine this number by using two basically different techniques: (1) analyzing and counting the number of nonadjacent positions for Al and Barb, or (2) subtracting the number of seating arrangements determined in Example 5 from the total number of ways that 5 people can be seated in 5 seats. Try doing this problem both ways and see if you agree with the answer of 72 ways.

As you apply the fundamental principle of counting, you may find that for certain problems, simply thinking about an appropriate tree diagram is helpful, even though the size of the problem may make it inappropriate to write out the diagram in detail. Consider the following example.

EXAMPLE 6

Suppose that the undergraduate students in three departments—geography, history, and psychology—are to be classified according to sex and year in school. How many categories are needed?

Solution Let's represent the various classifications symbolically as follows.

M:	Male	1.	Freshman	G:	Geography
F:	Female	2.	Sophomore	H:	History
		3.	Junior	P:	Psychology
		4.	Senior		

We can mentally picture a tree diagram such that each of the 2 sex classifications branches into 4 school-year classifications, which in turn branch into 3 department classifications. Thus, we have $(2)(4)(3) = 24$ different categories.

Another technique that works on certain problems involves what some people call the "backdoor" approach. For example, suppose we know that the classroom contains 50 seats. On some days, it may be easier to determine the number of students present by counting the number of empty seats and subtracting from 50 than by counting the number of students in attendance. (We suggested this backdoor approach as one way to count the nonadjacent seating arrangements in the discussion following Example 5.) The next example further illustrates this approach.

EXAMPLE 7

When rolling a pair of dice, in how many ways can we obtain a sum greater than 4?

Solution For clarification purposes, let's use a red die and a white die. (It is not necessary to use different colored dice, but it does help us analyze the different possible outcomes.) With a moment of thought you will see that there are more ways to get a sum greater than 4 than there are ways to get a sum of 4 or less. Therefore, let's determine the number of possibilities for getting a sum of 4 or less; then we'll subtract that number from the total number of possible outcomes when rolling a pair of dice.

First, we can simply list and count the ways of getting a sum of 4 or less.

Red die	White die
1	1
1	2
1	3
2	1
2	2
3	1

There are 6 ways of getting a sum of 4 or less.

Second, since there are 6 possible outcomes on the red die and 6 possible outcomes on the white die, there is a total of $(6)(6) = 36$ possible outcomes when rolling a pair of dice.

Therefore, subtracting the number of ways of getting 4 or less from the total number of possible outcomes, we obtain $36 - 6 = 30$ ways of getting a sum greater than 4.

Problem Set 11.1

1. If a woman has 2 skirts and 10 blouses, how many different skirt–blouse combinations does she have?

2. If a man has 8 shirts, 5 pairs of slacks, and 3 pairs of shoes, how many different shirt–slack–shoe combinations does he have?

3. In how many ways can 4 people be seated in a row of 4 seats?

4. How many numbers of two different digits can be formed by choosing from the digits 1, 2, 3, 4, 5, 6, and 7?

5. How many *even* numbers of three different digits can be formed by choosing from the digits 2, 3, 4, 5, 6, 7, 8, and 9?

6. How many *odd* numbers of four different digits can be formed by choosing from the digits 1, 2, 3, 4, 5, 6, 7, and 8?

7. Suppose that the students at a certain university are to be classified according to their college (College of Applied Science, College of Arts and Sciences, College of Business, College of Education, College of Fine Arts, College of Health and Physical Education), sex (female, male), and year in school (1, 2, 3, 4). How many categories are possible?

8. A medical researcher classifies subjects according to sex (female, male), smoking habits (smoker, nonsmoker), and weight (below average, average, above average). How many different combined classifications are used?

9. A pollster classifies voters according to sex (female, male), party affiliation (Democrat, Republican, Independent), and family income (below $10,000, $10,000–$19,999, $20,000–$29,999, $30,000–$39,999, $40,000–$49,999, $50,000 and above). How many combined classifications does the pollster use?

10. A couple is planning to have four children. How many ways can this happen relative to a boy–girl classification? (For example, *BBBG* indicates that the first three children are boys and the last is a girl.)

11. In how many ways can three officers—president, secretary, and treasurer—be selected from a club that has twenty members?

12. In how many ways can three officers—president, secretary, and treasurer—be selected from a club with 15 female and 10 male members, so that the President is female and the Secretary and Treasurer are males?

13. A disc jockey wants to play 6 songs once each in a half-hour program. How many different ways can he order these songs?

14. A state has agreed to have their automobile license plates consist of two letters followed by four digits. They do not want to repeat any letters or digits in any license number. How many different license plates will be available?

15. In how many ways can 6 people be seated in a row of 6 seats?

16. In how many ways can Al, Bob, Carl, Don, Ed, and Fern be seated in a row of 6 seats if Al and Bob want to sit side by side?

17. In how many ways can Amy, Bob, Cindy, Dan, and Elmer be seated in a row of 5 seats so that neither Amy nor Bob occupies an end seat?

18. In how many ways can Al, Bob, Carl, Don, Ed, and Fern be seated in a row of 6 seats if Al and Bob are not to be seated side by side? (Hint: Either Al and Bob will be seated side by side or they will not be seated side by side.)

19. In how many ways can Al, Bob, Carol, Dawn, and Ed be seated in a row of 5 chairs if Al is to be seated in the middle chair?

20. In how many ways can 3 letters be dropped in 5 mailboxes?

21. In how many ways can 5 letters be dropped in 3 mailboxes?

22. In how many ways can 4 letters be dropped in 6 mailboxes so that no two letters go in the same box?

23. In how many ways can 6 letters be dropped in 4 mailboxes so that no two letters go in the same box?

24. If 5 coins are tossed, in how many ways can they fall?

25. If 3 dice are tossed, in how many ways can they fall?

26. In how many ways can a sum less than 10 be obtained when tossing a pair of dice?

✓ **27.** In how many ways can a sum greater than 5 be obtained when tossing a pair of dice?

28. In how many ways can a sum greater than 4 be obtained when tossing *three* dice?

✓ **29.** If no number contains repeated digits, how many numbers greater than 400 can be formed by choosing from the digits 2, 3, 4, and 5? (Hint: Consider both 3-digit and 4-digit numbers.)

30. If no number contains repeated digits, how many numbers greater than 5000 can be formed by choosing from the digits 1, 2, 3, 4, 5, and 6?

✓ **31.** In how many ways can 4 boys and 3 girls be seated in a row of 7 seats so that boys and girls occupy alternate seats?

32. In how many ways can 3 different mathematics books and 4 different history books be exhibited on a shelf so that all of the books in a subject area are side-by-side?

33. In how many ways can a true–false test of 10 questions be answered?

34. If no number contains repeated digits, how many even numbers greater than 3000 can be formed by choosing from the digits 1, 2, 3, and 4?

✓ **35.** If no number contains repeated digits, how many odd numbers greater than 40,000 can be formed by choosing from the digits 1, 2, 3, 4, and 5?

36. In how many ways can Al, Bob, Carol, Don, Ed, Faye, and George be seated in a row of 7 seats so that Al, Bob, and Carol occupy consecutive seats in some order?

37. The license plates for a certain state consist of two letters followed by a four-digit number such that the first digit of the number is not zero. An example would be PK-2446.
(a) How many different license plates can be produced?
(b) How many different plates do not have a repeated letter?
(c) How many plates do not have any repeated digits in the number part of the plate?
(d) How many plates do not have a repeated letter and also do not have any repeated numbers?

11.2

Permutations and Combinations

As we develop the material in this section, **factorial notation** becomes very useful. The notation $n!$ (read "n factorial") is used with positive integers as follows.

$$1! = 1$$
$$2! = 2 \cdot 1 = 2$$
$$3! = 3 \cdot 2 \cdot 1 = 6$$
$$4! = 4 \cdot 3 \cdot 2 \cdot 1 = 24$$

Notice that the factorial notation refers to an *indicated product*. In general, we write

$$n! = n(n-1)(n-2) \cdots 3 \cdot 2 \cdot 1.$$

Now, as an introduction to the first concept of this section, let's consider a counting problem that closely resembles problems from the previous section.

EXAMPLE 1

In how many ways can the 3 letters, A, B, and C, be arranged in a row?

Solution A Certainly one approach to the problem is simply to list and count the arrangements.

ABC, ACB, BAC, BCA, CAB, CBA

There are 6 arrangements of the 3 letters.

Solution B Another approach, one that can be generalized for more difficult problems, uses the fundamental principle of counting. Since there are 3 choices for the first letter of an arrangement, 2 choices for the second letter, and 1 choice for the third letter, there are $(3)(2)(1) = 6$ different arrangements.

Ordered arrangements are called **permutations**. In general, a permutation of a set of n elements is an ordered arrangement of the n elements; we will use the symbol $P(n, n)$ to denote the number of such permutations. For example, from Example 1 we know that $P(3, 3) = 6$. Furthermore, by using the same basic approach as in Solution B of Example 1, we can obtain the following results.

$$P(1, 1) = 1 = 1!$$

$$P(2, 2) = 2 \cdot 1 = 2!$$

$$P(4, 4) = 4 \cdot 3 \cdot 2 \cdot 1 = 4!$$

$$P(5, 5) = 5 \cdot 4 \cdot 3 \cdot 2 \cdot 1 = 5!$$

In general, the following formula becomes evident.

$$P(n, n) = n!$$

Now suppose that we are interested in the number of 2-letter permutations that can be formed by choosing from the 4 letters A, B, C, and D. (Some examples of such permutations are AB, BA, AC, CA, BC, and CB.) In other words, we want to find the number of 2-element permutations that can be formed from a set of 4 elements. We denote this number by $P(4, 2)$. To find $P(4, 2)$, we can reason as follows: First, we can choose any one of the 4 letters to occupy the first position in the permutation, and then we can choose any one of the 3 remaining letters for the second position. Therefore, by the fundamental principle of counting, we have $(4)(3) = 12$ different 2-letter permutations; that is, $P(4, 2) = 12$. By using a similar line of reasoning, the following numbers can be determined. (Make sure that you agree with each of these.)

$$P(4, 3) = 4 \cdot 3 \cdot 2 = 24$$

$$P(5, 2) = 5 \cdot 4 = 20$$

$$P(6, 4) = 6 \cdot 5 \cdot 4 \cdot 3 = 360$$

$$P(7, 3) = 7 \cdot 6 \cdot 5 = 210$$

In general, we say that **the number of *r*-element permutations that can be formed from a set of *n* elements is given by**

$$P(n, r) = \underbrace{n(n - 1)(n - 2) \cdots}_{r \text{ factors}}$$

Notice that the indicated product for $P(n, r)$ begins with n. Thereafter each factor is one less than the previous one and there is a total of r factors. For example,

$$P(6, 2) = 6 \cdot 5 = 30,$$

$$P(8, 3) = 8 \cdot 7 \cdot 6 = 336,$$

$$P(9, 4) = 9 \cdot 8 \cdot 7 \cdot 6 = 3024.$$

Let's consider two examples illustrating the use of $P(n, n)$ and $P(n, r)$.

EXAMPLE 2

In how many ways can 5 students be seated in a row of 5 seats?

Solution The problem is asking for the number of 5-element permutations that can be formed from a set of 5 elements. Thus, we can apply $P(n, n) = n!$.

$$P(5, 5) = 5! = 5 \cdot 4 \cdot 3 \cdot 2 \cdot 1 = 120$$

EXAMPLE 3

Suppose that 7 people enter a swimming race. In how many ways can 1st, 2nd, and 3rd prizes be awarded?

Solution This problem is asking for the number of 3-element permutations that can be formed from a set of 7 elements. Therefore, using the formula for $P(n, r)$, we obtain

$$P(7, 3) = 7 \cdot 6 \cdot 5 = 210.$$

It should be evident that both Example 2 and Example 3 could have been solved by applying the fundamental principle of counting. In fact, it should be noted that the formulas for $P(n, n)$ and $P(n, r)$ do not really give us much additional problem-solving power. However, as we will see in a moment, they do provide the basis for developing a formula that is very useful as a problem-solving tool.

Permutations Involving Nondistinguishable Objects

Suppose we have two identical H's and one T in an arrangement such as HTH. If we switch the two identical H's, the newly formed arrangement, HTH, will not be distinguishable from the original. In other words, there are fewer distinguishable permutations of n elements when some of those elements are identical than there are when the n elements are distinctly different.

To see the effect of identical elements on the number of distinguishable permutations, let's look at some specific examples.

2 identical H's: 1 permutation (HH)
2 different letters: 2! permutations (HT, TH)
Therefore, having 2 different letters affects the number of permutations by a *factor of* 2!.

3 identical H's: 1 permutation (HHH)
3 different letters: 3! permutations
Therefore, having 3 different letters affects the number of permutations by a *factor of* 3!.

4 identical H's: 1 permutation (HHHH)
4 different letters: 4! permutations
Therefore, having 4 different letters affects the number of permutations by a *factor of* 4!.

Now let's solve a specific problem.

EXAMPLE 4

How many distinguishable permutations can be formed from 3 identical H's and 2 identical T's?

Solution If we had 5 distinctly different letters, we could form 5! permutations. But the 3 identical H's affect the number of distinguishable permutations by a factor of 3!, and the 2 identical T's affect the number of permutations by a factor of 2!. Therefore, we must divide 5! by 3! and 2!. Thus, we obtain

$$\frac{5!}{(3!)(2!)} = \frac{5 \cdot \overset{2}{4} \cdot 3 \cdot 2 \cdot 1}{3 \cdot 2 \cdot 1 \cdot 2 \cdot 1} = 10$$

distinguishable permutations of 3 H's and 2 T's.

The type of reasoning used in Example 4 leads us to the following general counting technique: If there are n elements to be arranged, where there are r_1 of one kind, r_2 of another kind, r_3 of another kind, . . . , r_k of a kth kind, then the total number of distinguishable permutations is given by the expression

$$\frac{n!}{(r_1!)(r_2!)(r_3!) \cdots (r_k!)}.$$

EXAMPLE 5

How many different 11-letter permutations can be formed from the 11 letters of the word MISSISSIPPI?

Solution Since there are 4 I's, 4 S's, and 2 P's we can form

$$\frac{11!}{(4!)(4!)(2!)} = \frac{11 \cdot 10 \cdot 9 \cdot 8 \cdot 7 \cdot 6 \cdot 5 \cdot 4 \cdot 3 \cdot 2 \cdot 1}{4 \cdot 3 \cdot 2 \cdot 1 \cdot 4 \cdot 3 \cdot 2 \cdot 1 \cdot 2 \cdot 1} = 34{,}650$$

distinguishable permutations.

Combinations (Subsets)

Permutations are *ordered* arrangements; however, frequently *order* is not a consideration. For example, suppose that we want to determine the number of 3-person committees that can be formed from the 5 people, Al, Barb, Carol, Dawn, and Eric. Certainly the committee consisting of Al, Barb, and Eric is the same as the committee consisting of Barb, Eric, and Al. In other words, the order in which we choose or list the members is not important. Therefore, we are really dealing with subsets; that is, we are looking for the number of 3-element subsets that can be formed from a set of 5 elements. Traditionally in this context, subsets have been called **combinations**. So, stated another way, we are looking for the number of combinations of 5 things taken 3 at a time. In general, *r*-element subsets taken from a set of *n* elements are called **combinations of n things taken r at a time**. The symbol $C(n, r)$ denotes the number of these combinations.

Now let's restate the previous committee problem and show a detailed solution that can be generalized to handle a variety of problems dealing with combinations.

EXAMPLE 6

How many 3-person committees can be formed from the 5 people, Al, Barb, Carol, Dawn, and Eric?

Solution Let's use the set $\{A, B, C, D, E\}$ to represent the 5 people. Consider one possible 3-person committee (subset), such as $\{A, B, C\}$; there are 3! permutations of these 3 letters. Now take another committee, such as $\{A, B, D\}$; there are also 3! permutations of these 3 letters. If we were to continue this process with all of the 3-letter subsets that can be formed from the 5 letters, we would be counting all possible 3-letter permutations of the 5 letters. That is, we would obtain $P(5, 3)$. Therefore, if we let $C(5, 3)$ represent the number of 3-element subsets, then

$$(3!) \cdot C(5, 3) = P(5, 3).$$

Solving this equation for $C(5, 3)$ yields

$$C(5, 3) = \frac{P(5, 3)}{3!} = \frac{5 \cdot 4 \cdot 3}{3 \cdot 2 \cdot 1} = 10.$$

So there are *ten* 3-person committees that can be formed from the 5 people.

In general, $C(n, r)$ times r! yields $P(n, r)$. Thus,

$$(r!) \cdot C(n, r) = P(n, r),$$

and solving this equation for $C(n, r)$ produces

$$C(n, r) = \frac{P(n, r)}{r!}.$$

In other words, we can find the number of *combinations* of n things taken r at a time by dividing by $r!$ the number of permutations of n things taken r at a time. The following examples illustrate this idea.

$$C(7, 3) = \frac{P(7, 3)}{3!} = \frac{7 \cdot 6 \cdot 5}{3 \cdot 2 \cdot 1} = 35$$

$$C(9, 2) = \frac{P(9, 2)}{2!} = \frac{9 \cdot 8}{2 \cdot 1} = 36$$

$$C(10, 4) = \frac{P(10, 4)}{4!} = \frac{10 \cdot 9 \cdot 8 \cdot 7}{4 \cdot 3 \cdot 2 \cdot 1} = 210$$

EXAMPLE 7

How many different 5-card hands can be dealt from a deck of 52 playing cards?

Solution Since the order in which the cards are dealt is not an issue, we are working with a combination (subset) problem. Thus, using the formula for $C(n, r)$, we obtain

$$C(52, 5) = \frac{P(52, 5)}{5!} = \frac{52 \cdot 51 \cdot 50 \cdot 49 \cdot 48}{5 \cdot 4 \cdot 3 \cdot 2 \cdot 1} = 2,598,960.$$

There are 2,598,960 different 5-card hands that can be dealt from a deck of 52 playing cards.

Some counting problems can be solved by using the fundamental principle of counting along with the combination formula, as the next example illustrates.

EXAMPLE 8

How many committees consisting of 3 women and 2 men can be formed from a group of 5 women and 4 men?

Solution Let's think of this problem in terms of two tasks.

Task 1. Choose a subset of 3 women from the 5 women. This can be done in

$$C(5, 3) = \frac{P(5, 3)}{3!} = \frac{5 \cdot 4 \cdot 3}{3 \cdot 2 \cdot 1} = 10 \text{ ways.}$$

Task 2. Choose a subset of 2 men from the 4 men. This can be done in

$$C(4, 2) = \frac{P(4, 2)}{2!} = \frac{4 \cdot 3}{2 \cdot 1} = 6 \text{ ways.}$$

Task 1 followed by Task 2 can be done in $(10)(6) = 60$ ways. Therefore, there are 60 committees consisting of 3 women and 2 men that can be formed.

Sometimes it takes a little thought to decide whether permutations or combinations should be used. Remember that **if order is to be considered, permutations should be used, but if order does not matter, then use combinations**. It is helpful to think of combinations as subsets.

EXAMPLE 9

A small accounting firm has 12 computer programmers. Three of these people are to be promoted to systems analysts. In how many ways can the firm select the 3 people to be promoted?

Solution Let's call the people A, B, C, D, E, F, G, H, I, J, K, and L. Suppose A, B, and C are chosen for promotion. Is this any different than choosing B, C, and A? Obviously not, and therefore order does not matter and we are being asked a question about combinations. More specifically, we need to find the number of combinations of 12 people taken 3 at a time. Thus, there are

$$C(12, 3) = \frac{P(12, 3)}{3!} = \frac{12 \cdot 11 \cdot 10}{3 \cdot 2 \cdot 1} = 220$$

different ways to choose the 3 people to be promoted.

EXAMPLE 10

A club is to elect 3 officers—president, secretary, and treasurer—from a group of 6 people, all of whom are willing to serve in any office. How many different ways can the officers be chosen?

Solution Let's call the candidates A, B, C, D, E, and F. Is electing A as president, B as secretary, and C as treasurer different than electing B as president, C as secretary, and A as treasurer? Obviously it is, and therefore we are working with permutations. Thus, there are

$$P(6, 3) = 6 \cdot 5 \cdot 4 = 120$$

different ways of filling the offices.

Problem Set 11.2

In Problems 1–12, evaluate each of the following.

1. $P(5, 3)$	**2.** $P(8, 2)$	**3.** $P(6, 4)$	**4.** $P(9, 3)$
5. $C(7, 2)$	**6.** $C(8, 5)$	**7.** $C(10, 5)$	**8.** $C(12, 4)$
9. $C(15, 2)$	**10.** $P(5, 5)$	**11.** $C(5, 5)$	**12.** $C(11, 1)$

13. How many permutations of the 4 letters A, B, C, and D can be formed by using all the letters in each permutation?

14. In how many ways can 6 students be seated in a row of 6 seats?

15. How many 3-person committees can be formed from a group of 9 people?

16. How many 2-card hands can be dealt from a deck of 52 playing cards?

17. How many 3-letter permutations can be formed from the first 8 letters of the alphabet if **(a)** repetitions are not allowed? **(b)** repetitions are allowed?

18. In a 7-team baseball league, in how many ways can the top 3 positions in the final standings be filled?

19. In how many ways can the manager of a baseball team arrange his batting order if he wants his 4 best hitters in the top 4 positions?

20. In a baseball league of 9 teams, how many games are needed to complete the schedule if each team plays 12 games with each other team?

21. How many committees consisting of 4 women and 4 men can be chosen from a group of 7 women and 8 men?

22. How many 3-element subsets containing one vowel and two consonants can be formed from the set $\{a, b, c, d, e, f, g, h, i\}$?

23. Five associate professors are being considered for promotion to the rank of full professor, but only 3 will be promoted. How many different ways are there of selecting the 3 to be promoted?

24. How many numbers of 4 different digits can be formed from the digits 1, 2, 3, 4, 5, 6, 7, 8, and 9, if each number must consist of 2 odd and 2 even digits?

25. How many 3-element subsets containing the letter A can be formed from the set $\{A, B, C, D, E, F\}$?

26. How many 4-person committees can be chosen from 5 women and 3 men if each committee must contain at least 1 man?

27. How many different 7-letter permutations can be formed from 4 identical H's and 3 identical T's?

28. How many different 8-letter permutations can be formed from 6 identical H's and 2 identical T's?

29. How many different 9-letter permutations can be formed from 3 identical A's, 4 identical B's, and 2 identical C's?

30. How many different 10-letter permutations can be formed from 5 identical A's, 4 identical B's, and one C?

31. How many different 7-letter permutations can be formed from the 7 letters of the word ALGEBRA?

32. How many different 11-letter permutations can be formed from the 11 letters of the word MATHEMATICS?

33. In how many ways can x^4y^2 be written without using exponents? (Hint: One way is *xxxxyy*.)

34. In how many ways can $x^3y^4z^3$ be written without using exponents?

35. Ten basketball players are going to divide into two teams of 5 players each for a game. In how many ways can this be done?

36. Ten basketball players are going to divide into 2 teams of 5 in such a way that the 2 best players are on opposite teams. In how many ways can this be done?

37. A box contains 9 good light bulbs and 4 defective bulbs. How many samples of 3 bulbs contain 1 defective bulb? How many samples of 3 bulbs contain *at least* 1 defective bulb?

38. How many 5-person committees consisting of 2 juniors and 3 seniors can be formed from a group of 6 juniors and 8 seniors?

39. In how many ways can 6 people be divided into 2 groups so that there are 4 in one group and 2 in the other? In how many ways can 6 people be divided into 2 groups of 3 each?

40. How many 5-element subsets containing both A and B can be formed from the set $\{A, B, C, D, E, F, G, H\}$?

41. How many 4-element subsets containing A or B but not both A and B can be formed from the set $\{A, B, C, D, E, F, G\}$?

42. How many different 5-person committees can be selected from 9 people if 2 of those people refuse to serve together on a committee?

43. How many different line segments are determined by 5 points? By 6 points? By 7 points? By n points?

44. (a) How many 5-card hands consisting of two kings and three aces can be dealt from a deck of 52 playing cards?
   (b) How many 5-card hands consisting of three kings and two aces can be dealt from a deck of 52 playing cards?
   (c) How many 5-card hands consisting of three cards of one face value and two cards of another face value can be dealt from a deck of 52 playing cards?

Miscellaneous Problems

45. In how many ways can 6 people be seated at a circular table? (Hint: Moving each person one place to the right (or left) does not create a new seating arrangement.)

46. The quantity $P(8, 3)$ can be expressed completely in factorial notation as follows.

$$P(8, 3) = \frac{P(8, 3) \cdot 5!}{5!} = \frac{(8 \cdot 7 \cdot 6)(5 \cdot 4 \cdot 3 \cdot 2 \cdot 1)}{5!} = \frac{8!}{5!}$$

Express each of the following in terms of factorial notation.
   (a) $P(7, 3)$ **(b)** $P(9, 2)$ **(c)** $P(10, 7)$
   (d) $P(n, r)$, $r \leq n$ and $0!$ is defined to be 1

47. Sometimes the formula

$$C(n, r) = \frac{n!}{r!(n - r)!}$$

is used to find the number of combinations of n things taken r at a time. Use the result from Problem 46(d) and develop this formula.

48. Compute $C(7, 3)$ and $C(7, 4)$. Compute $C(8, 2)$ and $C(8, 6)$. Compute $C(9, 8)$ and $C(9, 1)$. Now argue that $C(n, r) = C(n, n - r)$ for $r \leq n$.

11.3

Probability

In order to introduce some terminology and notation, let's consider a simple experiment of tossing a regular 6-sided die. There are 6 possible outcomes to this experiment: either the 1, the 2, the 3, the 4, the 5, or the 6 will land up. This set of possible outcomes is called a *sample space* and the individual elements of the sample space are called *sample points*. We will use S (sometimes with subscripts for identification purposes) to refer to a particular sample space of an experiment; then we will denote the number of sample points by $n(S)$. Thus, for the experiment of tossing a die, $S = \{1, 2, 3, 4, 5, 6\}$ and $n(S) = 6$.

In general, the set of all possible outcomes of a given experiment is called the **sample space** and the individual elements of the sample space are called **sample points**. (In this text we will be working only with sample spaces that are finite.)

Now suppose we are interested in some of the various possible outcomes in the die-tossing experiment. For example, we might be interested in the event, "An even number comes up." In this case we are satisfied if a 2, 4, or 6 appears on the top face of the die, and therefore the event "An even number comes up" is the subset $E = \{2, 4, 6\}$, where $n(E) = 3$. Perhaps, instead, we might be interested in the event, "A multiple of 3 comes up." This event determines the subset $F = \{3, 6\}$, where $n(F) = 2$.

In general, any subset of a sample space is called an **event** or an **event space**. If the event consists of exactly one element of the sample space, then it is called a **simple event**. Any nonempty event that is not simple is called a **compound event**. A compound event can be represented as the union of simple events.

It is now possible to give a very simple definition for *probability* as we want to use it in this text.

DEFINITION 11.1

In an experiment for which all possible outcomes in the sample space S are equally likely to occur, the **probability** of an event E is defined by

$$P(E) = \frac{n(E)}{n(S)},$$

where $n(E)$ denotes the number of elements in the event E and $n(S)$ denotes the number of elements in the sample space S.

Many probability problems can be solved by applying Definition 11.1. Such an approach requires that we be able to determine the number of elements in the sample space and the number of elements in the event space. For example, returning to the die-tossing experiment, the probability of getting an even number with one toss of a die is given by

$$P(E) = \frac{n(E)}{n(S)} = \frac{3}{6} = \frac{1}{2}.$$

Let's consider two examples where the number of elements in both the sample space and the event space are quite easy to determine.

EXAMPLE 1

A coin is tossed. Find the probability that a head turns up.

Solution Let the sample space be $S = \{H, T\}$; then $n(S) = 2$. The event of turning up a head is the subset $E = \{H\}$, so $n(E) = 1$. Therefore the probability of getting a head with one flip of a coin is given by

$$P(E) = \frac{n(E)}{n(S)} = \frac{1}{2}.$$

EXAMPLE 2

Two coins are tossed. What is the probability that *at least* one head will turn up?

Solution For clarification purposes, let the coins be a penny and a nickel. The possible outcomes of this experiment are (1) a head on both coins, (2) a head on the penny and a tail on the nickel, (3) a tail on the penny and a head on the nickel, or (4) a tail on both coins. Using ordered-pair notation, where the first entry of a pair represents the penny and the second entry the nickel, the sample space can be written

$$S = \{(H, H), (H, T), (T, H), (T, T)\}.$$

and $n(S) = 4$.

Let E be the event of getting at least one head. Thus, $E = \{(H, H), (H, T), (T, H)\}$ and $n(E) = 3$. Therefore, the probability of getting at least one head with one toss of two coins is

$$P(E) = \frac{n(E)}{n(S)} = \frac{3}{4}.$$

As you might expect, the counting techniques discussed in the first two sections of this chapter can frequently be used to solve probability problems.

EXAMPLE 3

Four coins are tossed. Find the probability of getting 3 heads and 1 tail.

Solution The sample space consists of the possible outcomes for tossing 4 coins. Since there are 2 things that can happen on each coin, by the fundamental principle of counting there are $2 \cdot 2 \cdot 2 \cdot 2 = 16$ possible outcomes for tossing 4 coins. So we know that $n(S) = 16$ without taking the time to list all of the elements. The event of getting 3 heads and 1 tail is the subset $E = \{(H, H, H, T), (H, H, T, H), (H, T, H, H), (T, H, H, H)\}$, where $n(E) = 4$. Therefore, the requested probability is

$$P(E) = \frac{n(E)}{n(S)} = \frac{4}{16} = \frac{1}{4}.$$

EXAMPLE 4

Al, Bob, Chad, Dawn, Eve, and Francis are randomly seated in a row of 6 chairs. What is the probability that Al and Bob are seated in the end seats?

Solution The sample space consists of all possible ways of seating 6 people in 6 chairs, or in other words, the permutations of 6 things taken 6 at a time. Thus, $n(S) = P(6, 6) = 6! = 6 \cdot 5 \cdot 4 \cdot 3 \cdot 2 \cdot 1 = 720.$

The event space consists of all possible ways of seating the 6 people so that Al and Bob both occupy end seats. The number of these possibilities can be counted as follows.

Task 1. Put Al and Bob in the end seats. This can be done in 2 ways since Al can be on the left end and Bob on the right end, or vice versa.

Task 2. Put the other 4 people in the remaining 4 seats. This can be done in $4! = 4 \cdot 3 \cdot 2 \cdot 1 = 24$ different ways.

Therefore, Task 1 followed by Task 2 can be done in $(2)(24) = 48$ different ways; so $n(E) = 48$. Thus, the requested probability is

$$P(E) = \frac{n(E)}{n(S)} = \frac{48}{720} = \frac{1}{15}.$$

Notice that in Example 3, by using the fundamental principle of counting to determine the number of elements in the sample space, we did not actually have to list all of the elements. For the event space, we listed the elements and counted them in the usual way. In Example 4 we used the permutation formula $P(n, n) = n!$ to determine the number of elements in the sample space, and then we used the fundamental principle to determine the number of elements in the event space. There are no definite rules about when to list the elements and when to apply some sort of counting technique. In general, we suggest that if you do not immediately see a counting pattern for a particular problem, you should begin the listing process. If a counting pattern then emerges as you are listing the elements, use the pattern at that time.

The combination (subset) formula we developed in Section 11.2, $C(n, r) = P(n, r)/r!$, is also a very useful tool for solving certain kinds of probability problems. The next three examples illustrate some problems of this type.

EXAMPLE 5

A committee of 3 people is randomly selected from Alice, Barb, Chad, Dee, and Eric. What is the probability that Alice is on the committee?

Solution The sample space, S, consists of all possible 3-person committees that can be formed from the 5 people. Therefore,

$$n(S) = C(5, 3) = \frac{P(5, 3)}{3!} = \frac{5 \cdot 4 \cdot 3}{3 \cdot 2 \cdot 1} = 10.$$

The event space, E, consists of all of the 3-person committees that have Alice as a member. Each of these committees contains Alice and 2 other people chosen from the 4 remaining people. Thus, the number of such committees is $C(4, 2)$. So we obtain

$$n(E) = C(4, 2) = \frac{P(4, 2)}{2!} = \frac{4 \cdot 3}{2 \cdot 1} = 6.$$

The requested probability is

$$P(E) = \frac{n(E)}{n(S)} = \frac{6}{10} = \frac{3}{5}.$$

EXAMPLE 6

A committee of 4 is chosen at random from a group of 5 seniors and 4 juniors. Find the probability that the committee will contain 2 seniors and 2 juniors.

Solution The sample space, S, consists of all possible 4-person committees that can be formed from the 9 people. Thus,

$$n(S) = C(9, 4) = \frac{P(9, 4)}{4!} = \frac{9 \cdot 8 \cdot 7 \cdot 6}{4 \cdot 3 \cdot 2 \cdot 1} = 126.$$

The event space, E, consists of all 4-person committees that contain 2 seniors and 2 juniors. They can be counted as follows.

Task 1. Choose 2 seniors from the 5 available seniors in $C(5, 2) = 10$ ways.

Task 2. Choose 2 juniors from the 4 available juniors in $C(4, 2) = 6$ ways.

Therefore, there are $10 \cdot 6 = 60$ committees consisting of 2 seniors and 2 juniors. The requested probability is

$$P(E) = \frac{n(E)}{n(S)} = \frac{60}{126} = \frac{10}{21}.$$

EXAMPLE 7

Eight coins are tossed. Find the probability of getting 2 heads and 6 tails.

Solution Since two things can happen on each coin, the total number of possible outcomes, $n(S)$, is $2^8 = 256$.

We can select 2 coins, which are to fall heads, in $C(8, 2) = 28$ ways. For each of these ways, there is only one way to select the other 6 coins that are to fall tails. Therefore, there are $28 \cdot 1 = 28$ ways of getting 2 heads and 6 tails; so $n(E) = 28$. The requested probability is

$$P(E) = \frac{n(E)}{n(S)} = \frac{28}{256} = \frac{7}{64}.$$

Problem Set 11.3

For Problems 1–4, *two* coins are tossed. Find the probability of tossing each of the following events.

1. One head and one tail

2. Two tails

3. At least one tail

4. No tails

For Problems 5–8, *three* coins are tossed. Find the probability of tossing each of the following events.

5. Three heads

6. Two heads and a tail

7. At least one head

8. Exactly one tail

For Problems 9–12, *four* coins are tossed. Find the probability of tossing each of the following events.

9. Four heads

10. Three heads and a tail

11. Two heads and two tails

12. At least one head

For Problems 13–16, *one* die is tossed. Find the probability of rolling each of the following events.

13. A multiple of 3

14. A prime number

15. An even number

16. A multiple of 7

For Problems 17–22, *two* dice are tossed. Find the probability of rolling each of the following events.

17. A sum of 6

18. A sum of 11

19. A sum less than 5

20. A 5 on exactly one die

21. A 4 on at least one die

22. A sum greater than 4

For Problems 23–26, *one* card is drawn from a standard deck of 52 playing cards. Find the probability of each of the following events.

23. A heart is drawn.

24. A king is drawn.

25. A spade or a diamond is drawn.

26. A red jack is drawn.

For Problems 27–30, suppose that *25* slips of paper numbered 1 to 25, inclusive, are put in a hat and then one is drawn out at random. Find the probability of each of the following events.

27. The slip with the 5 on it is drawn.

28. A slip with an even number on it is drawn.

29. A slip with a prime number on it is drawn.

30. A slip with a multiple of 6 on it is drawn.

For Problems 31–34, suppose that a committee of 2 boys is to be chosen at random from the 5 boys, Al, Bill, Carl, Dan, and Elmer. Find the probability of each of the following events.

31. Dan is on the committee.

32. Dan and Elmer are both on the committee.

33. Bill and Carl are not both on the committee.

34. Dan or Elmer but not both of them are on the committee.

For Problems 35–38, suppose that a 5-person committee is selected at random from the 8 people, Al, Barb, Chad, Dawn, Eric, Fern, George, and Harriet. Find the probability of each of the following events.

35. Al and Barb are both on the committee.

36. George is not on the committee.

37. Either Chad or Dawn, but not both, is on the committee.

38. Neither Al nor Barb is on the committee.

For Problems 39–41, suppose that a box of 10 items from a manufacturing company is known to contain 2 defective and 8 nondefective items. If a sample of 3 items is selected at random, find the probabilities of each of the following events.

39. The sample contains all nondefective items.

40. The sample contains 1 defective and 2 nondefective items.

41. The sample contains 2 defective and 1 nondefective items.

Solve the following problems.

42. A building has 5 doors. Find the probability that 2 people, entering the building at random, will choose the same door.

43. Bill, Carol, and Alice are seated at random in a row of 3 seats. Find the probability that Bill and Carol will be seated side by side.

44. April, Bill, Carl, and Denise are to be seated at random in a row of 4 chairs. What is the probability that April and Bill will occupy the end seats?

45. A committee of 4 girls is to be chosen at random from the 5 girls, Alice, Becky, Candy, Dee, and Elaine. Find the probability that Elaine is not on the committee.

46. Three boys and 2 girls are randomly seated in a row of 5 seats. What is the probability that the boys and girls will be in alternate seats?

47. Four different mathematics books and 5 different history books are randomly placed on a shelf. What is the probability that all of the books on a subject are side by side?

48. Each of 3 letters is to be mailed in any one of 5 different mailboxes. What is the probability that all are mailed in the same mail box?

49. Randomly form a 4-digit number by using the digits 2, 3, 4, and 6 once each. What is the probability that the number formed is greater than 4000?

50. Randomly select one of the 120 permutations of the letters a, b, c, d, and e. Find the probability that in the chosen permutation, the letter a precedes the b (the a is to the left of the b).

51. A committee of 4 is chosen at random from a group of 6 women and 5 men. Find the probability that the committee contains 2 women and 2 men.

52. A committee of 3 is chosen at random from a group of 4 women and 5 men. Find the probability that the committee will contain at least 1 man.

53. Al, Bob, Carl, Dan, Ed, Frank, Gino, Harry, Jerry, and Mike are randomly divided into two 5-man teams for a basketball game. What is the probability that Al, Bob, and Carl are on the same team?

54. Seven coins are tossed. Find the probability of getting 4 heads and 3 tails.

55. Nine coins are tossed. Find the probability of getting 3 heads and 6 tails.

56. Six coins are tossed. Find the probability of getting at least 4 heads.

57. Five coins are tossed. Find the probability of getting no more than 3 heads.

58. Each arrangement of the 11 letters of the word MISSISSIPPI is put on a slip of paper and placed in a hat. One slip is drawn at random from the hat. Find the probability that the slip contains an arrangement of the letters with the 4 S's at the beginning.

59. Each arrangement of the 7 letters of the word OSMOSIS is put on a slip of paper and placed in a hat. One slip is drawn at random from the hat. Find the probability that the slip contains an arrangement of the letters with an O at the beginning and an O at the end.

60. Consider all possible arrangements of 3 identical H's and 3 identical T's. Suppose that one of these arrangements is selected at random. What is the probability that the selected arrangement has the 3 H's in consecutive positions?

Miscellaneous Problems

In Example 7 of Section 11.2, we found that there are 2,598,960 different 5-card hands that can be dealt from a deck of 52 playing cards. Therefore, probabilities for certain kinds of 5-card poker hands can be calculated using 2,598,960 as the number of elements in the sample space. For Problems 61–69, determine the number of different 5-card poker hands of the indicated type that can be obtained.

61. A straight flush (five cards in sequence and of the same suit; aces are both low and high, so A2345 and 10JQKA are both acceptable)

62. Four of a kind (4 of the same face value, such as 4 kings)

63. A full house (3 cards of one face value and 2 cards of another face value)

64. A flush (5 cards of the same suit but not in sequence)

65. A straight (5 cards in sequence but not all of the same suit)

66. Three of a kind (3 cards of one face value and 2 cards of two different face values)

67. Two pairs   **68.** Exactly one pair   **69.** No pairs

11.4

Some Properties of Probability; Expected Values

There are several basic properties that are useful in the study of probability from both a theoretical and a computational viewpoint. We will discuss two of these properties at this time and some additional ones in the next section. The first property may seem to state the obvious, but it still needs to be mentioned.

PROPERTY 11.1

For all events E,

$$0 \le P(E) \le 1.$$

Property 11.1 simply states that probabilities must fall in the range from 0 to 1, inclusive. This should seem reasonable since $P(E) = n(E)/n(S)$, and E is a subset of S. The next two examples illustrate circumstances for which $P(E) = 0$ and $P(E) = 1$.

EXAMPLE 1

Toss a regular 6-sided die. What is the probability of getting a 7?

Solution The sample space is $S = \{1, 2, 3, 4, 5, 6\}$, where $n(S) = 6$. The event space is $E = \emptyset$, so $n(E) = 0$. Therefore, the probability of getting a 7 is

$$P(E) = \frac{n(E)}{n(S)} = \frac{0}{6} = 0.$$

EXAMPLE 2

What is the probability of getting a head or a tail with one flip of a coin?

Solution The sample space is $S = \{H, T\}$ and the event space is $E = \{H, T\}$. Therefore, $n(S) = n(E) = 2$ and

$$P(E) = \frac{n(E)}{n(S)} = \frac{2}{2} = 1.$$

An event that has a probability of 1 is sometimes called **certain success**, and an event with a probability of zero is called **certain failure**.

It should also be mentioned that Property 11.1 serves as a check for reasonableness of answers. In other words, when computing probabilities, we know that our answer must fall between 0 and 1, inclusive. Any other probability answer is simply not reasonable.

Complementary Events

Complementary events are complementary sets such that S, the sample space, serves as the universal set. The following examples illustrate this idea.

Sample space	Event space	Complement of event space
$S = \{1, 2, 3, 4, 5, 6\}$	$E = \{1, 2\}$	$E' = \{3, 4, 5, 6\}$
$S = \{H, T\}$	$E = \{T\}$	$E' = \{H\}$
$S = \{2, 3, 4, \ldots, 12\}$	$E = \{2, 3, 4\}$	$E' = \{5, 6, 7, \ldots, 12\}$
$S = \{1, 2, 3, \ldots, 25\}$	$E = \{3, 4, 5, \ldots, 25\}$	$E' = \{1, 2\}$

In each case, note that E' (the complement of E) consists of all elements of S that are *not* in E. Thus, E and E' are called *complementary events*. Also note that for each example, $P(E) + P(E') = 1$. We can state the following general property.

PROPERTY 11.2

If E is any event of a sample space S, and E' is the complementary event, then

$$P(E) + P(E') = 1.$$

From a computational viewpoint, Property 11.2 provides us with a double-barreled attack to some probability problems. That is to say, once we compute either $P(E)$ or $P(E')$, then the other one is determined simply by subtracting from 1. For example, suppose that for a particular problem we can determine that $P(E) = \frac{3}{13}$. Then we immediately know that $P(E') = 1 - P(E) = 1 - \frac{3}{13} = \frac{10}{13}$. The following examples further illustrate the usefulness of Property 11.2.

EXAMPLE 3

Two dice are tossed. Find the probability of getting a sum greater than 3.

Solution Let S be the familiar sample space of ordered pairs for this problem, where $n(S) = 36$. Let E be the event of obtaining a sum greater than 3. Then E' is the event of obtaining a sum less than or equal to 3, that is, $E' = \{(1, 1), (1, 2), (2, 1)\}$. Thus,

$$P(E') = \frac{n(E')}{n(S)} = \frac{3}{36} = \frac{1}{12}.$$

From this we conclude that

$$P(E) = 1 - P(E') = 1 - \tfrac{1}{12} = \tfrac{11}{12}.$$

EXAMPLE 4

Toss three coins and find the probability of getting at least one head.

Solution The sample space, S, consists of all possible outcomes for tossing three coins. Using the fundamental principle of counting, we know that there are $(2)(2)(2) = 8$ outcomes, so $n(S) = 8$. Let E be the event of getting at least one head. Then E' is the complementary event of not getting any heads. The set E' is easy to list, namely, $E' = \{(T, T, T)\}$. Thus, $n(E') = 1$ and $P(E') = \tfrac{1}{8}$. From this, $P(E)$ can be determined to be

$$P(E) = 1 - P(E') = 1 - \tfrac{1}{8} = \tfrac{7}{8}.$$

EXAMPLE 5

A 3-person committee is chosen at random from a group of 5 women and 4 men. Find the probability that the committee contains at least one woman.

Solution Let the sample space, S, be the set of all possible 3-person committees that can be formed from 9 people. There are $C(9, 3) = 84$ such committees; therefore, $n(S) = 84$.

Let E be the event, "The committee contains at least one woman." Then E' is the complementary event, "The committee contains all men." Thus, E' consists of all 3-man committees that can be formed from 4 men. There are $C(4, 3) = 4$ such committees; thus $n(E') = 4$. So we have

$$P(E') = \frac{n(E')}{n(S)} = \frac{4}{84} = \frac{1}{21}$$

which determines $P(E)$ to be

$$P(E) = 1 - P(E') = 1 - \tfrac{1}{21} = \tfrac{20}{21}.$$

$E \cap F$

$E \cup F$

(Figure 11.1 Continues)

The concepts of *set intersection* and *set union* play an important role in the study of probability. If E and F are two events in a sample space S, then $E \cap F$ is the event consisting of all sample points of S that are in both E and F. Likewise, $E \cup F$ is the event consisting of all sample points of S that are in E *or* F, or both.

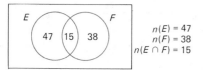

$n(E) = 47$
$n(F) = 38$
$n(E \cap F) = 15$

Figure 11.1

(Continued from previous page)

In Figure 11.1 there are 47 sample points in E, 38 sample points in F, and 15 sample points in $E \cap F$. How many sample points are there in $E \cup F$? Simply adding the number of points in E and F would result in counting the 15 points in $E \cap F$ twice. Therefore, 15 must be subtracted from the total number of points in E and F, yielding $47 + 38 - 15 = 70$ points in $E \cup F$. We can state the following general counting property:

$$n(E \cup F) = n(E) + n(F) - n(E \cap F).$$

If we divide both sides of this equation by $n(S)$, we obtain the following probability property.

PROPERTY 11.3

For events E and F of a sample space S,

$$P(E \cup F) = P(E) + P(F) - P(E \cap F).$$

EXAMPLE 6

What is the probability of getting an odd number or a prime number with one toss of a die?

Solution Let $S = \{1, 2, 3, 4, 5, 6\}$ be the sample space, $E = \{1, 3, 5\}$ the event of getting an odd number, and $F = \{2, 3, 5\}$ the event of getting a prime number. Then $E \cap F = \{3, 5\}$ and, using Property 11.3, we obtain

$$P(E \cup F) = \tfrac{3}{6} + \tfrac{3}{6} - \tfrac{2}{6} = \tfrac{4}{6} = \tfrac{2}{3}.$$

EXAMPLE 7

Toss three coins. What is the probability of getting at least two heads or exactly one tail?

Solution Using the fundamental principle of counting, we know that there are $2 \cdot 2 \cdot 2 = 8$ possible outcomes of tossing three coins; thus $n(S) = 8$. Let

$$E = \{(H, H, H), (H, H, T), (H, T, H), (T, H, H)\}$$

be the event of getting at least two heads and let

$$F = \{(H, H, T), (H, T, H), (T, H, H)\}$$

be the event of getting exactly one tail. Then

$$E \cap F = \{(H, H, T), (H, T, H), (T, H, H)\}$$

and we can compute $P(E \cup F)$ as follows.

$$P(E \cup F) = P(E) + P(F) - P(E \cap F)$$
$$= \tfrac{4}{8} + \tfrac{3}{8} - \tfrac{3}{8}$$
$$= \tfrac{4}{8} = \tfrac{1}{2}$$

In Property 11.3, if $E \cap F = \varnothing$, then the events E and F are said to be **mutually exclusive**. In other words, mutually exclusive events are events that cannot occur at the same time. For example, when rolling a die, the event of getting a 4 is mutually exclusive of the event of getting a 5; they cannot both happen on the same roll. If $E \cap F = \varnothing$, then $P(E \cap F) = 0$ and Property 11.3 becomes $P(E \cup F) = P(E) + P(F)$ for **mutually exclusive events**.

EXAMPLE 8

Suppose we have a jar containing 5 white, 7 green, and 9 red marbles. If one marble is drawn at random from the jar, find the probability that it is white or green.

Solution The events of "drawing a white marble" and "drawing a green marble" are mutually exclusive. Therefore, the probability of drawing a white or green marble is

$$\tfrac{5}{21} + \tfrac{7}{21} = \tfrac{12}{21} = \tfrac{4}{7}.$$

Note in the solution for Example 8 that we did not explicitly name and list the elements of the sample space or event spaces. It was obvious that the sample space contained 21 elements (21 marbles in the jar) and the event spaces contained 5 elements (5 white marbles) and 7 elements (7 green marbles). Thus, it was unnecessary to name and list the sample space and event spaces.

EXAMPLE 9

Suppose that the data in the following table represent a survey of 1000 drivers after a holiday weekend.

	RAIN (R)	NO RAIN (R')	TOTAL
ACCIDENT (A)	35	10	45
NO ACCIDENT (A')	450	505	955
TOTAL	485	515	1000

If a person is selected at random, what is the probability that the person was in an accident or that it rained?

Solution First, let's form a **probability table** by dividing each entry by 1000, the total number surveyed.

	RAIN (R)	NO RAIN (R')	TOTAL
ACCIDENT (A)	.035	.010	.045
NO ACCIDENT (A')	.450	.505	.955
TOTAL	.485	.515	1.000

Now we can use Property 11.3 and compute $P(A \cup R)$.

$$\begin{aligned} P(A \cup R) &= P(A) + P(R) - P(A \cap R) \\ &= .045 + .485 - .035 \\ &= .495 \end{aligned}$$

Expected Value

Suppose we toss a coin 500 times. We would expect to get approximately 250 heads. In other words, since the probability of getting a head with one toss of a coin is $\frac{1}{2}$, then in 500 tosses we should get approximately $500(\frac{1}{2}) = 250$ heads. The word "approximately" conveys a key idea. As we know from experience, it is possible to toss a coin several times and get all heads. However, with a large number of tosses, things should "average out," so that we get about an equal number of heads and tails.

As another example, consider the fact that the probability of getting a sum of 6 with one toss of a pair of dice is $\frac{5}{36}$. Therefore, if a pair of dice is tossed 360 times, we should expect to get a sum of 6 approximately $360(\frac{5}{36}) = 50$ times. Let us now define the concept of *expected value*.

DEFINITION 11.2

> If the k possible outcomes of an experiment are assigned the values x_1, $x_2, x_3, \ldots, x_k$ and occur with probabilities of $p_1, p_2, p_3, \ldots, p_k$, respectively, then the **expected value** of the experiment (E_v) is given by
>
> $$E_v = x_1 p_1 + x_2 p_2 + x_3 p_3 + \cdots + x_k p_k.$$

The concept of expected value (also called **mathematical expectation**) is used in a variety of probability situations dealing with such things as "fairness of games" and "decision-making" in business ventures. Let's consider some examples.

EXAMPLE 10

Suppose that you buy one ticket in a lottery for which 1000 tickets are sold. Furthermore, suppose that 3 prizes are awarded: one of $500, one of $300, and one of $100. What is your mathematical expectation?

Solution Since you bought one ticket, your probability of winning $500 is $\frac{1}{1000}$, that of winning $300 is $\frac{1}{1000}$, and that of winning $100 is $\frac{1}{1000}$. Multiplying each of these probabilities times the corresponding prize money and then adding the results yields your mathematical expectation.

$$E_v = \$500(\tfrac{1}{1000}) + \$300(\tfrac{1}{1000}) + \$100(\tfrac{1}{1000})$$
$$= \$.50 + \$.30 + \$.10$$
$$= \$.90$$

In Example 10, if you pay more than $.90 for a ticket, then it is not a **fair game** from your standpoint. If the price of the game is included in the calculation of the expected value, then a fair game is defined to be one for which the expected value is zero.

EXAMPLE 11

A player pays $5 to play a game for which the probability of winning is $\frac{1}{5}$ and the probability of losing is $\frac{4}{5}$. If the player wins the game, he receives $25. Is this a fair game for the player?

Solution Using Definition 11.2, let $x_1 = \$20$, which represents the $25 won minus the \$5 paid to play, and let $x_2 = -\$5$, the amount paid to play the game. We also are given that $p_1 = \frac{1}{5}$ and $p_2 = \frac{4}{5}$. Thus, the expected value is

$$E_v = \$20(\tfrac{1}{5}) + (-\$5)(\tfrac{4}{5})$$
$$= \$4 - \$4$$
$$= 0.$$

Since the expected value is zero, it is a fair game.

EXAMPLE 12

Suppose you are interested in insuring a diamond ring for $2000 against theft. An insurance company charges a premium of \$25 per year, claiming that there is a probability of .01 that the ring will be stolen during the year. What is your expected gain or loss if you take out the insurance?

Solution Using Definition 11.2, let $x_1 = \$1975$, which represents the $2000 minus the cost of the premium, \$25, and let $x_2 = -\$25$. We also are given that $p_1 = .01$ and so $p_2 = 1 - .01 = .99$. Thus, the expected value is

$$E_v = \$1975(.01) + (-\$25)(.99)$$
$$= \$19.75 - \$24.75$$
$$= -\$5.00.$$

This means that if you insure with this company over many years and the circumstances remain the same, you will have an average net loss of $5 per year.

Problem Set 11.4

For Problems 1–4, *two* dice are tossed. Find the probability of rolling each of the following events.

1. A sum of 6
2. A sum greater than 2
3. A sum less than 8
4. A sum greater than 1

For Problems 5–8, *three* dice are tossed. Find the probability of rolling each of the following events.

5. A sum of 3
6. A sum greater than 4
7. A sum less than 17
8. A sum greater than 18

For Problems 9–12, *four* coins are tossed. Find the probability of getting each of the following events.

9. Four heads
10. Three heads and a tail
11. At least one tail
12. At least one head

For Problems 13–16, *five* coins are tossed. Find the probability of each of the following events.

13. Five tails
14. Four heads and a tail
15. At least one tail
16. At least two heads

Solve the following problems.

17. Toss a pair of dice. What is the probability of not getting a double?

18. The probability that a certain horse will win the Kentucky Derby is $\frac{1}{20}$. What is the probability that it will lose the race?

19. One card is randomly drawn from a deck of 52 playing cards. What is the probability that it is not an ace?

20. Six coins are tossed. Find the probability of getting at least two heads.

21. A subset of 2 letters is chosen at random from the set $\{a, b, c, d, e, f, g, h, i\}$. Find the probability that the subset will contain at least one vowel.

22. A 2-person committee is chosen at random from a group of 4 men and 3 women. Find the probability that the committee contains at least one man.

23. A 3-person committee is chosen at random from a group of 7 women and 5 men. Find the probability that the committee contains at least one man.

For Problems 24–27, one die is tossed. Find the probability of rolling each of the following events.

24. A 3 or an odd number

25. A 2 or an odd number

26. An even number or a prime number

27. An odd number or a multiple of 3

For Problems 28–31, two dice are tossed. Find the probability of rolling each of the following events.

28. A double or a sum of 6

29. A sum of 10 or a sum greater than 8

30. A sum of 5 or a sum greater than 10

31. A double or a sum of 7

Solve the following problems.

32. Two coins are tossed. Find the probability of getting exactly one head or at least one tail.

33. Three coins are tossed. Find the probability of getting at least 2 heads or exactly one tail.

34. A jar contains 7 white, 6 blue, and 10 red marbles. If one marble is drawn at random from the jar, find the probability that (a) the marble is white or blue; (b) the marble is white or red; (c) the marble is blue or red.

35. A coin and a die are tossed. Find the probability of getting a head on the coin or a 2 on the die.

36. A card is randomly drawn from a deck of 52 playing cards. Find the probability that it is a red card or a face card. (Jacks, queens, and kings are the face cards.)

37. The data in the following table represent a survey of 1000 drivers after a holiday weekend.

	RAIN (R)	NO RAIN (R')	TOTAL
ACCIDENT (A)	45	15	60
NO ACCIDENT (A')	350	590	940
TOTAL	395	605	1000

If a person is selected at random from those surveyed, find the probability of each of the following events. (Express the probabilities in decimal form.)

(a) The person was in an accident or it rained.

(b) The person was not in an accident or it rained.

(c) The person was not in an accident or it did not rain.

38. One hundred people were surveyed and one question pertained to their educational background. The results of this question are given in the following table.

	FEMALE (F)	MALE (F')	TOTAL
COLLEGE DEGREE (D)	30	20	50
NO COLLEGE DEGREE (D')	15	35	50
TOTAL	45	55	100

If a person is selected at random from those surveyed, find the probability of each of the following events. Express the probabilities in decimal form.
(a) The person is female or has a college degree.
(b) The person is male or does not have a college degree.
(c) The person is female or does not have a college degree.

39. In a recent election there were 1000 eligible voters. They were asked to vote on two issues, A and B. The results were as follows: 300 people voted for A, 400 people voted for B, and 175 people voted for both A and B. If one person is chosen at random from the 1000 eligible voters, find the probability that the person voted for A or B.

40. A company has 500 employees of which 200 are females, 15 are high-level executives, and 7 of the high-level executives are females. If one of the 500 employees is chosen at random, find the probability that the person chosen is female or a high-level executive.

41. A die is tossed 360 times. How many times would you expect to get a 6?

42. Two dice are tossed 360 times. How many times would you expect to get a sum of 5?

43. Two dice are tossed 720 times. How many times would you expect to get a sum greater than 9?

44. Four coins are tossed 80 times. How many times would you expect to get 1 head and 3 tails?

45. Four coins are tossed 144 times. How many times would you expect to get 4 tails?

46. Two dice are tossed 300 times. How many times would you expect to get a double?

47. Three coins are tossed 448 times. How many times would you expect to get 3 heads?

48. Suppose 5000 tickets are sold in a lottery. There are 3 prizes: the first is $1000, the second is $500, and the third is $100. What is the mathematical expectation of winning?

49. Your friend challenges you with the following game: You are to roll a pair of dice and he will give you $5 if you roll a sum of 2 or 12, $2 if you roll a sum of 3 or 11, $1 if you roll a sum of 4 or 10; otherwise, you are to pay him $1. Should you play the game?

50. A contractor bids on a building project. There is a probability of .8 that he can show a profit of $30,000 and a probability of .2 that he will have to absorb a loss of $10,000. What is his mathematical expectation?

51. Suppose a person tosses 2 coins and receives $5 if 2 heads come up, $2 if 1 head and 1 tail come up, and has to pay $2 if 2 tails come up. Is it a fair game for him?

52. A "wheel of fortune" is divided into four colors: red, white, blue, and yellow. The probabilities of the spinner landing on each of the colors and the money received is given by the following chart. The price to spin the wheel is $1.50. Is it a fair game?

COLOR	PROBABILITY OF LANDING ON THE COLOR	MONEY RECEIVED FOR LANDING ON THE COLOR
red	$\frac{4}{10}$	$.50
white	$\frac{3}{10}$	1.00
blue	$\frac{2}{10}$	2.00
yellow	$\frac{1}{10}$	5.00

53. A contractor estimates a probability of .7 of making $20,000 on a building project and a probability of .3 of losing $10,000 on the project. What is his mathematical expectation?

54. A farmer estimates his corn crop at 30,000 bushels. Based on past experience, he also estimates a probability of $\frac{3}{5}$ that he will make a profit of $.50 per bushel and a probability of $\frac{1}{5}$ of losing $.30 per bushel. What is his expected income from the corn crop?

55. Bill finds that the annual premium for insuring a stereo system for $2500 against theft is $75. If the probability that the set will be stolen during the year is .02, what is Bill's expected gain or loss by taking out the insurance?

56. Sandra finds that the annual premium for a $2000 insurance policy against the theft of a painting is $100. If the probability that the painting will be stolen during the year is .01, what is Sandra's expected gain or loss by taking out the insurance?

Miscellaneous Problems

The term **odds** is sometimes used to express a probability statement. For example, we might say, "The odds in favor of the Cubs winning the pennant are 5 to 1," or "The odds against the Mets winning the pennant are 50 to 1." "Odds in favor" and "odds against" for equally likely outcomes can be defined as follows.

$$\text{odds in favor} = \frac{\text{number of favorable outcomes}}{\text{number of unfavorable outcomes}}$$

$$\text{odds against} = \frac{\text{number of unfavorable outcomes}}{\text{number of favorable outcomes}}$$

We have used the fractional form to define odds; however, in practice, the "to" vocabulary is commonly used. Thus, the odds in favor of rolling a 4 with one roll of a die are usually stated as "1 to 5" instead of $\frac{1}{5}$. The odds against rolling a 4 are stated as 5 to 1.

The "odds in favor of" statement about the Cubs means that there are 5 favorable outcomes compared to 1 unfavorable, or a total of 6 possible outcomes. So the "5 to 1 in favor of" statement also means that the probability of the Cubs winning the pennant is $\frac{5}{6}$. Likewise, the "50 to 1 against" statement about the Mets means that the probability that the Mets will not win the pennant is $\frac{50}{51}$.

Odds are usually stated in reduced form. For example, odds of 6 to 4 are usually stated as 3 to 2. Likewise, a fraction representing probability is reduced before changing to a statement about odds.

57. What are the odds in favor of getting 3 heads with a toss of 3 coins?

58. What are the odds against getting 4 tails with a toss of 4 coins?

59. What are the odds against getting 3 heads and 2 tails with a toss of 5 coins?

60. What are the odds in favor of getting 4 heads and 2 tails with a toss of 6 coins?

61. What are the odds in favor of getting a sum of 5 with one toss of a pair of dice?

62. What are the odds against getting a sum greater than 5 with one toss of a pair of dice?

63. Suppose that one card is drawn at random from a deck of 52 playing cards. Find the odds against drawing a red card.

64. Suppose that one card is drawn at random from a deck of 52 playing cards. Find the odds in favor of drawing an ace or a king.

65. If $P(E) = \frac{4}{7}$ for some event E, find the odds in favor of E happening.

66. If $P(E) = \frac{5}{9}$ for some event E, find the odds against E happening.

67. Suppose that there is a predicted 40% chance of freezing rain. State the prediction in terms of the odds against getting freezing rain.

68. Suppose that there is a predicted 20% chance of thunderstorms. State the prediction in terms of the odds in favor of getting thunderstorms.

69. If the odds against an event happening are 5 to 2, find the probability that the event will occur.

70. The odds against Belly Dancer winning the 5th race are 20 to 9. What is the probability of Belly Dancer winning the 5th race?

71. The odds in favor of the Mets winning the pennant are stated as 7 to 5. What is the probability of the Mets winning the pennant?

 72. The following chart contains some poker-hand probabilities. Complete the column "Odds against being dealt this hand." Notice that fractions are reduced before being changed to odds.

5-CARD HAND	PROBABILITY OF BEING DEALT THIS HAND	ODDS AGAINST BEING DEALT THIS HAND
straight flush	$\dfrac{40}{2,598,960} = \dfrac{1}{64,974}$	64,973 to 1
four of a kind	$\dfrac{624}{2,598,960} =$	
full house	$\dfrac{3744}{2,598,960} =$	
flush	$\dfrac{5108}{2,598,960} =$	
straight	$\dfrac{10,200}{2,598,960} =$	
three of a kind	$\dfrac{54,912}{2,598,960} =$	
two pairs	$\dfrac{123,552}{2,598,960} =$	
one pair	$\dfrac{1,098,240}{2,598,960} =$	
no pairs	$\dfrac{1,302,540}{2,598,960} =$	

11.5

Conditional Probability; Dependent and Independent Events

Two events are often related in a way that the probability of one of them may vary depending upon whether or not the other event has occurred. For example, the probability of rain may change drastically if additional information is obtained indicating a front moving through the area. Mathematically, the additional information about the front changes the sample space for the probability of rain.

In general, the probability of the occurrence of an event E, given the occurrence of another event F, is called a **conditional probability** and is denoted $P(E|F)$. Let's look at a simple example and use it to motivate a definition for conditional probability.

What is the probability of rolling a prime number in one roll of a die? Let $S = \{1, 2, 3, 4, 5, 6\}$, so $n(S) = 6$; and let $E = \{2, 3, 5\}$, so $n(E) = 3$. Therefore,

$$P(E) = \frac{n(E)}{n(S)} = \frac{3}{6} = \frac{1}{2}.$$

Next, what is the probability of rolling a prime number in one roll of a die, *given that an odd number has turned up?* Let $F = \{1, 3, 5\}$ be the new sample space of odd numbers. Then $n(F) = 3$. We are now interested in only that part of E (rolling a prime number) that is also in F, in other words, $E \cap F$. Therefore, since $E \cap F = \{3, 5\}$, the probability of E given F is

$$P(E|F) = \frac{n(E \cap F)}{n(F)} = \frac{2}{3}.$$

If the numerator and denominator of $n(E \cap F)/n(F)$ are both divided by $n(S)$, we obtain

$$\frac{\dfrac{n(E \cap F)}{n(S)}}{\dfrac{n(F)}{n(S)}} = \frac{P(E \cap F)}{P(F)}.$$

Therefore, we can state the following general definition of the conditional probability of E given F for arbitrary events E and F.

DEFINITION 11.3
Conditional Probability

$$P(E|F) = \frac{P(E \cap F)}{P(F)}, \qquad P(F) \neq 0$$

In an example of the previous section, the following probability table was formed relative to car accidents and weather conditions on a holiday weekend.

	RAIN (R)	NO RAIN (R')	TOTAL
ACCIDENT (A)	.035	.010	.045
NO ACCIDENT (A')	.450	.505	.955
TOTAL	.485	.515	1.000

Some conditional probabilities that can be calculated from the table are as follows.

$$P(A\,|\,R) = \frac{P(A \cap R)}{P(R)} = \frac{.035}{.485} = \frac{35}{485} = \frac{7}{97}$$

$$P(A'\,|\,R) = \frac{P(A' \cap R)}{P(R)} = \frac{.450}{.485} = \frac{450}{485} = \frac{90}{97}$$

$$P(A\,|\,R') = \frac{P(A \cap R')}{P(R')} = \frac{.010}{.515} = \frac{10}{515} = \frac{2}{103}$$

Note that the probability of an accident given that it was raining ($P(A\,|\,R)$) is greater than the probability of an accident given that it was not raining ($P(A\,|\,R')$). This seems reasonable.

EXAMPLE 1

A die is tossed. Find the probability that a 4 came up if it is known that an even number turned up.

Solution Let E be the event of rolling a 4 and let F be the event of rolling an even number. Therefore, $E = \{4\}$ and $F = \{2, 4, 6\}$, from which we obtain $E \cap F = \{4\}$. Using Definition 11.3, we obtain

$$P(E\,|\,F) = \frac{P(E \cap F)}{P(F)} = \frac{\frac{1}{6}}{\frac{3}{6}} = \frac{1}{3}.$$

EXAMPLE 2

Suppose the probability that a student will enroll in a mathematics course is .45, the probability that he (she) will enroll in a science course is .38, and the probability that he (she) will enroll in both courses is .26. Find the probability that a student will enroll in a mathematics course, given that he (she) is also enrolled in a science course. Also, find the probability that a student will enroll in a science course, given that he (she) is enrolled in mathematics.

Solution Let M be the event "Will enroll in mathematics" and S be the event "Will enroll in science." Therefore, using Definition 11.3, we obtain

$$P(M\,|\,S) = \frac{P(M \cap S)}{P(S)} = \frac{.26}{.38} = \frac{26}{38} = \frac{13}{19}$$

and

$$P(S\,|\,M) = \frac{P(S \cap M)}{P(M)} = \frac{.26}{.45} = \frac{26}{45}.$$

Independent and Dependent Events

Suppose that, when computing a conditional probability, we find that

$$P(E|F) = P(E).$$

This means that the probability of E is not affected by the occurrence or non-occurrence of F. In such a situation we say that event E is *independent* of event F. It can be shown that if event E is independent of event F, then F is also independent of E; thus, E and F are referred to as **independent events**. Furthermore, from the equations

$$P(E|F) = \frac{P(E \cap F)}{P(F)} \qquad \text{and} \qquad P(E|F) = P(E),$$

we see that

$$\frac{P(E \cap F)}{P(F)} = P(E),$$

which can be written

$$P(E \cap F) = P(E)P(F).$$

Therefore, we state the following general definition.

DEFINITION 11.4

> Two events E and F are said to be **independent** if and only if
>
> $$P(E \cap F) = P(E)P(F).$$
>
> Two events that are not independent are called **dependent events**.

In the probability table preceding Example 1, we see that $P(A) = .045$, $P(R) = .485$, and $P(A \cap R) = .035$. Since

$$P(A)P(R) = (.045)(.485) = .021825$$

and this does not equal $P(A \cap R)$, the events A (have a car accident) and R (rainy conditions) are not independent. This is not too surprising, since we would certainly expect rainy conditions and automobile accidents to be related.

EXAMPLE 3

Suppose we roll a white die and a red die. If we let E be the event "We roll a 4 on the white die" and F the event "We roll a 6 on the red die," are E and F independent events?

Solution The sample space for rolling a pair of dice has $(6)(6) = 36$ elements. Using ordered-pair notation, where the first entry represents the white die and the second entry the red die, we can list events E and F as follows.

$$E = \{(4, 1), (4, 2), (4, 3), (4, 4), (4, 5), (4,6)\}$$

$$F = \{(1, 6), (2, 6), (3, 6), (4, 6), (5, 6), (6, 6)\}$$

Therefore, $E \cap F = \{4, 6\}$. Since $P(E) = \frac{1}{6}$, $P(F) = \frac{1}{6}$, and $P(E \cap F) = \frac{1}{36}$, we see that $P(E \cap F) = P(E)P(F)$, and the events E and F are independent.

EXAMPLE 4

Two coins are tossed. Let E be the event "Toss not more than one head" and F the event "Toss at least one of each face." Are these independent events?

Solution The sample space has $(2)(2) = 4$ elements. The events E and F can be listed as follows.

$$E = \{(H, T), (T, H), (T, T)\}$$

$$F = \{(H, T), (T, H)\}$$

Therefore, $E \cap F = \{(H, T), (T, H)\}$. Since $P(E) = \frac{3}{4}$, $P(F) = \frac{1}{2}$, and $P(E \cap F) = \frac{1}{2}$, we see that $P(E \cap F) \neq P(E)P(F)$, so the events E and F are dependent.

Sometimes the independence issue can be decided by the physical nature of the events in the problem. For instance, in Example 3 it should seem evident that rolling a 4 on the white die is not affected by rolling a 6 on the red die. However, as in Example 4, the description of the events may not clearly indicate whether the events are dependent.

From a problem-solving viewpoint, the following two statements are very helpful.

1. If E and F are independent events, then

$$P(E \cap F) = P(E)P(F).$$

(This property generalizes to any finite number of independent events.)

2. If E and F are dependent events, then

$$P(E \cap F) = P(E)P(F \mid E).$$

Let's analyze some problems using these ideas.

EXAMPLE 5

A die is rolled 3 times. (This is equivalent to rolling 3 dice once each.) What is the probability of getting a 6 all 3 times?

Solution The events of a 6 on the first roll, a 6 on the second roll, and a 6 on the third roll are independent events. Therefore, the probability of getting three 6's is

$$\left(\tfrac{1}{6}\right)\left(\tfrac{1}{6}\right)\left(\tfrac{1}{6}\right) = \tfrac{1}{216}.$$

EXAMPLE 6

A jar contains 5 white, 7 green, and 9 red marbles. If 2 marbles are drawn in succession *without replacement*, find the probability that both marbles are white.

Solution Let E be the event of drawing a white marble on the first draw and F the event of drawing a white marble on the second draw. Since the marble drawn first is not to be replaced before drawing the second marble, we have dependent events. Therefore,

$$P(E \cap F) = P(E)P(F \mid E)$$
$$= (\tfrac{5}{21})(\tfrac{4}{20}) = \tfrac{20}{420} = \tfrac{1}{21}.$$

$P(F \mid E)$ means the probability of drawing a white marble on the second draw given that a white marble was obtained on the first draw.

The concept of *mutually exclusive events* may also enter the picture when working with independent or dependent events. Our final examples of this section illustrate this idea.

EXAMPLE 7

A coin is tossed 3 times. Find the probability of getting 2 heads and 1 tail.

Solution Two heads and 1 tail can be obtained in three different ways: (1) *HHT* (head on first toss, head on second toss, and tail on third toss), (2) *HTH*, and (3) *THH*. Thus we have three *mutually exclusive* events, each of which can be broken into *independent* events: "first toss," "second toss," and "third toss." Therefore, the probability can be computed as follows.

$$(\tfrac{1}{2})(\tfrac{1}{2})(\tfrac{1}{2}) + (\tfrac{1}{2})(\tfrac{1}{2})(\tfrac{1}{2}) + (\tfrac{1}{2})(\tfrac{1}{2})(\tfrac{1}{2}) = \tfrac{3}{8}$$

EXAMPLE 8

A jar contains 5 white, 7 green, and 9 red marbles. If 2 marbles are drawn in succession *without replacement*, find the probability that one of them is white and the other is green.

Solution The drawing of a white marble and a green marble can occur in two different ways: (1) by drawing a white first and then a green, or (2) by drawing a green first and then a white. Thus, we have two mutually exclusive events, each of which is broken into two *dependent* events: "first draw" and "second draw." Therefore, the probability can be computed as follows.

$$(\tfrac{5}{21})(\tfrac{7}{20}) \qquad + \qquad (\tfrac{7}{21})(\tfrac{5}{20}) = \tfrac{70}{420} = \tfrac{1}{6}$$

white on first draw   green on second draw   green on first draw   white on second draw

EXAMPLE 9

Two cards are drawn in succession *with replacement* from a deck of 52 playing cards. Find the probability of drawing a jack and a queen.

Solution Drawing a jack and a queen can occur two different ways: (1) a jack on the first draw and a queen on the second, or (2) a queen on the first draw and a jack on the second. Thus (1) and (2) are mutually exclusive events and each is broken into the *independent* events of first draw and second draw with replacement. Therefore, the probability can be computed as follows.

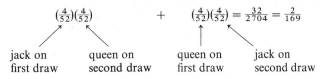

$$\left(\tfrac{4}{52}\right)\left(\tfrac{4}{52}\right) \qquad + \qquad \left(\tfrac{4}{52}\right)\left(\tfrac{4}{52}\right) = \tfrac{32}{2704} = \tfrac{2}{169}$$

jack on first draw   queen on second draw   queen on first draw   jack on second draw

Problem Set 11.5

1. A die is tossed. Find the probability that a 5 came up if it is known that an odd number came up.

2. A die is tossed. Find the probability that a prime number was obtained, given that an even number came up. Also find the probability that an even number came up, given that a prime number was obtained.

3. Two dice are rolled and someone indicates that the two numbers that come up are different. Find the probability that the sum of the two numbers is 6.

4. Two dice are rolled and someone indicates that the two numbers that come up are identical. Find the probability that the sum of the two numbers is 8.

5. One card is randomly drawn from a deck of 52 playing cards. Find the probability that it is a jack, given that the card is a face card. (We are considering jacks, queens, and kings as face cards.)

6. One card is randomly drawn from a deck of 52 playing cards. Find the probability that it is a spade, given the fact that it is a black card.

7. A coin and a die are tossed. Find the probability of getting a 5 on the die, given that a head comes up on the coin.

8. A family has 3 children. Assume that each child is as likely to be a boy as it is to be a girl. Find the probability that the family has 3 girls if it is known that they have at least 1 girl.

9. The probability that a student will enroll in a mathematics course is .7, the probability that he (she) will enroll in a history course is .3, and the probability that he (she) will enroll in both mathematics and history is .2. Find the probability that a student will enroll in mathematics, given that he (she) is also enrolled in history. Also find the probability that a student will enroll in history, given that he (she) is also enrolled in mathematics.

10. The following probability table contains data relative to car accidents and weather conditions on a holiday weekend.

	RAIN (R)	NO RAIN (R')	TOTAL
ACCIDENT (A)	.025	.015	.040
NO ACCIDENT (A')	.400	.560	.960
TOTAL	.425	.575	1.000

Find the probability that a person chosen at random from the survey was in an accident, given that it was raining. Also find the probability that a person was not in an accident, given that it was not raining.

11. One hundred people were surveyed and one question pertained to their educational background. The results of this question are given in the following table.

	FEMALE (F)	MALE (F')	TOTAL
COLLEGE DEGREE (D)	30	20	50
NO COLLEGE DEGREE (D')	15	35	50
TOTAL	45	55	100

Find the probability that a person chosen at random from the survey has a college degree, given that the person is female. Also find the probability that a person chosen is male, given that the person has a college degree.

12. In a recent election there were 1000 eligible voters. They were asked to vote on two issues, A and B. The results were as follows: 200 people voted for A, 400 people voted for B, and 50 people voted for both A and B. If one person is chosen at random from the 1000 eligible voters, find the probability that the person voted for A, given that he (she) voted for B. Also find the probability that the person voted for B, given that he (she) voted for A.

13. A small company has 100 employees of which 75 are males, 7 are administrators, and 5 of the administrators are males. If a person is chosen at random from the 100 employees, find the probability that the person is an administrator, given that he is a male. Also find the probability that the person chosen is a female, given that she is an administrator.

14. A survey claims that 80 percent of the households in a certain town have a color TV, 10 percent have a microwave oven, and 2 percent have both a color TV and a microwave oven. Find the probability that a randomly selected household will have a microwave oven, given that it has a color TV.

15. Consider a family of 3 children. Let E be the event "The first child is a boy" and let F be the event "They have exactly one boy." Are events E and F dependent or independent?

16. Roll a white die and a green die. Let E be the event "Roll a 2 on the white die" and let F be the event "Roll a 4 on the green die." Are E and F dependent or independent events?

17. Toss 3 coins. Let E be the event "Toss not more than one head" and let F be the event "Toss at least one of each face." Are E and F dependent or independent events?

18. A card is drawn at random from a standard deck of 52 playing cards. Let E be the event "The card is a 2" and let F be the event "The card is a 2 or 3." Are the events E and F dependent or independent?

19. A coin is tossed 4 times. Find the probability of getting 3 heads and 1 tail.

20. A coin is tossed 5 times. Find the probability of getting 4 heads and 1 tail.

21. Toss a pair of dice 3 times. Find the probability that a double is obtained on all three tosses.

22. Toss a pair of dice 3 times. Find the probability that each toss will produce a sum of 4.

For Problems 23–26, suppose that 2 cards are drawn in succession *without replacement* from a deck of 52 playing cards. Find the probability of each of the following events.

23. Both cards are 4's.

24. One card is an ace and one card is a king.

25. One card is a spade and one card is a diamond.

26. Both cards are black.

For Problems 27–30, suppose that 2 cards are drawn in succession *with replacement* from a deck of 52 playing cards. Find the probability of each of the following events.

27. Both cards are spades.

28. One card is an ace and one card is a king.

29. One card is the ace of spades and one card is the king of spades.

30. Both cards are red.

31. A person holds 3 kings from a deck of 52 playing cards. If the person draws two cards without replacement from the 49 cards remaining in the deck, find the probability of drawing the 4th king.

32. A person removes 2 aces and a king from a deck of 52 playing cards and draws, without replacement, 2 more cards from the deck. Find the probability that the person will draw 2 aces, or 2 kings, or an ace and a king.

For Problems 33–36, a bag contains 5 red and 4 white marbles. Two marbles are drawn in succession *with replacement*. Find the probability of each of the following events.

33. Both marbles drawn are red.

34. Both marbles drawn are white.

35. The first marble is red and the second marble is white.

36. At least one marble is red.

For Problems 37–40, a bag contains 5 white, 4 red, and 4 blue marbles. Two marbles are drawn in succession *with replacement*. Find the probability of each of the following events.

37. Both marbles drawn are white.

38. Both marbles drawn are red.

39. One red and one blue marble are drawn.

40. One white and one blue marble are drawn.

For Problems 41–44, a bag contains 1 red and 2 white marbles. Two marbles are drawn in succession *without replacement*. Find the probability of each of the following events.

41. One marble drawn is red and one marble drawn is white.

42. The first marble is red and the second one is white.

43. Both marbles drawn are white.

44. Both marbles drawn are red.

For Problems 45–48, a bag contains 5 red and 12 white marbles. Two marbles are drawn in succession *without replacement*. Find the probability of each of the following events.

45. Both marbles drawn are red.

46. Both marbles drawn are white.

47. One red and one white marble are drawn.

48. At least one marble drawn is red.

For Problems 49–52, a bag contains 2 red, 3 white, and 4 blue marbles. Two marbles are drawn in succession *without replacement*. Find the probability of each of the following events.

49. Both marbles drawn are white.

50. One marble drawn is white and one is blue.

51. Both marbles drawn are blue.

52. At least one red marble is drawn.

For Problems 53–56, a bag contains 5 white, 1 blue, and 3 red marbles. *Three* marbles are drawn in succession *with replacement*. Find the probability of each of the following events.

53. All three marbles drawn are blue.

54. One marble of each color is drawn.

55. One white and 2 red marbles are drawn.

56. One blue and 2 white marbles are drawn.

For Problems 57–60, a bag contains 4 white, 1 red, and 2 blue marbles. *Three* marbles are drawn in succession *without replacement*. Find the probability of each of the following events.

57. All three marbles drawn are white.

58. One red and 2 blue marbles are drawn.

59. One marble of each color is drawn.

60. One white and 2 red marbles are drawn.

61. Two boxes with red and white marbles are shown here. A marble is drawn at random from Box 1 and then a second marble is drawn from Box 2. Find the probability that both marbles drawn are white. Find the probability that both marbles drawn are red. Find the probability that one red and one white marble are drawn.

3 red		2 red
4 white		1 white
Box 1		*Box 2*

62. Three boxes containing red and white marbles are shown here. Randomly draw a marble from Box 1 and put it in Box 2. Then draw a marble from Box 2 and put it in Box 3. Then draw a marble from Box 3. What is the probability that the last marble drawn, from Box 3, is red? What is the probability that it is white?

2 red		3 red		
2 white		1 white		3 white
Box 1		*Box 2*		*Box 3*

11.6

The Binomial Theorem

In Chapter 1 we developed a pattern for expanding binomials, using Pascal's triangle to determine the coefficients of each term. Now we will be more precise and develop a general formula, called the *binomial formula*. In other words, we want to develop a formula that will allow us to expand $(x + y)^n$, where n is any positive integer.

Let's begin, as we did in Chapter 1, by looking at some specific expansions, which can be verified by direct multiplication.

$$(x + y)^0 = 1$$

$$(x + y)^1 = x + y$$

$$(x + y)^2 = x^2 + 2xy + y^2$$

$$(x + y)^3 = x^3 + 3x^2y + 3xy^2 + y^3$$

$$(x + y)^4 = x^4 + 4x^3y + 6x^2y^2 + 4xy^3 + y^4$$

$$(x + y)^5 = x^5 + 5x^4y + 10x^3y^2 + 10x^2y^3 + 5xy^4 + y^5$$

First, note the pattern of the exponents for x and y on a term-by-term basis. The exponents of x begin with the exponent of the binomial and decrease by 1, term by term, until the last term has x^0, which is 1. The exponents of y begin with zero ($y^0 = 1$) and increase by 1, term by term, until the last term contains y to the power of the binomial. In other words, the variables in the expansion of $(x + y)^n$ have the following pattern.

$$x^n, \quad x^{n-1}y, \quad x^{n-2}y^2, \quad x^{n-3}y^3, \ldots, xy^{n-1}, \quad y^n$$

Notice that for each term, the sum of the exponents of x and y is n.

Now let's look for a pattern for the coefficients by looking specifically at the expansion of $(x + y)^5$.

$$(x + y)^5 = x^5 + 5x^4y^1 + 10x^3y^2 + 10x^2y^3 + 5x^1y^4 + 1y^5$$

$$\uparrow \qquad \uparrow \qquad \uparrow \qquad \uparrow \qquad \uparrow$$

$$C(5, 1) \quad C(5, 2) \quad C(5, 3) \quad C(5, 4) \quad C(5, 5)$$

As indicated by the arrows, the coefficients are numbers that arise as different-sized combinations of 5 things. To see why this happens, consider the coefficient for the term containing x^3y^2. The two y's (for y^2) come from two of the factors of $(x + y)$ and therefore the three x's (for x^3) must come from the other three factors of $(x + y)$. In other words, the coefficient is $C(5, 2)$.

We can now state a general expansion formula for $(x + y)^n$; this formula is often called the **binomial theorem**. But, before stating it, let's make a small switch in notation. Instead of $C(n, r)$ we shall write $\binom{n}{r}$, which will prove to be a little more convenient at this time. The symbol $\binom{n}{r}$ still refers to the number of combinations of n things taken r at a time, but in this context it is often called a **binomial coefficient**.

Binomial Theorem

For any binomial $(x + y)$ and any natural number n,

$$(x + y)^n = x^n + \binom{n}{1}x^{n-1}y + \binom{n}{2}x^{n-2}y^2 + \cdots + \binom{n}{n}y^n.$$

The binomial theorem can be proved by mathematical induction, but we will not do that in this text. Instead we'll consider a few examples that put the binomial theorem to work.

EXAMPLE 1

Expand $(x + y)^7$.

Solution

$$(x + y)^7 = x^7 + \binom{7}{1}x^6y + \binom{7}{2}x^5y^2 + \binom{7}{3}x^4y^3 + \binom{7}{4}x^3y^4$$

$$+ \binom{7}{5}x^2y^5 + \binom{7}{6}xy^6 + \binom{7}{7}y^7$$

$$= x^7 + 7x^6y + 21x^5y^2 + 35x^4y^3 + 35x^3y^4 + 21x^2y^5 + 7xy^6 + y^7$$

EXAMPLE 2

Expand $(x - y)^5$.

Solution We shall treat $(x - y)^5$ as $[x + (-y)]^5$.

$$[x + (-y)]^5 = x^5 + \binom{5}{1}x^4(-y) + \binom{5}{2}x^3(-y)^2 + \binom{5}{3}x^2(-y)^3$$

$$+ \binom{5}{4}x(-y)^4 + \binom{5}{5}(-y)^5$$

$$= x^5 - 5x^4y + 10x^3y^2 - 10x^2y^3 + 5xy^4 - y^5$$

EXAMPLE 3

Expand $(2a + 3b)^4$.

Solution Let $x = 2a$ and $y = 3b$ in the binomial theorem.

$$(2a + 3b)^4 = (2a)^4 + \binom{4}{1}(2a)^3(3b) + \binom{4}{2}(2a)^2(3b)^2$$

$$+ \binom{4}{3}(2a)(3b)^3 + \binom{4}{4}(3b)^4$$

$$= 16a^4 + 96a^3b + 216a^2b^2 + 216ab^3 + 81b^4$$

EXAMPLE 4

Expand $\left(a + \dfrac{1}{n}\right)^5$.

Solution

$$\left(a + \frac{1}{n}\right)^5 = a^5 + \binom{5}{1}a^4\left(\frac{1}{n}\right) + \binom{5}{2}a^3\left(\frac{1}{n}\right)^2 + \binom{5}{3}a^2\left(\frac{1}{n}\right)^3$$

$$+ \binom{5}{4}a\left(\frac{1}{n}\right)^4 + \binom{5}{5}\left(\frac{1}{n}\right)^5$$

$$= a^5 + \frac{5a^4}{n} + \frac{10a^3}{n^2} + \frac{10a^2}{n^3} + \frac{5a}{n^4} + \frac{1}{n^5}$$

EXAMPLE 5

Expand $(x^2 - 2y^3)^6$.

Solution

$$[x^2 + (-2y^3)]^6 = (x^2)^6 + \binom{6}{1}(x^2)^5(-2y^3) + \binom{6}{2}(x^2)^4(-2y^3)^2$$

$$+ \binom{6}{3}(x^2)^3(-2y^3)^3 + \binom{6}{4}(x^2)^2(-2y^3)^4$$

$$+ \binom{6}{5}(x^2)(-2y^3)^5 + \binom{6}{6}(-2y^3)^6$$

$$= x^{12} - 12x^{10}y^3 + 60x^8y^6 - 160x^6y^9 + 240x^4y^{12}$$

$$- 192x^2y^{15} + 64y^{18}$$

Finding Specific Terms

Sometimes it is convenient to be able to write down the specific term of a binomial expansion without writing out the entire expansion. For example, suppose that we want the sixth term of the expansion $(x + y)^{12}$. We can proceed as follows: The sixth term will contain y^5. (Note in the binomial theorem that **the exponent of y is always one less than the number of the term**.) Since the sum of the exponents for x and y must be 12 (the exponent of the binomial), the sixth term will also contain x^7. The coefficient is $\binom{12}{5}$, where the 5 agrees with the exponent of y^5. Therefore, the sixth term of $(x + y)^{12}$ is

$$\binom{12}{5}x^7y^5 = 792x^7y^5.$$

EXAMPLE 6

Find the fourth term of $(3a + 2b)^7$.

Solution The fourth term will contain $(2b)^3$ and therefore it will also contain $(3a)^4$. The coefficient is $\binom{7}{3}$. Thus, the fourth term is

$$\binom{7}{3}(3a)^4(2b)^3 = (35)(81a^4)(8b^3) = 22{,}680a^4b^3.$$

Problem Set 11.6

Expand and simplify each of the following binomials.

1. $(x + y)^8$ 2. $(x + y)^9$ 3. $(x - y)^6$

4. $(x - y)^4$ 5. $(a + 2b)^4$ 6. $(3a + b)^4$

7. $(x - 3y)^5$ 8. $(2x - y)^6$ 9. $(2a - 3b)^4$

10. $(3a - 2b)^5$ 11. $(x^2 + y)^5$ 12. $(x + y^3)^6$

13. $(2x^2 - y^2)^4$ 14. $(3x^2 - 2y^2)^5$ 15. $(x + 3)^6$

16. $(x + 2)^7$ 17. $(x - 1)^9$ 18. $(x - 3)^4$

19. $\left(1 + \dfrac{1}{n}\right)^4$ **20.** $\left(2 + \dfrac{1}{n}\right)^5$ **21.** $\left(a - \dfrac{1}{n}\right)^6$

22. $\left(2a - \dfrac{1}{n}\right)^5$ **23.** $(1 + \sqrt{2})^4$ **24.** $(2 + \sqrt{3})^3$

25. $(3 - \sqrt{2})^5$ **26.** $(1 - \sqrt{3})^4$

Write the first four terms of each of the following expansions.

27. $(x + y)^{12}$ **28.** $(x + y)^{15}$ **29.** $(x - y)^{20}$

30. $(a - 2b)^{13}$ **31.** $(x^2 - 2y^3)^{14}$ **32.** $(x^3 - 3y^2)^{11}$

33. $\left(a + \dfrac{1}{n}\right)^9$ **34.** $\left(2 - \dfrac{1}{n}\right)^6$ **35.** $(-x + 2y)^{10}$

36. $(-a - b)^{14}$

Find the specified term for each of the following binomial expansions.

37. The 4th term of $(x + y)^8$ **38.** The 7th term of $(x + y)^{11}$

39. The 5th term of $(x - y)^9$ **40.** The 4th term of $(x - 2y)^6$

41. The 6th term of $(3a + b)^7$ **42.** The 3rd term of $(2x - 5y)^5$

43. The 8th term of $(x^2 + y^3)^{10}$ **44.** The 9th term of $(a + b^3)^{12}$

45. The 7th term of $\left(1 - \dfrac{1}{n}\right)^{15}$ **46.** The 8th term of $\left(1 - \dfrac{1}{n}\right)^{13}$

Miscellaneous Problems

Expand and simplify each of the following complex numbers.

47. $(1 + 2i)^5$ **48.** $(2 + i)^6$ **49.** $(2 - i)^6$ **50.** $(3 - 2i)^5$

Chapter 11 Summary

This chapter can be summarized around three main topics: counting techniques, probability, and the binomial theorem.

Counting Techniques

The *fundamental principle of counting* states that if a first task can be accomplished in x ways and, following this task, a second task can be accomplished in y ways, then task 1 followed by task 2 can be accomplished in $x \cdot y$ ways. The principle extends to any finite number of tasks. As you solve problems involving the fundamental principle of counting, it is often helpful to analyze the problem in terms of the tasks to be completed.

Ordered arrangements are called **permutations**. The number of permutations of n things taken n at a time is given by

$$P(n, n) = n!.$$

The number of r-element permutations that can be formed from a set of n elements is given by

$$P(n, r) = \underbrace{n(n - 1)(n - 2) \cdots}_{r \text{ factors}}.$$

If there are n elements to be arranged, where there are r_1 of one kind, r_2 of another kind, r_3 of another kind, . . . , r_k of a kth kind, then the number of distinguishable permutations is given by

$$\frac{n!}{(r_1!)(r_2!)(r_3!)\cdots(r_k!)}.$$

Combinations are subsets: the order in which the elements appear does not make a difference. The number of r-element combinations (subsets) that can be formed from a set of n elements is given by

$$C(n, r) = \frac{P(n, r)}{r!}.$$

"Does the order in which the elements appear make any difference?" This is a key question to consider when trying to decide whether a particular problem involves permutations or combinations. If the answer to the question is yes, then it is a permutation problem; if the answer is no, then it is a combination problem. Don't forget that combinations are subsets.

Probability

In an experiment for which all possible outcomes in the sample space S are equally likely to occur, the **probability** of an event E is defined by

$$P(E) = \frac{n(E)}{n(S)},$$

where $n(E)$ denotes the number of elements in the event E and $n(S)$ denotes the number of elements in the sample space S. The numbers $n(E)$ and $n(S)$ can often be determined by using one or more of the previously listed counting techniques. For all events E, it is always true that $0 \leq P(E) \leq 1$. That is to say, all probabilities fall in the range from 0 to 1, inclusive.

If E and E' are **complementary events**, then $P(E) + P(E') = 1$. Therefore, if we can calculate either $P(E)$ or $P(E')$, then we can find the other one by subtracting from 1.

For two events E and F, the probability of "E or F" is given by

$$P(E \cup F) = P(E) + P(F) - P(E \cap F).$$

If $E \cap F = \varnothing$, then E and F are **mutually exclusive events**.

The probability that an event E occurs, given that another event F has already occurred, is called **conditional probability**, and it is given by the equation

$$P(E \mid F) = \frac{P(E \cap F)}{P(F)}.$$

Two events E and F are said to be **independent** if and only if

$$P(E \cap F) = P(E)P(F).$$

Two events that are not independent are called **dependent** events, and the probability of two dependent events is given by

$$P(E \cap F) = P(E)P(F\,|\,E).$$

The Binomial Theorem

For any binomial $(x + y)$ and any natural number n,

$$(x + y)^n = x^n + \binom{n}{1}x^{n-1}y + \binom{n}{2}x^{n-2}y^2 + \cdots + \binom{n}{n}y^n.$$

Note the following patterns in a binomial expansion.

1. In each term, the sum of the exponents of x and y is n.

2. The exponents of x begin with the exponent of the binomial and decrease by 1, term by term, until the last term has x^0, which is 1. The exponents of y begin with zero ($y^0 = 1$) and increase by 1, term by term, until the last term contains y to the power of the binomial.

3. The coefficient of any term is given by $\binom{n}{r}$, where the value of r agrees with the exponent of y for that term. For example, if the term contains y^3, then the coefficient of that term is $\binom{n}{3}$.

4. The expansion of $(x + y)^n$ contains $n + 1$ terms.

Chapter 11 Review Problem Set

Problems 1–14 are counting-type problems.

1. How many different arrangements of the letters A, B, C, D, E, and F can be made?

2. How many different 9-letter arrangements can be formed from the 9 letters of the word APPARATUS?

3. How many odd numbers of 3 different digits each can be formed by choosing from the digits 1, 2, 3, 5, 7, 8, and 9?

4. In how many ways can Arlene, Brent, Cindy, Dave, Ernie, Frank, and Gladys be seated in a row of 7 seats so that Arlene and Cindy are side by side?

5. In how many ways can a committee of 3 people be chosen from 6 people?

6. How many committees consisting of 3 men and 2 women can be formed from 7 men and 6 women?

7. How many different 5-card hands consisting of all hearts can be formed from a deck of 52 playing cards?

8. If no number contains repeated digits, how many numbers greater than 500 can be formed by choosing from the digits 2, 3, 4, 5, and 6?

9. How many 3-person committees can be formed from 4 men and 5 women so that each committee contains at least one man?

10. How many different 4-person committees can be formed from 8 people if 2 particular people refuse to serve together on a committee?

11. How many 4-element subsets containing A or B but not both A and B can be formed from the set {A, B, C, D, E, F, G, H}?

12. How many different 6-letter permutations can be formed from 4 identical H's and 2 identical T's?

13. How many 4-person committees consisting of 2 seniors, 1 sophomore, and 1 junior can be formed from 3 seniors, 4 juniors, and 5 sophomores?

14. In a baseball league of 6 teams, how many games are needed to complete a schedule if each team plays 8 games with each other team?

Problems 15–35 pose some probability questions.

15. If 3 coins are tossed, find the probability of getting 2 heads and 1 tail.

16. If 5 coins are tossed, find the probability of getting 3 heads and 2 tails.

17. What is the probability of getting a sum of 8 with one roll of a pair of dice?

18. What is the probability of getting a sum more than 5 with one roll of a pair of dice?

19. Amy, Brenda, Chuck, Dave, and Elmer are randomly seated in a row of 5 seats. Find the probability that Amy and Chuck are not seated side by side.

20. Four girls and 3 boys are randomly seated in a row of 7 seats. Find the probability that the girls and boys will be seated in alternate seats.

21. Six coins are tossed. Find the probability of getting at least 2 heads.

22. Two cards are randomly chosen from a deck of 52 playing cards. What is the probability that 2 jacks are drawn?

23. Each arrangement of the 6 letters of the word CYCLIC is put on a slip of paper and placed in a hat. One slip is drawn at random. Find the probability that the slip contains an arrangement with the Y at the beginning.

24. A committee of 3 is randomly chosen from 1 man and 6 women. What is the probability that the man is not on the committee?

25. A 4-person committee is selected at random from the 8 people, Alice, Bob, Carl, Dee, Edna, Fred, Gina, and Hilda. Find the probability that Alice or Bob, but not both, is on the committee.

26. A committee of 3 is chosen at random from a group of 5 men and 4 women. Find the probability that the committee contains 2 men and 1 woman.

27. A committee of 4 is chosen at random from a group of 6 men and 7 women. Find the probability that the committee contains at least one woman.

28. A bag contains 5 red and 8 white marbles. Two marbles are drawn in succession with replacement. What is the probability that at least one red marble is drawn?

29. A bag contains 4 red, 5 white, and 3 blue marbles. Two marbles are drawn in succession with replacement. Find the probability that one red and one blue marble are drawn.

30. A bag contains 4 red and 7 blue marbles. Two marbles are drawn in succession without replacement. Find the probability of drawing one red and one blue marble.

31. A bag contains 3 red, 2 white, and 2 blue marbles. Two marbles are drawn in succession without replacement. Find the probability of drawing at least one red marble.

32. Each of 3 letters is to be mailed in any one of 4 different mailboxes. What is the probability that all 3 letters are mailed in the same mailbox?

33. The probability that a customer in a department store will buy a blouse is .15, the probability that she will buy a pair of shoes is .10, and the probability that she will buy both a blouse and a pair of shoes is .05. Find the probability that the customer will buy a blouse, given that she has already purchased a pair of shoes. Also find

the probability that she will buy a pair of shoes, given that she has already purchased a blouse.

34. A survey of 500 employees of a company produced the following information.

EMPLOYMENT LEVEL	COLLEGE DEGREE	NO COLLEGE DEGREE
Managerial	45	5
Nonmanagerial	50	400

Find the probability that an employee chosen at random **(a)** is working in a managerial position, given that he (she) has a college degree; and **(b)** has a college degree, given that he (she) is working in a managerial position.

35. From a survey of 1000 college students, it was found that 450 of them owned cars, 700 of them owned stereos, and 200 of them owned both a car and a stereo. If a student is chosen at random from the 1000 students, find the probability that **(a)** he (she) owns a car, given the fact that he (she) owns a stereo; and **(b)** he (she) owns a stereo, given the fact that he (she) owns a car.

For Problems 36–41, expand each binomial and simplify.

36. $(x + 2y)^5$

37. $(x - y)^8$

38. $(a^2 - 3b^3)^4$

39. $\left(x + \dfrac{1}{n}\right)^6$

40. $(1 - \sqrt{2})^5$

41. $(-a + b)^3$

42. Find the 4th term of the expansion of $(x - 2y)^{12}$.

43. Find the 10th term of the expansion of $(3a + b^2)^{13}$.

Answers to Odd-numbered Exercises
and All Review Problems

CHAPTER 1

Problem Set 1.1 (page 10)

1. True **3.** False **5.** False **7.** True
9. False **11.** $\{46\}$ **13.** $\{0, -14, 46\}$
15. $\{\sqrt{5}, -\sqrt{2}, -\pi\}$ **17.** $\{0, -14\}$ **19.** $\subseteq$
21. $\subseteq$ **23.** $\nsubseteq$ **25.** $\subseteq$ **27.** $\nsubseteq$ **29.** $\subseteq$
31. $\nsubseteq$ **33.** $\{1\}$ **35.** $\{0, 1, 2, 3\}$
37. $\{\ldots, -2, -1, 0, 1\}$ **39.** $\varnothing$ **41.** $\{0, 1, 2\}$
43. **(a)** 18 **(c)** 39 **(e)** 35
45. Commutative property of multiplication
47. Identity property of multiplication
49. Multiplication property of negative one
51. Distributive property
53. Commutative property of multiplication
55. Distributive property
57. Associative property of multiplication
59. -22 **61.** 100 **63.** -21 **65.** 8
67. 19 **69.** 66 **71.** -75 **73.** 34 **75.** 1
77. 11 **79.** 4

Problem Set 1.2 (page 19)

1. $\frac{1}{8}$ **3.** $-\frac{1}{1000}$ **5.** 27 **7.** 4
9. $-\frac{27}{8}$ **11.** 1 **13.** $\frac{16}{25}$ **15.** 4
17. $\frac{1}{100}$ or 0.01 **19.** $\frac{1}{100000}$ or 0.00001 **21.** 81
23. $\frac{1}{16}$ **25.** $\frac{3}{4}$ **27.** $\frac{256}{25}$ **29.** $\frac{16}{25}$ **31.** $\frac{64}{81}$
33. 64 **35.** $\frac{1}{100000}$ or 0.00001 **37.** $\frac{17}{72}$ **39.** $\frac{1}{6}$
41. $\frac{48}{19}$ **43.** $\frac{1}{x^4}$ **45.** $\frac{1}{a^2}$ **47.** $\frac{1}{a^6}$ **49.** $\frac{y^4}{x^3}$

51. $\frac{c^3}{a^3 b^6}$ **53.** $\frac{y^2}{4x^4}$ **55.** $\frac{x^4}{y^6}$ **57.** $\frac{9a^2}{4b^4}$
59. $\frac{1}{x^3}$ **61.** $\frac{a^3}{b}$ **63.** $\frac{6}{x^3 y}$ **65.** $\frac{6}{a^2 y^3}$ **67.** $\frac{4x^3}{y^5}$
69. $-\frac{5}{a^2 b}$ **71.** $\frac{1}{4x^2 y^4}$ **73.** $\frac{x+1}{x^2}$ **75.** $\frac{y - x^2}{x^2 y}$
77. $\frac{3b^3 + 2a^2}{a^2 b^3}$ **79.** $\frac{y^2 - x^2}{xy}$ **81.** $12x^{3a+1}$
83. 1 **85.** x^{2a} **87.** $-4y^{6b+2}$ **89.** x^b

Problem Set 1.3 (page 26)

1. $14x^2 + x - 6$ **3.** $-x^2 - 4x - 9$ **5.** $6x - 11$
7. $6x^2 - 5x - 7$ **9.** $-x - 34$
11. $12x^3 y^2 + 15x^2 y^3$ **13.** $30a^4 b^3 - 24a^5 b^3 + 18a^4 b^4$
15. $x^2 + 20x + 96$ **17.** $n^2 - 16n + 48$
19. $sx + sy - tx - ty$ **21.** $6x^2 + 7x - 3$
23. $12x^2 - 37x + 21$ **25.** $x^2 + 8x + 16$
27. $4n^2 + 12n + 9$ **29.** $x^3 + x^2 - 14x - 24$
31. $6x^3 - x^2 - 11x + 6$ **33.** $x^3 + 2x^2 - 7x + 4$
35. $t^3 - 1$ **37.** $6x^3 + x^2 - 5x - 2$
39. $x^4 + 8x^3 + 15x^2 + 2x - 4$ **41.** $25x^2 - 4$
43. $x^4 - 10x^3 + 21x^2 + 20x + 4$ **45.** $4x^2 - 9y^2$
47. $x^3 + 15x^2 + 75x + 125$ **49.** $8x^3 + 12x^2 + 6x + 1$
51. $64x^3 - 144x^2 + 108x - 27$
53. $125x^3 - 150x^2 y + 60xy^2 - 8y^3$
55. $a^7 + 7a^6 b + 21a^5 b^2 + 35a^4 b^3 + 35a^3 b^4$
$\qquad\qquad\qquad\qquad + 21a^2 b^5 + 7ab^6 + b^7$
57. $x^5 - 5x^4 y + 10x^3 y^2 - 10x^2 y^3 + 5xy^4 - y^5$
59. $x^4 + 8x^3 y + 24x^2 y^2 + 32xy^3 + 16y^4$

61. $64a^6 - 192a^5b + 240a^4b^2 - 160a^3b^3$
$\qquad + 60a^2b^4 - 12ab^5 + b^6$

63. $x^{14} + 7x^{12}y + 21x^{10}y^2 + 35x^8y^3 + 35x^6y^4$
$\qquad + 21x^4y^5 + 7x^2y^6 + y^7$

65. $32a^5 - 240a^4b + 720a^3b^2 - 1080a^2b^3$
$\qquad + 810ab^4 - 243b^5$

67. $3x^2 - 5x$ **69.** $-5a^4 + 4a^2 - 9a$

71. $5ab + 11a^2b^4$ **73.** $x^{2a} - y^{2b}$

75. $x^{2b} - 3x^b - 28$ **77.** $6x^{2b} + x^b - 2$

79. $x^{4a} - 2x^{2a} + 1$ **81.** $x^{3a} - 6x^{2a} + 12x^a - 8$

Problem Set 1.4 (page 33)

1. $2xy(3 - 4y)$ **3.** $(z + 3)(x + y)$

5. $(x + y)(3 + a)$ **7.** $(x - y)(a - b)$

9. $(3x + 5)(3x - 5)$ **11.** $(1 + 9n)(1 - 9n)$

13. $(x + 4 + y)(x + 4 - y)$

15. $(3s + 2t - 1)(3s - 2t + 1)$

17. $(x - 7)(x + 2)$ **19.** $(5 + x)(3 - x)$

21. Not factorable **23.** $(3x - 5)(x - 2)$

25. $(5x + 1)(2x - 7)$ **27.** $(x - 2)(x^2 + 2x + 4)$

29. $(4x + 3y)(16x^2 - 12xy + 9y^2)$ **31.** $4(x^2 + 4)$

33. $x(x + 3)(x - 3)$ **35.** $(3a - 7)^2$

37. $2n(n^2 + 3n + 5)$ **39.** $(5x - 3)(2x + 9)$

41. $(6a - 1)^2$ **43.** $(4x - y)(2x + y)$

45. Not factorable **47.** $2n(n^2 + 7n - 10)$

49. $4(x + 2)(x^2 - 2x + 4)$ **51.** $(x + 3)(x - 3)(x^2 + 5)$

53. $2y(x + 4)(x - 4)(x^2 + 3)$

55. $(a + b + c + d)(a + b - c - d)$

57. $(x + 4 + y)(x + 4 - y)$ **59.** $(x + y + 5)(x - y - 5)$

61. $(x^a + 4)(x^a - 4)$ **63.** $(x^n - y^n)(x^{2n} + x^ny^n + y^{2n})$

65. $(x^a + 4)(x^a - 7)$ **67.** $(2x^n - 5)(x^n + 6)$

69. $(x^{2n} + y^{2n})(x^n + y^n)(x^n - y^n)$

71. (a) $(x + 32)(x + 3)$ **(c)** $(x - 21)(x - 24)$
(e) $(x + 28)(x + 32)$

Problem Set 1.5 (page 42)

1. $\dfrac{2x}{3}$ **3.** $\dfrac{7y^3}{9x}$ **5.** $\dfrac{a + 4}{a - 9}$ **7.** $\dfrac{x(2x + 7)}{y(x + 9)}$

9. $\dfrac{x^2 + xy + y^2}{x + 2y}$ **11.** $-\dfrac{2}{x + 1}$ **13.** $\dfrac{x}{2y^3}$

15. $-\dfrac{8x^3y^3}{15}$ **17.** $\dfrac{14}{27a}$ **19.** $5y$ **21.** $\dfrac{5(a + 3)}{a(a - 2)}$

23. $\dfrac{(x + 6y)^2(2x + 3y)}{y^3(x + 4y)}$ **25.** $\dfrac{3xy}{4(x + 6)}$ **27.** $\dfrac{x - 9}{42x^2}$

29. $\dfrac{8x + 5}{12}$ **31.** $\dfrac{7x}{24}$ **33.** $\dfrac{35b + 12a^3}{80a^2b^2}$

35. $\dfrac{12 + 9n - 10n^2}{12n^2}$ **37.** $\dfrac{9y + 8x - 12xy}{12xy}$

39. $\dfrac{13x + 14}{(2x + 1)(3x + 4)}$ **41.** $\dfrac{7x + 21}{x(x + 7)}$ **43.** $\dfrac{1}{a - 2}$

45. $\dfrac{1}{x + 1}$ **47.** $\dfrac{9x + 73}{(x + 3)(x + 7)(x + 9)}$

49. $\dfrac{3x^2 + 30x - 78}{(x + 1)(x - 1)(x + 8)(x - 2)}$ **51.** $\dfrac{-x^2 - x + 1}{(x + 1)(x - 1)}$

53. $\dfrac{-8}{(n^2 + 4)(n + 2)(n - 2)}$ **55.** $\dfrac{5x^2 + 16x + 5}{(x + 1)(x - 4)(x + 7)}$

57. (a) $\dfrac{5}{x - 1}$ **(c)** $\dfrac{5}{a - 3}$ **(e)** $x + 3$

59. $\dfrac{5y^2 - 3xy^2}{x^2y + 2x^2}$ **61.** $\dfrac{x + 1}{x - 1}$ **63.** $\dfrac{n - 1}{n + 1}$

65. $\dfrac{-6x - 4}{3x + 9}$ **67.** $\dfrac{x^2 + x + 1}{x + 1}$ **69.** $\dfrac{a^2 + 4a + 1}{4a + 1}$

Problem Set 1.6 (page 51)

1. 9 **3.** 5 **5.** $\frac{6}{7}$ **7.** $-\frac{3}{2}$ **9.** $2\sqrt{6}$

11. $4\sqrt{7}$ **13.** $-6\sqrt{11}$ **15.** $\dfrac{3\sqrt{5}}{2}$ **17.** $2x\sqrt{3}$

19. $8x^2y^3\sqrt{y}$ **21.** $\dfrac{9y^3\sqrt{5x}}{7}$ **23.** $4\sqrt[3]{2}$

25. $2x\sqrt[3]{2x}$ **27.** $2x\sqrt[4]{3x}$ **29.** $\dfrac{2\sqrt{3}}{5}$ **31.** $\dfrac{\sqrt{14}}{4}$

33. $\dfrac{4\sqrt{15}}{5}$ **35.** $\dfrac{3\sqrt{2}}{7}$ **37.** $\dfrac{\sqrt{15}}{6x^2}$ **39.** $\dfrac{2\sqrt{15a}}{5ab}$

41. $\dfrac{3\sqrt[3]{2}}{2}$ **43.** $\dfrac{\sqrt[3]{18x^2y}}{3x}$ **45.** $12\sqrt{3}$ **47.** $3\sqrt{7}$

49. $\dfrac{11\sqrt{3}}{6}$ **51.** $-\dfrac{89\sqrt{2}}{30}$ **53.** $48\sqrt{6}$

55. $10\sqrt{6} + 8\sqrt{30}$ **57.** $3x\sqrt{6y} - 6\sqrt{2xy}$

59. $13 + 7\sqrt{3}$ **61.** $30 + 11\sqrt{6}$ **63.** 16

65. $x + 2\sqrt{xy} + y$ **67.** $a - b$ **69.** $3\sqrt{5} - 6$

71. $\sqrt{7} + \sqrt{3}$ **73.** $\dfrac{-2\sqrt{10} + 3\sqrt{14}}{43}$ **75.** $\dfrac{x + \sqrt{x}}{x - 1}$

77. $\dfrac{x - \sqrt{xy}}{x - y}$ **79.** $\dfrac{6x + 7\sqrt{xy} + 2y}{9x - 4y}$

Problem Set 1.7 (page 56)

1. 7 **3.** 8 **5.** -4 **7.** 2 **9.** 64
11. 0.001 **13.** $\frac{1}{32}$ **15.** 2 **17.** $15x^{7/12}$

19. $y^{5/12}$ **21.** $64x^{3/4}y^{3/2}$ **23.** $4x^{4/15}$ **25.** $\dfrac{7}{a^{1/12}}$

27. $\dfrac{16x^{4/3}}{81y}$ **29.** $\dfrac{y^{3/2}}{x}$ **31.** $8a^{9/2}x^2$ **33.** $\sqrt[4]{8}$

35. $\sqrt[12]{x^7}$ **37.** $xy\sqrt[4]{xy^3}$ **39.** $a\sqrt[12]{a^5b^{11}}$ **41.** $4\sqrt[6]{2}$

43. $\sqrt[6]{2}$ **45.** $\sqrt{2}$ **47.** $x\sqrt[12]{x^7}$ **49.** $\dfrac{5\sqrt[3]{x^2}}{x}$

51. $\dfrac{\sqrt[6]{x^3y^4}}{y}$ **53.** $\dfrac{\sqrt[20]{x^{15}y^8}}{y}$ **55.** $\dfrac{5\sqrt[12]{x^9y^8}}{4x}$

57. (a) $\sqrt[6]{2}$ **(c)** $\sqrt{x}$
61. (a) 13.391   **(c)** 2.702   **(e)** 4.304

Problem Set 1.8 (page 63)

1. $13 + 8i$ **3.** $3 + 4i$ **5.** $-11 + i$
7. $-1 - 2i$ **9.** $-\frac{3}{20} + \frac{5}{12}i$ **11.** $\frac{7}{10} - \frac{11}{12}i$
13. 4 **15.** $3i$ **17.** $i\sqrt{19}$ **19.** $\frac{2}{3}i$ **21.** $2i\sqrt{2}$
23. $3i\sqrt{3}$ **25.** $3i\sqrt{6}$ **27.** $18i$ **29.** $12i\sqrt{2}$
31. -8 **33.** $-\sqrt{6}$ **35.** $-2\sqrt{5}$ **37.** $-2\sqrt{15}$
39. $-2\sqrt{14}$ **41.** 3 **43.** $\sqrt{6}$ **45.** -21
47. $8 + 12i$ **49.** $0 + 26i$ **51.** $53 - 26i$
53. $10 - 24i$ **55.** $-14 - 8i$ **57.** $-7 + 24i$
59. $-3 + 4i$ **61.** $113 + 0i$ **63.** $13 + 0i$
65. $-\frac{8}{13} + \frac{12}{13}i$ **67.** $1 - \frac{2}{3}i$ **69.** $0 - \frac{3}{2}i$
71. $\frac{22}{41} - \frac{7}{41}i$ **73.** $-1 + 2i$ **75.** $-\frac{17}{10} + \frac{1}{10}i$
77. $\frac{5}{13} - \frac{1}{13}i$

Chapter 1 Review Problem Set (page 66)

1. $\frac{1}{125}$ **2.** $-\frac{1}{81}$ **3.** $\frac{16}{9}$ **4.** $\frac{1}{9}$ **5.** -8
6. $\frac{3}{2}$ **7.** $-\frac{1}{2}$ **8.** $\frac{1}{6}$ **9.** 4 **10.** -8
11. $12x^2y$ **12.** $-30x^{7/6}$ **13.** $\dfrac{48}{a^{1/6}}$ **14.** $\dfrac{27y^{3/5}}{x^2}$
15. $\dfrac{4y^5}{x^5}$ **16.** $\dfrac{8y}{x^{7/12}}$ **17.** $\dfrac{16x^6}{y^6}$ **18.** $-\dfrac{a^3b^1}{3}$
19. $4x - 1$ **20.** $-3x + 8$ **21.** $12a - 19$
22. $20x^2 - 11x - 42$ **23.** $-12x^2 + 17x - 6$
24. $-35x^2 + 22x - 3$ **25.** $x^3 + x^2 - 19x - 28$
26. $2x^3 - x^2 + 10x + 6$ **27.** $25x^2 - 30x + 9$
28. $9x^2 + 42x + 49$ **29.** $8x^3 - 12x^2 + 6x - 1$
30. $27x^3 + 135x^2 + 225x + 125$
31. $x^4 + 2x^3 - 6x^2 - 22x - 15$
32. $2x^4 + 11x^3 - 16x^2 - 8x + 8$ **33.** $-4x^2y^3 + 8xy^2$
34. $-7y + 9xy^2$ **35.** $(3x + 2y)(3x - 2y)$
36. $3x(x + 5)(x - 8)$ **37.** $(2x + 5)^2$
38. $(x - y + 3)(x - y - 3)$ **39.** $(x - 2)(x - y)$
40. $(4x - 3y)(16x^2 + 12xy + 9y^2)$
41. $(3x - 4)(5x + 2)$ **42.** $3(x^3 + 12)$

43. Not factorable   **44.** $3(x + 2)(x^2 - 2x + 4)$
45. $(x + 3)(x - 3)(x + 2)(x - 2)$

46. $(2x - 1 - y)(2x - 1 + y)$ **47.** $\dfrac{2}{3y}$ **48.** $\dfrac{-5a^2}{3}$

49. $\dfrac{3x + 5}{x}$ **50.** $\dfrac{2(3x - 1)}{x^2 + 4}$ **51.** $\dfrac{29x - 10}{12}$

52. $\dfrac{x - 38}{15}$ **53.** $\dfrac{-6n + 15}{5n^2}$ **54.** $\dfrac{-3x - 16}{x(x + 7)}$

55. $\dfrac{3x^2 - 8x - 40}{(x + 4)(x - 4)(x - 10)}$ **56.** $\dfrac{8x - 4}{x(x + 2)(x - 2)}$

57. $\dfrac{3xy - 2x^2}{5y + 7x^2}$ **58.** $\dfrac{3x - 2}{4x + 3}$ **59.** $20\sqrt{3}$

60. $6x\sqrt{6x}$ **61.** $2xy\sqrt[3]{4xy^2}$ **62.** $\sqrt{3}$ **63.** $\dfrac{\sqrt{10x}}{2y}$

64. $\dfrac{15 - 3\sqrt{2}}{23}$ **65.** $\dfrac{24 - 4\sqrt{6}}{15}$ **66.** $\dfrac{3x + 6\sqrt{xy}}{x - 4y}$

67. $\sqrt[6]{5^5}$ **68.** $\sqrt[12]{x^{11}}$ **69.** $x^2\sqrt[6]{x^5}$

70. $x\sqrt[10]{xy^9}$ **71.** $\sqrt[6]{5}$ **72.** $\dfrac{\sqrt[12]{x^{11}}}{x}$

73. $-11 - 6i$ **74.** $-1 - 2i$ **75.** $1 - 2i$
76. $21 + 0i$ **77.** $26 - 7i$ **78.** $-25 + 15i$
79. $-14 - 12i$ **80.** $29 + 0i$ **81.** $0 - \frac{5}{3}i$
82. $-\frac{6}{25} + \frac{17}{25}i$ **83.** $0 + i$ **84.** $-\frac{12}{29} - \frac{30}{29}i$
85. $10i$ **86.** $2i\sqrt{10}$ **87.** $16i\sqrt{5}$ **88.** -12
89. $-4\sqrt{3}$ **90.** $2\sqrt{2}$

CHAPTER 2

Problem Set 2.1 (page 74)

1. $\{-2\}$ **3.** $\{-\frac{1}{2}\}$ **5.** $\{7\}$ **7.** $\{-\frac{3}{2}\}$
9. $\{-\frac{10}{3}\}$ **11.** $\{-10\}$ **13.** $\{-17\}$
15. $\{-\frac{21}{16}\}$ **17.** $\{\frac{3}{5}\}$ **19.** $\{-14\}$ **21.** $\{9\}$
23. $\{\frac{10}{7}\}$ **25.** $\{-10\}$ **27.** $\{1\}$ **29.** $\{-12\}$
31. $\{27\}$ **33.** $\{\frac{159}{5}\}$ **35.** $\{3\}$ **37.** $\{0\}$
39. $\{-\frac{2}{3}\}$ **41.** $\{\frac{1}{2}\}$ **43.** 41   **45.** 7 and 23
47. 18, 19, and 20   **49.** 31, 33, and 35
51. 6, 7, and 8
53. $24,000 for Renee, $20,000 for Kelly, $16,000 for Nina
55. 48 pennies, 21 nickels, and 11 dimes
57. 17 females and 26 males   **59.** 13 years old
61. Brad is 29 and Pedro is 23.

Problem Set 2.2 (page 85)

1. $\{1\}$ **3.** $\{9\}$ **5.** $\{\frac{10}{3}\}$ **7.** $\{4\}$ **9.** $\{14\}$
11. $\{9\}$ **13.** $\{\frac{1}{2}\}$ **15.** $\{\frac{1}{4}\}$ **17.** $\{\frac{2}{3}\}$

19. $\{-8\}$ **21.** $\varnothing$ **23.** $\{12\}$ **25.** $\{300\}$

27. $\{275\}$ **29.** $\{-\frac{66}{37}\}$ **31.** $\{6\}$

33. $w = \dfrac{P - 2l}{2}$ **35.** $h = \dfrac{A - 2\pi r^2}{2\pi r}$

37. $F = \dfrac{9C + 160}{5}$ or $F = \dfrac{9C}{5} + 32$

39. $T = \dfrac{NC - NV}{C}$ **41.** $T = \dfrac{I + klt}{kl}$

43. $R_n = \dfrac{R_1 R_2}{R_1 + R_2}$ **45.** 17 and 81   **47.** 9

49. \$900 and \$1350   **51.** 37 teachers and 740 students

53. \$65   **55.** \$950 per month   **57.** \$32.20

59. \$75   **61.** \$30   **63.** 14 nickels and 29 dimes

65. 15 dimes, 45 quarters, and 10 half-dollars

67. \$2000 at 9% and \$3500 at 10%

69. \$3500   **71.** 6 centimeters by 10 centimeters

Problem Set 2.3 (page 96)

1. $\{-4, 7\}$ **3.** $\{-3, \frac{4}{3}\}$ **5.** $\{0, \frac{3}{2}\}$ **7.** $\left\{\pm\dfrac{2\sqrt{3}}{3}\right\}$

9. $\left\{\dfrac{-1 \pm 2\sqrt{5}}{2}\right\}$ **11.** $\{-\frac{5}{3}, \frac{2}{5}\}$ **13.** $\{2 \pm 2i\}$

15. $\{-\frac{7}{2}, \frac{1}{5}\}$ **17.** $\{4, 6\}$ **19.** $\{-5 \pm 3\sqrt{3}\}$

21. $\left\{\dfrac{3 \pm \sqrt{5}}{2}\right\}$ **23.** $\{-2 \pm i\sqrt{2}\}$ **25.** $\left\{\dfrac{-6 \pm \sqrt{46}}{2}\right\}$

27. $\{-16, 18\}$ **29.** $\left\{\dfrac{-5 \pm \sqrt{37}}{6}\right\}$ **31.** $\{-6, 9\}$

33. $\{-5, -\frac{1}{3}\}$ **35.** $\{1 \pm \sqrt{5}\}$ **37.** $\left\{\dfrac{3 \pm \sqrt{7}}{2}\right\}$

39. $\left\{\dfrac{3 \pm i\sqrt{19}}{2}\right\}$ **41.** $\{4 \pm 2\sqrt{3}\}$ **43.** $\{\frac{1}{2}\}$

45. $\{-\frac{3}{2}, \frac{1}{4}\}$ **47.** $\{-14, 12\}$ **49.** $\left\{\dfrac{3 \pm i\sqrt{47}}{4}\right\}$

51. $\{-1, \frac{5}{3}\}$ **53.** $\left\{\dfrac{-1 \pm \sqrt{2}}{2}\right\}$ **55.** $\{8 \pm 5\sqrt{2}\}$

57. $\{-10 \pm 5\sqrt{5}\}$ **59.** $\left\{\dfrac{1 \pm \sqrt{6}}{5}\right\}$

61. (a) One real solution   **(c)** One real solution
(e) Two complex but nonreal solutions
(g) Two unequal real solutions

63. 11 and 12   **65.** 8 and 14

67. 10 meters and 24 meters   **69.** 8 in. by 14 in.

71. 7 meters wide and 18 meters long

73. 1 meter   **75.** 7 in. by 11 in.

77. (a) $r = \dfrac{\sqrt{A\pi}}{\pi}$ **(c)** $t = \dfrac{\sqrt{2gs}}{g}$ **(e)** $y = \dfrac{b\sqrt{x^2 - a^2}}{a}$

Problem Set 2.4 (page 104)

1. $\{-12\}$ **3.** $\{\frac{37}{15}\}$ **5.** $\{-2\}$ **7.** $\{-8, 1\}$

9. $\{\frac{6}{29}\}$ **11.** $\{n \mid n \neq \frac{3}{2}$ and $n \neq 3\}$ **13.** $\{-4, \frac{4}{3}\}$

15. $\{-\frac{1}{4}\}$ **17.** $\varnothing$ **19.** $\{-1\}$

21. 9 rows and 14 trees per row   **23.** $4\frac{1}{2}$ hours

25. 50 miles

27. 50 mph for the freight and 70 mph for the express

29. 3 liters

31. 3.5 liters of the 50% solution and 7 liters of the 80% solution

33. 5 quarts   **35.** $2\frac{2}{5}$ hours   **37.** 60 minutes

39. 9 minutes for Pat and 18 minutes for Mike

41. 60 words per minute for Amelia and 40 words per minute for Paul

43. 7 golf balls   **45.** 60 hours

Problem Set 2.5 (page 111)

1. $\{-2, -1, 2\}$ **3.** $\{\pm i, \frac{3}{2}\}$ **5.** $\left\{-\frac{5}{4}, 0, \pm\dfrac{\sqrt{2}}{2}\right\}$

7. $\{0, 16\}$ **9.** $\{1\}$ **11.** $\{6\}$ **13.** $\{3\}$

15. $\varnothing$ **17.** $\{-15\}$ **19.** $\{9\}$ **21.** $\{\frac{2}{3}, 1\}$

23. $\{5\}$ **25.** $\{7\}$ **27.** $\{-2, -1\}$ **29.** $\{0\}$

31. $\{6\}$ **33.** $\{0, 4\}$ **35.** $\{\pm 1, \pm 2\}$

37. $\left\{\pm\dfrac{\sqrt{2}}{2}, \pm 2\right\}$ **39.** $\{\pm i\sqrt{5}, \pm\sqrt{7}\}$

41. $\{\pm\sqrt{2 + \sqrt{3}}, \pm\sqrt{2 - \sqrt{3}}\}$ **43.** $\{-125, 8\}$

45. $\{-\frac{8}{27}, \frac{27}{8}\}$ **47.** $\{-\frac{1}{6}, \frac{1}{2}\}$ **49.** $\{25, 36\}$ **51.** $\{4\}$

Problem Set 2.6 (page 117)

1. $(-\infty, -2]$
-2

3. $(1, 4)$
$1 \quad 4$

5. $(0, 2)$
$0 \quad 2$

7. $[-2, -1]$
$-2 \ -1$

9. $(-\infty, -2)$ **11.** $[-\frac{5}{3}, \infty)$ **13.** $[7, \infty)$

15. $(-\infty, \frac{17}{5}]$ **17.** $(-\infty, \frac{7}{3})$ **19.** $[-20, \infty)$

21. $(300, \infty)$ **23.** $(\frac{1}{5}, \frac{7}{5})$ **25.** $[1, 5]$

27. $(-4, 1)$ **29.** $(-\infty, -3) \cup (5, \infty)$
31. $[-1, 2]$ **33.** $(-\infty, -4) \cup (\frac{1}{3}, \infty)$ **35.** $[\frac{2}{5}, \frac{4}{3}]$
37. $(-\infty, \frac{1}{2}) \cup (\frac{1}{2}, \infty)$ **39.** $(-2, -1)$
41. $(-\infty, \frac{22}{3}]$ **43.** $(-4, 1) \cup (2, \infty)$
45. $(-\infty, -2] \cup [\frac{1}{2}, 5]$ **47.** $[-4, 0] \cup [6, \infty)$
49. $(-3, 2) \cup (2, \infty)$ **51.** Greater than 12%
53. 98 or better **55.** Between 5°C and 15°C, inclusive
57. $8.8 \le M \le 15.4$ **59.** More than 250 miles
61. **(a)** $(-\infty, -5) \cup (1, \infty)$ **(c)** $(-4, 3)$
(e) $(-4, -1) \cup (2, \infty)$

Problem Set 2.7 (page 124)

1. $(-\infty, -1) \cup (5, \infty)$ **3.** $(-2, \frac{1}{2})$ **5.** $(\frac{1}{3}, 3]$
7. $[-3, -2)$ **9.** $(-\infty, -5) \cup (-2, \infty)$
11. $(-\infty, -5)$ **13.** $(-3, 2)$ **15.** $\{-4, 8\}$
17. $\{-\frac{13}{20}, \frac{3}{20}\}$ **19.** $\{-3, 4\}$ **21.** $\{-3, \frac{1}{3}\}$
23. $\varnothing$ **25.** $\{-\frac{10}{3}, 2\}$ **27.** $\{\frac{1}{4}, \frac{7}{4}\}$ **29.** $\{-\frac{2}{5}, 4\}$
31. $\{-2, 0\}$ **33.** $\{-1\}$ **35.** $(-6, 6)$
37. $(-\infty, -2) \cup (8, \infty)$ **39.** $[-3, 4]$
41. $(-\infty, -\frac{11}{3}) \cup (\frac{7}{3}, \infty)$ **43.** $\varnothing$ **45.** $(-\frac{1}{2}, \frac{7}{2})$
47. $(-\infty, -9] \cup [7, \infty)$ **49.** $(-1, 3)$
51. $(-\infty, -6) \cup (-2, \infty)$ **53.** $(-\infty, \frac{5}{4}) \cup (\frac{7}{2}, \infty)$
55. $(-\infty, -3) \cup (-3, -1)$ **57.** $[-\frac{2}{5}, 0) \cup (0, \frac{2}{3}]$
59. $(-\infty, \frac{2}{5}] \cup [\frac{2}{3}, \infty)$

Chapter 2 Review Problem Set (page 126)

1. $\{-14\}$ **2.** $\{-19\}$ **3.** $\{\frac{10}{7}\}$ **4.** $\{200\}$
5. $\{-1, \frac{5}{3}\}$ **6.** $\{\frac{5}{4}, 6\}$ **7.** $\{3 \pm i\}$
8. $\{-22, 18\}$ **9.** $\{-\frac{2}{5}, 0, \frac{1}{3}\}$ **10.** $\{-5\}$
11. $\{\frac{1}{2}, 6\}$ **12.** $\{\pm 3i, \pm \sqrt{5}\}$ **13.** $\left\{\pm \frac{\sqrt{5}}{5}, \pm \sqrt{2}\right\}$
14. $\left\{-1, 2, \frac{-5 \pm \sqrt{33}}{2}\right\}$ **15.** $\{2\}$ **16.** $\{-1, \frac{1}{2}\}$
17. $\{0\}$ **18.** $\{-\frac{6}{5}, \frac{8}{5}\}$ **19.** $\{\frac{2}{5}, 12\}$ **20.** $\{\frac{1}{4}, \frac{7}{4}\}$
21. $(-8, \infty)$ **22.** $[-\frac{65}{4}, \infty)$ **23.** $(-\infty, -\frac{9}{2})$
24. $(-\infty, 400]$ **25.** $[-2, 1]$ **26.** $(-\frac{2}{3}, 2)$
27. $(-3, 6)$ **28.** $(-\infty, -2] \cup [7, \infty)$
29. $(-\infty, -2) \cup (1, 4)$ **30.** $[-4, \frac{3}{2})$
31. $(-\infty, \frac{1}{5}) \cup (2, \infty)$ **32.** $[-7, -3)$ **33.** $(-\infty, 4)$
34. $(-\infty, -\frac{1}{2}) \cup (2, \infty)$ **35.** $[-\frac{19}{3}, 3]$ **36.** $(-\frac{9}{2}, \frac{3}{2})$
37. $(-1, 0) \cup (0, \frac{1}{3})$ **38.** $(-\frac{3}{2}, \infty)$ **39.** 21, 23, and 25
40. 9 and 65 **41.** 7 cm by 12 cm
42. 13 nickels, 39 dimes, and 36 quarters
43. $20 **44.** 20 gallons
45. Rosie is 14 years old and her mother is 33 years old.
46. $350 at 9% and $450 at 12% **47.** 95 or better
48. $10\frac{10}{11}$ minutes **49.** $26\frac{2}{3}$ minutes

50. 40 shares at $15/share
51. 54 mph for Mike and 52 mph for Larry
52. Cindy 4 hours and Bill 6 hours

CHAPTER 3

Problem Set 3.1 (page 136)

1. 10 **3.** -5 **5.** 6 **7.** 15 **9.** 7
11. $\frac{1}{3}$ **13.** -7 **15.** $10; (6, 4)$ **17.** $\sqrt{13}; (2, -\frac{5}{2})$
19. $3\sqrt{2}; (\frac{15}{2}, -\frac{11}{2})$ **21.** $\frac{\sqrt{74}}{6}; \left(\frac{1}{12}, \frac{11}{12}\right)$ **23.** $(3, 5)$
25. $(2, 5)$ **27.** $(\frac{17}{8}, -7)$ **29.** $(4, \frac{25}{4})$
35. $15 + 9\sqrt{5}$ **39.** 3 or -7 **41.** $(3, 8)$

Problem Set 3.2 (page 142)

1. 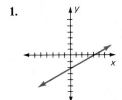 **3.**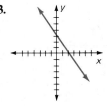

5. 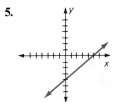 **7.**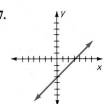

9. 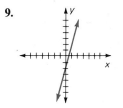 **11.**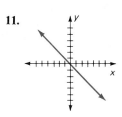

13. The graph is the y-axis. **15.**

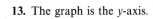

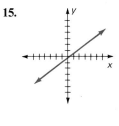

17.

19.

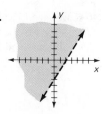

21.

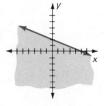

23.

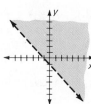

25.

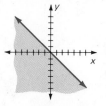

27.

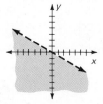

29.

31.

33.

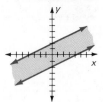

35.

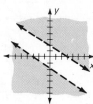

37.

39.

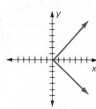

41.

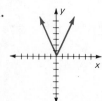

Problem Set 3.3 (page 149)

1. $\frac{3}{4}$ **3.** -5 **5.** 0 **7.** $-b/a$ **9.** $x = 7$
11. $x = -2$ **13–18.** Answers will vary.
19. $x - 3y = -10$ **21.** $2x - y = 0$
23. $2x + 3y = -1$ **25.** $y = -2$
27. $5x - 7y = -11$ **29.** $5x + 6y = 37$
31. $x + 5y = 14$ **33.** $y = -3$ **35.** $x - 2y = -6$
37. $3x + 7y = 14$ **39.** $8x - 2y = -3$
41. $10x + 12y = 3$ **43.** $5x - 4y = 20$
45. $x = -4$ **47.** $5x + 2y = 14$
49. $4x + y = -2$ **51.** Parallel   **53.** Perpendicular
55. Intersecting lines that are not perpendicular
57. Perpendicular   **59.** $m = \frac{2}{3}, b = -\frac{4}{3}$
61. $m = \frac{1}{2}, b = -\frac{7}{2}$ **63.** $m = -3, b = 0$
65. $m = \frac{7}{5}, b = -\frac{12}{5}$

67. (a)

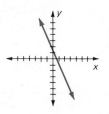

(c)

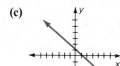

(e)

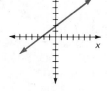

73. $x - y = -7,\ x + 5y = -19,\ 2x + y = 16$
75. $9x + 8y = -2,\ 6x - 7y = 11,\ 15x + y = 9$
77. (a) 32 cm
79. (a) $3x - y = 9$ **(c)** $2x + 7y = 0$
83. (a) $2x - y = 4$ **(c)** $5x + 2y = 2$

Problem Set 3.4 (page 156)

1. $(4, -3); (-4, 3); (-4, -3)$ **3.** $(-6, 1); (6, -1); (6, 1)$
5. $(0, -4); (0, 4); (0, -4)$ **7.** y-axis   **9.** x-axis
11. x-axis, y-axis, and origin   **13.** None
15. Origin   **17.** None   **19.** y-axis
21.

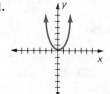

23.

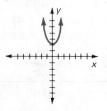

25.
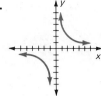

27.
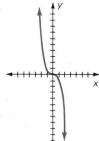

29.
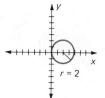

31.

29.
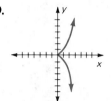

31.
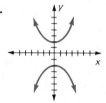

33.
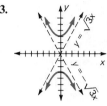

35.
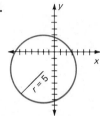

33.
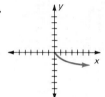

35.
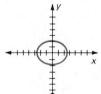

37.
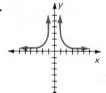

39.
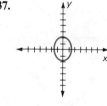

37.
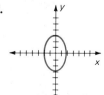

39.
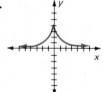

41.
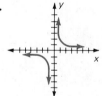

43.
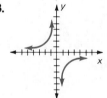

45. The x- and y-axes

47. (a) $(1, 4)$; $r = 3$ **(c)** $(-6, -4)$; $r = 8$
(e) $(0, 6)$; $r = 9$

Problem Set 3.5 (page 162)

1. $x^2 + y^2 - 4x - 6y - 12 = 0$
3. $x^2 + y^2 + 2x + 10y + 17 = 0$ **5.** $x^2 + y^2 - 6x = 0$
7. $x^2 + y^2 = 49$ **9.** $(3, 5)$; $r = 2$
11. $(-5, -7)$; $r = 1$ **13.** $(5, 0)$; $r = 5$
15. $(0, 0)$; $r = 2\sqrt{2}$ **17.** $(\frac{1}{2}, 1)$; $r = 2$
19. $x^2 + y^2 - 6x - 6y - 67 = 0$
21. $x^2 + y^2 - 14x + 14y + 49 = 0$
23. $x^2 + y^2 + 6x - 10y + 9 = 0$ and $x^2 + y^2 + 6x + 10y + 9 = 0$

25.
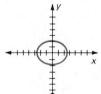

27.
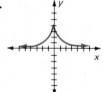

Chapter 3 Review Problem Set (page 165)

1. 5   **2.** -5 **3.** $(9, \frac{1}{3})$ **4.** $(-2, 6)$
7. x-axis   **8.** None   **9.** x-axis, y-axis, and origin
10. y-axis   **11.** Origin   **12.** y-axis
13.
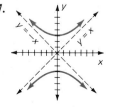

14.
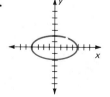

15.

16.

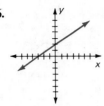

17.

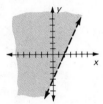

18.

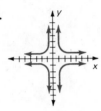

19.

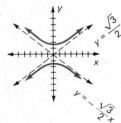

20.

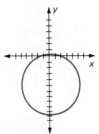

21.

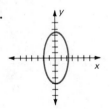

22.

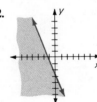

23. -5 **24.** $\frac{5}{7}$ **25.** $3x + 4y = 29$
26. $2x - y = -4$ **27.** $4x + 3y = -4$
28. $x + 2y = 3$ **29.** $x^2 + y^2 - 10x + 12y + 60 = 0$
30. $x^2 + y^2 - 4x - 6y - 4 = 0$
31. $x^2 + y^2 + 10x - 24y = 0$
32. $x^2 + y^2 + 8x + 8y + 16 = 0$

Chapters 1, 2, and 3 Cumulative Review Problem Set (page 166)

1. $\frac{1}{27}$ **2.** $-\frac{1}{16}$ **3.** $\frac{9}{4}$ **4.** $-\frac{2}{3}$ **5.** 9

6. $\frac{9}{16}$ **7.** $\frac{20}{x^2 y^3}$ **8.** $-\frac{56a}{b}$ **9.** $4x^4 y^2$

10. $\frac{5y^2}{x^4}$ **11.** $\frac{x^{1/3}}{17y^{7/4}}$ **12.** $\frac{4a^8}{b^{14}}$ **13.** $-30\sqrt{2}$

14. $6xy\sqrt{3x}$ **15.** $2xy^2\sqrt[3]{7xy}$ **16.** $\frac{3\sqrt{6}}{10}$

17. $\frac{\sqrt{21xy}}{7y}$ **18.** $-\frac{5(\sqrt{2} + 3)}{7}$ **19.** $\frac{6\sqrt{14} + 3\sqrt{42}}{2}$

20. $\frac{4x - 12\sqrt{xy}}{x - 9y}$ **21.** $\frac{3x^3 y^2}{8}$ **22.** $-\frac{3b^3}{8a^3}$

23. $\frac{5x + 1}{x}$ **24.** $\frac{21x + 5}{24}$ **25.** $\frac{10 - 3n}{6n^2}$

26. $\frac{5x^2 + 18x + 27}{(x + 9)(x - 3)(x + 3)}$ **27.** $\{-\frac{23}{4}\}$ **28.** $\{3\}$

29. $\{\frac{3}{7}\}$ **30.** $\{\pm\frac{2}{3}\}$ **31.** $\{-4, 0, 2\}$ **32.** $\{\frac{3}{7}, 4\}$

33. $\{\pm 4i, \pm 1\}$ **34.** $\{-\frac{1}{5}, 1\}$ **35.** $\left\{\frac{3 \pm \sqrt{17}}{4}\right\}$

36. $\left\{\frac{-13 \pm \sqrt{205}}{2}\right\}$ **37.** $\{1\}$ **38.** $\left\{\frac{1 \pm 2i}{2}\right\}$

39. $(-\infty, \frac{1}{8})$ **40.** $[-\frac{5}{9}, \infty)$ **41.** $(-\infty, 250]$
42. $(-\infty, -8) \cup (3, \infty)$ **43.** $(-\frac{3}{2}, \frac{1}{3})$
44. $(-3, \frac{1}{2}) \cup (4, \infty)$ **45.** $(-1, \frac{2}{3}]$
46. $(1, 7]$ **47.** $(-\infty, -\frac{4}{3}) \cup (2, \infty)$ **48.** $(-\frac{9}{5}, 3)$

49.

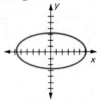

50.

51.

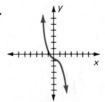

52.

53.

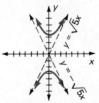

54.

55. $(-7, 4)$; $r = 3$ **56.** $3x - 4y = -26$
57. $(-2, 6)$ **58.** $4x + 3y = 25$

CHAPTER 4

Problem Set 4.1 (page 174)

1. $f(3) = -1$; $f(5) = -5$; $f(-2) = 9$
3. $g(3) = -20$; $g(-1) = -8$; $g(-4) = -41$
5. $h(3) = \frac{5}{4}$; $h(4) = \frac{23}{12}$; $h(-\frac{1}{2}) = -\frac{13}{12}$
7. $f(5) = 3$; $f(\frac{1}{2}) = 0$; $f(23) = 3\sqrt{5}$ **9.** 4
11. $2a + h - 3$ **13.** $4a + 2h + 7$ **15.** $3a^2 + 3ah + h^2$
17. Yes   **19.** No   **21.** Yes   **23.** No
25. $D = \{x \mid x \geq 0\}$; $R = \{f(x) \mid f(x) \geq 0\}$
27. $D = \{x \mid x \text{ is any real number}\}$; $R = \{f(x) \mid f(x) \geq 1\}$
29. The domain and the range both consist of the set of all real numbers.
31. $D = \{x \mid x \text{ is any real number}\}$; $R = \{f(x) \mid f(x) \text{ is any nonnegative real number}\}$
33. $D = \{x \mid x \neq 4\}$ **35.** $D = \{x \mid x \neq 2 \text{ and } x \neq -3\}$
37. $D = \{x \mid x \geq -\frac{1}{5}\}$
39. $D = \{x \mid x \neq -2 \text{ and } x \neq -3\}$
41. $D = \{x \mid x \neq 0 \text{ and } x \neq -4\}$
43. $(-\infty, -1] \cup [1, \infty)$ **45.** $(-\infty, \infty)$
47. $(-\infty, -4] \cup [6, \infty)$ **49.** $(-\infty, -\frac{3}{4}] \cup [\frac{2}{3}, \infty)$
51. -1500; 600; 1000; 900   **53.** 97.3; 93.9; 55.8
55. \$74; \$98; \$122; \$258   **57.** 8; $4\sqrt{3}$; 0   **59.** Even
61. Even   **63.** Neither   **65.** Odd   **67.** Odd
69. 4; 10; 9; 25   **71.** 6; 10; 6; 10   **73.** 1; 0; 0; -1

Problem Set 4.2 (page 182)

1.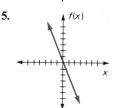
3.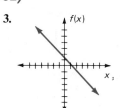
5.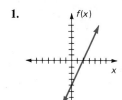
7.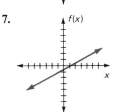
9.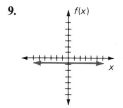
11.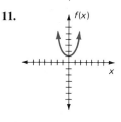
13.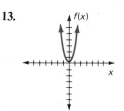
15.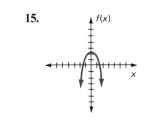
17.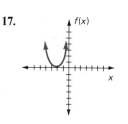
19.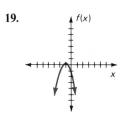
21.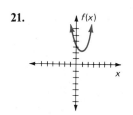
23.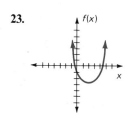
25.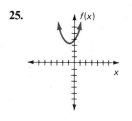
27.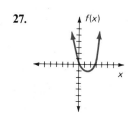
29.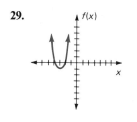
31.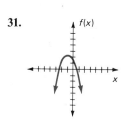
33.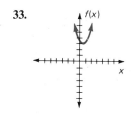
35.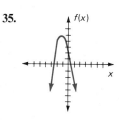

37.

39.
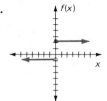

13.

15.
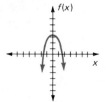

41.
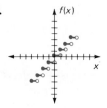

43.
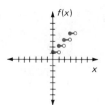

17.

19.
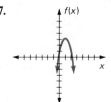

45.
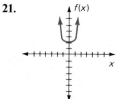

21.
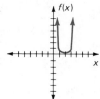

23.
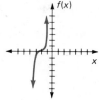

Problem Set 4.3 (page 189)

1.

3.
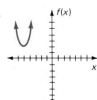

25.
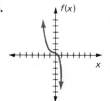

27.
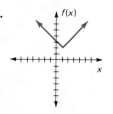

5.

7.
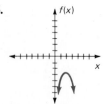

29.
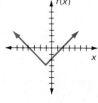

31.
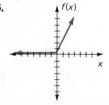

9.
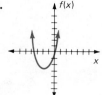

11.
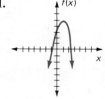

33.
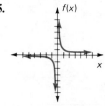

35.

37.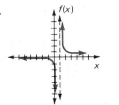

39.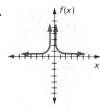

41. 70   **43.** 144 feet   **45.** 25 and 25
47. 60 meters by 60 meters
49. 1100 subscribers at \$13.75 per month
51. $10\sqrt{2}$ feet   **53.** 62.5 feet

Problem Set 4.4 (page 194)

1. $8x - 2;\; -2x - 6;\; 15x^2 - 14x - 8;\; \dfrac{3x - 4}{5x + 2}$

3. $x^2 - 7x + 3;\; x^2 - 5x + 5;\; -x^3 + 5x^2 + 2x - 4;$
$\dfrac{x^2 - 6x + 4}{-x - 1}$

5. $2x^2 + 3x - 6;\; -5x + 4;\; x^4 + 3x^3 - 10x^2 + x + 5;$
$\dfrac{x^2 - x - 1}{x^2 + 4x - 5}$

7. $\sqrt{x - 1} + \sqrt{x};\; \sqrt{x - 1} - \sqrt{x};\; \sqrt{x^2 - x};\; \dfrac{\sqrt{x(x - 1)}}{x}$

9. $(f \circ g)(x) = 6x - 2;\quad D = \{\text{all reals}\}$
$(g \circ f)(x) = 6x - 1;\quad D = \{\text{all reals}\}$
11. $(f \circ g)(x) = 10x + 2;\quad D = \{\text{all reals}\}$
$(g \circ f)(x) = 10x - 5;\quad D = \{\text{all reals}\}$
13. $(f \circ g)(x) = 3x^2 + 7;\quad D = \{\text{all reals}\}$
$(g \circ f)(x) = 9x^2 + 24x + 17;\quad D = \{\text{all reals}\}$
15. $(f \circ g)(x) = 3x^2 + 9x - 16;\quad D = \{\text{all reals}\}$
$(g \circ f)(x) = 9x^2 - 15x;\quad D = \{\text{all reals}\}$

17. $(f \circ g)(x) = \dfrac{1}{2x + 7};\quad D = \{x \mid x \neq -\tfrac{7}{2}\}$

$(g \circ f)(x) = \dfrac{7x + 2}{x};\quad D = \{x \mid x \neq 0\}$

19. $(f \circ g)(x) = \sqrt{3x - 3};\quad D = \{x \mid x \geq 1\}$
$(g \circ f)(x) = 3\sqrt{x - 2} - 1;\quad D = \{x \mid x \geq 2\}$

21. $(f \circ g)(x) = \dfrac{x}{2 - x};\quad D = \{x \mid x \neq 0 \text{ and } x \neq 2\}$

$(g \circ f)(x) = 2x - 2;\quad D = \{x \mid x \neq 1\}$

23. 4; 50   **25.** 9; 0   **27.** $\sqrt{11}$; 5
35. $b + 2a = 2$

Problem Set 4.5 (page 200)

1. Yes   **3.** No   **5.** Yes
7. Domain of f: $\{1, 2, 5\}$
Range of f: $\{5, 9, 21\}$
$f^{-1} = \{(5, 1), (9, 2), (21, 5)\}$
Domain of f^{-1}: $\{5, 9, 21\}$
Range of f^{-1}: $\{1, 2, 5\}$
9. Domain of f: $\{0, 2, -1, -2\}$
Range of f: $\{0, 8, -1, -8\}$
$f^{-1} = \{(0, 0), (8, 2), (-1, -1), (-8, -2)\}$
Domain of f^{-1}: $\{0, 8, -1, -8\}$
Range of f^{-1}: $\{0, 2, -1, -2\}$
19. No   **21.** Yes   **23.** No   **25.** Yes
27. Yes   **29.** $f^{-1}(x) = x + 4$

31. $f^{-1}(x) = \dfrac{-x - 4}{3}$ **33.** $f^{-1}(x) = \dfrac{12x + 10}{9}$

35. $f^{-1}(x) = -\tfrac{3}{2}x$ **37.** $f^{-1}(x) = x^2$ for $x \geq 0$
39. $f^{-1}(x) = \sqrt{x - 4}$ for $x \geq 4$
41. $f^{-1}(x) = \tfrac{1}{3}x$

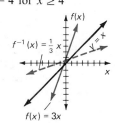

43. $f^{-1}(x) = \dfrac{x - 1}{2}$

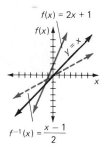

45. $f^{-1}(x) = \dfrac{x + 2}{x}$ for $x > 0$

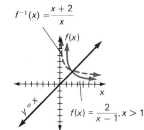

47. $f^{-1}(x) = \sqrt{x+4}$ for $x \geq -4$

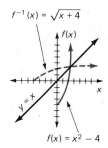

51. (a) $\dfrac{x+7}{6}$ **(c)** $\dfrac{x+7}{6}$

Problem Set 4.6 (page 207)

1. $y = kx^3$ **3.** $A = klw$ **5.** $V = \dfrac{k}{P}$

7. $V = khr^2$ **9.** 24 **11.** $\frac{22}{7}$ **13.** $\frac{1}{2}$ **15.** 7

17. 6 **19.** 8 **21.** 96 **23.** 5 hours

25. 2 seconds **27.** 24 days **29.** 28

31. \$2400 **33. (a)** \$210 **(c)** \$1050

35. 3560.76 cubic meters **37.** 0.048

Chapter 4 Review Problem Set (page 211)

1. 7; 4; 32

2. (a) -5 **(b)** $4a + 2h - 1$ **(c)** $-6a - 3h + 2$

3. The domain is the set of all real numbers and the range is the set of real numbers greater than or equal to 5.

4. The domain is the set of all real numbers except $\frac{1}{2}$ and -4.

5. $(-\infty, 2] \cup [5, \infty)$

6.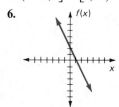

7.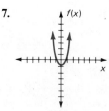

8.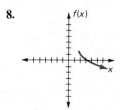

9.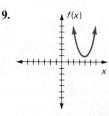

10.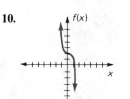

11.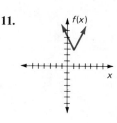

12.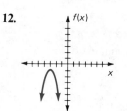

13.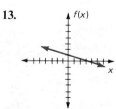

14.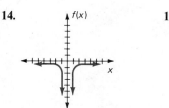

15.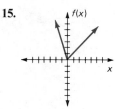

16. $x^2 - 2x$; $-x^2 + 6x + 6$; $2x^3 - 5x^2 - 18x - 9$;

$\dfrac{2x+3}{x^2 - 4x - 3}$

17. $(f \circ g)(x) = -6x + 12$; $D = \{\text{all reals}\}$
$(g \circ f)(x) = -6x + 25$; $D = \{\text{all reals}\}$

18. $(f \circ g)(x) = 25x^2 - 40x + 11$; $D = \{\text{all reals}\}$
$(g \circ f)(x) = 5x^2 - 29$; $D = \{\text{all reals}\}$

19. $(f \circ g)(x) = \sqrt{x-3}$; $D = \{x \mid x \geq 3\}$
$(g \circ f)(x) = \sqrt{x-5} + 2$; $D = \{x \mid x \geq 5\}$

20. $(f \circ g)(x) = \dfrac{x+2}{-3x-5}$; $D = \{x \mid x \neq -2 \text{ and } x \neq -\frac{5}{3}\}$

$(g \circ f)(x) = \dfrac{x-3}{2x-5}$; $D = \{x \mid x \neq 3 \text{ and } x \neq \frac{5}{2}\}$

21. 1; 5

22. The domain of f equals the range of g, and the range of f equals the domain of g. Furthermore, $(f \circ g)(x) = x$ and $(g \circ f)(x) = x$.

23. Yes **24.** No **25.** Yes **26.** Yes

27. $f^{-1}(x) = \dfrac{x-5}{4}$ **28.** $f^{-1}(x) = \dfrac{-x-7}{3}$

29. $f^{-1}(x) = \dfrac{6x+2}{5}$ **30.** $f^{-1}(x) = \sqrt{-2-x}$

31. 112 students **32.** $k = 9$ **33.** 441

34. 128 pounds

CHAPTER 5

Problem Set 5.1 (page 218)

1. $\{3\}$ **3.** $\{3\}$ **5.** $\{4\}$ **7.** $\{2\}$
9. $\{-2\}$ **11.** $\{\frac{5}{3}\}$ **13.** $\{\frac{3}{2}\}$ **15.** $\{\frac{4}{9}\}$

17. 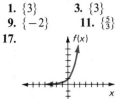 **19.**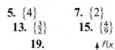

21. 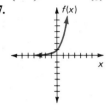 **23.**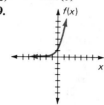

25. 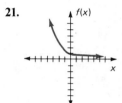 **27.**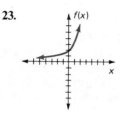

29. 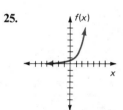 **31.**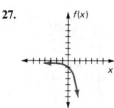

33. 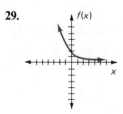 **35.**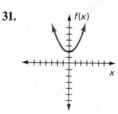

37. 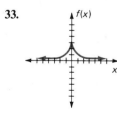 **?**

Problem Set 5.2 (page 223)

1. (a) $.67$ **(c)** $\$2.31$ **(e)** $\$12,623$ **(g)** $\$803$
3. $\$384.66$ **5.** $\$480.31$ **7.** $\$2479.35$

9. $\$1816.70$ **11.** $\$1356.59$ **13.** $\$745.88$
15. $\$2174.40$ **17.** $\$4416.52$

19.

	8%	10%	12%	14%
COMPOUNDED ANNUALLY	$2159	2594	3106	3707
COMPOUNDED SEMIANNUALLY	2191	2653	3207	3870
COMPOUNDED QUARTERLY	2208	2685	3262	3959
COMPOUNDED MONTHLY	2220	2707	3300	4022
COMPOUNDED CONTINUOUSLY	2225	2718	3320	4055

21.

	8%	10%	12%	14%
5 YEARS	$1492	1649	1822	2014
10 YEARS	2225	2718	3320	4055
15 YEARS	3320	4481	6049	8164
20 YEARS	4952	7388	11020	16440
25 YEARS	7388	12179	20079	33103

23. 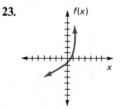 **25.**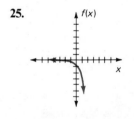

27. 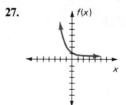 **29.** 8243; 22,405; 100,396

31. 203 grams; 27 grams; 0.5 grams
33. 82,887; 87,136; 96,299 **35.** 500 grams; 210 grams
37. $-\frac{1}{25}$

39. 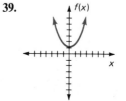 **41.**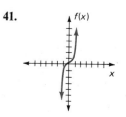

Problem Set 5.3 (page 232)

1. $\log_3 9 = 2$ **3.** $\log_5 125 = 3$ **5.** $\log_2(\frac{1}{16}) = -4$
7. $\log_{10} 0.01 = -2$ **9.** $2^6 = 64$ **11.** $10^{-1} = 0.1$
13. $2^{-4} = \frac{1}{16}$ **15.** 2   **17.** -1 **19.** 1
21. $\frac{1}{2}$ **23.** $\frac{1}{2}$ **25.** $-\frac{1}{8}$ **27.** 7   **29.** 0
31. $\{25\}$ **33.** $\{32\}$ **35.** $\{9\}$ **37.** $\{1\}$
39. 1.1461   **41.** 0.6020   **43.** 2.5353   **45.** 0.1505
47. 1.1268   **49.** 1.4471   **51.** 1.9912   **53.** 2.3010
55. 3.1461   **57.** $\log_b x + \log_b y + \log_b z$
59. $2\log_b x + 3\log_b y$ **61.** $\frac{1}{2}\log_b x + \frac{1}{2}\log_b y$

63. $\frac{1}{2}\log_b x - \frac{1}{2}\log_b y$ **65.** $\log_b\left(\dfrac{xy}{z}\right)$

67. $\log_b\left(\dfrac{x}{yz}\right)$ **69.** $\log_b(x\sqrt{y})$ **71.** $\log_b\left[\dfrac{x^2\sqrt{x-1}}{(2x+5)^4}\right]$

73. $\{2\}$ **75.** $\{2\}$ **77.** $\{\frac{2}{9}\}$ **79.** $\{6\}$

Problem Set 5.4 (page 240)

1.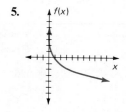

3.

5.

7. Same graph as Problem 1
9. Same graph as Problem 5

11.

13.

15. .9754   **17.** 1.5393   **19.** 3.6741
21. $-.2132$ **23.** -2.3279 **25.** .9405
27. 1.7551   **29.** 2.8500   **31.** $.6920 + (-1)$
33. $.7226 + (-3)$ **35.** 5.8825   **37.** 33.597
39. 8580.3   **41.** 3.5620   **43.** 0.71581

45. 0.0022172   **47.** 3.55   **49.** 30.4   **51.** 983
53. 640,000   **55.** 0.542   **57.** 0.00731
59. 0.179   **61.** 0.6931   **63.** 3.0634   **65.** 6.0210
67. -1.1394 **69.** -2.6381 **71.** -7.1309
73. 1.3083   **75.** 4.3567   **77.** 6.4297
79. -0.8675 **81.** -4.7677

Problem Set 5.5 (page 248)

1. $\{3.17\}$ **3.** $\{1.95\}$ **5.** $\{1.81\}$ **7.** $\{1.41\}$
9. $\{1.41\}$ **11.** $\{3.10\}$ **13.** $\{1.82\}$ **15.** $\{7.84\}$
17. $\{10.32\}$ **19.** $\{2\}$ **21.** $\{\frac{29}{8}\}$

23. $\left\{\dfrac{-1+\sqrt{65}}{2}\right\}$ **25.** $\{\sqrt{2}\}$ **27.** $\{6\}$

29. $\{1, 100\}$ **31.** 2.402   **33.** .461   **35.** 2.657
37. 1.211   **39.** 7.9 years   **41.** 12.2 years
43. 11.8%   **45.** 6.6 years   **47.** 1.5 hours
49. 34.7 years   **51.** $k = 0.17$ **55.** $\{1.13\}$
57. $x = \ln(y + \sqrt{y^2 + 1})$

Problem Set 5.6 (page 254)

1. .6362   **3.** 2.1103   **5.** $-1 + .9426$
7. $-3 + .7148$ **9.** 2.945   **11.** 155.8
13. 0.3181   **15.** 20930   **17.** 5.749   **19.** 1,549,000
21. 614.1   **23.** 0.9614   **25.** 13.79   **27.** 15.91

Chapter 5 Review Problem Set (page 257)

1. 32   **2.** -125 **3.** 81   **4.** 3   **5.** -2
6. $\frac{1}{3}$ **7.** $\frac{1}{4}$ **8.** -5 **9.** 1   **10.** 12
11. $\{5\}$ **12.** $\{\frac{1}{9}\}$ **13.** $\{\frac{7}{2}\}$ **14.** $\{3.40\}$
15. $\{8\}$ **16.** $\{\frac{1}{11}\}$ **17.** $\{1.95\}$ **18.** $\{1.41\}$
19. $\{1.56\}$ **20.** $\{20\}$ **21.** $\{10^{100}\}$ **22.** $\{2\}$
23. $\{\frac{11}{2}\}$ **24.** $\{0\}$ **25.** .3680   **26.** 1.3222
27. 1.4313   **28.** .5634
29. (a) $\log_b x - 2\log_b y$ (b) $\frac{1}{4}\log_b x + \frac{1}{2}\log_b y$
(c) $\frac{1}{2}\log_b x - 3\log_b y$

30. (a) $\log_b x^3 y^2$ (b) $\log_b\left(\dfrac{\sqrt{y}}{x^4}\right)$ (c) $\log_b\left(\dfrac{\sqrt{xy}}{z^2}\right)$

31. 1.58   **32.** 0.63   **33.** 3.79   **34.** -2.12

35.   **36.**

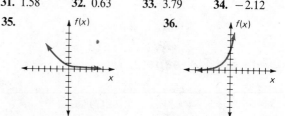

37.

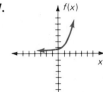

38.

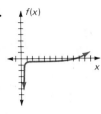

39.

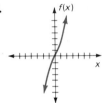

40.

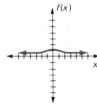

41.

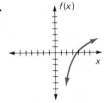

42.

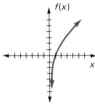

43. $2219.91   **44.** $4797.55   **45.** $15,999.31
46. Approximately 5.3 years
47. Approximately 12.1 years
48. Approximately 8.7%   **49.** 61070; 67493; 74591
50. Approximately 4.8 hours

CHAPTER 6

Problem Set 6.1 (page 264)

1. Q: $4x + 5$ **3.** Q: $t^2 + 2t - 4$
    R: 0 R: 0
5. Q: $2x + 5$ **7.** Q: $3x - 4$
    R: 1 R: $3x - 1$
9. Q: $5y - 1$ **11.** Q: $4a + 6$
    R: $-8y - 2$ R: $7a - 19$
13. Q: $3x + 4y$ **15.** Q: $3x + 4$ **17.** Q: $x + 6$
    R: 0 R: 0 R: 14
19. Q: $4x - 3$ **21.** Q: $x^2 - 1$
    R: 2 R: 0
23. Q: $3x^3 - 4x^2 + 6x - 13$ **25.** Q: $x^2 - 2x - 3$
    R: 12 R: 0
27. Q: $x^3 + 7x^2 + 21x + 56$ **29.** Q: $x^2 + 3x + 2$
    R: 167 R: 0
31. Q: $x^4 + x^3 + x^2 + x + 1$
    R: 0
33. Q: $x^4 + x^3 + x^2 + x + 1$ **35.** Q: $2x^2 + 2x - 3$
    R: 2 R: $\frac{9}{2}$

37. Q: $4x^3 + 2x^2 - 4x - 2$
    R: 0

Problem Set 6.2 (page 268)

1. $f(2) = -2$ **3.** $f(-4) = -105$
5. $f(-2) = 9$ **7.** $f(6) = 74$ **9.** $f(3) = 200$
11. $f(-1) = 5$ **13.** $f(7) = -5$ **15.** $f(-2) = -27$
17. $f(\frac{1}{2}) = -2$ **19.** Yes   **21.** Yes   **23.** No
25. Yes   **27.** Yes   **29.** $(x + 2)(x + 6)(x - 1)$
31. $(x - 3)(2x - 1)(3x + 2)$ **33.** $(x + 1)^2(x - 4)$
35. $k = 6$ **37.** $k = -30$
39. Let $f(x) = x^{12} - 4096$; then $f(-2) = 0$. Therefore
    $x + 2$ is a factor of $f(x)$.
41. Let $f(x) = x^n - 1$. Since $1^n = 1$ for all positive
    integers n, then $f(1) = 0$ and $x - 1$ is a factor.
43. **(a)** Let $f(x) = x^n - y^n$. Therefore $f(y) = y^n - y^n = 0$
        and $x - y$ is a factor of $f(x)$.
    (c) Let $f(x) = x^n + y^n$. Therefore $f(-y) = (-y)^n + y^n =$
        $-y^n + y^n = 0$ when n is odd, and $x - (-y) =$
        $x + y$ is a factor of $f(x)$.
45. $f(1 + i) = 2 + 6i$
49. **(a)** $f(4) = 137; f(-5) = 11; f(7) = 575$
    (c) $f(4) = -79; f(5) = -162; f(-3) = 110$

Problem Set 6.3 (page 278)

1. $\{-2, -1, 2\}$ **3.** $\{-\frac{3}{2}, \frac{1}{3}, 1\}$ **5.** $\{-7, \frac{2}{3}, 2\}$
7. $\{-1, 4\}$ **9.** $\{-3, 1, 2, 4\}$ **11.** $\{-2, 1 \pm \sqrt{7}\}$
13. $\{-\frac{2}{3}, 1, \pm\sqrt{2}\}$ **15.** $\{-\frac{4}{3}, 0, \frac{1}{2}, 3\}$
17. $\{-1, 2, 1 \pm i\}$ **19.** $\{-1, \frac{3}{2}, 2, \pm i\}$
27. **(a)** $\{-4, -2, 1\}$ **(c)** $\{-4, -2, \frac{3}{2}\}$
29. 2 positive or 2 nonreal complex solutions
31. 1 negative and 2 nonreal complex solutions
33. 1 positive and 2 negative solutions OR 1 positive
    and 2 nonreal complex solutions
35. 1 negative, 2 positive, and 2 nonreal complex solutions
    OR 1 negative and 4 nonreal complex solutions
37. 1 positive, 1 negative, and 4 nonreal complex solutions
41. **(a)** An upper bound of 3 and a lower bound of -1
    (c) An upper bound of 3 and a lower bound of -6
    (e) An upper bound of 5 and a lower bound of -3

Problem Set 6.4 (page 285)

1.

3.

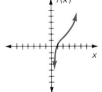

5.
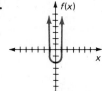

7.

9.
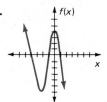

11.
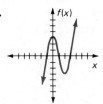

13.
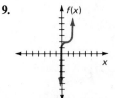

15.
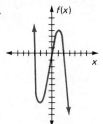

17.
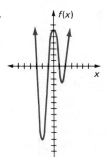

19.
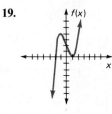

21.
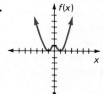

23.
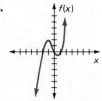

25.
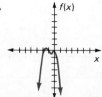

27.
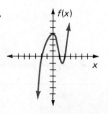

29.
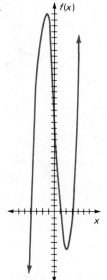

31.
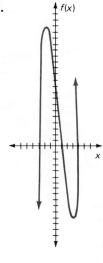

33.
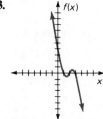

35. (a) 60   **(c)** $f(x) > 0$ for $(-4, 3) \cup (5, \infty)$;
$f(x) < 0$ for $(-\infty, -4) \cup (3, 5)$

37. (a) 432   **(c)** $f(x) > 0$ for $(-3, 4) \cup (4, \infty)$;
$f(x) < 0$ for $(-\infty, -3)$

39. (a) 8
(c) $f(x) > 0$ for $(-\infty, -2) \cup (-2, 1) \cup (2, \infty)$;
$f(x) < 0$ for $(1, 2)$

41. (a) 512   **(c)** $f(x) > 0$ for $(-2, 4) \cup (4, \infty)$;
$f(x) < 0$ for $(-\infty, -2)$

Problem Set 6.5 (page 292)

1.
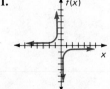

3.
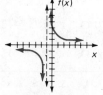

5.

7.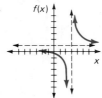

9.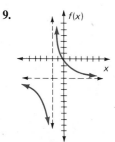

11.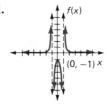
(0, −1)

13.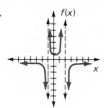

15.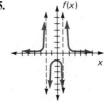

17.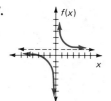

19.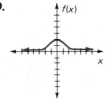

21.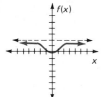

23. (a)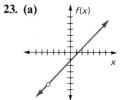

(c)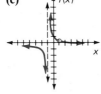

Problem Set 6.6 (page 296)

1.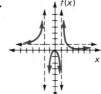

3.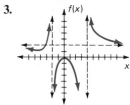

5.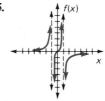

7.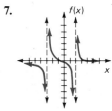

9.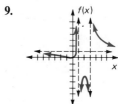

11.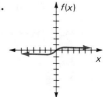

13.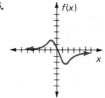

15.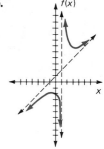

17.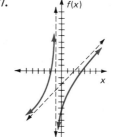

19.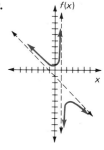

Problem Set 6.7 (page 302)

1. $\dfrac{4}{x-2} + \dfrac{7}{x+1}$ **3.** $\dfrac{3}{x+1} - \dfrac{5}{x-1}$

5. $\dfrac{1}{3x-1} + \dfrac{6}{2x+3}$ **7.** $\dfrac{2}{x-1} + \dfrac{3}{x+2} - \dfrac{4}{x-3}$

9. $\dfrac{-1}{x} + \dfrac{2}{2x-1} - \dfrac{3}{4x+1}$ **11.** $\dfrac{2}{x-2} + \dfrac{5}{(x-2)^2}$

13. $\dfrac{4}{x} + \dfrac{7}{x^2} - \dfrac{10}{x+3}$ **15.** $\dfrac{-3}{x^2+1} - \dfrac{2}{x-4}$

17. $\dfrac{3}{x+2} - \dfrac{2}{(x+2)^2} + \dfrac{1}{(x+2)^3}$ **19.** $\dfrac{2}{x} + \dfrac{3x+5}{x^2-x+3}$

21. $\dfrac{2x}{x^2+1} + \dfrac{3-x}{(x^2+1)^2}$

Chapter 6 Review Problem Set (page 304)

1. Q: $3x^2 - 5x + 4$
R: 4
2. Q: $2a - 1$
R: 5
3. Q: $3x^2 - x + 5$
R: 3
4. Q: $5x^2 - 3x - 3$
R: 16
5. Q: $-2x^3 + 9x^2 - 38x + 151$
R: -605
6. Q: $-3x^3 - 9x^2 - 32x - 96$ **7.** $f(1) = 1$
R: -279
8. $f(-3) = -197$ **9.** $f(-2) = 20$ **10.** $f(8) = 0$
11. Yes   **12.** No   **13.** Yes   **14.** Yes
15. $\{-3, 1, 5\}$ **16.** $\{-\frac{7}{2}, -1, \frac{5}{4}\}$
17. $\{1, 2, 1 \pm 5i\}$ **18.** $\{-2, 3 \pm \sqrt{7}\}$
19. 2 positive and 2 negative solutions OR 2 positive and 2 nonreal complex solutions OR 2 negative and 2 nonreal complex solutions OR 4 nonreal complex solutions
20. 1 negative and 4 nonreal complex solutions

21.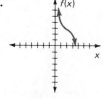

22.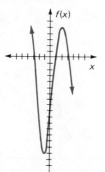

23.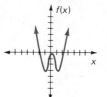

24.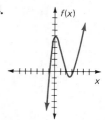

25.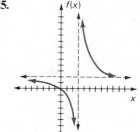

26.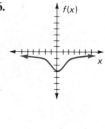

27.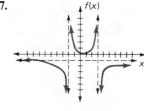

28.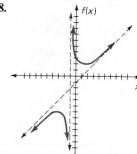

29. $\dfrac{1}{x} - \dfrac{2}{x^2} + \dfrac{4}{x+2}$ **30.** $\dfrac{3x+1}{x^2+4} - \dfrac{5}{2x-1}$

CHAPTER 7

Problem Set 7.1 (page 312)

1. $\{(7, 9)\}$ **3.** $\{(-4, 7)\}$ **5.** $\{(6, 3)\}$
7. $a = -3$ and $b = -4$
9. $\{(k, \frac{2}{3}k - \frac{4}{3}) \,|\, k$ is a real number$\}$, a dependent system
11. $u = 5$ and $t = 7$ **13.** $\{(2, -5)\}$
15. $\varnothing$, an inconsistent system   **17.** $\{(-\frac{3}{4}, -\frac{6}{5})\}$
19. $\{(3, -4)\}$ **21.** $\{(2, 8)\}$ **23.** $\{(-1, -5)\}$
25. $\varnothing$, an inconsistent system   **27.** $a = \frac{5}{27}$ and $b = -\frac{26}{27}$
29. $s = -6$ and $t = 12$ **31.** $\{(-\frac{1}{2}, \frac{1}{3})\}$
33. $\{(\frac{13}{22}, \frac{2}{11})\}$ **35.** $\{(-4, 2)\}$ **37.** $\{(5, 5)\}$
39. $\varnothing$, an inconsistent system   **41.** $\{(12, -24)\}$
43. $t = 8$ and $u = 3$ **45.** $\{(200, 800)\}$
47. $\{(400, 800)\}$ **49.** $\{(3.5, 7)\}$ **51.** 17 and 36

53. 5 and 15   **55.** 72   **57.** 34
59. 8 single rooms and 15 double rooms
61. 2500 student tickets and 500 nonstudent tickets
63. $500 at 9% and $1500 at 11%
65. 8 liters of the 40% solution and 12 liters of the 60% solution
67. $1.25 per tennis ball and $1.75 per golf ball
69. 30 five-dollar bills and 18 ten-dollar bills
71. $\{(4,6)\}$ **73.** $\{(2,-3)\}$ **75.** $\{(\frac{1}{4},-\frac{2}{3})\}$

Problem Set 7.2 (page 321)

1. $\{(3,1,2)\}$ **3.** $\{(-1,3,5)\}$ **5.** $\{(-2,-1,3)\}$
7. $\{(0,2,4)\}$ **9.** $\{(4,-1,-2)\}$ **11.** $\{(-4,0,-1)\}$
13. $\{(2,2,-3)\}$ **15.** 8, 15, and 20
17. 7 nickels, 13 dimes, and 22 quarters
19. $40°$, $60°$, and $80°$
21. $500 at 12%, $1000 at 13%, and $1500 at 14%
23. 50 of type A, 75 of type B, and 150 of type C
25. $x^2 + y^2 - 4x - 20y - 2 = 0$

Problem Set 7.3 (page 330)

1. Yes   **3.** Yes   **5.** No   **7.** No   **9.** Yes
11. $\{(-1,-5)\}$ **13.** $\{(3,-6)\}$ **15.** $\varnothing$
17. $\{(-2,-9)\}$ **19.** $\{(-1,-2,3)\}$
21. $\{(3,-1,4)\}$ **23.** $\{(0,-2,4)\}$
25. $\{(-7k+8,-5k+7,k)\}$ **27.** $\{(-4,-3,-2)\}$
29. $\{(4,-1,-2)\}$ **31.** $\{(1,-1,2,-3)\}$
33. $\{(2,1,3,-2)\}$ **35.** $\{(-2,4,-3,0)\}$
37. $\varnothing$ **39.** $\{(-3k+5,-1,-4k+2,k)\}$
41. $\{(-3k+9,k,2,-3)\}$ **43.** $\{(17k-6,10k-5,k)\}$
45. $\{(-\frac{1}{2}k+\frac{34}{11},\frac{1}{2}k-\frac{5}{11},k)\}$ **47.** $\varnothing$

Problem Set 7.4 (page 340)

1. 22   **3.** -29 **5.** 20   **7.** 5   **9.** -2
11. $-\frac{2}{3}$ **13.** -25 **15.** 58   **17.** 39
19. -12 **21.** -41 **23.** -8 **25.** 1088
27. -140 **29.** 81   **31.** 146   **33.** Property 7.3
35. Property 7.2   **37.** Property 7.4
39. Property 7.3   **41.** Property 7.5

Problem Set 7.5 (page 346)

1. $\{(1,4)\}$ **3.** $\{(3,-5)\}$ **5.** $\{(2,-1)\}$ **7.** $\varnothing$
9. $\{(-\frac{1}{4},\frac{2}{3})\}$ **11.** $\{(\frac{2}{17},\frac{52}{17})\}$ **13.** $\{(9,-2)\}$
15. $\{(2,-\frac{5}{7})\}$ **17.** $\{(0,2,-3)\}$ **19.** $\{(2,6,7)\}$
21. $\{(4,-4,5)\}$ **23.** $\{(-1,3,-4)\}$
25. Infinitely many solutions   **27.** $\{(-2,\frac{1}{2},-\frac{2}{3})\}$
29. $\{(3,\frac{1}{2},-\frac{1}{3})\}$ **31.** $\{(-4,6,0)\}$ **35.** $\{(0,0,0)\}$
37. Infinitely many solutions

Chapter 7 Review Problem Set (page 350)

1. $\{(3,-7)\}$ **2.** $\{(-1,-3)\}$ **3.** $\{(0,-4)\}$
4. $\{(\frac{23}{3},-\frac{14}{3})\}$ **5.** $\{(4,-6)\}$ **6.** $\{(-\frac{6}{7},-\frac{15}{7})\}$
7. $\{(-1,2,-5)\}$ **8.** $\{(2,-3,-1)\}$ **9.** $\{(5,-4)\}$
10. $\{(2,7)\}$ **11.** $\{(-2,2,-1)\}$ **12.** $\{(0,-1,2)\}$
13. $\{(-3,-1)\}$ **14.** $\{(4,6)\}$ **15.** $\{(2,-3,-4)\}$
16. $\{(-1,2,-5)\}$ **17.** $\{(5,-5)\}$ **18.** $\{(-12,12)\}$
19. $\{(\frac{5}{7},\frac{4}{7})\}$ **20.** $\{(-10,-7)\}$ **21.** $\{(1,1,-4)\}$
22. $\{(-4,0,1)\}$ **23.** $\varnothing$ **24.** $\{(-2,-4,6)\}$
25. -34 **26.** 13   **27.** -40 **28.** 16
29. 51   **30.** 125   **31.** 72
32. $900 at 10% and $1600 at 12%
33. 20 nickels, 32 dimes, and 54 quarters
34. $25°$, $45°$, and $110°$

CHAPTER 8

Problem Set 8.1 (page 358)

1. $\begin{bmatrix} 3 & -5 \\ 8 & 3 \end{bmatrix}$ **3.** $\begin{bmatrix} -2 & 21 \\ -7 & 2 \end{bmatrix}$ **5.** $\begin{bmatrix} -2 & 1 \\ -3 & 19 \end{bmatrix}$

7. $\begin{bmatrix} -1 & -5 \\ 2 & 3 \end{bmatrix}$ **9.** $\begin{bmatrix} -12 & -14 \\ -18 & -20 \end{bmatrix}$

11. $\begin{bmatrix} 2 & -11 \\ -7 & 0 \end{bmatrix}$

13. $AB = \begin{bmatrix} 4 & -6 \\ 8 & -12 \end{bmatrix}$, $BA = \begin{bmatrix} -5 & 5 \\ 3 & -3 \end{bmatrix}$

15. $AB = \begin{bmatrix} -5 & -18 \\ -4 & 42 \end{bmatrix}$, $BA = \begin{bmatrix} 19 & -39 \\ -16 & 18 \end{bmatrix}$

17. $AB = \begin{bmatrix} 14 & -28 \\ 7 & -14 \end{bmatrix}$, $BA = \begin{bmatrix} 0 & 0 \\ 0 & 0 \end{bmatrix}$

19. $AB = \begin{bmatrix} -14 & -7 \\ -12 & -1 \end{bmatrix}$, $BA = \begin{bmatrix} -2 & -3 \\ -32 & -13 \end{bmatrix}$

21. $AB = \begin{bmatrix} 1 & 0 \\ 0 & 1 \end{bmatrix}$, $BA = \begin{bmatrix} 1 & 0 \\ 0 & 1 \end{bmatrix}$

23. $AB = \begin{bmatrix} 0 & -\frac{5}{3} \\ \frac{17}{6} & -3 \end{bmatrix}$, $BA = \begin{bmatrix} 0 & -\frac{17}{6} \\ \frac{5}{3} & -3 \end{bmatrix}$

25. $AB = \begin{bmatrix} 1 & 0 \\ 0 & 1 \end{bmatrix}$, $BA = \begin{bmatrix} 1 & 0 \\ 0 & 1 \end{bmatrix}$

27. $AB = \begin{bmatrix} 3 & -2 \\ 4 & 5 \end{bmatrix}$, $BA = \begin{bmatrix} 5 & 4 \\ -2 & 3 \end{bmatrix}$

29. $AD = \begin{bmatrix} 1 & 1 \\ 9 & 9 \end{bmatrix}$, $DA = \begin{bmatrix} 3 & 7 \\ 3 & 7 \end{bmatrix}$

45. $A^2 = \begin{bmatrix} -1 & -4 \\ 8 & 7 \end{bmatrix}$, $A^3 = \begin{bmatrix} -9 & -11 \\ 22 & 13 \end{bmatrix}$

Problem Set 8.2 (page 364)

1. $\begin{bmatrix} 3 & -7 \\ -2 & 5 \end{bmatrix}$ **3.** $\begin{bmatrix} -5 & 8 \\ 2 & -3 \end{bmatrix}$ **5.** $\begin{bmatrix} -\frac{2}{5} & \frac{1}{5} \\ \frac{3}{10} & \frac{1}{10} \end{bmatrix}$

7. Inverse does not exist.   **9.** $\begin{bmatrix} -\frac{5}{7} & \frac{2}{7} \\ -\frac{4}{7} & \frac{3}{7} \end{bmatrix}$

11. $\begin{bmatrix} -\frac{3}{5} & \frac{1}{5} \\ 1 & 0 \end{bmatrix}$ **13.** $\begin{bmatrix} -\frac{4}{5} & \frac{3}{5} \\ \frac{1}{5} & -\frac{2}{5} \end{bmatrix}$ **15.** $\begin{bmatrix} 2 & -\frac{5}{3} \\ 1 & -\frac{2}{3} \end{bmatrix}$

17. $\begin{bmatrix} \frac{1}{2} & \frac{1}{2} \\ \frac{1}{2} & -\frac{1}{2} \end{bmatrix}$ **19.** $\begin{bmatrix} 30 \\ 36 \end{bmatrix}$ **21.** $\begin{bmatrix} 0 \\ 5 \end{bmatrix}$ **23.** $\begin{bmatrix} -4 \\ 13 \end{bmatrix}$

25. $\begin{bmatrix} -4 \\ -13 \end{bmatrix}$ **27.** $\{(2, 3)\}$ **29.** $\{(-2, 5)\}$

31. $\{(0, -1)\}$ **33.** $\{(-1, -1)\}$ **35.** $\{(4, 7)\}$
37. $\{(-\frac{1}{3}, \frac{1}{2})\}$ **39.** $\{(-9, 20)\}$

Problem Set 8.3 (page 370)

1. $A + B = \begin{bmatrix} 1 & 3 & -3 \\ 3 & -6 & 7 \end{bmatrix}$;

$A - B = \begin{bmatrix} 3 & -5 & 11 \\ -7 & 6 & 3 \end{bmatrix}$;

$2A + 3B = \begin{bmatrix} 1 & 10 & -13 \\ 11 & -18 & 16 \end{bmatrix}$;

$4A - 2B = \begin{bmatrix} 10 & -12 & 30 \\ -18 & 12 & 16 \end{bmatrix}$

3. $A + B = \begin{bmatrix} -1 & -7 & 13 & 7 \end{bmatrix}$;
$A - B = \begin{bmatrix} 5 & 5 & -5 & 17 \end{bmatrix}$;
$2A + 3B = \begin{bmatrix} -5 & -20 & 35 & 9 \end{bmatrix}$;
$4A - 2B = \begin{bmatrix} 14 & 8 & -2 & 58 \end{bmatrix}$

5. $A + B = \begin{bmatrix} 8 & -3 & -2 \\ 9 & 2 & -3 \\ 7 & 5 & 21 \end{bmatrix}$;

$A - B = \begin{bmatrix} -2 & -1 & 4 \\ -11 & 6 & -11 \\ -7 & 5 & -3 \end{bmatrix}$;

$2A + 3B = \begin{bmatrix} 21 & -7 & -7 \\ 28 & 2 & -2 \\ 21 & 10 & 54 \end{bmatrix}$;

$4A - 2B = \begin{bmatrix} 2 & -6 & 10 \\ -24 & 20 & -36 \\ -14 & 20 & 12 \end{bmatrix}$

7. $A + B = \begin{bmatrix} 0 & 2 \\ -1 & 10 \\ 1 & -9 \\ 2 & 9 \end{bmatrix}$;

$A - B = \begin{bmatrix} -2 & -2 \\ 5 & -4 \\ -11 & 1 \\ -16 & 13 \end{bmatrix}$;

$2A + 3B = \begin{bmatrix} 1 & 6 \\ -5 & 27 \\ 8 & -23 \\ 13 & 16 \end{bmatrix}$;

$4A - 2B = \begin{bmatrix} -6 & -4 \\ 14 & -2 \\ -32 & -6 \\ -46 & 48 \end{bmatrix}$

9. $AB = \begin{bmatrix} 11 & -8 & 14 \\ 4 & -16 & 8 \\ -28 & 22 & -36 \end{bmatrix}$; $BA = \begin{bmatrix} -20 & 21 \\ 8 & -21 \end{bmatrix}$

11. $AB = \begin{bmatrix} 22 & -8 & 1 & 3 \\ -42 & 36 & -26 & -20 \end{bmatrix}$; BA does not exist.

13. $AB = \begin{bmatrix} -12 & 5 & -5 \\ 14 & -2 & 4 \\ -10 & 13 & -5 \end{bmatrix}$; $BA = \begin{bmatrix} -1 & 0 & -6 \\ 10 & -2 & 16 \\ -8 & 5 & -16 \end{bmatrix}$

15. $AB = \begin{bmatrix} -9 \end{bmatrix}$; $BA = \begin{bmatrix} -2 & 1 & -3 & -4 \\ -6 & 3 & -9 & -12 \\ 4 & -2 & 6 & 8 \\ -8 & 4 & -12 & -16 \end{bmatrix}$

17. AB does not exist; $BA = \begin{bmatrix} 20 \\ 2 \\ -30 \end{bmatrix}$

19. $AB = \begin{bmatrix} 9 & -12 \\ -12 & 16 \\ 6 & -8 \end{bmatrix}$; BA does not exist.

21. $\begin{bmatrix} -\frac{1}{5} & \frac{3}{10} \\ \frac{2}{5} & -\frac{1}{10} \end{bmatrix}$ **23.** $\begin{bmatrix} 4 & -1 \\ -7 & 2 \end{bmatrix}$

25. $\begin{bmatrix} -\frac{4}{5} & -\frac{1}{5} \\ -\frac{3}{5} & -\frac{2}{5} \end{bmatrix}$ **27.** $\begin{bmatrix} \frac{7}{2} & -3 & \frac{1}{2} \\ -\frac{1}{2} & 0 & \frac{1}{2} \\ -\frac{1}{2} & 1 & -\frac{1}{2} \end{bmatrix}$

29. $\begin{bmatrix} -50 & -9 & 11 \\ -23 & -4 & 5 \\ 5 & 1 & -1 \end{bmatrix}$ **31.** Inverse does not exist.

33. $\begin{bmatrix} \frac{4}{7} & -1 & -\frac{9}{7} \\ -\frac{3}{14} & \frac{1}{2} & \frac{6}{7} \\ \frac{2}{7} & 0 & -\frac{1}{7} \end{bmatrix}$ **35.** $\begin{bmatrix} \frac{1}{2} & 0 & 0 \\ 0 & \frac{1}{4} & 0 \\ 0 & 0 & \frac{1}{10} \end{bmatrix}$

37. $\{(-3,2)\}$ **39.** $\{(2,5)\}$ **41.** $\{(-1,-2,1)\}$
43. $\{(-2,3,5)\}$ **45.** $\{(-4,3,0)\}$
47. (a) $\{(-1,2,3)\}$ **(c)** $\{(-5,0,-2)\}$
 (e) $\{(1,-1,-1)\}$
49. (a) y-axis reflection
 (c) 90° counterclockwise rotation

Problem Set 8.4 (page 379)

1.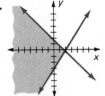

3.

5.

7.

9.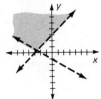

11.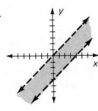

13.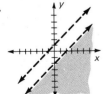

15.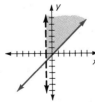

17. $\varnothing$

19.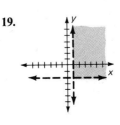

21.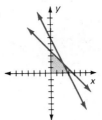

23.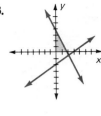

25. Minimum of 8 and maximum of 52
27. Minimum of 0 and maximum of 28
29. 63   **31.** 340   **33.** 2   **35.** 98
37. $5000 at 9% and $5000 at 12%
39. 300 of type A and 200 of type B
41. 12 units of A and 16 units of B

Chapter 8 Review Problem Set (page 385)

1. $\begin{bmatrix} 7 & -5 \\ -3 & 10 \end{bmatrix}$ **2.** $\begin{bmatrix} 3 & 3 \\ 3 & -6 \end{bmatrix}$ **3.** $\begin{bmatrix} 2 & 1 \\ -6 & 8 \\ -2 & 2 \end{bmatrix}$

4. $\begin{bmatrix} 19 & -11 \\ -6 & 22 \end{bmatrix}$ **5.** $\begin{bmatrix} 7 & 1 \\ -14 & 20 \\ 1 & -2 \end{bmatrix}$

6. $\begin{bmatrix} -11 & -3 & 15 \\ 24 & 2 & -20 \\ -40 & -5 & 38 \end{bmatrix}$ **7.** $\begin{bmatrix} 16 & -26 \\ 0 & 13 \end{bmatrix}$

8. $\begin{bmatrix} 26 & -36 \\ -15 & 32 \end{bmatrix}$ **9.** $\begin{bmatrix} -27 \\ 26 \end{bmatrix}$

10. EF does not exist.   **14.** $\begin{bmatrix} 4 & -5 \\ -7 & 9 \end{bmatrix}$

15. $\begin{bmatrix} -3 & 4 \\ 7 & -9 \end{bmatrix}$ **16.** $\begin{bmatrix} -\frac{3}{8} & \frac{1}{8} \\ \frac{1}{4} & \frac{1}{4} \end{bmatrix}$

17. Inverse does not exist.   **18.** $\begin{bmatrix} \frac{5}{7} & -\frac{3}{7} \\ -\frac{4}{7} & \frac{1}{7} \end{bmatrix}$

19. $\begin{bmatrix} \frac{2}{7} & \frac{1}{7} \\ -\frac{1}{3} & 0 \end{bmatrix}$ **20.** $\begin{bmatrix} \frac{39}{8} & -\frac{17}{8} & -\frac{1}{8} \\ 2 & -1 & 0 \\ \frac{1}{8} & \frac{1}{8} & \frac{1}{8} \end{bmatrix}$

21. $\begin{bmatrix} 8 & -8 & 5 \\ -3 & 2 & -1 \\ -1 & -1 & 1 \end{bmatrix}$ **22.** Inverse does not exist.

23. $\begin{bmatrix} -\frac{20}{3} & -\frac{7}{3} & \frac{1}{3} \\ -\frac{1}{3} & -\frac{2}{3} & -\frac{1}{3} \\ -\frac{5}{3} & -\frac{1}{3} & \frac{1}{3} \end{bmatrix}$ **24.** $\{(-2,6)\}$ **25.** $\{(4,-1)\}$

26. $\{(2,-3,-1)\}$ **27.** $\{(-3,2,5)\}$ **28.** $\{(-4,3,4)\}$

29.

30.

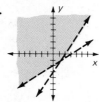

31.

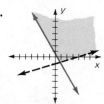

32.

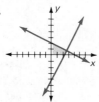

33. 37 **34.** 56 **35.** 57 **36.** 1700
37. 75 one-gallon and 175 two-gallon freezers

CHAPTER 9

Problem Set 9.1 (page 393)

1. $V(0,0)$, $F(2,0)$, $x = -2$ **3.** $V(0,0)$, $F(0,-3)$, $y = 3$

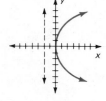

5. $V(0,0)$, $F = (-\frac{1}{2}, 0)$, $x = \frac{1}{2}$ **7.** $V(0,0)$, $F(0, \frac{3}{2})$, $y = -\frac{3}{2}$

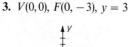

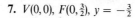

9. $V(0,2)$, $F(0,3)$, $y = 1$ **11.** $V(0,-2)$, $F(0,-4)$, $y = 0$

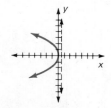

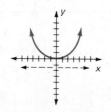

13. $V(2,0)$, $F(5,0)$, $x = -1$ **15.** $V(1,2)$, $F(1,3)$, $y = 1$

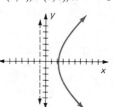

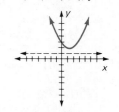

17. $V(-3,1)$, $F(-3,-1)$, $y = 3$ **19.** $V(3,1)$, $F(0,1)$, $x = 6$

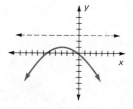

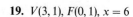

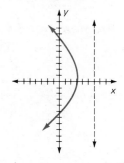

21. $V(-2,-3)$, $F(-1,-3)$, $x = -3$

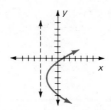

23. $x^2 = 12y$ **25.** $y^2 = -4x$
27. $x^2 + 12y - 48 = 0$ **29.** $x^2 - 6x - 12y + 21 = 0$
31. $y^2 - 10y + 8x + 41 = 0$ **33.** $3y^2 = -25x$
35. $y^2 = 10x$ **37.** $x^2 - 14x - 8y + 73 = 0$
39. $y^2 + 6y - 12x + 105 = 0$
41. $x^2 + 18x + y + 80 = 0$ **43.** $x^2 = 750(y - 10)$

Problem Set 9.2 (page 399)

For Problems 1–22, the foci are indicated above the graph, and the vertices and endpoints of the minor axes are indicated on the graph.

1. $F(\sqrt{3}, 0)$, $F'(-\sqrt{3}, 0)$ **3.** $F(0, \sqrt{5})$, $F'(0, -\sqrt{5})$

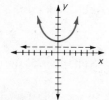

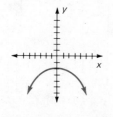

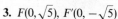

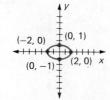

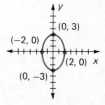

5. $F(0, \sqrt{6})$, $F'(0, -\sqrt{6})$

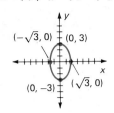

7. $\cdot F(\sqrt{15}, 0)$, $F'(-\sqrt{15}, 0)$

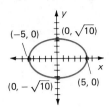

21. $F(0, 4)$, $F'(-6, 4)$

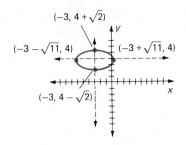

9. $F(0, \sqrt{33})$, $F'(0, -\sqrt{33})$

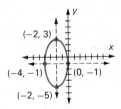

11. $F(2, 0)$, $F'(-2, 0)$

23. $16x^2 + 25y^2 = 400$

25. $36x^2 + 11y^2 = 396$ **27.** $x^2 + 9y^2 = 9$

29. $100x^2 + 36y^2 = 225$ **31.** $7x^2 + 3y^2 = 75$

33. $3x^2 - 6x + 4y^2 - 8y - 41 = 0$

35. $9x^2 + 25y^2 - 50y - 200 = 0$ **37.** $3x^2 + 4y^2 = 48$

39. $\dfrac{10\sqrt{5}}{3}$ feet

13. $F(1 + \sqrt{5}, 2)$, $F'(1 - \sqrt{5}, 2)$

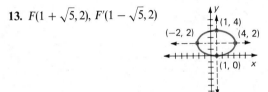

Problem Set 9.3 (page 406)

For Problems 1–22, the foci and the equations of the asymptotes are indicated above the graphs. The vertices are given on the graphs.

1. $F(\sqrt{13}, 0)$, $F'(-\sqrt{13}, 0)$
$y = \pm\frac{2}{3}x$

3. $F(0, \sqrt{13})$, $F'(0, -\sqrt{13})$
$y = \pm\frac{2}{3}x$

15. $F(-2, -1 + 2\sqrt{3})$, $F'(-2, -1 - 2\sqrt{3})$

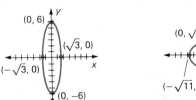

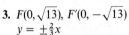

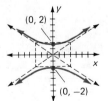

5. $F(0, 5)$, $F'(0, -5)$
$y = \pm\frac{4}{3}x$

7. $F(3\sqrt{2}, 0)$, $F'(-3\sqrt{2}, 0)$
$y = \pm x$

17. $F(3 + \sqrt{3}, 0)$, $F'(3 - \sqrt{3}, 0)$

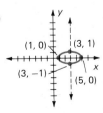

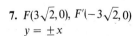

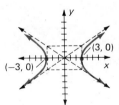

19. $F(4, -1 + \sqrt{7})$, $F'(4, -1 - \sqrt{7})$

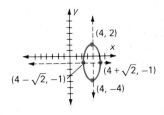

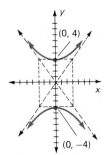

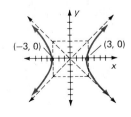

9. $F(0, \sqrt{30}), F'(0, -\sqrt{30})$

$$y = \pm \frac{\sqrt{5}}{5} x$$

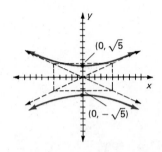

11. $F(\sqrt{10}, 0), F'(-\sqrt{10}, 0)$

$$y = \pm 3x$$

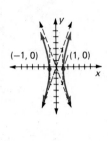

19. $F(0, -5 + \sqrt{10}), F'(0, -5 - \sqrt{10})$

$$3x - y = 5 \quad \text{and} \quad 3x + y = -5$$

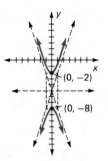

13. $F(3 + \sqrt{13}, -1), F'(3 - \sqrt{13}, -1)$

$$2x - 3y = 9 \quad \text{and} \quad 2x + 3y = 3$$

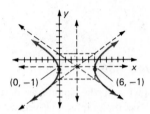

21. $F(-2 + \sqrt{2}, -2), F'(-2 - \sqrt{2}, -2)$

$$x - y = 0 \quad \text{and} \quad x + y = -4$$

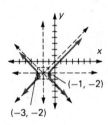

15. $F(-3, 2 + \sqrt{5}), F'(-3, 2 - \sqrt{5})$

$$2x - y = -8 \quad \text{and} \quad 2x + y = -4$$

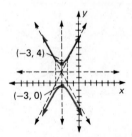

23. $5x^2 - 4y^2 = 20$ **25.** $16y^2 - 9x^2 = 144$

27. $3x^2 - y^2 = 3$ **29.** $4y^2 - 3x^2 = 12$

31. $7x^2 - 16y^2 = 112$

33. $5x^2 - 40x - 4y^2 - 24y + 24 = 0$

35. $3y^2 - 30y - x^2 - 6x + 54 = 0$

37. $5x^2 - 20x - 4y^2 = 0$ **39.** Circle

41. Straight line   **43.** Ellipse

45. Hyperbola   **47.** Parabola

Problem Set 9.4 (page 411)

1. $\{(1, 2)\}$ **3.** $\{(1, -5), (-5, 1)\}$

5. $\{(2 + i\sqrt{3}, -2 + i\sqrt{3}), (2 - i\sqrt{3}, -2 - i\sqrt{3})\}$

7. $\{(-6, 7), (-2, -1)\}$ **9.** $\{(-3, 4)\}$

11. $\left\{ \left(\dfrac{-1 + i\sqrt{3}}{2}, \dfrac{-7 - i\sqrt{3}}{2} \right), \left(\dfrac{-1 - i\sqrt{3}}{2}, \dfrac{-7 + i\sqrt{3}}{2} \right) \right\}$

13. $\{(-1, 2)\}$ **15.** $\{(-6, 3), (-2, -1)\}$ **17.** $\{(5, 3)\}$

19. $\{(1, 2), (-1, 2)\}$ **21.** $\{(-3, 2)\}$ **23.** $\{(2, 0), (-2, 0)\}$

25. $\{(\sqrt{2}, \sqrt{3}), (\sqrt{2}, -\sqrt{3}), (-\sqrt{2}, \sqrt{3}), (-\sqrt{2}, -\sqrt{3})\}$

27. $\{(1, 1), (1, -1), (-1, 1), (-1, -1)\}$

29. (a) $\{(2i, 2\sqrt{2}), (2i, -2\sqrt{2}), (-2i, 2\sqrt{2}), (-2i, -2\sqrt{2})\}$

17. $F(2 + \sqrt{6}, 0), F'(2 - \sqrt{6}, 0)$

$$\sqrt{2}x - y = 2\sqrt{2} \quad \text{and} \quad \sqrt{2}x + y = 2\sqrt{2}$$

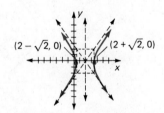

Chapter 9 Review Problem Set (page 413)

1. $F(4,0), F'(-4,0)$ **2.** $F(-3,0)$

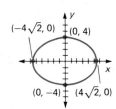

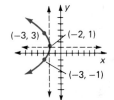

3. $F(0, 2\sqrt{3}), F'(0, -2\sqrt{3})$

$y = \pm\dfrac{\sqrt{3}}{3} x$

4. $F(\sqrt{15}, 0), F'(-\sqrt{15}, 0)$

$y = \pm\dfrac{\sqrt{6}}{3} x$

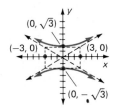

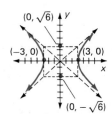

5. $F(0, \sqrt{6}), F'(0, -\sqrt{6})$ **6.** $F(0, \tfrac{1}{2})$

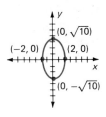

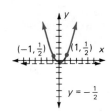

7. $F(4 + \sqrt{6}, 1), F'(4 - \sqrt{6}, 1)$

$\sqrt{2}x - 2y = 4\sqrt{2} - 2$ and $\sqrt{2}x + 2y = 4\sqrt{2} + 2$

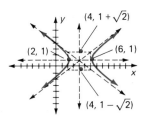

8. $F(3, -2 + \sqrt{7}), F'(3, -2 - \sqrt{7})$

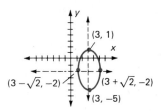

9. $F(-3, 1), x = -1$ **10.** $F(-1, -5), y = -1$

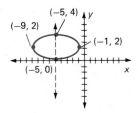

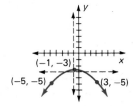

11. $F(-5 + 2\sqrt{3}, 2), F'(-5 - 2\sqrt{3}, 2)$

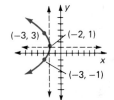

12. $F(-2, -2 + \sqrt{10}), F'(-2, -2 - \sqrt{10})$

$\sqrt{6}x - 3y = 6 - 2\sqrt{6}$ and $\sqrt{6}x + 3y = -6 - 2\sqrt{6}$

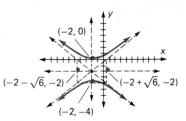

13. $y^2 = -20x$ **14.** $y^2 + 16x^2 = 16$

15. $25x^2 - 2y^2 = 50$ **16.** $4x^2 + 3y^2 = 16$

17. $3x^2 = 2y$ **18.** $9y^2 - x^2 = 9$

19. $9x^2 - 108x + y^2 - 8y + 331 = 0$

20. $y^2 + 4y - 8x + 36 = 0$

21. $3y^2 + 24y - x^2 - 10x + 20 = 0$

22. $x^2 + 12x - y + 33 = 0$ **23.** $4x^2 + 40x + 25y^2 = 0$

24. $4x^2 - 32x - y^2 + 48 = 0$ **25.** $\{(-1, 4)\}$

26. $\{(3, 1)\}$ **27.** $\{(-1, -2), (-2, -3)\}$

28. $\left\{ \left(\dfrac{4\sqrt{2}}{3}, \dfrac{4}{3}i \right), \left(\dfrac{4\sqrt{2}}{3}, -\dfrac{4}{3}i \right), \left(-\dfrac{4\sqrt{2}}{3}, \dfrac{4}{3}i \right), \right.$
$\left. \left(-\dfrac{4\sqrt{2}}{3}, -\dfrac{4}{3}i \right) \right\}$

29. $\{(0, 2), (0, -2)\}$

30. $\left\{ \left(\dfrac{\sqrt{15}}{5}, \dfrac{2\sqrt{10}}{5} \right), \left(\dfrac{\sqrt{15}}{5}, -\dfrac{2\sqrt{10}}{5} \right), \left(-\dfrac{\sqrt{15}}{5}, \dfrac{2\sqrt{10}}{5} \right), \right.$
$\left. \left(-\dfrac{\sqrt{15}}{5}, -\dfrac{2\sqrt{10}}{5} \right) \right\}$

CHAPTER 10

Problem Set 10.1 (page 420)

1. $-4, -1, 2, 5, 8$ **3.** $2, 0, -2, -4, -6$
5. $2, 11, 26, 47, 74$ **7.** $0, 2, 6, 12, 20$
9. $4, 8, 16, 32, 64$ **11.** $a_{15} = -79; a_{30} = -154$
13. $a_{25} = 1; a_{50} = -1$ **15.** $a_n = 2n + 9$

17. $a_n = -3n + 5$ **19.** $a_n = \dfrac{n + 2}{2}$

21. $a_n = 4n - 2$ **23.** $a_n = -3n$
25. 73 **27.** 334 **29.** 35 **31.** 7 **33.** 86
35. 2700 **37.** 3200 **39.** -7950 **41.** 637.5
43. 4950 **45.** 1850 **47.** -2030 **49.** 3591
51. $40{,}000$ **53.** $58{,}250$ **55.** 2205
57. -1325 **59.** 5265 **61.** -810 **63.** 1276
65. 660 **67.** 55 **69.** 431
71. $3, 3, 7, 7, 11, 11$ **73.** $4, 7, 10, 13, 17, 21$
75. $4, 12, 36, 108, 324, 972$ **77.** $1, 1, 2, 3, 5, 8$
79. $3, 1, 4, 9, 25, 256$

Problem Set 10.2 (page 429)

1. $a_n = 3(2)^{n-1}$ **3.** $a_n = 3^n$ **5.** $a_n = (\tfrac{1}{2})^{n+1}$
7. $a_n = 4^n$ **9.** $a_n = (0.3)^{n-1}$ **11.** $a_n = (-2)^{n-1}$
13. 64 **15.** $\tfrac{1}{9}$ **17.** -512 **19.** $\tfrac{1}{4374}$
21. $\tfrac{2}{3}$ **23.** 2 **25.** 1023 **27.** $19{,}682$
29. $394\tfrac{1}{16}$ **31.** 1364 **33.** 1089 **35.** $7\tfrac{511}{512}$
37. -547 **39.** $127\tfrac{3}{4}$ **41.** 540 **43.** $2\tfrac{61}{64}$
45. 4 **47.** 3 **49.** No sum   **51.** $\tfrac{27}{4}$ **53.** 2
55. $\tfrac{16}{3}$ **57.** $\tfrac{1}{3}$ **59.** $\tfrac{26}{99}$ **61.** $\tfrac{41}{333}$ **63.** $\tfrac{4}{15}$
65. $\tfrac{106}{495}$ **67.** $\tfrac{7}{3}$

Problem Set 10.3 (page 434)

1. \$24,200   **3.** 11,550 students
5. 7320 students   **7.** 125 liters   **9.** 512 gallons

11. \$116.25   **13.** \$163.84; \$327.67   **15.** \$24,900
17. 1936 feet   **19.** $\tfrac{15}{16}$ gram   **21.** 2910 feet
23. 325 logs   **25.** 5.9%   **27.** $\tfrac{5}{64}$ gallon

Problem Set 10.4 (page 440)

These problems are proofs by mathematical induction and require class discussion.

Chapter 10 Review Problem Set (page 442)

1. $a_n = 6n - 3$ **2.** $a_n = 3^{n-2}$ **3.** $a_n = 5 \cdot 2^n$
4. $a_n = -3n + 8$ **5.** $a_n = 2n - 7$ **6.** $a_n = 3^{3-n}$
7. $a_n = -(-2)^{n-1}$ **8.** $a_n = 3n + 9$

9. $a_n = \dfrac{n + 1}{3}$ **10.** $a_n = 4^{n-1}$ **11.** 73 **12.** 106

13. $\tfrac{1}{32}$ **14.** $\tfrac{4}{9}$ **15.** -92 **16.** $\tfrac{1}{16}$ **17.** -5
18. 85 **19.** $\tfrac{5}{9}$ **20.** 2 or -2 **21.** $121\tfrac{40}{81}$
22. 7035 **23.** $-10{,}725$ **24.** $31\tfrac{31}{32}$ **25.** $32{,}015$
26. 4757 **27.** $85\tfrac{21}{64}$ **28.** $37{,}044$ **29.** $12{,}726$
30. -1845 **31.** 225 **32.** 255 **33.** 8244
34. $85\tfrac{1}{3}$ **35.** $\tfrac{4}{11}$ **36.** $\tfrac{41}{90}$ **37.** \$750
38. \$46.50   **39.** \$3276.70   **40.** 10,935 gallons

CHAPTER 11

Problem Set 11.1 (page 448)

1. 20 **3.** 24 **5.** 168 **7.** 48 **9.** 36
11. 6840 **13.** 720 **15.** 720 **17.** 36
19. 24 **21.** 243 **23.** Impossible   **25.** 216
27. 26 **29.** 36 **31.** 144 **33.** 1024
35. 30 **37. (a)** $6{,}084{,}000$ **(c)** $3{,}066{,}336$

Problem Set 11.2 (page 456)

1. 60 **3.** 360 **5.** 21 **7.** 252 **9.** 105
11. 1 **13.** 24 **15.** 84 **17. (a)** 336
19. 2880 **21.** 2450 **23.** 10 **25.** 10
27. 35 **29.** 1260 **31.** 2520 **33.** 15
35. 126 **37.** $144; 202$ **39.** $15; 10$ **41.** 20

43. $10; 15; 21; \dfrac{n(n-1)}{2}$ **45.** 120

Problem Set 11.3 (page 462)

1. $\dfrac{1}{2}$ **3.** $\dfrac{3}{4}$ **5.** $\dfrac{1}{8}$ **7.** $\dfrac{7}{8}$ **9.** $\dfrac{1}{16}$

11. $\dfrac{3}{8}$ 13. $\dfrac{1}{3}$ 15. $\dfrac{1}{2}$ 17. $\dfrac{5}{36}$ 19. $\dfrac{1}{6}$

21. $\dfrac{11}{36}$ 23. $\dfrac{1}{4}$ 25. $\dfrac{1}{2}$ 27. $\dfrac{1}{25}$ 29. $\dfrac{9}{25}$

31. $\dfrac{2}{5}$ 33. $\dfrac{9}{10}$ 35. $\dfrac{5}{14}$ 37. $\dfrac{15}{28}$ 39. $\dfrac{7}{15}$

41. $\dfrac{1}{15}$ 43. $\dfrac{2}{3}$ 45. $\dfrac{1}{5}$ 47. $\dfrac{1}{63}$ 49. $\dfrac{1}{2}$

51. $\dfrac{5}{11}$ 53. $\dfrac{1}{6}$ 55. $\dfrac{21}{128}$ 57. $\dfrac{13}{16}$ 59. $\dfrac{1}{21}$

61. 40 63. 3744 65. 10,200 67. 123,552
69. 1,302,540

Problem Set 11.4 (page 471)

1. $\dfrac{5}{36}$ 3. $\dfrac{7}{12}$ 5. $\dfrac{1}{216}$ 7. $\dfrac{53}{54}$ 9. $\dfrac{1}{16}$

11. $\dfrac{15}{16}$ 13. $\dfrac{1}{32}$ 15. $\dfrac{31}{32}$ 17. $\dfrac{5}{6}$ 19. $\dfrac{12}{13}$

21. $\dfrac{7}{12}$ 23. $\dfrac{37}{44}$ 25. $\dfrac{2}{3}$ 27. $\dfrac{2}{3}$ 29. $\dfrac{5}{18}$

31. $\dfrac{1}{3}$ 33. $\dfrac{1}{2}$ 35. $\dfrac{7}{12}$

37. (a) .410 (c) .955 39. .525 41. 60
43. 120 45. 9 47. 56 49. It is a fair game.
51. Yes 53. $11,000 55. −$25 57. 1 to 7
59. 11 to 5 61. 1 to 8 63. 1 to 1

65. 4 to 3 67. 3 to 2 69. $\dfrac{2}{7}$ 71. $\dfrac{7}{12}$

Problem Set 11.5 (page 481)

1. $\dfrac{1}{3}$ 3. $\dfrac{2}{15}$ 5. $\dfrac{1}{3}$ 7. $\dfrac{1}{6}$ 9. $\dfrac{2}{3};\dfrac{2}{7}$

11. $\dfrac{2}{3};\dfrac{2}{5}$ 13. $\dfrac{1}{15};\dfrac{2}{7}$ 15. Dependent

17. Independent 19. $\dfrac{1}{4}$ 21. $\dfrac{1}{216}$ 23. $\dfrac{1}{221}$

25. $\dfrac{13}{102}$ 27. $\dfrac{1}{16}$ 29. $\dfrac{1}{1352}$ 31. $\dfrac{2}{49}$

33. $\dfrac{25}{81}$ 35. $\dfrac{20}{81}$ 37. $\dfrac{25}{169}$ 39. $\dfrac{32}{169}$

41. $\dfrac{2}{3}$ 43. $\dfrac{1}{3}$ 45. $\dfrac{5}{68}$ 47. $\dfrac{15}{34}$ 49. $\dfrac{1}{12}$

51. $\dfrac{1}{6}$ 53. $\dfrac{1}{729}$ 55. $\dfrac{5}{27}$ 57. $\dfrac{4}{35}$ 59. $\dfrac{8}{35}$

61. $\dfrac{4}{21};\dfrac{2}{7};\dfrac{11}{21}$

Problem Set 11.6 (page 487)

1. $x^8 + 8x^7y + 28x^6y^2 + 56x^5y^3 + 70x^4y^4 + 56x^3y^5 + 28x^2y^6 + 8xy^7 + y^8$
3. $x^6 - 6x^5y + 15x^4y^2 - 20x^3y^3 + 15x^2y^4 - 6xy^5 + y^6$
5. $a^4 + 8a^3b + 24a^2b^2 + 32ab^3 + 16b^4$
7. $x^5 - 15x^4y + 90x^3y^2 - 270x^2y^3 + 405xy^4 - 243y^5$
9. $16a^4 - 96a^3b + 216a^2b^2 - 216ab^3 + 81b^4$
11. $x^{10} + 5x^8y + 10x^6y^2 + 10x^4y^3 + 5x^2y^4 + y^5$
13. $16x^8 - 32x^6y^2 + 24x^4y^4 - 8x^2y^6 + y^8$
15. $x^6 + 18x^5 + 135x^4 + 540x^3 + 1215x^2 + 1458x + 729$
17. $x^9 - 9x^8 + 36x^7 - 84x^6 + 126x^5 - 126x^4 + 84x^3 - 36x^2 + 9x - 1$

19. $1 + \dfrac{4}{n} + \dfrac{6}{n^2} + \dfrac{4}{n^3} + \dfrac{1}{n^4}$

21. $a^6 - \dfrac{6a^5}{n} + \dfrac{15a^4}{n^2} - \dfrac{20a^3}{n^3} + \dfrac{15a^2}{n^4} - \dfrac{6a}{n^5} + \dfrac{1}{n^6}$

23. $17 + 12\sqrt{2}$ 25. $843 - 589\sqrt{2}$
27. $x^{12} + 12x^{11}y + 66x^{10}y^2 + 220x^9y^3$
29. $x^{20} - 20x^{19}y + 190x^{18}y^2 - 1140x^{17}y^3$
31. $x^{28} - 28x^{26}y^3 + 364x^{24}y^6 - 2912x^{22}y^9$

33. $a^9 + \dfrac{9a^8}{n} + \dfrac{36a^7}{n^2} + \dfrac{84a^6}{n^3}$

35. $x^{10} - 20x^9y + 180x^8y^2 - 960x^7y^3$
37. $56x^5y^3$ 39. $126x^5y^4$ 41. $189a^2b^5$

43. $120x^6y^{21}$ 45. $\dfrac{5005}{n^6}$ 47. $41 - 38i$

49. $-117 - 44i$

Chapter 11 Review Problem Set (page 490)

1. 720 2. 30,240 3. 150 4. 1440
5. 20 6. 525 7. 1287 8. 264
9. 74 10. 55 11. 40 12. 15 13. 60

14. 120 15. $\dfrac{3}{8}$ 16. $\dfrac{5}{16}$ 17. $\dfrac{5}{36}$ 18. $\dfrac{13}{18}$

19. $\dfrac{3}{5}$ 20. $\dfrac{1}{35}$ 21. $\dfrac{57}{64}$ 22. $\dfrac{1}{221}$ 23. $\dfrac{1}{6}$

24. $\dfrac{4}{7}$ 25. $\dfrac{4}{7}$ 26. $\dfrac{10}{21}$ 27. $\dfrac{140}{143}$ 28. $\dfrac{105}{169}$

29. $\dfrac{1}{6}$ **30.** $\dfrac{28}{55}$ **31.** $\dfrac{5}{7}$ **32.** $\dfrac{1}{16}$ **33.** $\dfrac{1}{2}; \dfrac{1}{3}$

34. (a) $\dfrac{9}{19}$ (b) $\dfrac{9}{10}$ **35.** (a) $\dfrac{2}{7}$ (b) $\dfrac{4}{9}$

36. $x^5 + 10x^4y + 40x^3y^2 + 80x^2y^3 + 80xy^4 + 32y^5$

37. $x^8 - 8x^7y + 28x^6y^2 - 56x^5y^3 + 70x^4y^4 - 56x^3y^5$
$$+ 28x^2y^6 - 8xy^7 + y^8$$

38. $a^8 - 12a^6b^3 + 54a^4b^6 - 108a^2b^9 + 81b^{12}$

39. $x^6 + \dfrac{6x^5}{n} + \dfrac{15x^4}{n^2} + \dfrac{20x^3}{n^3} + \dfrac{15x^2}{n^4} + \dfrac{6x}{n^5} + \dfrac{1}{n^6}$

40. $41 - 29\sqrt{2}$ **41.** $-a^3 + 3a^2b - 3ab^2 + b^3$

42. $-1760x^9y^3$ **43.** $57915a^4b^{18}$

Index

Abscissa, 131
Absolute value:
  definition of, 6, 121
  equations involving, 121
  inequalities involving, 121
  properties of, 6
Addition
  of complex numbers, 58
  of functions, 191
  of matrices, 355, 365
  of polynomials, 21
  of radical expressions, 47
  of rational expressions, 37
  of rational numbers, 37
Addition property of equality, 70
Addition property of inequality, 113
Additive inverse property, 7
Algebraic equation, 69
Algebraic expression, 2
Algebraic fraction, 35
Algebraic identity, 69
Algebraic inequality, 113
Analytic geometry, 129
Antilogarithm, 237
Arithmetic sequence, 415
Associative property
  of addition of matrices, 354, 365
  of addition of real numbers, 7
  of multiplication of matrices, 358, 367
  of multiplication of real numbers, 7
Asymptotes, 161, 289, 295, 402
Augmented matrix, 323
Axis of a coordinate system, 131
Axis of symmetry, 178, 184

Base of a logarithm, 226
Base of a power, 13
Binomial, 20

Binomial expansion, 23, 484
Bounded subset, 377
Braces for set notation, 3

Cartesian coordinate system, 132
Center
  of a circle, 157
  of an ellipse, 395
  of a hyperbola, 401
Change of base formula, 247
Characteristic of a logarithm, 236
Checking
  solutions of equations, 70
  solutions of inequalities, 114
  solutions of word problems, 73
Circle, 157
Circle, equation of, 157
Closure property
  for addition, 7
  for multiplication, 7
Coefficient, numerical, 8
Cofactor, 334
Combination, 454
Common difference of an arithmetic
  sequence, 416
Common logarithm, 234
Common ratio of a geometric sequence,
  423
Commutative property
  of addition of matrices, 354, 364
  of addition of real numbers, 7
  of multiplication of real numbers, 7
Complementary events, 466
Completely factored form, 27
Completing the square, 89
Complex fraction, 41
Complex number, 57
Composition of functions, 192

Compound event, 459
Compound inequalities, 115
Compound interest, 219
Conditional equation, 69
Conditional probability, 476
Conic section, 387
Conjugate, 61
Conjugate axis of a hyperbola, 401
Consistent system of equations, 306
Constant of proportionality, 202
Constant of variation, 202
Constraints, 378
Coordinate geometry, 129
Coordinate of a point, 5, 130
Counting numbers, 3
Cramer's rule, 343, 345
Critical numbers, 116
Cross-multiplication property, 78
Cube root, 45

Decimals,
    nonrepeating, 4
    repeating, 4
    terminating, 4
Decreasing function, 183, 217
Degree
    of a monomial, 20
    of a polynomial, 20, 270
Denominator,
    least common, 38
    rationalizing a, 49
Dependent equations, 307
Dependent events, 478
Dependent variable, 174
Descartes's rule of signs, 276
Determinant, 332
Difference quotient, 171
Difference of squares, 23
Dimension of a matrix, 323
Direct variation, 202
Directrix of a parabola, 388
Discriminant, 93
Distance formula, 133
Distributive property, 7, 356, 358
Division
    of complex numbers, 59
    of functions, 191
    of polynomials, 25, 260
    of radical expressions, 48
    of rational expressions, 37
    of rational numbers, 37

Division algorithm for polynomials, 261
Domain of a function, 169

e, 221
Element of a set, 2
Elementary row operations, 323
Ellipse, 159, 394
Empty set, 3
Equal matrices, 354
Equality,
    addition property of, 70
    multiplication property of, 70
    reflexive property of, 69
    substitution property of, 69
    symmetric property of, 69
    transitive property of, 70
Equation(s),
    definition of, 69
    dependent, 307
    equivalent, 69
    exponential, 215, 241
    first-degree in one variable, 69
    first-degree in two variables, 138
    first-degree in three variables, 315
    inconsistent, 306
    linear, 70
    logarithmic, 243
    polynomial, 270
    quadratic, 88
    radical, 108
Equivalent equations, 69
Equivalent systems, 308
Evaluating algebraic expressions, 9
Event space, 459
Expansion of a binomial, 23, 484
Expansion of a determinant by minors, 333
Expected value, 470
Exponent(s),
    integers as, 13
    negative, 15
    properties of, 14, 215
    rational numbers as, 53
    zero as an, 15
Exponential equation, 215, 241
Exponential function, 216
Extraneous solution or root, 108

Factor, 27
Factor theorem, 267
Factorial notation, 450

Factoring,
  complete, 27
  difference of cubes, 32
  difference of squares, 28
  grouping, 28
  sum of cubes, 32
  trinomials, 30
Fair game, 470
Feasible solutions, 378
First-degree equation
  in one variable, 69
  in three variables, 315
  in two variables, 138
Focus
  of an ellipse, 394
  of a hyperbola, 401
  of a parabola, 388
Formulas, 79
Function(s),
  decreasing, 183, 217
  definition of, 169
  domain of a, 169
  even, 176
  exponential, 216
  graph of a, 170
  greatest integer, 183
  identity, 178
  increasing, 183, 217
  inverse of a, 196
  linear, 177
  logarithmic, 233
  one-to-one, 195
  polynomial, 280
  quadratic, 178
  range of a, 169
  rational, 287
Functional notation, 169
Fundamental principle of counting, 445

Gaussian elimination, 323
General term of a sequence, 416, 423
Geometric sequence, 423
Graph
  of an equation, 138
  of a function, 170
  of an inequality, 140
Graphing suggestions, 187

Half-life, 223
Horizontal asymptote, 289
Horizontal line test, 195
Hyperbola, 160, 400

i, 57
Identity element
  for addition of matrices, 354, 365
  for addition of real numbers, 7
  for multiplication of matrices, 360, 367
  for multiplication of real numbers, 7
Identity function, 178
Imaginary number, 57
Inconsistent equations, 306
Increasing function, 183, 217
Independent events, 478
Independent variable, 174
Index of a radical, 45
Index of summation, 419
Inequalities,
  equivalent, 113
  graphs of, 140
  involving absolute value, 119
  linear in one variable, 113
  linear in two variables, 140
  quadratic, 115
  solutions of, 113
Infinite geometric sequence, 427
Integers, 3
Interpolation, linear, 250
Intersection of sets, 467
Interval notation, 114
Inverse of a function, 196
Inverse variation, 203
Irrational numbers, 4

Joint variation, 205

Law of exponential growth, 222
Least common denominator, 38
Least common multiple, 38
Like terms, 9
Line, number, 5
Linear equations,
  definition of, 70, 138
  graph of, 139
  slope-intercept form, 146, 149
  standard form of, 149
Linear function,
  definition of, 177, 376
  graph of a, 177
Linear inequality, 140
Linear interpolation, 250
Linear programming, 376
Linear systems of equations, 306, 315

Literal factor, 8
Logarithm(s),
　base of a, 226
　characteristic of a, 236
　common, 234
　definition of, 226
　mantissa of a, 236
　natural, 238
　properties of, 228–230, 252
　table of common logarithms (back of
　　book)
　table of natural logarithms, 239
Logarithmic equations, 243
Logarithmic function, 233
Lower bound, 279

Major axis of an ellipse, 159, 395
Mantissa, 236
Mathematical expectation, 470
Mathematical induction, 436
Matrix, 322
Maximum value of a function, 185, 377
Midpoint formula, 135
Minimum value of a function, 185, 377
Minors, expansion of a determinant by,
　333
Minor axis of an ellipse, 159, 395
Mirror image, 197
Monomial(s),
　addition of, 21
　definition of, 20
　degree of, 20
　division of, 25
　multiplication of, 21
　subtraction of, 21
Multiple, least common, 38
Multiple roots, 271
Multiplication
　of complex numbers, 59
　of functions, 191
　of matrices, 356, 366
　of polynomials, 21
　of radical expressions, 47
　of rational expressions, 36
　of rational numbers, 36
Multiplication property
　of equality, 70
　of inequality, 113
　of negative one, 7
　of zero, 7
Multiplicative inverse property, 7

Multiplicity of roots, 271
Mutually exclusive events, 469

nth root, 46
Natural logarithms, 238
Natural numbers, 3
Notation,
　functional, 169
　scientific, 19
　set-builder, 3
　summation, 419
Null set, 3
Number(s),
　absolute value of, 6
　complex, 57
　counting, 3
　imaginary, 58
　integers, 3
　irrational, 4
　natural, 3
　rational, 3
　real, 4
　whole, 3
Numerical coefficient, 8
Numerical expression, 2
Numerical statement of equality, 69
Numerical statement of inequality, 113

Objective function, 378
Oblique asymptote, 295
Odds, 474
One, multiplication property of, 7
One-to-one correspondence, 130
One-to-one function, 195
Open sentence, 69
Operations, order of, 10
Ordinate, 131
Origin, 131
Origin symmetry, 154

Parabola, 178, 388
Partial fractions, 297
Pascal's triangle, 24
Perfect square trinomial, 23
Permutation, 451
Points of division of a line segment, 134
Point-slope form, 145
Polynomial(s),
　addition of, 21
　completely factored form of, 27
　definition of, 20

degree of a, 20
division of, 25
multiplication of, 21
subtraction of, 21
Polynomial equation, 270
Polynomial function, 280
Principal root, 45
Principle of mathematical induction,
   436
Probability, 459
Problem solving suggestions, 73, 80, 99,
   431
Properties of
   determinants, 336
   equality, 69
   exponents, 14, 215
   inequality, 113
   logarithms, 228–230, 252
   real numbers, 7
Proportion, 78
Pure imaginary number, 57
Pythagorean theorem, 95

Quadrant, 131
Quadratic equation(s),
   definition of, 88
   discriminant of a, 93
   nature of solutions of, 93
   standard form of, 88
Quadratic formula, 91
Quadratic function,
   definition of, 178
   graph of a, 178
Quadratic inequality, 115

Radical(s),
   addition of, 47
   changing form of, 46
   definition of, 45, 46
   division of, 49
   index of a, 46
   multiplication of, 47
   simplest form of, 49
   subtraction of, 47
Radical equation, 108
Radicand, 45
Radius of a circle, 157
Range of a function, 169
Ratio, 78
Ratio of a geometric sequence, 423
Rational exponent, 53

Rational expression, 35
Rational function, 287
Rational number, 3
Rational root theorem, 272
Rationalizing a denominator, 49
Real number, 4
Real number line, 5
Reciprocal, 8
Rectangular coordinate system, 131
Reduced echelon form, 325
Reducing fractions, 35
Reflexive property of equality, 69
Relation, 170
Remainder theorem, 266
Roots of an equation, 423

Sample points, 459
Sample space, 459
Scalar multiplication, 355
Scientific notation, 19
Sequence,
   arithmetic, 415
   definition of, 415
   general term of, 415
   geometric, 423
   infinite, 415
Set(s),
   element of a, 3
   empty, 3
   equal, 3
   intersection of, 467
   notation, 3
   null, 3
   solution, 69
   union of, 467
Set of feasible solutions, 378
Sign variations, 276
Similar terms, 9
Simple event, 459
Simplest radical form, 49
Slope, 143
Slope-intercept form, 146, 149
Solution(s),
   of equations, 69
   extraneous, 108
   of inequalities, 113
   of systems, 306
Solution set
   of an equation, 69
   of an inequality, 113
   of a system, 306

Square matrix, 332, 367
Square root, 44
Standard form
  of complex numbers, 57
  of equation of a circle, 157
  of equation of a straight line, 149
  of a quadratic equation, 88
Subset, 5
Substitution of functions, 192
Substitution property of equality, 69
Subtraction
  of complex numbers, 58
  of functions, 191
  of matrices, 355, 366
  of polynomials, 21
  of radical expressions, 47
  of rational expressions, 37
  of rational numbers, 37
Suggestions for solving word problems,
    73, 80, 99, 431
Sum
  of an arithmetic sequence, 417
  of a geometric sequence, 424
  of an infinite geometric sequence, 427
Summation notation, 419
Symmetric property of equality, 69
Symmetry, 153, 154, 178
Synthetic division, 262
System(s)
  of linear equations in two variables,
    306
  of linear equations in three variables,
    315
  of linear inequalities, 374
  of nonlinear equations, 407

Table of common logarithms (back of
    book)
Table of natural logarithms, 239
Term(s):
  addition of like, 9
  of an algebraic expression, 8, 20
  like, 9
  similar, 9
Test number, 116
Test point, 141
Transitive property of equality, 69
Transverse axis of a hyperbola, 401
Tree diagrams, 445
Triangular form of a system, 329
Trinomial, 20
Turning points, 283

Unbounded subset, 377
Union of sets, 467
Upper bound, 279

Variable, 2
Variation,
  constant of, 202
  direct, 202
  inverse, 203
  joint, 205
Vertex
  of an ellipse, 395
  of a hyperbola, 401
  of a parabola, 178, 388
Vertical asymptote, 289
Vertical line test, 170

Whole numbers, 3

x-axis symmetry, 154
x-intercept, 139

y-axis symmetry, 153
y-intercept, 139

Zero,
  addition property of, 7
  as an exponent, 15

Table of Common Logarithms

N	0	1	2	3	4	5	6	7	8	9
1.0	.0000	.0043	.0086	0.128	.0170	.0212	.0253	.0294	.0334	.0374
1.1	.0414	.0453	.0492	.0531	.0569	.0607	.0645	.0682	.0719	.0755
1.2	.0792	.0828	.0864	.0899	.0934	.0969	.1004	.1038	.1072	.1106
1.3	.1139	.1173	.1206	.1239	.1271	.1303	.1335	.1367	.1399	.1430
1.4	.1461	.1492	.1523	.1553	.1584	.1614	.1644	.1673	.1703	.1732
1.5	.1761	.1790	.1818	.1847	.1875	.1903	.1931	.1959	.1987	.2014
1.6	.2041	.2068	.2095	.2122	.2148	.2175	.2201	.2227	.2253	.2279
1.7	.2304	.2330	.2355	.2380	.2405	.2430	.2455	.2480	.2504	.2529
1.8	.2553	.2577	.2601	.2625	.2648	.2672	.2695	.2718	.2742	.2765
1.9	.2788	.2810	.2833	.2856	.2878	.2900	.2923	.2945	.2967	.2989
2.0	.3010	.3032	.3054	.3075	.3096	.3118	.3139	.3160	.3181	.3201
2.1	.3222	.3243	.3263	.3284	.3304	.3324	.3345	.3365	.3385	.3404
2.2	.3424	.3444	.3464	.3483	.3502	.3522	.3541	.3560	.3579	.3598
2.3	.3617	.3636	.3655	.3674	.3692	.3711	.3729	.3747	.3766	.3784
2.4	.3802	.3820	.3838	.3856	.3874	.3892	.3909	.3927	.3945	.3962
2.5	.3979	.3997	.4014	.4031	.4048	.4065	.4082	.4099	.4116	.4133
2.6	.4150	.4166	.4183	.4200	.4216	.4232	.4249	.4265	.4281	.4298
2.7	.4314	.4330	.4346	.4362	.4378	.4393	.4409	.4425	.4440	.4456
2.8	.4472	.4487	.4502	.4518	.4533	.4548	.4564	.4579	.4594	.4609
2.9	.4624	.4639	.4654	.4669	.4683	.4698	.4713	.4728	.4742	.4757
3.0	.4771	.4786	.4800	.4814	.4829	.4843	.4857	.4871	.4886	.4900
3.1	.4914	.4928	.4942	.4955	.4969	.4983	.4997	.5011	.5024	.5038
3.2	.5051	.5065	.5079	.5092	.5105	.5119	.5132	.5145	.5159	.5172
3.3	.5185	.5198	.5211	.5224	.5237	.5250	.5263	.5276	.5289	.5302
3.4	.5315	.5328	.5340	.5353	.5366	.5378	.5391	.5403	.5416	.5428
3.5	.5441	.5453	.5465	.5478	.5490	.5502	.5514	.5527	.5539	.5551
3.6	.5563	.5575	.5587	.5599	.5611	.5623	.5635	.5647	.5658	.5670
3.7	.5682	.5694	.5705	.5717	.5729	.5740	.5752	.5763	.5775	.5786
3.8	.5798	.5809	.5821	.5832	.5843	.5855	.5866	.5877	.5888	.5899
3.9	.5911	.5922	.5933	.5944	.5955	.5966	.5977	.5988	.5999	.6010
4.0	.6021	.6031	.6042	.6053	.6064	.6075	.6085	.6096	.6107	.6117
4.1	.6128	.6138	.6149	.6160	.6170	.6180	.6191	.6201	.6212	.6222
4.2	.6232	.6243	.6253	.6263	.6274	.6284	.6294	.6304	.6314	.6325
4.3	.6335	.6345	.6355	.6365	.6375	.6385	.6395	.6405	.6415	.6425
4.4	.6435	.6444	.6454	.6464	.6474	.6484	.6493	.6503	.6513	.6522
4.5	.6532	.6542	.6551	.6561	.6571	.6580	.6590	.6599	.6609	.6618
4.6	.6628	.6637	.6646	.6656	.6665	.6675	.6684	.6693	.6702	.6712
4.7	.6721	.6730	.6739	.6749	.6758	.6767	.6776	.6785	.6794	.6803
4.8	.6812	.6821	.6830	.6839	.6848	.6857	.6866	.6875	.6884	.6893
4.9	.6902	.6911	.6920	.6928	.6937	.6946	.6955	.6964	.6972	.6981
5.0	.6990	.6998	.7007	.7016	.7024	.7033	.7042	.7050	.7059	.7067
5.1	.7076	.7084	.7093	.7101	.7110	.7118	.7126	.7135	.7143	.7152
5.2	.7160	.7168	.7177	.7185	.7193	.7202	.7210	.7218	.7226	.7235
5.3	.7243	.7251	.7259	.7267	.7275	.7284	.7292	.7300	.7308	.7316
5.4	.7324	.7332	.7340	.7348	.7356	.7364	.7372	.7380	.7388	.7396